Lecture Notes in Statistics 209

For further volumes:
http://www.springer.com/series/694

Eric D. Feigelson • G. Jogesh Babu
Editors

Statistical Challenges in Modern Astronomy V

Springer

Editors
Eric D. Feigelson
Department of Astronomy
and Astrophysics
Pennsylvania State University
University Park, PA, USA

G. Jogesh Babu
Department of Statistics
Pennsylvania State University
University Park, PA, USA

ISBN 978-1-4614-3519-8 ISBN 978-1-4614-3520-4 (eBook)
DOI 10.1007/978-1-4614-3520-4
Springer New York Heidelberg Dordrecht London

Library of Congress Control Number: 2012939628

Printed on acid-free paper

Springer is part of Springer Science+Business Media (www.springer.com)

Preface

Twenty years ago, the first *Statistical Challenges in Modern Astronomy* (SCMA) conference was held at Penn State University. Serving as a gathering of two scholarly communities with common interests, SCMA meetings have been held every 5 years for cross-disciplinary discussions of methodological issues arising in astronomical research. These are the proceedings of the fifth SCMA conference held in June 2011. While some of the topics are the similar as those in the 1991 meeting, the level of sophistication and accomplishment has enormously increased. Astronomers and statisticians worldwide have developed collaborations to address some of the most challenging and important problems facing astronomy today. These involve data mining enormous datasets from widefield surveys obtained with major new telescope systems, fitting of cosmological and other astrophysical models to complex datasets, and studying the temporal behaviors of innumerable variable objects in the sky. Bayesian inference has gained considerable momentum in astrophysical model fitting. These advanced methods are gaining attention outside of the world of expert astrostatisticians, as the broad astronomical community realize that twenty-first century science goals can not be achieved with nineteenth and twentieth century statistical methods. At SCMA V, both young and experienced astrostatisticians presented work and engaged in discussions on how these problems can be best addressed.

The proceedings are divided into six sections; most invited talks are followed by invited commentaries by scholars in the other field. The volume begins with five talks on *Statistics in Cosmology* demonstrating significant recent accomplishments in this most-important field of astronomy and astrophysics. Modern accomplishments of modern quantitative cosmology rely heavily on sophisticated statistical analysis of large datasets. Topics reviewed include likelihood-free estimation of quasar luminosity functions (Schaefer and Freeman), estimation of galaxy photometric redshifts and quantification of voids in galaxy Large-Scale Structure (Wandelt), inference based on comparing data to cosmological simulations (Higdon), likelihood estimation of gravitational lensing of the cosmic microwave background (CMB) radiation (Anderes), and application of needlets to cosmic microwave background studies (Marinucci).

The second section provides a sampling of the growing applications of *Bayesian Analysis Across Astronomy*. Here we have both invited reviews by senior researchers, and a sampling of the many works by younger researchers. The reviews discuss Bayesian models constructed to model galaxy star formation histories (Weinberg), model selection within the consensus ΛCDM cosmological model family (Trotta), and measurement errors in astronomical regression and density estimation problems (Kelly). The shorter talks treat asteroseismology (Benomar), event detection in time series (Blocker and Protopapas), reverberation mapping in active galactic nuclei (Brewer), modeling of Poisson images (Guglielmetti et al.), treatment of instrument calibration errors (Kashyap et al.), modeling of Type Ia supernova data (Mandel), and faint source flux estimation (Switzer et al.). Advanced methods for hierarchical modeling and Monte Carlo Markov Chain computational techniques are discussed in many of these talks and associated commentaries.

The third section of the proceedings address the use of modern techniques techniques of *Data Mining and Astroinformatics* for the analysis of massive datasets emerging from many new observatories. Compressive sensing, an extension of wavelet analysis, is very promising for many problems (Starck). Diffusion maps can treat non-linear structures in high-dimensional datasets (Lee and Freeman). Nearest neighbor techniques are used for outlier detection in megadatasets (Borne and Vedachalam). Bayesian approaches can help cross-identification of sources between astronomical catalogs (Budavári). Likelihood-based data compression can assist parameter estimation in large datasets (Jimenez).

The fourth section considers challenges arising in astronomical *Image and Time Series Analysis*. Techniques of mathematical morphology are applied to classifying sunspots (Stenning et al.). Realistic images are simulated using knowledge of celestial populations and telescope characteristics (Connolly et al.). Structure recognition algorithms are discussed for three-dimensional astronomical datacubes (Rosolowsky). The problem of locating faint transient sources in multiepoch image datasets is addressed by controlling the False Discovery Rate (Clements et al.). Wavelets are a valuable tool for modeling irregularly spaced time series (Mondal and Percival).

The fifth section provides perspectives on *The Future of Astrostatistics*. The field is gaining a presence in international organizations (Hilbe). The public domain **R** statistical computing environment is a very promising new software environment to implement existing and develop new statistical analyses for astronomical research (Tierney). A Panel Discussion discusses various aspects of astrostatistical practice and research for the coming decade (van Dyk, Feigelson, Loredo, Scargle). The final section of the proceedings gives brief presentations of the contributed posters. Many fascinating problems and sophisticated statistical methods are described.

The work of many individuals and organizations contributed to the success of the SCMA V conference. The invited speakers and cross-disciplinary commentators were the central pillar of the conference, and we are grateful for their presentations and manuscripts. Staff in the Departments of Statistics and Astronomy and Astrophysics provided administrative support. Funding support for the conference was provided by the two departments, Penn State's Eberly College of Science, and the National Science Foundation through grant AST-1113001. Finally, we are appreciative of our families' support during the many phases of this conference organization.

Pennsylvania State University, PA, USA Eric D. Feigelson
Pennsylvania State University, PA, USA G. Jogesh Babu

Contents

Part II Bayesian Analysis Across Astronomy

Part III Data Mining and Astroinformatics

Part IV Image and Time Series Analysis

Contributors

Mark Allen Kavli Institute for Particle Astrophysics and Cosmology, Stanford University, Stanford, CA, USA

Steven W. Allen Kavli Institute for Particle Astrophysics and Cosmology, Stanford University, Stanford, CA, USA

Ethan Anderes Statistics Department, University of California, Davis, CA, USA

Keith Arnaud NASA Goddard Space Flight Center, Greenbelt, MD, USA

Zsolt Bagoly Eötvös University, Budapest, Hungary

Paul D. Baines Department of Statistics, University of California, Davis, CA, USA

Lajos G. Balázs Konkoly Observatory, Budapest, Hungary

Nicholas M. Ball National Research Council Herzberg Institute of Astrophysics, Victoria, BC, Canada

Iréne Balmés Laboratoire Univers et Théories (LUTh), UMR 8102 CNRS, Observatoire de Paris, Université Paris Diderot, Meudon, France

A.J. Banday Institut de Recherche in Astrophysique et Planétologie, Université de Toulouse, Toulouse, France

David Banks Department of Statistical Science, Duke University, Durham, NC, USA

Richard Baraniuk Department of Electrical and Computer Engineering, Rice University, Houston, TX, USA

D. Bard SLAC National Accelerator Laboratory, Menlo Park, CA, USA

Graham Barnes CoRA, NorthWest Research Associates, Boulder, CO, USA

Shantanu Basu Department of Physics and Astronomy, London, ON, Canada

Guillaume Belanger European Space Agency, Villanueva de la Canada, Spain

Othman Benomar School of Physics, University of Sydney, Sydney, Australia

Sharmodeep Bhattacharyya Department of Statistics, University of California, Berkeley, CA, USA

P. Bielewicz Centre d'Étude Spatiale des Rayonnements, Toulouse Cedex 4, France

Alexander W. Blocker Department of Statistics, Harvard University, Cambridge, MA, USA

Joshua S. Bloom Department of Astronomy, University of California, Berkeley, CA, USA

John Bochanski Department of Astronomy and Astrophysics, Penn State University, University Park, PA, USA

Hans Böhringer Max-Planck-Institut für extraterrestrische Physik, Garching, Germany

Kirk D. Borne Department of Computational and Data Sciences, George Mason University, Fairfax, VA, USA

Brendon J. Brewer Department of Physics, University of California, Santa Barbara, CA, USA

Tamara Broderick Department of Statistics, University of California, Berkeley, CA, USA

Tamás Budavári Department of Physics and Astronomy, Johns Hopkins University, Baltimore, MD, USA

Nathaniel R. Butler School of Earth and Space Exploration, Arizona State University, Tempe, AZ, USA

Yonk-Ik Byun Department of Astronomy, Yonsei University, Seoul, Korea

Seo-Won Chang Department of Astronomy, Yonsei University, Seoul, Korea

David Chernoff Cornell University, Ithaca, NY, USA

Jessica Cisewski Department of Statistics and Operations Research, University of North Carolina, Carrboro, NC, USA

Nicolle Clements Department of Statistics, Fox School of Business and Management, Temple University, Philadelphia, PA

Andrew Connolly Department of Astronomy, University of Washington, Seattle, WA, USA

Alanna Connors Eureka Scientific, Oakland CA, USA

Thomas M. Crawford The Kavli Institute for Cosmological Physics, The University of Chicago, Chicago, IL, USA

Istvan Csabai Department of Physics and Astronomy, Johns Hopkins University, Baltimore, MD, USA

Nathan Deg Department of Astronomy, Queens University, Kingston, ON, Canada

Stewart DeSoto Department of Physics, Wheaton College, Wheaton, IL, USA

Volker Dose Max-Planck-Institut für Plasmaphysik, Garching, Germany

David Donoho Statistics Department, Stanford University, Stanford, CA, USA

Jeremy Drake Harvard-Smithsonian Center for Astrophysics, Cambridge, MA, USA

Garcia Etchegaray Estefania, Department of Statistics, Baker Hall, Carnegie Mellon University, Pittsburgh, PA, USA

Benjamin Farr Dearborn Observatory, Northwestern University, Evanston, IL, USA

Krzysztof Findeisen Department of Astronomy, California Institute of Technology, Pasadena, CA, USA

Rainer Fischer Max-Planck-Institut für Plasmaphysik, Garching, Germany

Didier Fraix-Burnet Université Jules Fournier, Grenoble Cedex 9, France

Peter E. Freeman Department of Statistics, Carnegie Mellon University, Pittsburgh, PA, USA

Christopher Genovese Department of Statistics, Carnegie Mellon University, Pittsburgh, PA, USA

Lluís Gil Tavernes Blanques, University of València, València, Spain

Facundo Gomez Department of Physics and Astronomy, Michigan State University, East Lansing, MI, USA

Carlo Graziani Department of Astronomy, University of Chicago, Chicago, IL, USA

Philip C. Gregory Physics and Astronomy, University of British Columbia, Vancouver, BC, Canada

Dirk Grupe Department of Astronomy and Asrophysics, Penn State University, University Park, PA, USA

Fabrizia Guglielmetti Max Planck Institut für Exterterrestrischephysik, Giessenbachstr, Garching bei Muenchen, Germany

Wenge Guo Department of Mathematical Sciences, New Jersey Institute of Technology, University Heights, Newark, NJ

Jaegyoon Hahm Supercomputing Center, Korea Institute of Science and Technology Information, Daejeon, Korea

Kenji Hamaguchi NASA Goddard Space Flight Center, Greenbelt, MD, USA

Jiangang Hao Fermi National Laboratory, Batavia, IL, USA

Salman Habib Argonne National Laboratory, Argonne, IL, USA

Alan Heavens SUPA, Institute for Astronomy, Royal Observatory of Edinburgh, Edinburgh, UK

Katrin Heitmann Argonne National Laboratory, Argonne, IL 60439, USA

Martin A. Hendry School of Physics and Astronomy, Kelvin Building, University of Glasgow, Glasgow, UK

David Higdon Los Alamos National Laboratory, Los Alamos, NM, USA

Joseph M. Hilbe School of Mathematics and Statistical Sciences, Arizona State University, Tempe, AZ, USA

Darren Homrighausen Department of Statistics, Carnegie Mellon University, Pittsburgh, PA, USA

István Horváth Department of Physics, Bolyai Military University, Budapest, Hungary

Talvikki Hovatta Department of Physics, Purdue University, West Lafayette, IN, USA

R. Izbicki Department of Statistics, Carnegie Mellon University, Pittsburgh, PA, USA

Fadili M. Jalal Groupe de Recherche en Informatique, Image, Automatique et Instrumentation de Caen, École Nationale Superieure d'Ingéntiers, Caen, France

John Berian James Niels Bohr Institute, Copenhagen, Denmark

Jens Jasche Argelander-Institut für Astronomie, Bonn, Germany

Garrett Jernigan Space Sciences Laboratory, University California, Berkeley, CA, USA

Raul Jimenez Instituto de Ciencias del Cosmos, Facultad de Fisica, Universidad de Barcelona, Barcelona, Spain

Babu G. Jogesh Department of Statistics, Pennsylvania State University, PA, USA

Vinay L. Kashyap Harvard-Smithsonian Center for Astrophysics, Cambridge, MA, USA

Neal Katz Department of Astronomy, University of Massachusetts, Amherst, MA, USA

David Keitel Max-Planck-Institut für Gravitationsphysik, Hannover, Germany

Brandon C. Kelly Department of Physics, University of California, Santa Barbara CA, USA

Michael Kuhn Department of Astronomy and Astrophysics, Penn State University, University Park, PA, USA

Aleksandra Kurek Astronomical Observatory, Jagiellonian University, Krakow, Poland

Guilhem Lavaux Department of Physics and Astronomy, University of Waterloo, Waterloo, ON, Canada

Earl Lawrence Los Alamos National Laboratory, Los Alamos, NM, USA

Ann B. Lee Department of Statistics, Carnegie Mellon University, Pittsburgh, PA, USA

Thomas C.M. Lee Department of Statistics, University of Statistics, Davis, CA, USA

Adrienne Leonard IRFU Service d'Astrophysique, Centre d'Études Atomiques, Gif sur Yvette Cedex, France

James Long Department of Statistics, University of California, Berkeley, CA, USA

Thomas J. Loredo Department of Astronomy, Cornell University, Ithaca, NY, USA

Yu Lu Kavli Institute for Particle Astrophysics and Cosmology, Stanford University, Stanford, CA, USA

Daniel Pereira Machado Service d'Astrophysique IRFU, Centre d'Études Atomiques de Saclay, Paris, France

Barry Madore Carnegie Observatories, Pasadena, CA, USA

Kaisey S. Mandel Harvard-Smithsonian Center for Astrophysics, Cambridge, MA, USA

Adam Mantz NASA Goddard Space Flight Center, Greenbelt, MD, USA

Domenico Marinucci Department of Mathematics, University of Rome, Roma, Italy

Elizabeth Martinez-Gomez Center for Astrostatistics, Penn State University, University Park, PA, USA

Kushal Mehta Harvard-Smithsonian Center for Astrophysics, Cambridge, MA, USA

Attila Mészáros Faculty of Mathematics and Physics, Charles University, Prague 8, Czech Republic

Brendan Miller Department of Astronomy, University of Michigan, Ann Arbor, MI, USA

H.J. Mo Department of Astronomy, University of Massachusetts, Amherst, MA, USA

Debashis Mondal Department of Statistics, University of Chicago, Chicago, IL, USA

Adam N. Morgan Department of Astronomy, University of California, Berkeley, CA, USA

Kellen Murphy Department of Physics and Astronomy, Ohio University, Athens, OH, USA

J. Newman Department of Physics and Astronomy, University of Pittsburgh, Pittsburgh, PA, USA

Tim Naylor School of Physics, University of Exeter, Exeter, UK

Mark Neyrinck Department of Physics and Astronomy, Johns Hopkins University, Baltimore, MD, USA

Fabio Noviello CNRS and Université Paris-Sud, Orsay, France

Martin Paegert 6301 Stevenson Ctr, Vanderbilt University, Nashville, TN, USA

Paykari Paniez Service d'Astrophysique, Centre dÉtudes Atomiques de Saclay, Orme des Merisiers, Gif-sur-Yvette, France

M.A. Papa Max Planck Institute for Gravitational Physics, Potsdam, Germany

Donald B. Percival Department of Statistics, University of Washington, Seattle, WA, USA

John Peterson Department of Physics, Purdue University, West Lafayette, IN, USA

Rosati Piero European Southern Observatory, Garching, Germany

Andrew Pollock XMM-Newton Science Operations Center, European Space Astronomy Centre, Madrid, Spain

Adrian Pope Argonne National Laboratory, Santa Fe, NM, USA

Jennifer Posson-Brown Harvard-Smithsonian Center for Astrophysics, Cambridge MA, USA

R. Prix Max Planck Institute for Gravitational Physics, Hannover, Germany

Pavlos Protopapas Harvard-Smithsonian Center for Astrophysics, Harvard University, Cambridge, MA, USA

David Rapetti Dark Cosmology Centre, Niels Bohr Institute, University of Copenhagen, Copenhagen, Denmark

Kathy Reeves Harvard-Smithsonian Center for Astrophysics, Cambridge, MA, USA

Christian L. Reichardt Department of Physics, University of California, Berkeley, CA, USA

John Rice Department of Statistics, University of California, Berkeley, CA, USA

Joseph W. Richards Department of Statistics and Astronomy, University of California, Berkeley, CA, USA

Erik Rosolowsky University of British Columbia, Kelowna, BC, Canada

Gerald Ruch Department of Physics, University of St. Thomas, Saint Paul, MN, USA

David Ruppert School of Operations Research and Information Engineering, Cornell University, Ithaca, NY, USA

Claes-Erik Rydberg Department of Astronomy, Stockholm University, Stockholm, Sweden

Steve Saar Harvard-Smithsonian Center for Astrophysics, Cambridge MA, USA

Gendith Sardane Department of Physics and Astronomy, University of Pittsburgh, Pittsburgh, PA, USA

Sanat K. Sarkar Department of Statistics, Fox School of Business and Management, Temple University, Philadelphia, PA

Jeffrey D. Scargle Planetary Systems Branch, NASA Ames Research Center, Moffett Field, CA, USA

Chad M. Schafer Department of Statistics, Carnegie Mellon University, Pittsburgh, PA, USA

Sam Schmidt Department of Physics, University of California, Davis, CA, USA

P. Schneider Argelander-Institut für Astronomie, University of Bonn, Bonn, Germany

Carolyn Sealfon Department of Physics, West Chester University, Merion Science Center, West Chester, PA, USA

Marco Selig Max Planck Institut für Astronomie, Garching, Germany

Jonathan Sick Department of Physics, Queen's University, Kingston, ON, Canada

Jennifer Siegal-Gaskins Center for Cosmology and Astro-Particle Physics, Ohio State University, Columbus, OH, USA

Aneta Siemiginowska Harvard-Smithsonian Center for Astrophysics, Cambridge, MA, USA

Maham Siddiqi Max Planck Institute for Gravitational Physics, Hannover, Germany

Dejan Slepčev Department of Mathematics, Carnegie Mellon University, Pittsburgh, PA, USA

Kunlaya Soiaporn School of Operations Research and Information Engineering, Cornell University, Ithaca, NY, USA

Jean-Luc Starck Service d'Astrophysique, Centre d'Études Atomiques de Saclay, Orme des Merisiers, Gif-sur-Yvette, France

Dan L. Starr Astronomy Department, University of California, Berkeley, CA, USA

Nathan Stein Department of Statistics, Harvard University, Science Center, Cambridge, MA, USA

David Stenning Department of Statistics, University of California, Irvine, CA, USA

Eric R. Switzer Kavli Institute Cosmological Physics, University of Chicago, Chicago, IL, USA

Dorottya Szécsi Eötvös Loránd University, Budapest, Hungary

Ismael Tereno Department de Física, Universidade de Lisboa, Lisbon, Portugal

Luke Tierney Department of Statistics and Actuarial Science, University of Iowa, Iowa City, IA, USA

Paolo Tozzi Osservatorio Astronomico di Trieste, Trieste, Italy

Roberto Trotta Astrophysics Group, Blackett Laboratory, Imperial College London, London, UK

Irina S. Udaltsova Department of Statistics, University of California, Davis CA, USA

David A. van Dyk Department of Mathematics, Imperial College London, London, UK

Alexander van Engelen McGill University, Montréal, QC, Canada

Roland Vavrek European Space Astronomy Centre, Villanueva de la Cañada, Madrid, Spain

Arun Vedachalam Department of Computational and Data Sciences, George Mason University, Fairfax, MD, USA

Licia Verde Instituto de Ciencias del Cosmos, Facultad de Fisica, Universidad de Barcelona, Barcelona, Spain

Péter Veres Department of Astronomy & Astrophysics, Pennsylvania State University, University Park, PA, USA

Ricardo Vilalta Department of Computer Science, University of Houston, Houston, TX, USA

Benjamin D. Wandelt Institut d'Astrophysique, Paris, France

Ira Wasserman Department of Astronomy, Cornell University, Ithaca, NY, USA

Michael Way NASA Ames Research Center, Moffett Field, CA, USA

Martin D. Weinberg Department of Astronomy, Lederle Graduate Research Center, University of Massachusetts, Amherst, MA, USA

Philipp Wilking Institut für Astronomie, University of Bonn, Bonn, Germany

Jin Xu Department of Statistics, University of California, Irvine, CA, USA

Ilsang Yoon Department of Astronomy, University of Massachusetts, Amherst, MA, USA

C. Alex Young NASA Goddard Space Flight Center, Greenbelt, MD, USA

Andreas Zezas Department of Physics, University of Crete, Heraklion, Greece

Part I
Statistics in Cosmology

Chapter 1
Likelihood-Free Inference in Cosmology: Potential for the Estimation of Luminosity Functions

Chad M. Schafer and Peter E. Freeman

Abstract Statistical inference of cosmological quantities of interest is complicated by significant observational limitations, including heteroscedastic measurement error and irregular selection effects. These observational difficulties exacerbate challenges posed by the often-complex relationship between estimands and the distribution of observables; indeed, in some situations it is only possible to simulate realizations of observations under various assumed cosmological theories. When faced with these challenges, one is naturally led to consider utilizing repeated simulations of the full data generation process, and then comparing observed and simulated data sets to constrain the parameters. In such a scenario, one would not have a likelihood function relating the parameters to the observable data. This paper will present an overview of methods that allow a likelihood-free approach to inference, with emphasis on approximate Bayesian computation, a class of procedures originally motivated by similar inference problems in population genetics.

1.1 Introduction

The ever-increasing efforts to build catalogs of astronomical objects, and to measure key properties of these objects, is, in large part, motivated by the goal of inferring unknown constants that characterize the Universe. This paper seeks to present an example of such a problem, and to describe some of the features of the data and their collection that complicates what is otherwise a standard statistical inference problem. To an outsider of this field, it can be surprising the extent to which

C.M. Schafer (✉) • P.E. Freeman
Department of Statistics, Carnegie Mellon University, 5000 Forbes Avenue, Pittsburgh, PA 15213, USA
e-mail: cschafer@stat.cmu.edu; pfreeman@stat.cmu.edu

E.D. Feigelson and G.J. Babu (eds.), *Statistical Challenges in Modern Astronomy V*, Lecture Notes in Statistics 209, DOI 10.1007/978-1-4614-3520-4_1,

Table 1.1 Examples of key cosmological parameters

Parameter	Description	In Fig. 1.1[a]
Ω_m	Ratio of total matter density to that needed for a flat Universe	0.266
Ω_Λ	Similar to Ω_m, but for dark energy density	0.734
H_0	Hubble constant: the current expansion rate of the Universe	71.0 km/s/Mpc

[a] Estimates based on WMAP7 [2]

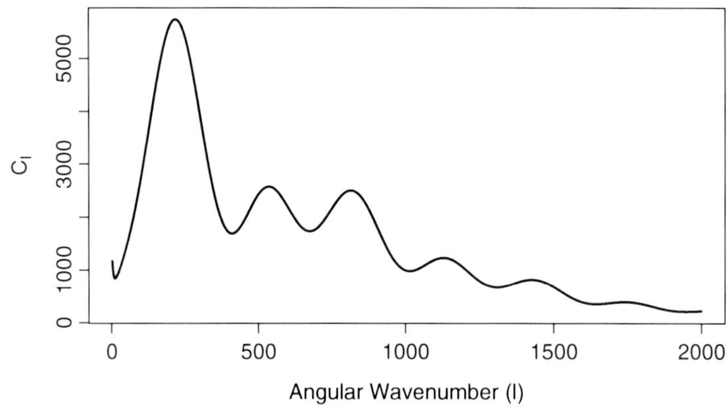

Fig. 1.1 Power spectrum, a function of cosmological parameters, of fluctuations in the temperature of photons that comprise the cosmic microwave background (CMB). The parameter values are fixed to those shown in Table 1.1

many questions regarding the nature of Universe have been boiled down to the estimation of a relatively small number of *cosmological parameters*. Table 1.1 gives some examples of these physical constants. Carefully-derived cosmological theory posits relationships between these parameters and the distribution of observables. In (relatively) simple situations, the distribution of the data is of a "standard" form, and the likelihood function can be derived. This allows for utilization of well-established methods of inference, including finding maximum likelihood estimates or exploring the posterior distribution of these parameters given the observed data.

One of the most important inference problems that fits into this framework is the estimation of cosmological parameters using fluctuations in the temperature of photons that comprise the cosmic microwave background (CMB). These photons are remnants of the time, only 300,000 years after the Big Bang, when the temperature of the Universe had cooled sufficiently for light to travel freely. The slight variation in the temperature of these photons encodes important information regarding the nature of the early Universe; the amount of correlation on different angular scales has been characterized as a function of cosmological parameters. Figure 1.1 shows the *power spectrum* that describes the Gaussian process on the sphere used to model the process; this power spectrum corresponds to the parameter values shown in Table 1.1. A succession of experiments has observed this background radiation to greater precision, and hence has achieved stronger constraints on the unknowns. The estimates in Table 1.1 are based on the recent WMAP 7 data release [2].

The relationship between the cosmological parameters and the power spectrum of the CMB fluctuations is complex: It is highly nonlinear, and there are strong degeneracies between some the parameters. The complexity of this relationship presents its own challenges. Bayesian methods dominate in cosmology, and MCMC is feasible in this situation; one only needs to make small steps in the cosmological parameter space, and the parameter vectors are mapped into the corresponding power spectrum, which in turn defines the likelihood function for the data. Schafer and Stark [3] presents a Monte Carlo method for constructing confidence regions of optimal expected size that is specifically motivated by this type of situation. Yet, both of these methods rely upon knowledge of the likelihood function of the data. Increasingly, we are faced with situations in which this is not a reasonable assumption. This may be because the distribution of the data is inherently complex, or it may be because of data corrupted by irregular truncation effects and/or heteroscedastic measurement error with complex dependence structure.

This paper describes *likelihood-free* approaches to inference, in particular, *approximate Bayesian computation* (ABC). The term "likelihood-free" is not intended to imply that a likelihood function does not exist in these applications; instead, it is the case that the likelihood function is too complex to admit a form that can be evaluated reliably for different values of the parameters of interest. These procedures will instead be built upon repeated simulation of the data-generating process (allowing for the incorporation of any complex computer models, data contamination, or selection effects) and then comparing simulated with observed data. Implementation of these approaches presents their own set of challenges. The difficulty of deriving an appropriate likelihood function is replaced with that of finding an approximate *sufficient statistic* for the parameter of interest. There are also computational challenges to implementing these procedures, but these can be mitigated via the design of efficient algorithms. This paper will present a brief introduction to some techniques and directions for addressing these challenges.

Another objective of this paper is to allow a reader familiar with statistical inference, but not with astronomy, the chance to learn some background on a relatively simple cosmological inference problem that possesses some of the aforementioned challenges. In the next section we will present two examples, with background information. The first is a stylized example of estimating cosmological parameters using observations of Type Ia supernovae. This example serves largely to introduce important concepts and methods. The second is the problem of estimating a bivariate *luminosity function*, the distribution of astronomical objects of interest as a function of their distance and the amount of light they emit. We will utilize the quasar catalog of [4] to motivate a promising approach to estimating the bivariate luminosity function which relies upon forward simulation of the full data generation process.

1.2 Examples and Astronomical Background

In this section we will present two examples of statistical inference using astronomical data. The first is relatively simple and will serve only to demonstrate basic likelihood-free techniques. The second application possesses the type of complications that motivate the consideration of these approaches. Both of these build upon the same astronomical background, including the following key quantities described below.

Key Quantities in the Examples

1. **Redshift** (often denoted z)—Because the Universe is expanding, light emitted by an astronomical object is shifted to longer wavelengths prior to reaching the observer: the ratio of the wavelength at which the light is observed to the wavelength when emitted equals $1+z$. Since the magnitude of this shift increases as a function of the time since the light was emitted, redshift is often taken as a (nonlinear) proxy for time (or distance). For the current epoch, $z = 0$; for quasars, $z \leq 7$; and for the CMB, the most distant structure yet observed in the Universe, $z \approx 1089$.

2. **Apparent magnitude** (m)—The brightness of the object as measured by the observer. Magnitudes are measured on a logarithmic scale such that *decreasing* the magnitude by five corresponds to changing the brightness by a factor of 100. The root of the magnitude system was the classification of stars by the Greek astronomer Hipparchus, who used one for the brightest stars and six for the faintest.

3. **Absolute magnitude** (M)—The apparent magnitude of that an object would have if it were located 10 pc (or about 32 light-years) from Earth. The relationship between m and M in a flat Universe can be written as

$$M = m - \frac{(1+z)}{c\,H_0} \int_0^z \left(\Omega_m (1+u)^3 + \Omega_\Lambda\right)^{-0.5} du, \qquad (1.1)$$

where c is the speed of light, and H_0, Ω_m, and Ω_Λ are among the cosmological parameters shown in Table 1.1.

Equation 1.1 establishes a relationship between a measurable property of astronomical objects (the apparent magnitude), and a scientifically useful quantity (the absolute magnitude). Note how this transformation depends not only on the redshift of the object, but on the values of unknown physical constants. In the examples that follow, this expression will be utilized in different ways. In the first case, Type Ia

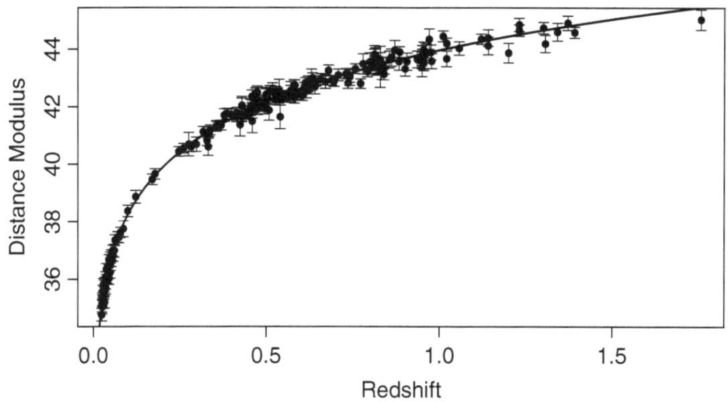

Fig. 1.2 Plot of distance modulus vs. redshift for a sample of 182 SNe Ia [5]. The curve is the predicted relationship when $H_0 = 72.76$ km/s/Mpc and $\Omega_m = 0.341$, the MLE under a simple model

supernovae, for which both M and m are known, are used in order to constrain the cosmological parameters. In the second example, values for these parameters are assumed in order transform m into M for a sample of quasars.

1.2.1 Demonstration Example: Estimation with Type Ia Supernovae

A white dwarf star that accumulates matter from a companion star will not remain stable once its mass exceeds the *Chandrasekhar limit* of approximately 1.38 times the mass of the sun. The resulting thermonuclear explosion is called a *Type Ia supernova* (SN Ia). The uniformity in mass of these stars at the time of their demise implies uniformity in their absolute magnitudes (M) and hence SNe Ia are approximate *standard candles*, in that variation in their apparent magnitude (m) (measured from Earth) is attributable primarily to variation in the differences in their distance from us. Thus, the *distance modulus* (denoted μ), defined to be the difference between the apparent and absolute magnitudes, is a proxy for the space-time distance to the SN Ia. Redshift (z) can also be considered a proxy for space-time distance and estimates of the redshifts are also available for each of the SNe Ia. Equation 1.1 establishes a direct relationship between distance modulus and redshift as a function of cosmological parameters H_0, Ω_m and Ω_Λ, and hence these observations can be used to constrain these parameters.

Figure 1.2 shows measurements of these quantities for each of 182 SNe Ia [5]. The error bars depicted for each distance modulus reflect the uncertainty in the magnitude measurements. These uncertainties are typically taken to be "known," derived from properties of the observing conditions and the scientific instrument in use.

For the purposes of this example, we will assume that the errors are normally distributed with mean zero, and are independent. We will also assume that $\Omega_m + \Omega_\Lambda = 1$. The result is a simple, two-parameter model, one for which it is not difficult to write out the full likelihood function.

The Statistical Model

Assume observe realizations of pairs (z_i, Y_i) for $i = 1, 2, \ldots, n$ such that

$$Y_i = \frac{(1+z_i)}{c\,H_0} \int_0^{z_i} \left(\Omega_m(1+u)^3 + (1-\Omega_m)\right)^{-0.5} du + \sigma_i \varepsilon_i$$

where the ε_i are independent, identically distributed standard normal random variables, and the σ_i are known.

In what follows we will use these data and this model to make comparisons between the between standard and likelihood-free methods for estimating H_0 and Ω_m. The solid line in Fig. 1.2 is the case where $H_0 = 72.76$ km/s/Mpc and $\Omega_m = 0.341$, the maximum likelihood estimate under this model. There are various ways in which these assumptions could be relaxed, and hence make the results of more scientific interest. As this is done, however, it will be increasingly difficult to derive the likelihood function, and one would start to see the appeal of taking a likelihood-free approach.

1.2.2 *Motivating Example: Luminosity Function Estimation*

Broadly stated, the *luminosity function* of a particular class of astronomical objects is the distribution of the absolute magnitudes of those objects. For example, one can seek to estimate the luminosity function of all galaxies, the luminosity function of galaxies that are of a particular type, the luminosity function of galaxies at redshift $z = 2.0$, and so forth. To a statistician, this is a familiar *density estimation* problem. From a cosmological perspective, it is of interest to study how the luminosity function evolves with redshift, setting up a bivariate density estimation problem in the (z, M) plane. The underlying goal is to compare predicted evolution under proposed theories with the observed evolution. Hence, we can view the luminosity function as an important cosmological unknown, and an accurate estimate of the luminosity function are of fundamental scientific interest. There are complications in this estimation, namely the presence of heteroscedastic measurement error in the key observables, and physical limitations on the objects we are able to view.

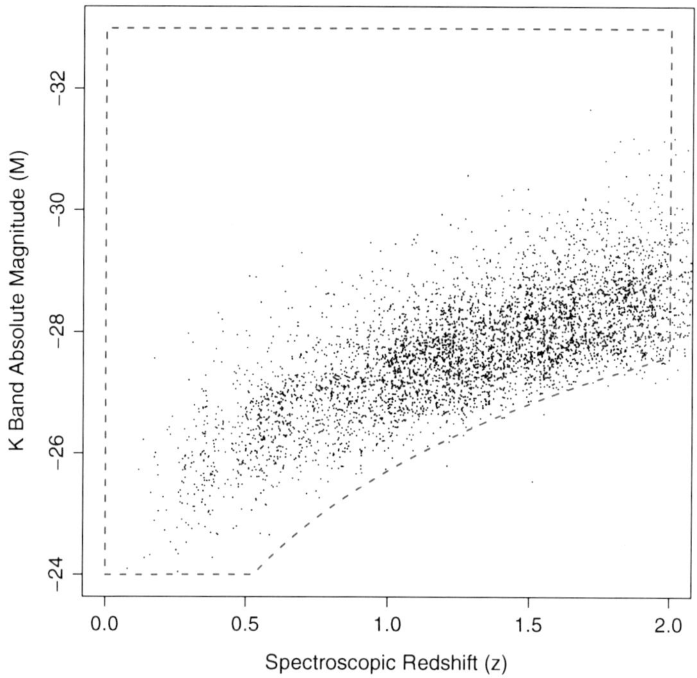

Fig. 1.3 Redshift and absolute magnitude measurements for a subset of the quasars in [4]

Here we will consider the specific problem of estimating the luminosity function of quasars. *Quasars* are ultra-luminous galactic nuclei powered by the infall of matter into supermassive black holes. Because of their compactness, they appear like stars, or "quasi-stellar," hence the name. The rate of matter infall into supermassive black holes, which dictates when a quasar is "on" or "off," is directly tied to the physics of galaxy formation and evolution. Thus the quasar luminosity function provides a means by which to constrain theoretical models of these processes. We will utilize a subset of 5,000 quasars taken from the catalog of [4]. The full catalog consists of over 130,000 quasars; for the purpose of demonstrating our methods, we will focus on the reduced sample. For the problem at hand, there are two key measured quantities for each quasar: the redshift and the apparent magnitude. One then calculates the absolute magnitude via (1.1).

Figure 1.3 shows the (z,M) pairs for each of the quasars in our sample. Outside of the dashed region, quasars in the sample are truncated because of the difficulty of observing quasars that are too dim. The curve in the truncation region arises because the limit is in terms of apparent magnitude; the depicted bound corresponds to truncating quasars with $m > 18.4$. As mentioned above, it is of interest to estimate the bivariate luminosity function (the bivariate density in (z,M) space).

1.2.2.1 Estimators for the Bivariate Luminosity Function

The irregular truncation boundary shown as the dashed line in Fig. 1.3 presents challenges to the fitting of a bivariate density to this sample, even without the presence of measurement error in the observations. If a well-motivated parametric form for the density exists, then maximum likelihood estimation would be a natural choice. But, lacking such a form, the focus has been on nonparametric estimators. Lynden-Bell [6] introduced in the astronomy literature the nonparametric maximum likelihood (NPMLE) estimator for the case of one-sided truncation of absolute magnitude and [7] derived some of the asymptotic properties of this estimator. Efron and Petrosian [8] extended the NPMLE to the case of double truncation of absolute magnitude. Each of these papers assumes that absolute magnitude and redshift are statistically independent (and, hence, that the luminosity function does not evolve with redshift.) The density estimate (or distribution function estimate) which results from a NPMLE procedure places all of the probability on observed data values, but even smoothing this estimate may not be sufficient to remove artifacts: An estimate can suffer from what [9] referred to as "large jumps," where lone data points can greatly influence the estimator. Efron and Petrosian [8] also developed a permutation test for independence of the two variables. Independence of absolute magnitude and redshift is a strong assumption, and not justified in most applications. In practice, one of these methods is applied to a narrow bin of observations in redshift.

In [10], a method is presented for fitting bivariate luminosity functions of the semiparametric form

$$\log(\phi(z,M)) = \mathbf{f}(z) + \mathbf{g}(M) + \tau z M. \tag{1.2}$$

Thus, the log density is additive in functions, estimated nonparametrically, of only z and M, plus a term that accounts for the evolution of the luminosity function with redshift. This first-order approximation to the true form for the evolution does appear to fit to observed data well; Schafer [10] makes comparisons between the results from the fitting procedure and those built on "binning," and there is good agreement; see Fig. 1.4. This form for the bivariate luminosity function will be a key ingredient to our likelihood-free approach.

1.2.2.2 A Further Complication: Redshift Estimation

Our reduced sample from [4] consists of 5,000 quasars which each have two estimates of the redshift. The first is the high-quality *spectroscopic* estimate of the redshift, constructed from the full emission spectrum of the quasar. Figure 1.5 shows such a spectrum; by matching this spectrum with a *template spectrum* of a quasar, one is able estimate to good accuracy the redshift of the observed quasar. Unfortunately, such spectroscopic data is difficult to obtain, and many experiments only provide *photometric* magnitudes accumulated over wide ranges of wavelength.

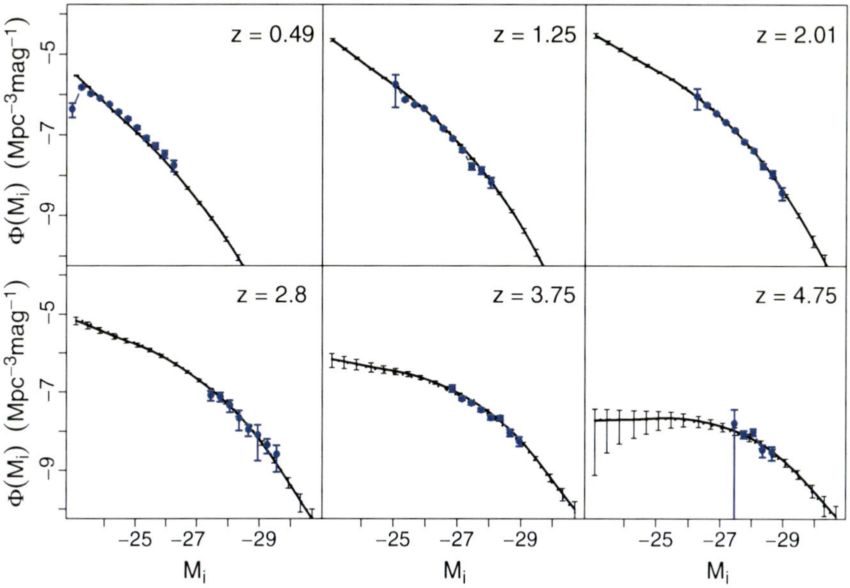

Fig. 1.4 A comparison of the evolution of the quasar luminosity function as estimated by Schafer [10] and that obtained by unbiased estimators built on binning

The left plot of Fig. 1.5 depicts the situation. Instead of observing the full spectrum, one can only observe the spectrum integrated against each of the five bands (Z, Y, J, H, and K). Then, estimation of redshift becomes a regression problem. There is a training set, consisting of quasars for which there are both spectra and photometric observations; these are used to fit a model relating the two. This model is applied to the quasars for which there are only photometric observations in order to predict their redshift. The relationship is highly nonlinear, and extensive work on this problem only has served to demonstrate the difficulty of the challenge. See the right plot of Fig. 1.5 for the results of performing such an analysis on the data of [4].

1.2.2.3 From True Bivariate Luminosity Function to Observable Data

Consider the aggregate effects of the use of photometric data:

1. Redshift (z) has *measurement error*
2. The distribution of this error depends on true redshift
3. Conversion from apparent to absolute magnitude (M) has error
4. There will be strong dependence between errors in z and M
5. The truncation will be performed on error-filled data

It would be difficult to construct an adequate likelihood function that takes into account the above features of the model for the observable data. When faced with

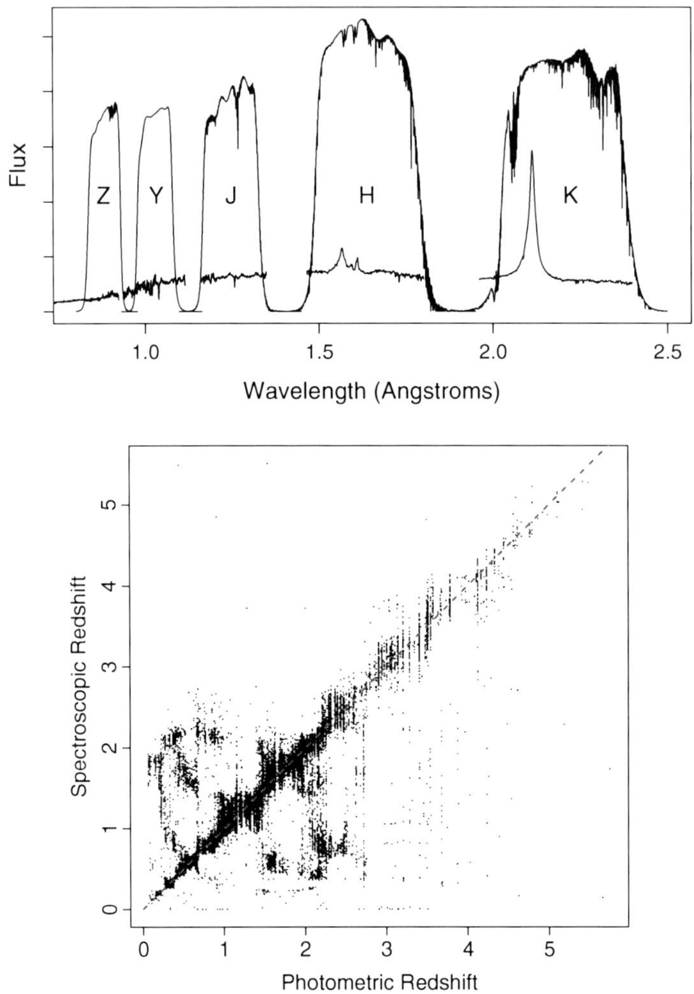

Fig. 1.5 *Left:* The spectrum of a quasar, with the filters of photometric bands superimposed. *Right:* Plot of spectroscopic redshift versus photometric redshift for 5,000 quasars in [4]

such a challenging situation, one is naturally led to consider the *forward process* that generated these data. If one is able to adequately simulate the individual steps, it would be possible to generate data sets under conditions similar to those that led to the observed data, varying only the parameters to be estimated. These simulations could then be compared to the observed data. This is the fundamental idea behind likelihood-free inference.

1.3 Likelihood-Free Inference

Standard techniques for statistical inference are built upon knowledge of (a good approximation to) the likelihood function for the data as a function of the parameters of interest. This relationship between parameters and distribution for the data, denoted $f_\theta(x)$, can be complex, but as long as one can evaluate this expression for different values of θ and x, proper implementations of well-established algorithms, such as MCMC, will lead to accurate constraints on the unknowns. A *likelihood-free* approach to inference is necessary when $f_\theta(x)$ is not available; as stated above, in this paper we concern ourselves with the case where the effect of contamination of the observations by measurement error makes (even approximate) derivation of the likelihood function impossible.

Frequentist likelihood-free approaches to inference are built upon the following, simple approximation: To estimate $f_\theta(x)$, the likelihood evaluated at data x when θ is the truth, sample B data values $x_1, x_2, \ldots, x_B$ under the model implied by θ. Then use

$$f_\theta(x) \approx K \sum_{i=1}^{B} \mathbf{1}_{\Delta(x,x_i) \leq \varepsilon} \tag{1.3}$$

for some $\varepsilon > 0$, constant K and choice of distance metric Δ. In other words, the proportion of simulated data values that are "close" to x (as measured by the metric Δ) is proportional to the likelihood function evaluated at the pair (x, θ). Diggle and Gratton (1984), for example, approximate the likelihood surface by applying nonparametric density estimators to likelihoods approximated in this way, and then proceed to find the maximum likelihood estimator. The primary challenge in such an approach is the difficulty encountered when θ is of high dimension.

Bayesian approaches are appealing because, just as with MCMC, one can generate a sample from the high-dimensional posterior and still estimate most integrals over the posterior, including marginal distributions for parameters, via Monte Carlo approximations. *Approximate Bayesian computation* (ABC) refers to a class of methods used to approximate the posterior distribution in cases where a functional form for the likelihood is not available. The development of these methods was motivated by estimation problems in population genetics, but recent work is expanding the areas of application. In this section we describe a simple algorithm utilized in this growing field of research.

The basic ABC algorithm is the *ABC Rejection Algorithm* outlined below.

The ABC Rejection Algorithm

First, define a distance metric Δ and a tolerance ε. Then, repeat the following until sample of size N is generated:

1. Choose θ^* from prior $\pi(\theta)$.
2. Generate $x_{\text{sim}} \sim f_{\theta^*}$.

(continued)

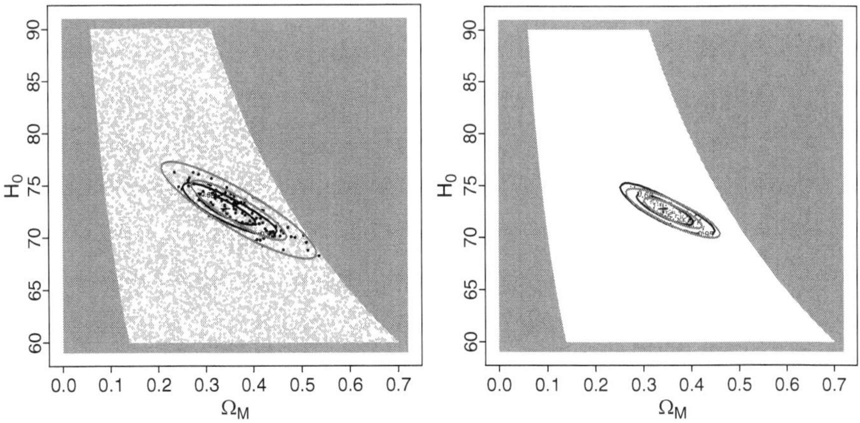

Fig. 1.6 *Left*: An application of the ABC Rejection method to the SNe example. *Gray* points correspond to rejected proposals, while *black* points are accepted. *Right*: Same situation, except using the ABC SMC method

(continued)

3. If $\Delta(x_{\text{sim}}, x_{\text{obs}}) > \varepsilon$, then return to step 1; otherwise, accept this θ^* into the posterior sample.

This algorithm works because the pair $(\theta^*, x_{\text{sim}})$ that results from steps one and two are a draw from the distribution with density $f_\theta(x)\pi(\theta)$ and, if this θ^* is accepted in step three, the probability of θ^* being in set A is

$$\int_A \int_{N(x_{\text{obs}},\varepsilon)} f_\theta(x)\pi(\theta)\,dx\,d\theta \approx K \int_A f_\theta(x_{\text{obs}})\pi(\theta)\,d\theta = \int_A \pi(\theta \mid x_{\text{obs}})\,d\theta$$

where $N(x_{\text{obs}}, \varepsilon)$ is the collection of all x values that are within ε of x_{obs}, and K is a constant that does not depend on θ or x_{obs}. Hence, the accepted θ^* is approximately distributed as a draw from the posterior $\pi(\theta \mid x_{\text{obs}})$. The left plot of Fig. 1.6 depicts the result of application of this method to the two parameter estimation problem using Type Ia SNe described above. One notes that in this case 5,633 proposed θ^* were rejected in order to generate a collection of 100 accepted parameter values, and yet the tolerance ε is still not sufficiently small for the posterior estimated from the draws (gray contours) to be a good approximation to the true posterior (black contours).

Thus, although conceptually and (typically) computationally simple, the ABC rejection algorithm can be incredibly inefficient, rejecting a high proportion of the proposed θ^*, especially if the parameter space is of high dimension. *Sequential Monte Carlo (SMC)* [11] methods were developed to address such challenges. These approaches migrate a family of N *particles* through a sequence of steps; at each step the target distribution for the particles is a little closer to the primary objective: the posterior. This allows one to start with a generous amount of tolerance, and hence not reject such a large proportion of the proposals, and then subsequently tighten the standards to the point where the distribution of the particles is similar to a sample from the posterior. In [12], a version of SMC was developed that operated in the absence of a likelihood function, again motivated by complex genetics models that did not yield a tractable form for the likelihood. This is described below.

The ABC SMC Algorithm [12]

First, define a distance metric Δ and a sequence $\varepsilon_0 > \varepsilon_1 > \cdots > \varepsilon_T$.
At main iteration $t = 0$, for each of $i = 1,2,\ldots,N$:

1. Choose θ_i^* from prior $\pi(\theta)$.
2. Generate $x_{\text{sim}} \sim f_{\theta_i^*}$.
3. If $\Delta(x_{\text{sim}}, x_{\text{obs}}) > \varepsilon_0$, then return to step 1; otherwise, accept this $\theta_i^{(t)}$.
4. Set $w_i = 1/N$.

At main iteration $t = 1,2,\ldots,T$, for each of $i = 1,2,\ldots,N$:

1. Choose θ_i^* from among the $\theta_j^{(t-1)}$ with probabilities $w_j^{(t-1)}$
2. Generate $\theta_i^{(t)}$ by perturbing θ_i^* using kernel $K(\theta_i^*,\cdot)$
3. Generate $x_{\text{sim}} \sim f_{\theta_i^{(t)}}$
4. If $\Delta(x_{\text{sim}}, x_{\text{obs}}) > \varepsilon_t$, then return to step 1; otherwise, accept this $\theta_i^{(t)}$
5. Calculate the new weight as

$$w_i^{(t)} = \frac{\pi\left(\theta_i^{(t)}\right)}{\sum_{j=1}^{N} w_j^{(t-1)} K\left(\theta_j^*, \theta_i^{(t)}\right)}$$

Note that, when using the algorithm, the $\theta_i^{(t)}$ are a sample from the distribution

$$g(\theta) = \sum_{j=1}^{N} w_j^{(t-1)} K(\theta_j^*, \theta) f_{\theta,\varepsilon_t}(x_{\text{obs}}).$$

The weights $w_i^{(t)}$ can be viewed as importance sampling weights

$$w_i^{(t)} = \frac{\pi\left(\theta_i^{(t)}\right) f_{\theta_i^{(t)},\varepsilon_t}(x_{\text{obs}})}{\sum_{j=1}^{N} w_j^{(t-1)} K\left(\theta_j^*, \theta_i^{(t)}\right) f_{\theta_i^{(t)},\varepsilon_t}(x_{\text{obs}})}.$$

This collection of parameter values can then be used as a sample from the (approximated) posterior, and then be used much in the same way as would the output of an MCMC implementation (with the small added complication of incorporating the weights). When applied in the SNe example, the improvement in the estimation of the posterior distribution can be seen in the right plot of Fig. 1.6.

1.3.1 Quantifying the Distance Between Data Sets

Both of the aforementioned algorithms are built upon the same crucial ingredient unique to the ABC approach: a distance metric Δ capable of assessing the degree of similarity between the observed data and a simulated data set. In practice, this comparison is not made between the raw data objects, but instead between *summary statistics*, either a smoothed version or a low-dimensional representation of the original data. The resulting compression is an important step; if done appropriately, the summary statistic will preserve the information useful for constraining the input parameters and throw out the useless ancillary information. Indeed, the better this summary statistic approximates a *minimal sufficient statistic,* the better the ABC procedure will mimic the results that would have been obtained with full knowledge of the likelihood function.

As a result, current research is focused on procedures for constructing such a statistic. A method for assessing the value of proposed summary statistics is proposed in [13]. In [14, 15], an approach of *indirect inference* is utilized. An auxiliary model is fit to the data that incorporates not only the parameters of interest θ, but also ancillary parameters that make the model flexible enough to fit to the real data. This model is chosen to take a sufficiently simple form that estimation of all of the parameters is feasible. The vector consisting of the MLE of these parameters serves as a summary statistic. In cosmology applications, however, it may not generally be feasible to construct such an auxiliary model. The general concept, however, is relevant: The amount of compression performed on the data to create the summary statistic should be equivalent to the compression performed when the MLE of θ is found.

For instance, in the SNe example, the summary statistic is chosen to be the fit of a smoother through the simulated redshift and absolute magnitude data. Ideally, the amount of smoothing would be equivalent to the smoothness of the set of curves found when varying H_0 and Ω_m. Of course, without knowledge of the likelihood function, one would need to utilize a more extensive set of simulations to explore the nature of how the distribution of the data changes as θ is varied. Returning again to the SNe example, repeated simulations of data sets for fixed θ would reveal the smooth relationship between redshift and the distance modulus; repeating this for many different values of θ would reveal that the shape of this curve does not change much over the parameter space. In this way, with enough simulations, one could uncover the true low-dimensional structure present in the relationship between θ and the distribution of the observable data. Such a procedure is described in [16]. This seems to be a very promising direction for the practical implementation of ABC approaches in cosmology.

1.3.2 Luminosity Function Estimation

Finally, we will briefly outline how we are implementing a likelihood-free approach to estimating luminosity functions, specifically to analyze the quasar sample of [4]. First, we assume that the true form of the bivariate luminosity function (i.e., the bivariate density) takes the form given in (1.2). As already mentioned, previous studies have justified this choice. It is further assumed that the functions $\mathbf{f}(\cdot)$ and $\mathbf{g}(\cdot)$ are quadratic; the result is that there are seven parameters in the model once τ is included. A normal prior is assumed for each of these parameters. Once values are chosen for each of these parameters, once can then run the "forward process" shown below to generate data that has been subjected to the same effects as the observed data.

- Draw true z and M values
- Convert to true apparent magnitude m
- Simulate photometric redshift by drawing from joint distribution
- Calculate estimated absolute magnitude M
- Apply truncation to error-filled observations

Once generated, a data set is converted into a "summary statistic" by fitting a bivariate density to the observations; this is again using the form given in (1.2). The distance is then calculated using simple L_2 distance between the two (observed and simulated) bivariate densities. Although this is a challenging implementation, some of the preliminary results are promising. Figure 1.7 show an estimated luminosity function when the ABC SMC method was applied to a case where the data were subjected to errors and truncation identical to those present in the sample of [4], but the truth was fixed and shown as the dashed line.

1.4 Conclusion

This article presents an overview of approaches to approximate Bayesian computation, which are likelihood-free statistical inference procedures. These could prove to be useful in a range of cosmological inference problems. Here, the framework for

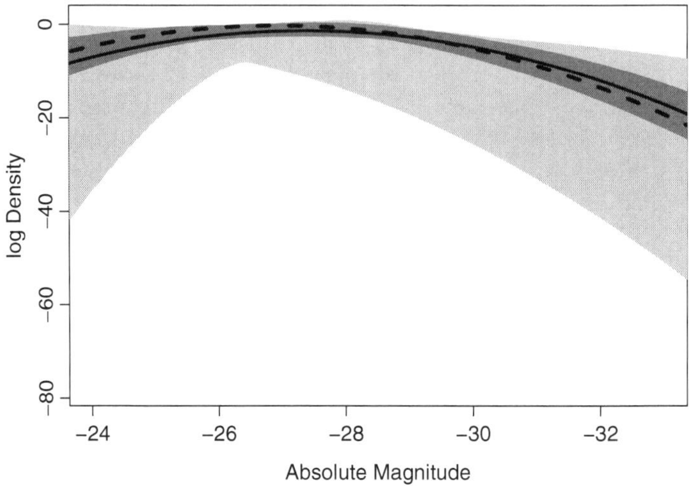

Fig. 1.7 Quasar luminosity function estimate based on simulations. The *dashed line* is the known truth. The inner band is the 68% credible region, while the larger, outer region is the 95% credible region

the application of these methods to luminosity function estimation is motivated. Of particular relevance is how these procedures could allow for adequate incorporation of the significant observation limitations that are present, including the reality of the limitations of photometric estimates of redshifts. If successful, these approaches will make full use of the flood of data to be gathered by photometric surveys.

References

1. P.J. Diggle, R.J. Gratton, J. Royal Stat. Soc., ser B **46**, 193–212 (1984)
2. D. Larson, J. Dunkley, G. Hinshaw, E. Komatsu, M.R. Nolta, C.L. Bennett, B. Gold, M. Halpern, R.S. Hill, N. Jarosik, A. Kogut, M. Limon, S.S. Meyer, N. Odegard, L. Page, K.M. Smith, D.N. Spergel, G.S. Tucker, J.L. Weiland, E. Wollack, E.L. Wright, Astrophys. J. Suppl. **192**(2), 16 (2011)
3. C. Schafer, P. Stark, J. Am. Stat. Assoc. **192**, 1080 (2009)
4. M.A. Peth, N.P. Ross, D.P. Schneider, Astron. J. **141**(4), 105 (2011)
5. A.G. Riess, L.G. Strolger, S. Casertano, H.C. Ferguson, B. Mobasher, B. Gold, P.J. Challis, A.V. Filippenko, S. Jha, W. Li, J. Tonry, R. Foley, R.P. Kirshner, M. Dickinson, E. MacDonald, D. Eisenstein, M. Livio, J. Younger, C. Xu, T. Dahln, D. Stern, Astrophys. J. **659**, 98 (2007)
6. D. Lynden-Bell, Month. Not. Royal Astron. Soc. **155**, 95 (1971)
7. M. Woodroofe, Ann. Stat. **13**(1), 163 (1985)
8. B. Efron, V. Petrosian, J. Am. Stat. Assoc. **94**(447), 824 (1999)
9. M. Woodroofe, in *Statistical Challenges in Modern Astronomy*, ed. by E. Feigelson, G. Babu (Spring-Verlag, New York, 1992), pp. 196–200
10. C. Schafer, Astrophys. J. **661**, 703 (2007)
11. P. Del Moral, A. Doucet, A. Jasra, J. Roy. Stat. Soc., Ser. B **68**(3), 411 (2006)

12. M.A. Beaumont, J.M. Cornuet, J.M. Marin, C.P. Robert, Biometrika pp. 983–990 (2009)
13. P. Joyce, P. Marjoram, Statistical Applications in Genetics and Molecular Biology **7**(1) (2008)
14. K. Heggland, A. Frigessi, J. Roy. Stat. Soc., Ser. B **66**(2), 447 (2004)
15. C.C. Drovandi, A.N. Pettitt, M.J. Faddy, J. Roy. Stat. Soc., Ser. C **60**(3), 317 (2011)
16. P. Fearnhead, D. Prangle, ArXiv e-prints (2010)

Chapter 2
Commentary: Likelihood-Free Inference in Cosmology: Potential for the Estimation of Luminosity Functions

Martin A. Hendry

Abstract The identification, diagnosis and removal of systematic biases, due to e.g. measurement errors and observational selection effects, has become a key challenge for the so-called 'era of precision cosmology'. In this commentary I will describe some specific examples of where and how this challenge may arise in the analysis of astronomical surveys, thus illustrating ways in which the construction of an explicit likelihood function is rendered complicated in this field. These various examples therefore provide further motivation for the potential usefulness of the likelihood-free inference approach which Schafer has proposed.

2.1 Introduction

The 20 years since the first SCMA conference have seen rapid growth in the reach and impact of astrostatistics—particularly in the field of cosmology. The application of physically well-motivated cosmological probes such as Type Ia supernovae (SNIe) and the cosmic microwave background radiation has placed strong constraints on the parameters which define our cosmological model, leading to the emergence of the so-called "Concordance Cosmology", supported by observations across a range of astrophysical phenomena. While there remain serious unresolved issues with the Concordance model, the quantity and quality of the data that emerged in the late 1990s prompted the label "the era of precision cosmology" to enter common use [1].

The appropriateness of this label is undermined, however, by the potential impact of systematic errors. These may arise for a variety of reasons, including instrumental or atmospheric effects, measurement errors and observational selection

---

M.A. Hendry (✉)
SUPA, School of Physics and Astronomy, University of Glasgow, Glasgow, G12 8QQ, UK
e-mail: Martin.Hendry@glasgow.ac.uk

E.D. Feigelson and G.J. Babu (eds.), *Statistical Challenges in Modern Astronomy V*, Lecture Notes in Statistics 209, DOI 10.1007/978-1-4614-3520-4_2,

due to e.g. truncation or censoring, and may be strongly correlated, non-Gaussian, non-stationary or otherwise problematic. Their identification and diagnosis can present significant challenges for the analysis of astronomical surveys via traditional likelihood-bases methods. In this brief commentary I will describe some specific examples of where and how these challenges may arise—thus providing further motivation for the potential usefulness of the likelihood-free inference approach which Schafer has proposed.

2.2 Systematic Effects in Astronomical Surveys

The surveying of astronomical populations is commonplace across a wide range of scales, from the statistics of nanoflares on the Sun to the demographics of distant quasars. As Schafer has noted in the preceding article, the approach adopted to date in studying astronomical populations has generally been likelihood based. For instance in estimating the galaxy luminosity function (LF) a range of maximum likelihood methods—both parametric and robust—has been developed, many of which explicitly account for the impact of observational selection (see [2] for a recent and comprehensive review) and the semi-parametric method of [3] is a powerful recent addition to these techniques.

In this context however, and as the preceding article also discusses, a significant complication in this field is the growing prevalence in very large survey datasets of photometric redshifts. These have hugely increased the volume and size of redshift surveys and the efficiency with which they may be carried out but at the cost of introducing a significant measurement error on the redshift of each source. The trend towards extremely large photometric redshift surveys is firmly set to continue as we approach the era of 'petascale' datasets promised by the Large Synoptic survey Telescope [4]. Consequently the impact of photometric redshift errors on likelihood-based approaches to survey analysis, and the exploration of alternative methodologies, appears to be an important future research direction—a conclusion which was also reached at SCMA4 in the context of the report presented there on the work of the astronomical surveys group within the 2006 Astrostatistics program at SAMSI [5]. This conclusion would appear to be equally relevant, if not more so, today.

A common feature shared by likelihood-based methods to probe survey luminosity functions is the adoption of a simple, approximate form for the sample selection function—for example a step function to describe the flux limit(s) of the survey [6]. While these approximations may be necessary to make the problem analytically tractable, the reality may be considerably more complicated, particularly when objects (such as distant SNIe or high redshift galaxies) are being detected in crowded fields, where issues of blended sources and source misclassification can be important [7]. These effects can render the flux limit of selected sources strongly dependent on environment, sky direction and 'seeing' conditions at the time of observation—all of which may not easily be reducible to a simple step function of flux alone.

Another common problem with flux limited surveys is where the sources are originally selected in the optical—based on a historical catalogue of e.g. B-band galaxy apparent magnitudes—but the survey involves observations made in another waveband, for example I-band photometry for the purpose of estimating galaxy distances and peculiar velocities via the Tully-Fisher relation [8]. In this situation the intrinsic correlation between galaxy luminosity and colour means that the B-band selection to which the original catalogue was subject will translate into an I-band selection function in the Tully-Fisher survey. However, since the correlation between B-band luminosity and $B-I$ colour is not perfect but has an appreciable scatter, the I-band selection function will be blurred even if the original B-band selection is well described by a sharp apparent magnitude limit [9].

A further complication when observing the very distant Universe is that surveys of e.g. quasars or high redshift galaxies may be subject to complex and poorly understood evolutionary effects (indeed probing this source evolution is often the main object of the survey in the first place!). In addition the application of so-called 'k-corrections' is required because the spectral energy distribution emitted by a high redshift source in its rest frame will be observed redshifted towards longer wavelengths by the expansion of the Universe [10, 11].

Other surveyed sources such as radio pulsars, gamma ray bursts or active galactic nuclei may be affected by geometrical selection effects, where the emitted radiation is strongly anisotropic [12]. These effects can impact significantly on the detectability of sources and influence their apparent brightness due to e.g. relativistic beaming, as well as introducing strong degeneracies between source parameters such as inferred distance and inclination to the line of sight. Similar issues are now being confronted in the nascent field of gravitational-wave astronomy [13], where the selection function of e.g. observed inspiralling binary neutron star sources will be the result of a complex interplay between the underlying cosmological model, the intrinsic star formation rate and a sky sensitivity pattern which is strongly dependent on direction, source orientation and frequency of the emitted gravitational waves [14].

Another very common and important source of systematic error in survey data is the effect of extinction: the wavelength dependent absorption of light by dust either in the environs of the source itself or within our own Milky Way galaxy. Extinction effects are often dealt with by carrying out multi-wavelength observations and correcting for their impact by fitting a (usually parametric) extinction law as a function of wavelength. This technique has been used extensively for example to infer extinction-free estimates of the distance to Cepheid variable stars in external galaxies observed by the Hubble Space Telescope [15].

Multiwavelength observations are also a key feature of the methodology used to harness SNIe as cosmological distance indicators. The multiwavelength approach is employed both to diagnose and correct for extinction and to improve the precision of the distance indicator itself by exploiting empirical correlations between the shape of the SNIe light curves and their intrinsic luminosity at different wavelengths. For more than 15 years advanced Bayesian methods have been applied for calibrating these relations to derive SNIe distance estimates [16]. Recently Mandel [17] has presented a sophisticated multilevel Bayesian model that addresses simultaneously extinction, intrinsic light curve shape, possible source evolution and cosmological parameter extraction. While this treatment is certainly 'state of the art' it shares with many of the other survey examples listed here the requirement of a complicated likelihood function, perhaps featuring a significant number of nuisance parameters, to fully capture the intrinsic characteristics of the source population and the observational selection effects to which they are subject.

2.3 The Case for a Likelihood-Free Approach

All of the complicating factors listed in the previous section—crowded fields, colour correlations, evolutionary effects, k-corrections, source orientation and beaming, extinction—are relatively straightforward to *simulate*, i.e. to model numerically via Monte Carlo simulation, but are not so easy to explicitly include in a likelihood model without potentially rendering that model unwieldy. In contrast, therefore, to the traditional methodology whereby adopts a likelihood function model that is as simple as possible and estimates the parameters of that model (see e.g. the VELMOD approach of [18] as a good archetype, in the area of peculiar velocity reconstruction), one can envisage instead a "forward modelling" approach in which one constructs sophisticated "mock" datasets that can simulate faithfully some or all of the above factors that would influence the journey of a real photon (or graviton!) from source to detectors. As described in the preceding article, one would draw inferences about the source population by comparing these mock datasets with the real survey data—analogous to the approach that has been adopted for many years in generating mock galaxy catalogues from high resolution n-body simulations of large scale structure [19].

As the preceding article has recognised, the key challenge in this approach is identifying a suitable metric for comparing the mock and real datasets, or some appropriate summary statistic constructed therefrom. The ABC algorithms which Schafer presents appear to offer a useful and practical solution to this challenge—particularly the sequential Monte Carlo algorithm which largely overcomes the problem of inefficient sampling of the Rejection algorithm. This is a crucial improvement since, as we have seen in Sect. 67.2, the complexity of simulations required to capture adequately the details of many future cosmological data sets may be considerable.

In a similar vein the preceding article underlines the importance of identifying and constructing useful summary statistics that *efficiently* measure the degree of similarity between the observed and simulated datasets. He proposes, for example, fitting a low-dimensional smoother through the real and simulated supernovae redshift and magnitude data to represent the luminosity distance-redshift relation. This is an approach that has already been explored—using a variety of different basis functions [20–22]—as an efficient method for representing non-parametrically the luminosity distance-redshift relation and its integral relationship to the cosmic equation of state. An approach of this form, applied to a variety of other cosmological datasets, would appear to hold promise for the efficient implementation of likelihood-free inference methods in the future.

References

1. M.S. Turner, arXiv: astro-ph/9811366
2. R.W.I. Johnston, Astron. & Astrophys. Reviews, **19**, 41 (2011)
3. C. Schafer, Astrophys. J. **661**, 703 (2007)
4. http://www.lsst.org. See also Z. Ivezic *et al.* arXiv: 0805.2366
5. T.J. Loredo, in 'Statistical Challenges in Modern Astronomy IV', ASP Conf. Ser. **371**, 121 (2007)
6. T.J. Loredo and M.A. Hendry, in 'Bayesian Methods in Cosmology', eds. A.R. Liddle et al. (Cambridge University Press), p245 (2010)
7. D.J. Mortlock, in 'Bayesian Methods in Cosmology', eds. A.R. Liddle *et al.* (Cambridge University Press), p193 (2010)
8. C. Springob, K.L. Masters, M.P. Haynes, R. Giovanelli and C. Marinoni, Astrophys. J. Supp. **172**, 599 (2007)
9. J.A. Willick, Astrophys. J. Supp. 92, 1 (1994)
10. E. Cameron and S.P. Driver, Astron. & Astrophys. **493**, 489 (2009)
11. A.L. O'Mill, F. Duplancic, G. Lambas and L. Sodr/'e Jr., Mon. Not. Royal Astron. Soc. **413**, 1395 (2011)
12. E. Berger *et al.*, Astrophys. J. **664**, 1000 (2007)
13. S. Nissanke, D.E. Holz, S.A. Hughes, N. Dalal, and J.L. Sievers, Astrophys. J. **725**, 396 (2010)
14. S.R. Taylor, J.R. Gair and I. Mandel, arXiv: gr-qc/1108.5161
15. W.L. Freedman and B.F. Madore, Annual Revs. Astron. & Astrophys. **48**, 673 (2010)
16. A.G. Riess, W.H. Press and R.P. Kirshner, Astrophys. J. **473**, 88 (1996)
17. K.S. Mandel, G. Narayan and R.P. Kirshner, Astrophys. J. **731**, 120 (2011)
18. J.A. Willick and M.A. Strauss, Astrophys. J. **507**, 64 (1998)
19. S. Cole, S. Hatton, D.H. Weinberg and C.S. Frenk, Astrophys. J. **300**, 945 (1998)
20. T.D. Saini, S. Raychaudhury, V. Sahni and A.A. Starobinsky, Phys. Rev. Lett. **85**, 1162 (2000)
21. A. Shafieloo, U. Alam, V. Sahni and A.A. Starobinsky, Mon. Not. Royal Astron. Soc. **366**, 1081 (2006)
22. C.A. Clarkson and C. Zunckel, Phys. Rev. Lett. **104**, 21 (2010)

Chapter 3
Robust, Data-Driven Inference in Non-linear Cosmostatistics

Benjamin D. Wandelt, Jens Jasche, and Guilhem Lavaux

Abstract We discuss two projects in non-linear cosmostatistics applicable to very large surveys of galaxies. The first is a Bayesian reconstruction of galaxy redshifts and their number density distribution from approximate, photometric redshift data. The second focuses on cosmic voids and uses them to construct *cosmic spheres* which allow reconstructing the expansion history of the Universe using the Alcock-Paczynski test. In both cases we find that non-linearities *enable* the methods or *enhance* the results: non-linear gravitational evolution creates voids and our photo-z reconstruction works best in the highest density (and hence most non-linear) portions of our simulations.

3.1 What is Cosmostatistics?

Cosmostatistics is the discipline of using the departures from homogeneity observed in astronomical surveys to distinguish between cosmological models. It therefore plays a central role in the cosmological agenda for the coming decade, which is to

B.D. Wandelt (✉)
UPMC Univ Paris 06, UMR7095, Institut d'Astrophysique de Paris, F-75014, Paris, France

CNRS, UMR7095, Institut d'Astrophysique de Paris, F-75014, Paris, France
e-mail: wandelt@iap.fr

J. Jasche
Argelander-Institut für Astronomie, Auf dem Hügel 71, D-53121 Bonn, Germany
e-mail: jasche@iap.fr

G. Lavaux
Department of Physics and Astronomy, University of Waterloo, 200 University Avenue West, Waterloo, ON N2L 3G1, Canada
e-mail: lavaux@uwaterloo.ca

E.D. Feigelson and G.J. Babu (eds.), *Statistical Challenges in Modern Astronomy V*, Lecture Notes in Statistics 209, DOI 10.1007/978-1-4614-3520-4_3,

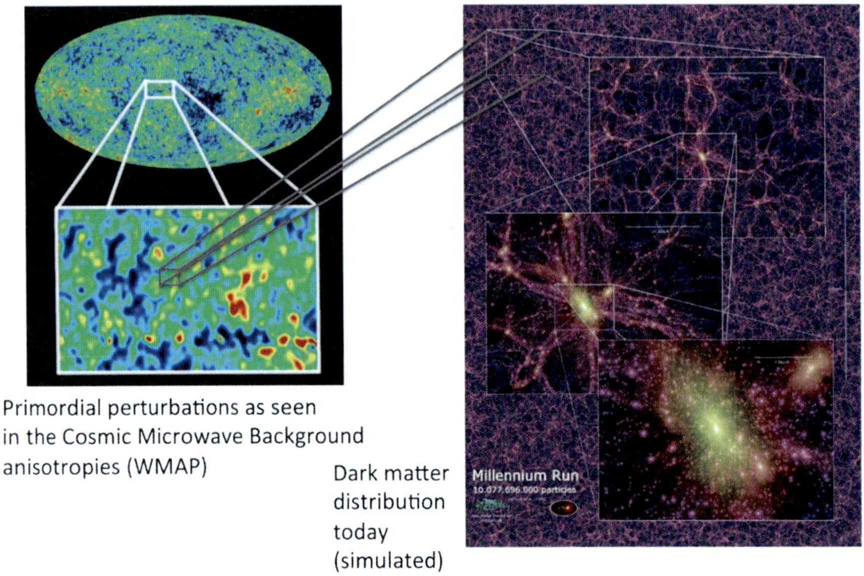

Fig. 3.1 Cosmostatistics uses the stochastic departures from homogeneity on all observable scales to distinguish between cosmological models

- Learn about the cosmic beginning;
- Understand the cosmic constituents, in particular Dark Matter and Dark Energy; and
- Understand cosmological evolution from initial seed perturbations to current observations

One of the challenges for cosmostatistics is that any given observable (maps of the cosmic microwave background, galaxy survey, etc.) is informative about all these goals in some way (Fig. 3.1).

We are fortunate to live in a time when the cosmic microwave background (CMB)is being mapped with high precision from space (by the WMAP [7] and Planck [9] missions), and ground-based and space-based missions are mapping out sizable fractions of the observable Universe in exquisite detail and in three dimensions, across large swaths of the electromagnetic spectrum. Between these two approaches we expect the CMB to have much more signal on very large scales, whereas in principle, probes of density *should* win overall, simply since there are vastly more modes in a three-dimensional data set which greatly reduces sample variance.

How do we realize the immense promise of large scale structure surveys for constraining cosmological models? A number of known and unknown systematics stand between where we are now and the dream of accessing the vast number of perturbation modes sampled by tracers of the underlying density field. Many of these systematics complicate the relationship between the distribution of tracers and the mass distribution we would actually like to probe.

These complications arise either due to the intricate physics of galaxy formation or through incomplete information in the data (e.g. having access only to approximate photometric redshift information instead of the much more expensive spectroscopic redshifts). In addition, the mass density has undergone non-linear dynamical evolution on length scales less than ∼20 Mpc/h, which has coupled the perturbation modes in ways that are non-trivial to model. Non-linear mode coupling erases information that the mode amplitudes carried about the state of the early Universe from whence they arose. On the largest scales the limits are set by causality and hence the finite volume of the observable Universe.

Most people would agree on the impracticality of incorporating fully non-linear gravitational evolution into cosmological inference, let alone a fully physical model of galaxy formation. So the challenge is to find ways of looking at the data that are robust to these systematics.

When it comes to dealing with incomplete information, the challenge is to produce a joint analysis with uncontroversial prior information that allows reconstituting some of the information that has not been captured in the data.

In this talk we will highlight two recent papers which give examples of these two approaches. In one case [3], we develop a Bayesian approach to improving photometric redshift estimates (and simultaneously estimate the density of the tracers). The prior information we assume to achieve this information recovery is local isotropy of the tracer distribution.

In the second paper [5] we define a new observable to prove the physical properties of dark energy: stacked voids. In this case we choose a very specific pre-processing step to extract features of the data which should be robust to galaxy bias and to non-linearity. The approach explicitly projects out the details of the tracer distribution in the non-linear density field to obtain nearly spherical objects that nearly co-move with the expansion which serve as the basis of a powerful and purely geometrical test of the expansion history of the Universe. Again, local isotropy underlies this approach which posits that underdense regions are not preferentially oriented with respect to an observer's line of sight.

3.2 Bayesian Inference from Photometric Redshift Surveys

The vast majority of ongoing and future surveys (CFHTLS, DES, Pan-STARRS, LSST) are or will be photometric. This is a simple consequence of the cost of taking a galaxy spectrum with current technology. Photometric redshift errors of Δz 0.03, the current state-of-the-art, translate into smearing along the line of sight on scales of ∼200 Mpc. Such errors are not detrimental to certain kinds of science but will cause any structure smaller than 100 Mpc to be wiped out, as illustrated in Fig. 3.2.

Looking at the trivial density estimate calculated binning photometric tracers shown in Fig. 3.2 it is immediately clear that the line-like finger-of-god artifacts introduced by photo-z smearing are very recognizable, since they break local isotropy, a core element of our cosmology. Since they stand out so visibly, we wondered if they could be removed.

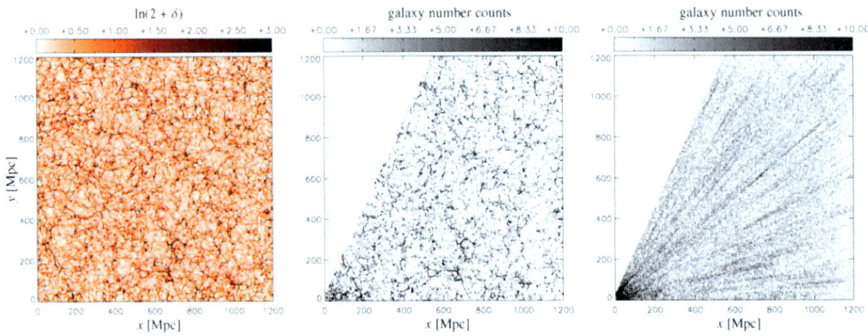

Fig. 3.2 From an n-body simulation to the simulated photo-z survey: the particle density in the simulation (*left*), after application of the mask (*center*), and after simulation of photo-z uncertainties (*right*)

In the following we will often refer to the tracers as galaxies, but the nature of the tracer is of no importance to the functioning or implementation of the algorithm.

3.2.1 A Simple Model of a Photo-z Catalogue

First we build a hierarchical model for the distribution of tracers. A simple approach is to consider the points an inhomogeneous Poisson process. The intensity function of the Poisson process is the underlying number density field, which in turn is a correlated, statistically isotropic, log-normal random field. For the purposes of this exercise we will assume that the correlation function (or equivalently the power spectrum $P(k)$) is known. Relaxing this assumption will be subject of a future study.

The third level in the model hierarchy: photo-z distortions modify the galaxy positions along the radial lines of sight. It is assumed that the redshift uncertainties are specified in terms of a pdf for each tracer. These photo-z pdfs are assumed to be the output of an earlier analysis step which uses any information available, except the spatial distribution of the tracers in the catalog. All photometric information for the galaxy including any morphological features that can be discerned in the images are fair game.

3.2.1.1 Implementation

This hierarchical model can be straightforwardly implemented. The challenge is to explore the posterior density in an efficient manner since the parameter space is enormous: approximately 16 million parameters for the number density and 20 million galaxy redshifts. We choose a block Gibbs sampling approach with the following steps:

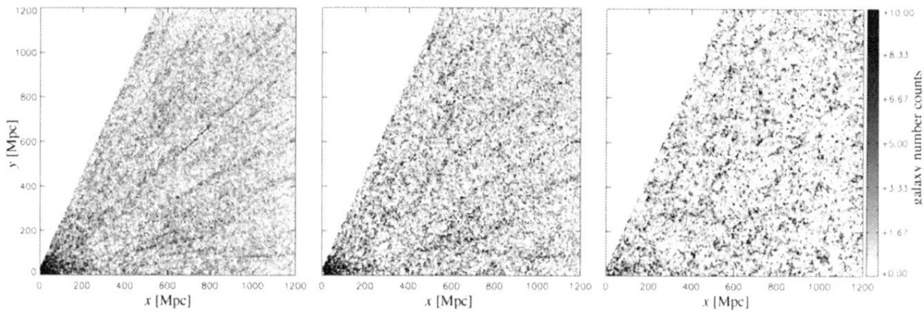

Fig. 3.3 Constrained realizations of the reconstructed density field. The data was simulated using an n-body simulation and the reconstruction assumes the Poisson-lognormal prior with isotropic correlations

Sample the number density given the current galaxy redshifts. We draw from the conditional posterior of the number density assuming that the current "guess" of the galaxy redshift is correct. This is a solved problem [4]; it uses a Hamiltonian sampling approach to update the number density field using the galaxy positions and incorporating the correlated log-normal prior.

Sample the galaxy redshifts given the number density. The redshift posteriors for the galaxies are conditionally independent given the number density field. This feature allows parallelizing this step over the number of galaxies. Each galaxy performs one step of a Metropolis-Hastings Markov Chain Monte Carlo along the line of sight. The conditional posterior for each galaxy is the product of the input photo-z pdf for this galaxy and the number density.

Conditional independence is the key feature that allows this algorithm to scale to tens of millions of galaxies. From the perspective of the message passing paradigm of Bayesian inference, the number density field communicates the information about all the other galaxies to each individual one.

3.2.2 Results

Figures 3.3 and 3.4 illustrate our approach. Even within a few steps the samples of the number density isotropize. As the sampler progresses, individual galaxies explore along their line of sight in a number density field which in turn fluctuates in response to the changing galaxy positions.

Figures 3.3 and 3.4 illustrate our approach. The first figure shows that even within a few steps the samples of the number density become isotropized. In the second figure we track the redshift of an example galaxy as the sampler explores the range of possible reconstructions. The galaxies explore along their line of sight in a number density field that, in turn, fluctuates in response to the changing galaxy positions.

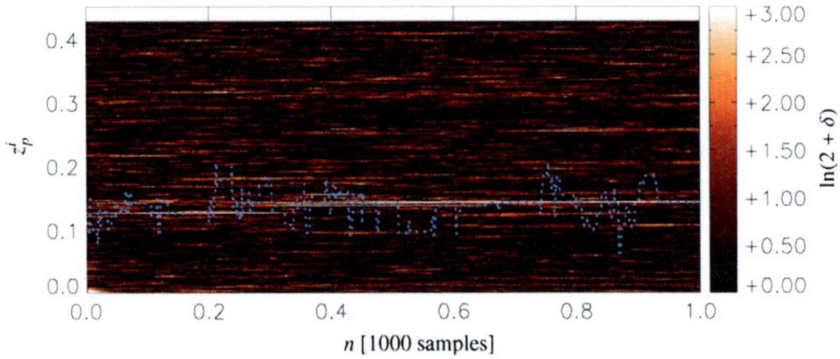

Fig. 3.4 Constrained realizations of the reconstructed density field. The data was simulated using an n-body simulation and the reconstruction assumes the Poisson-lognormal prior with isotropic correlations

The results are encouraging. In high density regions galaxy redshift uncertainties reduce by a factor of several. When a galaxy could reside in one of several concentrations lying along the line of sight the output pdf is multi-modal. Even so, the reconstructed redshift posteriors of the galaxies are generally far more informative than the inputs coming from photometric redshift estimators.

In order to summarize the result of the reconstruction we form the posterior mean estimator, the average of the number density field realizations that are explored by the sampler. We can compare this reconstruction to assess its capability to reproduce features of the input map. Figure 3.5 shows the k-space cross-correlations between the reconstructed and the input field. It is clear that the method is very successful in the high density parts of the sky.

3.2.3 Discussion and Conclusions

The first main point of this talk is that we demonstrated the technical achievement of running a fully Bayesian analysis of a simulated data set with tens of millions of galaxies, and density fields represented on tens of millions of grid zones. The scale of this application corresponds to that of the current generation of available surveys, so it should be feasible to apply this approach to existing data.

The second key issue is to test whether our analysis is sensitive to model misspecification, since the real data will not follow the correlated log-normal Poisson model. Our initial tests (of code correctness) used simulations that were consistent with the prior assumptions. These tests were passed. We do not show these tests here because the prior produces density fields that clearly not realistic, missing much of the filamentary structure which is characteristic of the cosmic web.

The work we present in this talk (and described in detail in Jasche and Wandelt) uses simulated from an n-body simulation. Our results demonstrate that the reconstruction is successful in spite of using an approximate model.

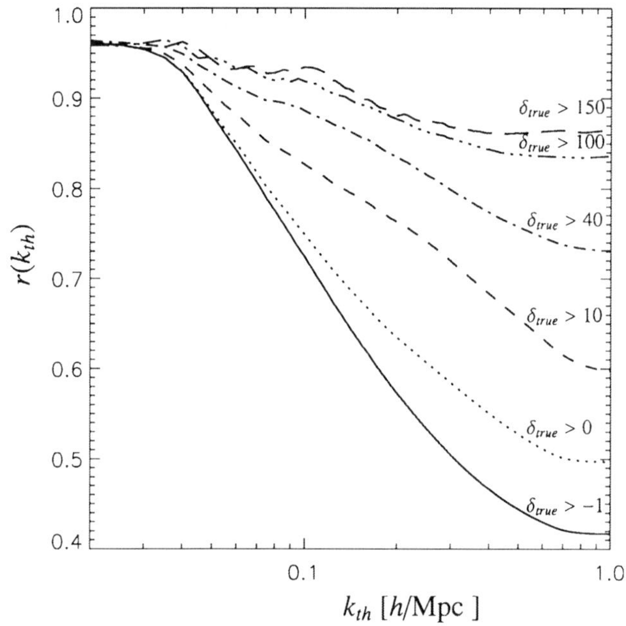

Fig. 3.5 The reconstructed density recovers the small scale features of the input density very well in high density regions. The figure shows the cross-correlation between the input field and the reconstructed density as a function of wave number. Different lines correspond to different thresholds of overdensity

The key feature underlying the reconstruction is clearly the ability to build in the prior assumption of isotropic correlations in the underlying cosmological number density field of the tracers. A secondary feature is the assumption of the shape of the correlations. What we show is that modeling those two aspects of the data results in acceptable reconstructions, that improve the redshift information for each galaxy significantly. It is also true that a better model including the morphological features of realistic gravitationally evolved number density would likely improve upon our results, since the differences between a correlated Poisson log-normal sample and a physical sample drawn from an n-body simulation are easily visible by eye. But it is clear that the reconstructions are not highly sensitive to the details of the assumed prior as long as two salient features of correlation and isotropy are included for the density field and we posit a simple statistical relationship of the tracers to the underlying density, in this case the inhomogeneous Poisson model.

Our approach is completely independent of and complementary to the means by which the photometric redshift is derived. The method is ready for tests on realistic data where the photoz pdfs will be specified in terms of a different pdf for each galaxy.

As a consequence the method will be able to benefit from those tracers whose redshifts are better determined that others. In particular we can merge the advantages of a large number of galaxies in photometric samples and the accuracy of spectroscopic samples! We will explore this idea further in follow-up studies.

This inference problem is of particular interest because it is an example where combining millions of noisy measurements with a physical prior, namely the assumption of isotropic correlations produces a decisive gain in information.

In the second part of the talk we will see another application of the notion of statistical isotropy—this time to the construction of an estimator for the expansion history of the Universe.

3.3 Precision Cosmography with Cosmic Voids

Understanding the physical properties of dark energy is a major goal of modern cosmology. There are essentially two distinct approaches to reaching this goal: cosmography and tracing structure formation.

Cosmography. The cosmography approach, which constrains dark energy properties using precision measurements of the expansion geometry of the Universe. Einstein's equation relates the geometrical properties of our Universe to its content. Since "dark energy" is just a placeholder for the terms in Einstein's equation that drive the observed accelerated expansion of the Universe, precision comographical measurements can tell us about the time dependence of these terms and hence about the value, and rate of change of the equation of state parameter.

Tracing structure formation. The expansion of the universe has an impact on the rate at which primordial perturbations amplify. These perturbations then form structures through non-linear gravitational evolution, galaxy formation etc. Observing the statistical properties (number, size etc) of these structures as a function of redshift constrains the growth of structure, and hence the expansion history, which is informative about the properties of dark energy.

It is clear from this description that geometrical approaches are more direct. In addition, approaches relying on the statistical measures of the amount of structure in the universe inevitably require a detailed understanding of the processes that relate the formed structures to the underlying perturbation amplitude. These processes (e.g. galaxy formation) can be highly complex and deeply non-linear and are research areas in themselves.

Geometrical approaches function by constructing standards out of observables (or combinations of observables) that can be modeled reliably such as standard candles (as in the case of type Ia supernovae), standard rulers (as in the case of Baryon Acoustic Oscillations (BAO)) or time standards (such as the (differences of) ages of galaxies).

3.3.1 The Stacked Voids Alcock-Paczynski Test

The Alcock-Paczynski (AP) test [2] requires a different standard: "standard, co-expanding spheres." One way to construct such standard spheres is through appealing to the statistical isotropy of the cosmological perturbations. In that case, correlations should depend only on the length, but not the direction of the vector connecting the two points being correlated. If the tracers that are being correlated did not move, any anisotropy in the correlation function could be interpreted as being due to the cosmological expansion at the redshift of the correlated objects.

The key difficulty in constructing standard spheres are peculiar velocity effects. Any tracers that happen to lie in gravitationally bound structure will have velocities of the order of the depth of the gravitational potential well of the structure. For clusters or groups of galaxies the resulting finger of god effect in redshift space dominates the cosmic expansion signal by an order of magnitude. To construct an Alcock Paczyski test would therefore require a separate high precision measurement of the depth and shape of the potential well of any structures whose parts were used in the construction of the test.

So far the main work-around has been to only use very long range correlations of order $100\,h^{-1}$Mpc where peculiar velocity effects become sub-dominant compared to cosmic expansion effect and where the baryon sound speed at radiation drag leads to a peak in the correlation function. The downside of this limiting oneself to such large scales is that the statistical constraints will depend on the number of independent correlation volumes in survey volume, which limits the number of perturbation modes that can be used to arrive at the dark energy constraints and therefore leads one to consider extremely large surveys.

In this talk we propose a new way of constructing standard spheres: stacking cosmic voids. While the AP test had been discussed for especially spherical individual voids [10] stacking many voids guarantees spherical symmetry since isotropy prevents cosmic voids from pointing at us (or away from us) preferentially. Finding voids in redshift shells, extracting them from the survey, co-centering them and stacking them, therefore gives rise to spherically symmetric underdensities.

There are several advantages to using cosmic voids:

- Voids are simple: peculiar velocities in and around voids are small compared to the cosmic expansion. We find that they give a 16% systematic effect on our reconstructed Hubble diagram, with a very mild dependence on void size and redshift.
- Voids are small: A typical void size is $10\,h^{-1}$ Mpc—for a dense enough survey the number of voids per unit volume that can be detected is therefore of order 1,000 times larger than the number of BAO correlation volumes.
- Voids remember: we find that voids have a well-ordered phase space—all they do is empty themselves out.

We use the term cosmic voids not to describe regions that are entirely empty, but regions that are underdense basins of repulsion in the cosmic density field.

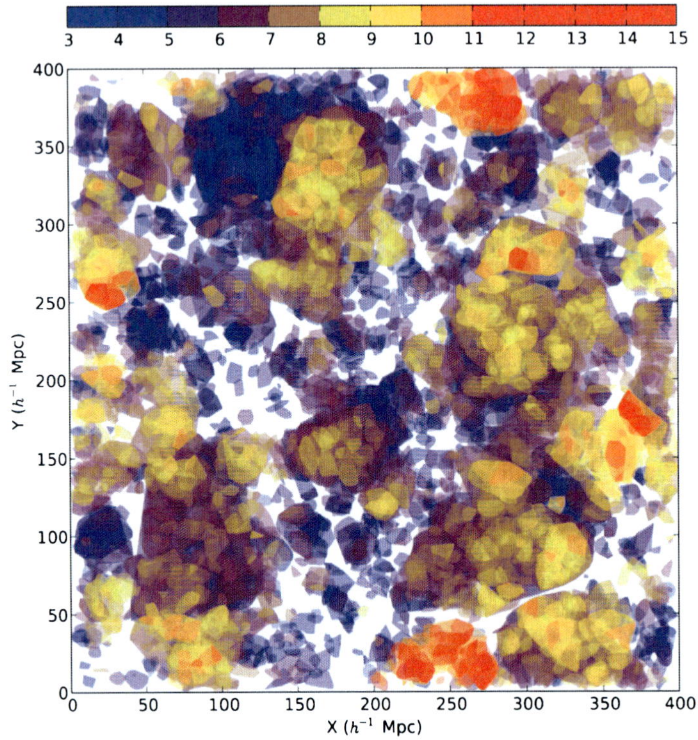

Fig. 3.6 The results of our void finder in a slice of an n-body simulation. The void finder constructs a hierarchical structure of voids. Each patch is a void, colored according to the level in the void hierarchy. When collecting voids in a size bin during the stacking procedure the algorithm traverses the tree in a depth first algorithm and marks and returns the first void it finds which satisfies the size criterion

In order to demonstrate the promise of stacked voids for constructing a powerful AP test we solved the following problems:

1. Create a suitable void definition: a modified ZOBOV algorithm [8] (see Fig. 3.6);
2. Define a method to add voids into stacks labeled by size and redshift, which both enhances signal to noise and sphericalizes them (see Fig. 3.7);
3. Determine the number of voids that would be available to this method in an observed cosmological volume (see Fig. 3.8); and
4. Measure their stretch along the line of sight in order to obtain the expansion history of the universe (see Figs. 3.9 and 3.10).

Details can be found in our main paper [5].

We tested these methods in a series of three pure dark matter N-body simulations with different realizations of the initial conditions but the same cosmology. The volume of each simulation is given by a cube of side $L = 500\ \mathrm{h}^{-1}\ \mathrm{M}_{\odot}$. Each simulation had $N = 512^3$ particles. We adopted a ΛCDM-WMAP7 cosmology with

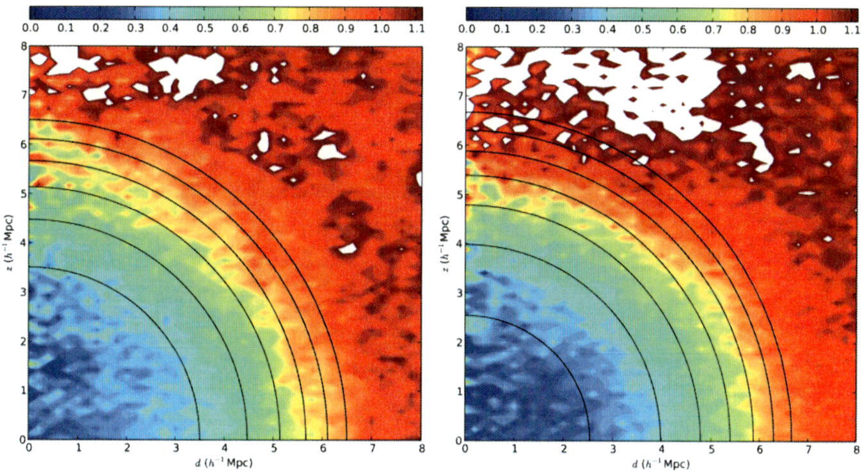

Fig. 3.7 A void stack for $8\,\mathrm{h}^{-1}$ Mpc voids. *Left*: the stack after fitting removing the cosmic expansion effect, but without including peculiar velocities in the simulation. We find our profile agrees well with that found in [13]. *Right*: The stack when peculiar velocities are included. The same cosmic expansion has been removed as in the *left panel*. Careful inspection shows that peculiar velocities lead to a small net compression of the void stack along the line of sight

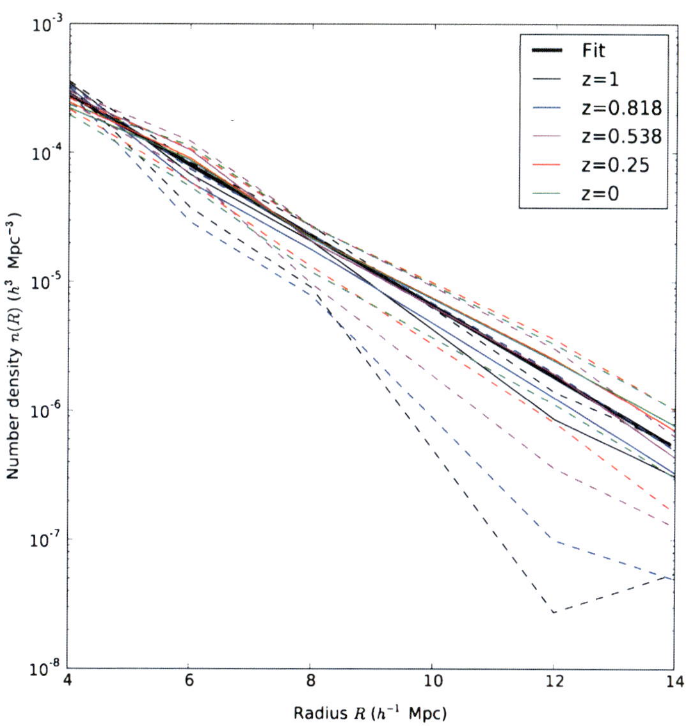

Fig. 3.8 Our simulation results for numbers densities of cosmic voids as a function of redshift for voids of different sizes. These simulation results agree with the model described in [11]

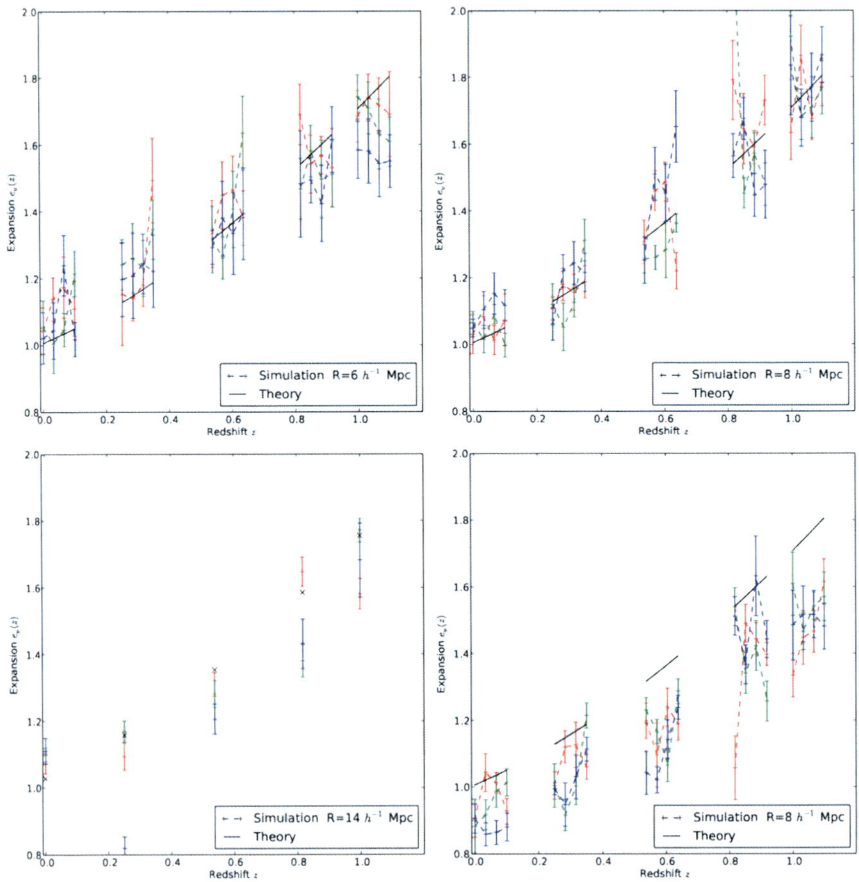

Fig. 3.9 The measured void stretch as a function of redshift for voids of 6, 8, and 14 h^{-1}Mpc (from *left* to *right* and *top* to *bottom*) for three simulations. The long-*dashed line* shows the result for the simulated cosmology. No peculiar velocities were included in the mock catalogs used for these plots. The *lower right panel* shows the result for 8 h^{-1} Mpc voids for mocks with peculiar velocities and without any correction for peculiar velocity effect. The lack of redshift dependence of the resulting bias is clear. The same plot after debiasing is shown in Fig. 3.10

the following parameters: $\Omega_b h^2 = 0.02258$, $\Omega_c h^2 = 0.1108$, $H = 71\,\mathrm{km\,s^{-1}\,Mpc^{-1}}$, $w = -1$, $n_S = 1$, $A_S = 2.34 \times 10^{-9}$. This corresponds to $\Omega_b = 0.045$, $\Omega_M = 0.264$, $\sigma_8 = 0.84$. Each particle had a mass $m_p = 2.05\,10^{11}\,\mathrm{h}^{-1}\,M_\odot$. The transfer function for density fluctuations for this cosmology was computed using CAMB [6]. The initial conditions were generated using ICGEN,[1] a code which uses the transfer function to generate a density field from the primordial power spectrum.

[1] Available from http://www.iap.fr/users/lavaux/

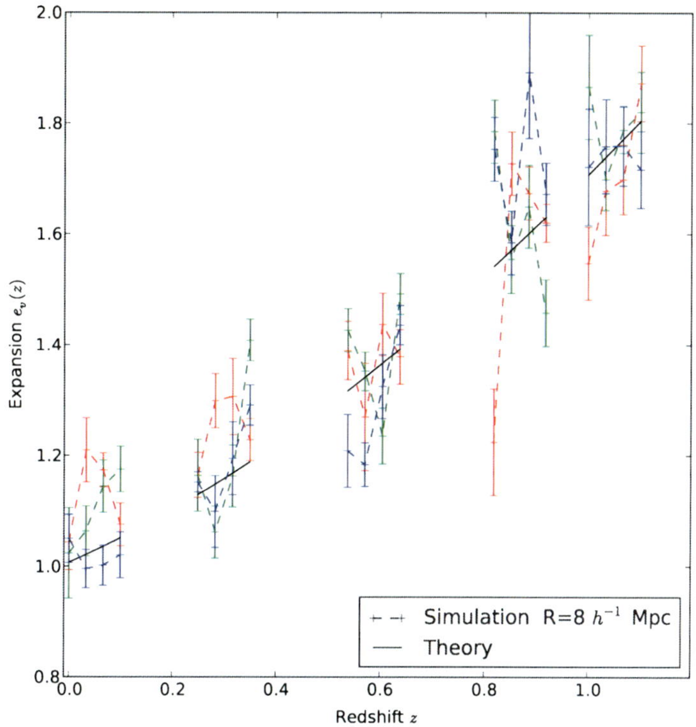

Fig. 3.10 Stretch inferred for 8 h^{-1} Mpc voids after the correction of a peculiar velocity bias. There is no evidence for residual bias at the level of our simulations

3.3.2 Discussion and Conclusion

Based on these results we performed a Fisher matrix forecast of the statistical constraints on dark energy equation of state parameter w_a and its rate of change w_p that we would expect from Euclid. We quantify the answer in terms of the figure of merit defined by the Dark Energy Task Force [1], i.e. the relative reduction in the area of the uncertainty ellipse for these two quantities. The result is exciting—we find that the stacked void Alcock-Paczynski test has the potential significantly to enhance the power of the proposed (and now selected) Euclid space craft to constrain dark energy phenomenology.

On the fact of it cosmic voids have the potential to provide a far more powerful constraint on dark energy than measurements of the Baryonic Acoustic Oscillation scale, by up to an order of magnitude. This large increase of information is easily understood in comparing the number of modes probed by voids compared to BAOs, which scales roughly as the third power of the ratio of the BAO scale to the scale of the smallest usable voids $\sim$1,000. The area of parameter constraints scales as the square root of the number of modes $\sim$30. When projected into the w_a, w_p plane using the Fisher matrix formalism for the EUCLID wide survey, we find the improvement over BAO on those parameters by a factor of $\sim$10.

We expect our stacked void shape measurements to be robust to galaxy bias as it is purely geometrical and relies on the topology of the density field [12]. In fact, it is possible that biased tracers of the density enhance the contrast of voids and therefore enhance the void detection rate. These expectations remains to verified on more realistic mock catalogs and real data.

Based on our Fisher matrix forecasts, the stacked voids technique promises a remarkable increase to the figure of merit from EUCLID when compared to the combined results from all other probes using EUCLID data (BAO, weak lensing, type Ia supernovae, cluster counts). The Alcock-Paczinsky test using stacked voids is therefore potentially a significant addition to the portfolio of major dark energy probes which merits further detailed studies focused on additional real-world systematics and optimal survey design.

Acknowledgements The authors wish to thank Eric Feigelson and Jogesh Babu for the invitation to speak at this highly enjoyable and fruitful meeting. We thank Laird Thompson, Joseph Silk, Mark Neyrinck, Thierry Sousbie, Miguel Aragón-Calvo, and Stéphane Colombi for useful discussions and e owe a special debt of gratitude to the respondent, Christopher Genovese for his insightful and valuable commentary.

The authors acknowledge financial support from NSF Grants AST 07-08849, AST 09-08693 and from BDW's Chaire d'Excellence granted by the Agence Nationale de Recherche. GL acknowledges support from CITA National Fellowship and financial support from the Government of Canada Post-Doctoral Research Fellowship. This research was supported by the National Science Foundation through TeraGrid resources provided by NCSA under grant number TG-AST100029.

We thank Francisco S. Kitaura, Torsten A. Enßlin and Simon D. White for useful discussions; Cristiano Porciani for the simulated density field and Nina Roth for help with handling the simulation data; Rainer Moll and Björn M. Schäfer for useful discussions and support with many valuable numerical gadgets. This work has been supported by the Deutsche Forschungsgemeinschaft within the Priority Programme 1177 under the project PO 1454/1-1.

Research at Perimeter Institute is supported by the Government of Canada through Industry Canada and by the Province of Ontario through the Ministry of Research and Innovation.

References

1. Albrecht, A., et al. 2006, ArXiv Astrophysics e-prints
2. Alcock, C., & Paczynski, B. 1979, Nature, 281, 358
3. Jasche, J., & Wandelt, B. D. 2011, e-print arXiv: 1106.2757
4. Jasche J., Kitaura F. S., 2010, MNRAS, 407, 29
5. Lavaux, G., & Wandelt, B. D. 2011, e-print, arXiv: 1110.0345
6. Lewis, A., Challinor, A., & Lasenby, A. 2000, Astrophys. J., 538, 473
7. Komatsu, E., et al. 2011, ApJS, 192, 18
8. Neyrinck, M. C. 2008, MNRAS, 386, 2101
9. Planck Collaboration, A&A accepted, eprint arXiv:1101.2022
10. Ryden, B. S. 1995, ApJ, 452, 25
11. Sheth, R. K., & van de Weygaert, R. 2004, MNRAS, 350, 517
12. Springel, V., et al. 1998, MNRAS, 298, 1169
13. van de Weygaert, R., & van Kampen, E. 1993, MNRAS, 263, 481

Chapter 4
Simulation-Aided Inference in Cosmology

David Higdon, Earl Lawrence, Katrin Heitmann, and Salman Habib

Abstract In this paper we describe two Bayesian statistical approaches for combining large-scale computational models with physical observations to make inferences about cosmological parameters. The first method is a Bayesian calibration approach adapted from Kennedy and O'Hagan (J R Stat Soc B 68:425–464, 2001) and Higdon et al. (J Am Stat Assoc 103:570–583, 2008). It makes use of a response surface model that approximates the simulation output at untried input settings. The second approach uses the ensemble Kalman filter (Evensen, IEEE Control Syst Mag 29:83–104, 2009), which makes use of an ensemble of simulations and physical observations to update the prior parameter distribution using standard equations from Kalman filtering. We apply these methods to large-scale structure simulations and observations from the Sloan Digital Sky Survey.

4.1 Introduction

In this paper we combine computationally intensive simulation results with measurements from the Sloan Digital Sky Survey (SDSS) to infer a subset of the parameters that control the ΛCDM model, cosmology's standard model. We describe two Bayesian approaches for carrying out this analysis. First, we describe a statistical framework adapted from Kennedy and O'Hagan [7] and Higdon et al. [4] to determine a posterior distribution for these cosmological parameters given the simulation output and the physical observations. Second, we show how to use

D. Higdon (✉) • E. Lawrence
Los Alamos National Laboratory, Los Alamos, NM 87545, USA
e-mail: dhigdon@lanl.gov; earl@lanl.gov

K. Heitmann • S. Habib
Argonne National Laboratory, Argonne, IL 60439, USA
e-mail: heitmann@anl.gov; habib@anl.gov

E.D. Feigelson and G.J. Babu (eds.), *Statistical Challenges in Modern Astronomy V*, Lecture Notes in Statistics 209, DOI 10.1007/978-1-4614-3520-4_4,

the ensemble Kalman filter [1] to estimate these cosmological parameters. We briefly contrast these two basic approaches for model calibration (i.e. parameter estimation).

4.2 Simulations and Physical Observations

The SDSS, shown in the left panel of Fig. 4.1 maps out the spatial location of galaxies around the Milky Way Galaxy. A key feature of the spatial distribution of galaxies is the combination of voids and high density filaments of matter. This

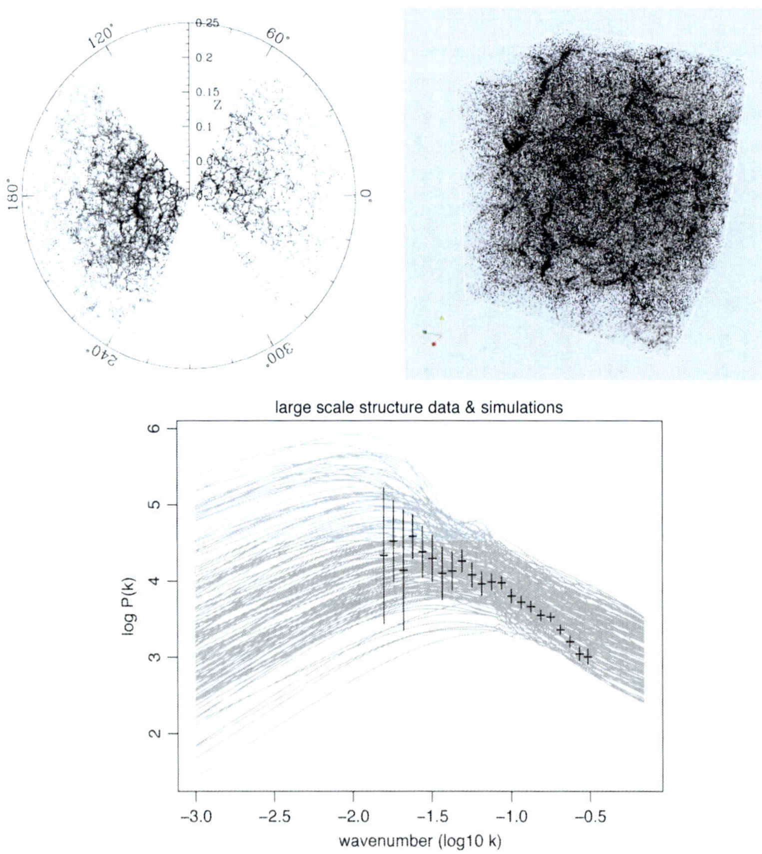

Fig. 4.1 *Top left*: Physical observations from the Sloan Digital Sky Survey (Credit: Sloan Digital Sky Survey). *Top right*: Simulation results from an *N*-body simulation. *Bottom*: Power spectra for the Matter density fields. The *gray lines* are from 128 simulations; the *black lines* give spectrum estimates derived from the physical observations

Table 4.1 ΛCDM parameters with their lower and upper bounds

Param	Explanation	Lower	Upper
n	Spectral index	0.8	1.4
h	Hubble constant	0.5	1.1
σ_8	Galaxy fluctuation amplitude	0.6	1.6
Ω_{CDM}	Dark matter density	0.0	0.6
Ω_{B}	Baryonic matter density	0.02	0.12

peculiarity is a result of the cumulative effect of gravity (and other forces) acting on slight matter density fluctuations present shortly after the big bang, as evidenced by the cosmic microwave background (CMB).

Predicting the current spatial distribution of matter in the universe, given the parameters of the ΛCDM model, requires substantial computing effort. For a given parameter setting, a very large-scale N-body simulation is carried out. The simulation initializes dark matter tracer particles according to the CMB and then propagates them according to gravity and other forces up to the present time. The result of one such simulation is shown in the middle frame of Fig. 4.1. Different cosmologies (i.e. cosmological parameter settings) yield simulations with different spatial structure. We would like to determine which cosmologies are consistent with physical observations of our universe, such as the power spectra in the right frame of Fig. 4.1.

It is difficult to directly compare the simulation output and the SDSS data. The simulations move dark matter particles over a periodic cube of space, while the SDSS data give a censored, local snapshot of the large scale structure of the universe. We can simplify the comparison by summarizing the simulation output and physical observations with their power spectra, describing the spatial distribution of matter density at a wide range of length scales, shown in the right frame of Fig. 4.1. Note that the wave number k on the x-axis of these spectra is given in h/Mpc. A megaparsec (Mpc) is a length scale; two galaxies are separated by about 1 Mpc on average. The gray lines in right hand plot of Fig. 4.1 show a number of matter power spectra produced by carrying out simulations using different cosmological parameter settings.

Computing the matter power spectrum is trivial for the simulation output since the output resides on a periodic, cubic lattice. Determining the matter power spectrum from the SDSS data has many difficulties: nonstandard survey geometry, redshift space distortions, luminosity bias and noise, just to name a few. Because of these challenges, we use the data and likelihood of Tegmark et al. [16], which is summarized in right hand plot of Fig. 4.1. This is chosen for demonstration purposes only as the spectra from the dark matter simulations are not directly comparable with the spectrum computed from luminous red galaxies. These data correspond to 22 independent pairs (y_i, k_i) with the two standard deviation bars shown in Fig. 4.1.

For the N-body simulations, we consider five ΛCDM parameters show in Table 4.1. Since we assume a flat universe and a constant dark energy equation of state, we expect that any variation in the unused ΛCDM parameters will not affect the resulting matter power spectra.

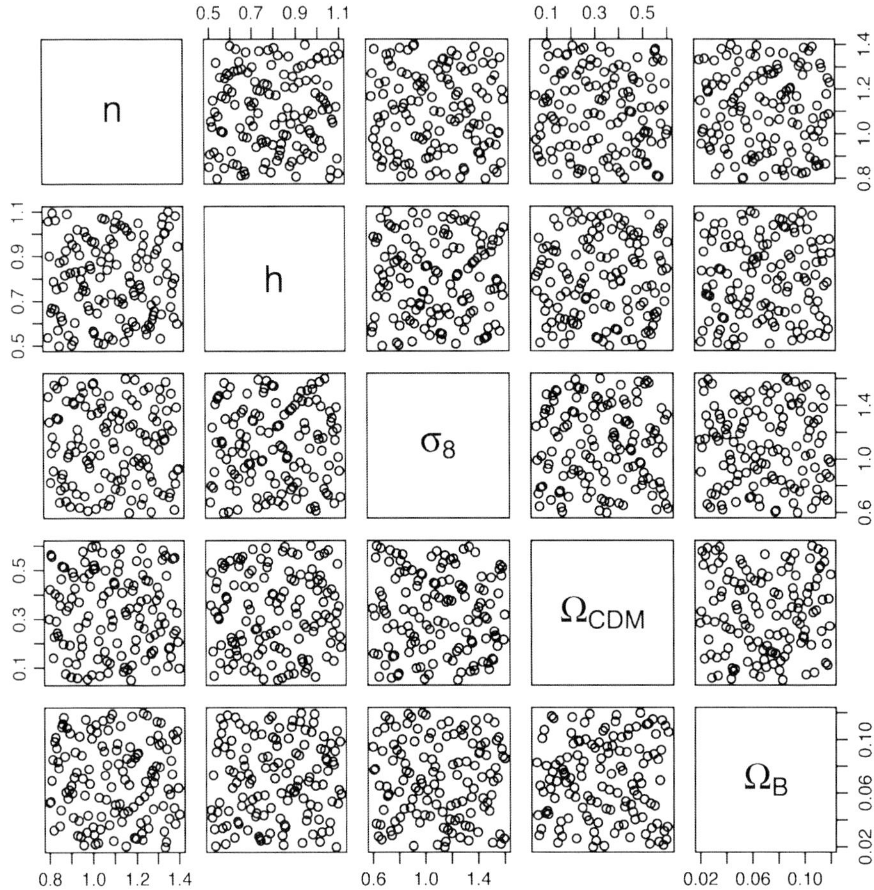

Fig. 4.2 128 input parameter settings over the 5-dimensional parameter space

The dark matter simulations are computationally demanding, requiring the computation of force interactions for over two million particles. Simulation accuracy is particularly important for the smaller length scales ($k \geq 0.2\,\mathrm{h\ Mpc^{-1}}$), where the gravitational effects become strongly nonlinear. For this demonstration, we use $m = 128$ simulations. For the Bayesian computer model calibration (BCMC) approach, a response surface is built to estimate power spectra at untried input settings. Experience indicates a preference for spreading the 128 inputs to fill in the 5-dimensional parameter space (see Fig. 4.2). For a survey of statistical designs for computer experiments, see Santner et al. [14], Chaps. 5 and 6. For the ensemble Kalman filter (EnKF) approach, this simulation output is treated as a sample from the prior distribution for the cosmological parameter settings.

4.3 Statistical Formulation

In this section we describe the statistical methodology for combining physical observations and simulation output to infer unknown model parameters. We use observations y from the matter power spectrum (Fig. 4.1) and matter power spectra derived from physical simulations.

Generally, the simulation models requires p-vector t of input parameters to produce a matter power spectrum $\eta(t)$. The simplest model to consider is that the vector of physical observations y is a noisy version of a simulation $\eta(\theta)$ at the true setting θ

$$y = \eta(\theta) + \varepsilon, \tag{4.1}$$

where the observation error vector is normal, with mean 0 and variance Σ_y. Given a prior distribution $\pi(\theta)$ for the true parameter vector θ, the resulting posterior distribution $\pi(\theta|y)$ for θ is given by

$$\pi(\theta|y) \propto L(y|\eta(\theta)) \cdot \pi(\theta), \tag{4.2}$$

where $L(y|\eta(\theta))$ comes from the normal sampling model for the data

$$L(y|\eta(\theta)) = \exp\left\{\frac{1}{2}(y-\eta(\theta))'\Sigma_y^{-1}(y-\eta(\theta))\right\} \tag{4.3}$$

and $\pi(\theta)$ is uniform over the 5-dimensional rectangle C given by the lower and upper bounds in Table 4.1. Note that we use the notation t to represent a generic input vector and the notation θ to represent the value or distribution of values for the input at which the simulator best matches physical observations.

This basic Bayesian formulation is the starting point for both the BCMC and EnKF approaches. If the computational model could be evaluated quickly, it could be directly incorporated in the likelihood and the posterior distribution could be explored via MCMC. However, each simulation requires hours or days of computation, thus a direct MCMC-based approach is infeasible.

Note that here we consider Σ_y to be known, accounting for the error in the physical observations. More generally, Σ_y could also incorporate error due to the mismatch between computational model and reality. This paper does not discuss the important topic of modeling this discrepancy, but more information can be found in Kennedy and O'Hagan [7], Kaipio and Somersalom [5] and Goldstein and Rougier [2], along with their accompanying discussions.

4.3.1 *Bayesian Computer Model Calibration*

The BCMC approach deals with the computational bottleneck by treating $\eta(\cdot)$ as an unknown function to be estimated from a fixed collection of simulations

$\eta(t_1),\ldots,\eta(t_m)$ carried out at input settings $t_1,\ldots,t_m$. This approach requires a prior distribution for the unknown function $\eta(\cdot)$, and treats the simulation output $\eta^* = (\eta(t_1),\ldots,\eta(t_m))'$ as data for the analysis. Because we are trying to estimate the function, as well as the input settings, there is an additional component of the likelihood obtained from the sampling model for η^* by $L(\eta^*|\eta(\cdot))$.

For this case, the resulting posterior distribution has the general form

$$\pi(\theta,\eta(\cdot)|y,\eta^*) \propto L(y|\eta(\theta)) \cdot L(\eta^*|\eta(\cdot)) \cdot \pi(\eta(\cdot)) \cdot \pi(\theta), \tag{4.4}$$

which has traded direct evaluations of the simulator model for a more complicated form which depends strongly on the prior model for the function $\eta(\cdot)$. Under this model, the marginal distribution for the cosmological parameters θ will be affected by uncertainty regarding $\eta(\cdot)$.

In the following subsections, we describe a particular formulation of (4.4) in the context of this large scale structure application. This formulation has been useful in a variety of physics and engineering applications which combine field observations with detailed simulation models for inference. We start with a description of the how to build an *emulator*, the model for $\eta(\cdot)$ at untried parameter settings. We then describe how the observed data is combined with the simulations and the emulator to give the posterior distribution.

4.3.1.1 Emulating the Simulator Output

In this section, we describe the probability model, which we call an emulator, for the simulator output at untried settings. For a given input t in the standardized input space $[0,1]^p$, the simulator produces a matter power spectrum of length n_η, as shown in Fig. 4.1. The emulator models the simulation output using a q-dimensional basis representation:

$$\eta(t) = \sum_{i=1}^{q} \phi_i w_i(t) + \varepsilon,\ t \in [0,1]^p, \tag{4.5}$$

where $\{\phi_1,\ldots,\phi_q\}$ is a collection of orthogonal, n_η-dimensional basis vectors, the $w_i(t)$ are weights depending on the input, and ε is an n_η-dimensional error term. This formulation reduces the problem of building an emulator that maps $[0,1]^p$ to R^{n_η} to building q independent, univariate models for each $w_i(t)$. Separate Gaussian processes (GP) are used to model each of the weight functions. The details of this model specification are given below.

Output from each of the m simulation runs prescribed by the input parameter design results in n_η-dimensional vectors which we denote by $\eta_1,\ldots,\eta_m$. Since the simulation outputs have no missing data, they can be efficiently represented via principal components [12]. We first center the simulations by subtracting the mean $(\frac{1}{m}\sum_{j=1}^{m}\eta_j)$ from each output vector. Depending on the application, some alternative standardization may be preferred. Whatever the choice of the standardization, the same standardization is also applied to the experimental data.

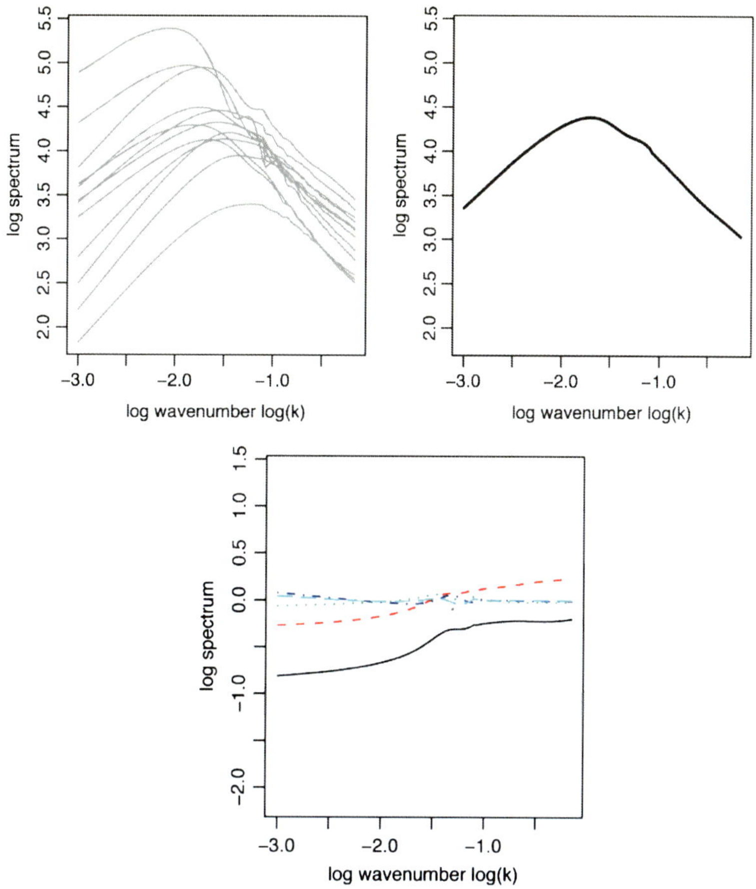

Fig. 4.3 Simulations (*top left*), mean (*top right*), and the first five principal component bases (*bottom*) derived from the simulation output

We define Ξ to be the $n_\eta \times m$ matrix ($n_\eta >> m$) obtained by column-binding the (standardized) output vectors from the simulations We apply the singular value decomposition (SVD) to the simulation output matrix Ξ giving

$$\Xi = [\eta_1; \cdots ; \eta_m] = UDV', \tag{4.6}$$

where U is a $n_\eta \times m$ orthogonal matrix, D is a diagonal $m \times m$ matrix holding the singular values, and V is a $m \times m$ orthonormal matrix. To construct a q-dimensional representation of the simulation output, we define the principal component (PC) basis matrix Φ_η to be the first q columns of $[UD\sqrt{m}]$. For the matter power spectrum application we take $q = 5$; the basis functions $\phi_1, \ldots, \phi_5$ are shown in Fig. 4.3.

Note that the ϕ_i are functions of log wave number.

We use the basis representation of (4.5) to model the n_η-dimensional simulator output over the input space. Each vector of basis weights $w_i(t)$, $i = 1,\dots,q$, is modeled as a zero mean GP

$$w_i(t) \sim \mathrm{N}(0, \lambda_{wi}^{-1} R(t;\rho_i)), \tag{4.7}$$

where λ_{wi} is the marginal precision of the process and $R(t;\rho_i)$ is a correlation matrix with entries dependent on the inputs and a set of parameters given by the correlation function

$$\mathrm{Corr}(w_i(t), w_i(t')) = \prod_{k=1}^{p} \rho_{ik}^{4(t_k - t_k')^2} \tag{4.8}$$

This is the Gaussian covariance function, which gives very smooth realizations, and has been used previously by Kennedy and O'Hagan [7] and Sacks et al. [13] to model computer simulation output. An advantage of the product form is that only a single additional parameter is required per additional input dimension, but the fitted GP response still allows for rather general interactions between inputs. We use the Gaussian form for the covariance function because the simulators we handle tend to respond very smoothly to changes in the inputs. The parameter ρ_{ik} controls the spatial range for the kth input dimension of the process w_i. Under this parameterization, ρ_{ik} gives the correlation between $w_i(t)$ and $w_i(t')$ when the input conditions t and t' are identical, except for a difference of 0.5 in the kth component. Note that this interpretation makes use of the standardization of the input space to $[0,1]^p$.

Restricting to the m input design settings , we define the m-vector w_i to be $w_i = (w_i(t_1),\dots,w_i(t_m))'$ for $i = 1,\dots,q$. In addition we define $R(t;\rho_i)$ to be the $m \times m$ correlation matrix resulting from applying (4.8) to each pair of input settings in the design. The p-vector ρ_i gives the correlation distances for each of the input dimensions. At the m simulation input settings, the mq-vector $w = (w_1',\dots,w_q')'$ then has prior distribution

$$\begin{pmatrix} w_1 \\ \vdots \\ w_q \end{pmatrix} \sim N\left(\begin{pmatrix} 0 \\ \vdots \\ 0 \end{pmatrix}, \begin{pmatrix} \lambda_{w1}^{-1} R(t;\rho_1) & 0 & 0 \\ 0 & \ddots & 0 \\ 0 & 0 & \lambda_{wq}^{-1} R(t;\rho_q) \end{pmatrix} \right), \tag{4.9}$$

which is controlled by q precision parameters held in λ_w and $q \cdot p$ spatial correlation parameters held in ρ. The prior above can be written more compactly as $w \sim N(0, \Sigma_w)$, where Σ_w, controlled by parameter vectors λ_w and ρ, is given by the block diagonal covariance matrix in (4.9).

We specify independent Gamma priors for each λ_{wi} and independent Beta priors for the ρ_{ik},

$$\begin{aligned} \pi(\lambda_{wi}) &\propto \lambda_{wi}^{a_w - 1} e^{-b_w \lambda_{wi}}, \ i = 1,\dots,q, \\ \pi(\rho_{ik}) &\propto \rho_{ik}^{a_\rho - 1} (1 - \rho_{ik})^{b_\rho - 1}, \ i = 1,\dots,q, k = 1,\dots,p. \end{aligned} \tag{4.10}$$

We expect the marginal variance for each $w_i(\cdot)$ process to be close to one due to the scaling of the basis functions. For this reason we specify that $a_w = b_w = 5$, encouraging each λ_{wi} to be close to 1. In addition, this informative prior helps stabilize the resulting posterior distribution for the correlation parameters which can trade off with the marginal precision parameter. Because we expect only a subset of the inputs to influence the simulator response, our prior for the correlation parameters reflects this expectation of *effect sparcity*. Under the parameterization in (4.8), input k is inactive for PC i if $\rho_{ik} = 1$. Choosing $a_\rho = 1$ and $0 < b_\rho < 1$ will give a density with substantial prior mass near one. We take $b_\rho = 0.1$, which makes $\Pr(\rho_{ik} < 0.98) \approx \frac{1}{3}$ a priori. In general, the selection of these hyperparameters should depend on how many of the p inputs are expected to be active. Alternatively, the prior could be specified to have some point mass at one as in Linkletter et al. [8].

Define $\eta = \text{vec}(\Xi)$, where $\text{vec}(\Xi)$ produces a vector by stacking the columns of matrix Ξ. Taking the error vector in (4.5) to be independent Gaussian with common precision λ_η, we get the sampling model for η:

$$\eta | w, \lambda_\eta \sim N\left(\Phi w, \frac{1}{\lambda_\eta} I\right), \tag{4.11}$$

where $\Phi = [I_m \otimes \phi_1; \cdots; I_m \otimes \phi_q]$, and the ϕ_i are the q basis vectors previously computed via SVD. A Gamma prior with parameters (a_η, b_η) is specified for the error precision λ_η.

Multiplying (4.9)–(4.11) and the Gamma prior for λ_η yields the posterior. After integrating out w, the posterior distribution for the unknown parameters becomes

$$\begin{aligned}
&\pi(\lambda_\eta, \lambda_w, \rho | \eta) \propto \\
&\left|(\lambda_\eta \Phi'\Phi)^{-1} + \Sigma_w\right|^{-\frac{1}{2}} \exp\{-\tfrac{1}{2}\hat{w}'([\lambda_\eta \Phi'\Phi]^{-1} + \Sigma_w)^{-1}\hat{w}\} \times \\
&\lambda_\eta^{a_\eta^* - 1} e^{-b_\eta^* \lambda_\eta} \times \prod_{i=1}^{q} \lambda_{wi}^{a_w - 1} e^{-b_w \lambda_{wi}} \times \prod_{i=1}^{q}\prod_{j=1}^{p} (1 - \rho_{ij})^{b_\rho - 1},
\end{aligned} \tag{4.12}$$

where

$$\begin{aligned}
a_\eta^* &= a_\eta + \frac{m(n_\eta - q)}{2}, \\
b_\eta^* &= b_\eta + \tfrac{1}{2}\eta'(I - \Phi(\Phi'\Phi)^{-1}\Phi')\eta, \text{ and} \\
\hat{w} &= (\Phi'\Phi)^{-1}\Phi'\eta.
\end{aligned} \tag{4.13}$$

This posterior distribution is a milepost on the way to the complete formulation incorporating experimental data. However, it is worth considering this intermediate posterior distribution for the simulator response. It can be explored via MCMC using standard Metropolis updates and we can view a number of posterior quantities to illuminate features of the simulator. Oakley and O'Hagan [10] use posterior of the simulator response to investigate formal sensitivity measures of a univariate sim-

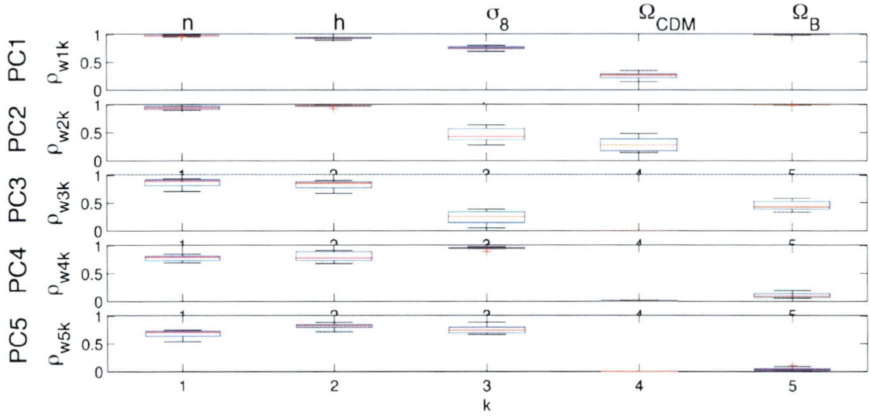

Fig. 4.4 Boxplots of posterior samples for each ρ_{ik} for the large scale structure application

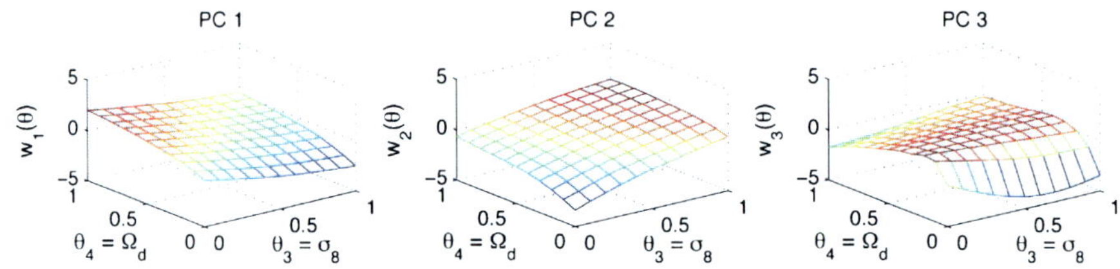

Fig. 4.5 Posterior mean surfaces for $w_i(\theta)$, $i = 1,2,3$. Here the other three parameters were held at their midpoints as σ_8 and Ω_{CDM} vary over the design range

ulator; Sacks et al. [13] consider sensitivity from a non-Bayesian perspective. For example, Fig. 4.4 shows boxplots of the posterior distributions for the components of ρ. From this figure it is apparent that PCs 1 and 2 are most influenced by σ_8 and Ω_{CDM}. Figure 4.5 shows the resulting posterior mean surfaces for $w_1(\cdot)$, $w_2(\cdot)$ and $w_3(\cdot)$ as a function of σ_8 and Ω_{CDM}.

Given the posterior realizations from (4.12), one can generate realizations from the process $\eta(\cdot)$ at any input setting $t^\star$. Since

$$\eta(t^\star) = \sum_{i=1}^{q} \phi_i w_i(t^\star), \tag{4.14}$$

realizations from the $w_i(t^\star)$ processes need to be drawn given the MCMC output. For a given draw $(\lambda_\eta, \lambda_w, \rho)$ a draw of $w^\star = (w_1(t^\star), \ldots, w_q(t^\star))'$ can be produced by using the fact

$$\begin{pmatrix} \hat{w} \\ w^\star \end{pmatrix} \sim N\left(\begin{pmatrix} 0 \\ 0 \end{pmatrix}, \left[\begin{pmatrix} (\lambda_\eta \boldsymbol{\Phi}'\boldsymbol{\Phi})^{-1} & 0 \\ 0 & 0 \end{pmatrix} + \Sigma_{w,w^\star}(\lambda_w, \rho) \right] \right), \tag{4.15}$$

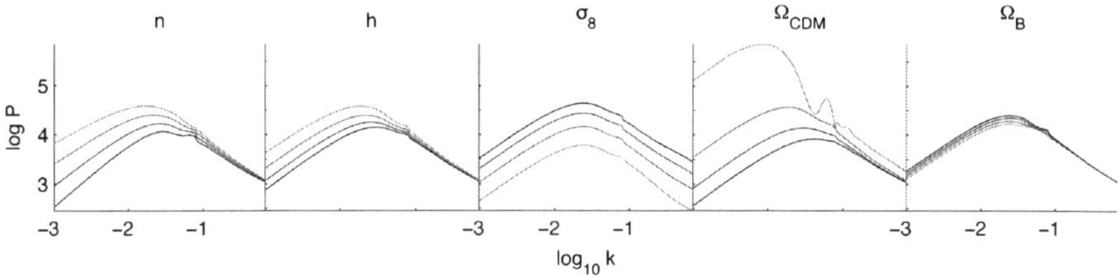

Fig. 4.6 Changes to the posterior mean simulator predictions obtained by varying one input, while holding others at their central values, i.e. at the midpoint of their range. The light to *dark lines* correspond to the smallest parameter setting to the biggest, for each parameter

where $\Sigma_{w,w^\star}$ is obtained by applying the prior covariance rule from (4.8) to the augmented input settings that include the original design and the new input setting $t^\star$. Recall that $\hat{w} = (\Phi'\Phi)^{-1}\Phi'\eta$. Application of the conditional normal rules then gives

$$w^\star|\hat{w} \sim N(V_{21}V_{11}^{-1}\hat{w}, V_{22} - V_{21}V_{11}^{-1}V_{12}), \tag{4.16}$$

where

$$V = \begin{pmatrix} V_{11} & V_{12} \\ V_{21} & V_{22} \end{pmatrix} = \left[\begin{pmatrix} (\lambda_\eta \Phi'\Phi)^{-1} & 0 \\ 0 & 0 \end{pmatrix} + \Sigma_{w,w^\star}(\lambda_w, \rho)\right] \tag{4.17}$$

is a function of the parameters produced by the MCMC output. Hence, for each posterior realization of $(\lambda_\eta, \lambda_w, \rho)$, a realization of $w^\star$ can be produced. The above recipe easily generalizes to give predictions over many input settings at once.

Figure 4.6 shows posterior means for the simulator response η where each of the inputs is varied over its prior (standardized) range of $[0,1]$ while the other four inputs are held at their midpoints. The posterior mean response conveys an idea of how the different parameters affect the highly multivariate simulation output. Other marginal functionals of the simulation response can also be calculated such as sensitivity indicies or estimates of the Sobol decomposition [10, 13]. Note that a simplified emulator can be constructed by taking plug in estimates for $(\lambda_\eta, \lambda_w, \rho)$.

4.3.1.2 Incorporating Physical Data

Given the model specifications for the simulator $\eta(\cdot)$, we can now consider the sampling model for the experimentally observed data. The data are contained in an n_y-vector y. For the matter power spectrum application $n_y = 22$, corresponding to different wave numbers as shown in Fig. 4.1. As previously stated, the data are modeled as a noisy version of the simulated spectrum $\eta(\theta)$ run at the true, but unknown, parameter setting θ. Thus

$$y = \eta(\theta) + \varepsilon, \tag{4.18}$$

where the errors are assumed to be $N(0,\Sigma_y)$. For notational convenience we represent the precision Σ_y^{-1} as $\lambda_y W_y$, leaving open the option to estimate a scaling of the error covariance with λ_y^{-1}. Using the basis representation for the simulator this equation becomes

$$y = \Phi_y w(\theta) + \varepsilon \tag{4.19}$$

where $w(\theta)$ is the q-vector $(w_1(\theta),\ldots,w_q(\theta))'$. Because the wave number support of y is not necessarily contained in the support of the simulation output, the basis vectors in Φ_y may have to be interpolated over wave number from the columns of Φ. Since the simulation output over wave number is quite dense, this interpolation is straightforward.

We specify a Gamma prior with parameters (a_y,b_y) for the precision parameter λ_y resulting in a normal-gamma form for the data model

$$y|w(\theta),\lambda_y \sim N(\Phi_y w(\theta),(\lambda_y W_y)^{-1}),\ \lambda_y \sim Ga(a_y,b_y). \tag{4.20}$$

The observation precision W_y is fairly well-known for the SDSS data, so we encourage λ_y to be near one with informative prior parameters $a_y = b_y = 5$.

We can now write out the entire posterior distribution for all of the parameters and the best fitting inputs θ. First, let

$$\begin{aligned}
\hat{w}_y &= (\Phi_y' W_y \Phi_y)^{-1}\Phi_y' W_y y,\\
a_y^* &= a_y + \tfrac{1}{2}(n-q),\\
b_y^* &= b_y + \tfrac{1}{2}(y-\Phi_y\hat{w}_y)' W_y (y-\Phi_y\hat{w}_y),\\
\Lambda_y &= \lambda_y \Phi_y' W_y \Phi_y,\\
\Lambda_\eta &= \lambda_\eta \Phi'\Phi, \qquad\qquad (4.21)\\
I_q &= q\times q \text{ identity matrix},\\
\Sigma_{w_y w} &= \begin{pmatrix} \lambda_{w1}^{-1}R(\theta,\theta^*;\rho_1) & 0 & 0\\ 0 & \ddots & 0\\ 0 & 0 & \lambda_{wq}^{-1}R(\theta,\theta^*;\rho_q)\end{pmatrix},\\
\hat{z} &= \begin{pmatrix}\hat{w}_y\\ \hat{w}\end{pmatrix},\\
\Sigma_{\hat{z}} &= \begin{pmatrix}\Lambda_y^{-1} & 0\\ 0 & \Lambda_\eta^{-1}\end{pmatrix} + \begin{pmatrix} I_q & \Sigma_{w_y w}\\ \Sigma_{w_y w}' & \Sigma_w\end{pmatrix}.
\end{aligned}$$

The posterior distribution has the form

$$\pi(\lambda_\eta, \lambda_w, \rho, \lambda_y, \theta|\hat{z}) \propto$$

$$|\Sigma_{\hat{z}}|^{-\frac{1}{2}} \exp\{-\tfrac{1}{2}\hat{z}'\Sigma_{\hat{z}}^{-1}\hat{z}\} \times \lambda_\eta^{a_\eta^*-1} e^{-b_\eta^*\lambda_\eta} \times \prod_{i=1}^{q} \lambda_{wi}^{a_w-1} e^{-b_w\lambda_{wi}} \times$$

$$\prod_{i=1}^{q}\prod_{k=1}^{p} \rho_{ik}^{a_\rho-1}(1-\rho_{ik})^{b_\rho-1} \times \lambda_y^{a_y^*-1} e^{-b_y^*\lambda_y} \times I[\theta \in C], \quad (4.22)$$

where C denotes the p-dimensional rectangle defined in Table 4.1.

Realizations from the posterior distribution are produced using standard, single site MCMC. Metropolis updates [9] are used for the components of ρ and θ with a uniform proposal distribution centered at the current value of the parameter. The precision parameters λ_η, λ_w and λ_y are sampled using Hastings updates [3]. Here the proposals are uniform draws, centered at the current parameter values, with a width that is proportional to the current parameter value. In a given application the candidate proposal width can be tuned for optimal performance.

The resulting posterior distribution estimate for θ is shown in Fig. 4.7 on the original scale. The posterior values can also be propagated through the emulator to produce realizations of the posterior spectrum. The right hand plot of Fig. 4.7 shows the posterior mean and pointwise 90% ranges for the power spectrum.

4.3.2 Ensemble Kalman Filter for Parameter Estimation

The ensemble Kalman filter (EnKF), a Monte Carlo extension of the Kalman filter, uses an ensemble of model runs that are updated as additional data are made available [1]. Unlike the Kalman filter [6], the EnKF does not require a linear model and doesn't assume Gaussian distributions. The EnKF can be easily extended to estimate model parameters by appending the parameter vector as an unobserved part of the state vector. To date, this approach has primarily been used in applications in oil recovery [11, 15], even though it seems applicable to a wide variety of inverse problems.

Below we briefly describe two basic variants of the EnKF for parameter estimation, differing in how they use the ensemble of model runs to approximate the resulting posterior distribution. One estimates the joint prior distribution for the states and parameters by computing a multivariate normal approximation to the ensemble of model runs and then uses the traditional Kalman updates to the mean and covariance to compute the posterior. The other uses the ensemble directly with EnKF updates to each ensemble member. In both cases, an ensemble of draws from the prior distribution of the model parameters θ are paired with the resulting simulation output to produce an ensemble of $(\eta(\theta), \theta)$ pairs, from which the sample covariance is used to produce an approximation to the posterior distribution. Hence

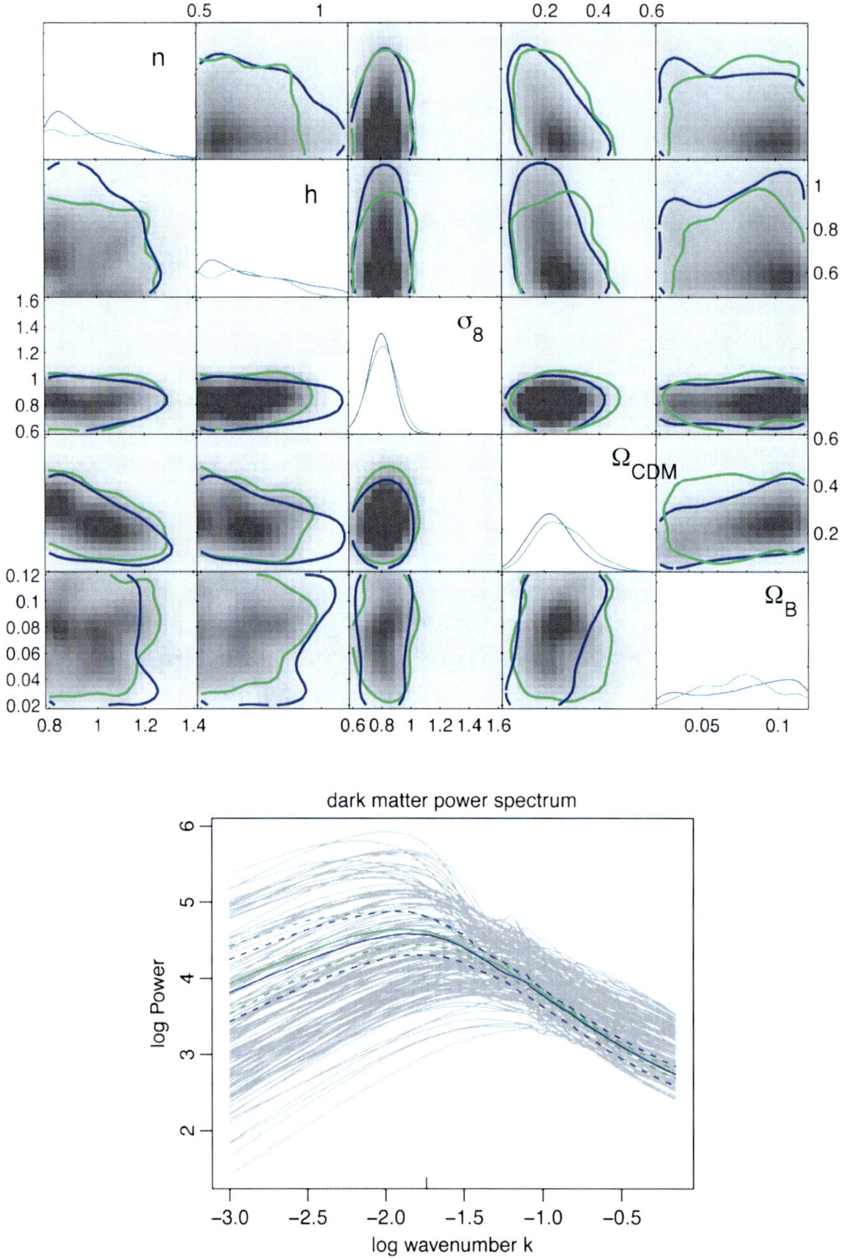

Fig. 4.7 Comparison of posteriors between the Bayesian computer model calibration (BCMC) approach (*blue*) and the ensemble Kalman filter (EnKF, *green*). *Top*: Estimated posterior distribution of the parameters $\theta = (n, h, \sigma_8, \Omega_{\mathrm{CDM}}, \Omega_B)$. The diagonal shows the estimated marginal posterior pdf for each parameter; the off-diagonal images give estimates of bivariate marginals; the contour lines show estimated 95% hpd regions. The *lower triangle* and *green lines* give the posterior under the EnKF approach; The *upper triangle* and *blue lines* give the posterior under the BCMC approach. *Bottom*: Posterior median and 95% uncertainty bounds for the posterior power spectrum. *Green lines* correspond to EnKF; *blue lines* correspond to BCMC

we treat the input parameter settings $t_1,\dots,t_m$ as m draws from the prior distribution $\pi(\theta)$. Note that even though the distribution of the simulator response $\eta(\theta)$ is completely determined by the distribution for θ, the covariance estimate used by the EnKF ignores this.

4.3.2.1 Gaussian Prior Approximation

The first approach fits a multivariate normal distribution to the prior ensemble for $(\eta(\theta),\theta)$. Implicitly, it uses a linear approximation for $\eta(\theta)$ to produce the posterior distribution for θ. The recipe:

1. For each of the $m = 128$ simulations form the $p_\eta + p$-vector

$$\begin{pmatrix} \eta(t_k) \\ t_k \end{pmatrix}, k = 1,\dots,m. \tag{4.23}$$

   Here $n_\eta = 88$ and $p = 5$. With these m vectors, compute the sample mean vector μ_{pr} and the $(n_\eta + p)\times(n_\eta + p)$ sample covariance matrix Σ_{pr}. Treat $(\eta(\theta),\theta)'$ as though it has $N(\mu_{\text{pr}},\Sigma_{\text{pr}})$ prior distribution.
2. In our large-scale structure example, the physical observations y correspond to an interpolation of the n_η elements of $\eta(\theta)$. Let H be the matrix for the that interpolates $\eta(\theta)$ and ignores θ in the combined state-parameter vector. In this case the likelihood can be rewritten as

$$L(y|\eta(\theta)) \propto \exp\left\{-\frac{1}{2}\left(y - H\begin{pmatrix}\eta(\theta)\\ \theta\end{pmatrix}\right)' \Sigma_y^{-1}\left(y - H\begin{pmatrix}\eta(\theta)\\ \theta\end{pmatrix}\right)\right\}. \tag{4.24}$$

3. Combining the normal approximation to the prior with the normal likelihood results in an updated, or posterior, distribution for $(\eta(\theta)),\theta)$ for which

$$\begin{pmatrix}\eta(\theta)\\ \theta\end{pmatrix} |y \sim N(\mu_{\text{post}},\Sigma_{\text{post}}), \tag{4.25}$$

   where

$$\Sigma_{\text{post}}^{-1} = \Sigma_{\text{pr}}^{-1} + H'\Sigma_y^{-1}H \tag{4.26}$$

   and

$$\mu_{\text{post}} = \Sigma_{\text{post}}\left(\Sigma_{\text{pr}}^{-1}\mu_{\text{pr}} + H'\Sigma_y^{-1}y\right). \tag{4.27}$$

   Note that the posterior mean can be rewritten in form more commonly used in Kalman filtering

$$\mu_{\text{post}} = \mu_{\text{pr}} + \Sigma_{\text{pr}}H'(H\Sigma_{\text{pr}}H' + H\Sigma_y H')^{-1}(y - H\mu_{\text{pr}}) \tag{4.28}$$

   where $\Sigma_{\text{pr}}H'(H\Sigma_{\text{pr}}H' + H\Sigma_y H')^{-1}$ is the Kalman gain matrix.

The joint normal computations used here effectively assume a linear plus Gaussian noise relationship between $\eta(\theta)$ and θ, inducing a normal posterior for θ.

4.3.2.2 Ensemble Representation

The second approach is basically the usual EnKF for one time step. The goal is to perturb member each of the ensemble $(\eta(t_k), t_k)$, in order to produce an updated member (η_k^*, θ_k^*) which is an approximate draw from the posterior distribution. This updated member is not produced with the simulator so that η_k^* will not be equal to the simulator evaluated at updated parameter value $\eta(\theta_k^*)$. The general recipe:

1. Construct the $(n_\eta + p) \times (n_\eta + p)$ sample covariance matrix Σ_{pr} as in Step 1 of the previous algorithm.
2. For $k = 1, \ldots, m$ do:

   (a) Draw a perturbed data value $y_k \sim N(y, \Sigma_y)$.
   (b) Produce the perturbed ensemble member

$$\begin{pmatrix} \eta_k^* \\ \theta_k^* \end{pmatrix} = \Sigma_{\text{post}} \left(\Sigma_{\text{pr}}^{-1} \begin{pmatrix} \eta(t_k) \\ t_k \end{pmatrix} + H' \Sigma_y^{-1} y_k \right). \tag{4.29}$$

   where Σ_{pr} and Σ_{post} are defined in the previous algorithm. Note this perturbation of the ensemble member can be equivalently written using the more standard Kalman gain update:

$$\begin{pmatrix} \eta_k^* \\ \theta_k^* \end{pmatrix} = \begin{pmatrix} \eta(t_k) \\ t_k \end{pmatrix} + \Sigma_{\text{pr}} H' (H \Sigma_{\text{pr}} H' + H \Sigma_y H')^{-1} (y_k - \eta(t_k)) \tag{4.30}$$

3. Treat this $m = 128$ member ensemble

$$\begin{pmatrix} \eta_k^* \\ \theta_k^* \end{pmatrix}, k = 1, \ldots, m. \tag{4.31}$$

   as draws from the updated, posterior distribution for $(\eta(\theta), \theta)$.

Note that this approach uses an update of two normal forms, just like the previous version, but updated separately for each ensemble member. Only, here the normal prior is centered at the ensemble member, and the normal likelihood is centered at the perturbed data value, rather than at the ensemble mean and the actual data value.

This produces a posterior ensemble for the distribution of θ, along with a posterior ensemble for the power spectrum. The green lines in the left panel of Fig. 4.7 show the posterior densities for the parameters—a kernel density estimator was used to produce the density plots. The green lines in the right panel show the posterior median and 95% uncertainty bounds for the power spectrum. For comparison, the blue lines show the same quantities estimated using the BCMC approach.

4.4 Discussion

The two methods yield somewhat similar results for the posterior distributions of the parameters and the spectrum, but there are differences. The EnKF uses a Gaussian simplifying assumption in order to include data, which basically uses a linear plus noise (i.e. regression) relationship between θ and η. As such, the BCMC approach is likely to produce more accurate results for both the parameters and the predicted spectrum since it gives a more accurate representation of the simulator response. Another advantage of the BCMC approach is that the emulator can be used for secondary purposes such as assessing parameter sensitivity. These advantages come at a cost. The BCMC approach requires considerably more computation than the EnKF's simple linear updating equation. Further, for high-dimensional parameter spaces the BCMC approach may experience difficulty with estimating the response surface without huge numbers of runs. In this case, the assumptions and the efficiency of the EnKF may produce a superior result.

References

1. Evensen, G. (2009) The ensemble Kalman filter for combined state and parameter estimation, *Control Systems Magazine, IEEE*, 29(3), 83–104
2. Goldstein, M. & Rougier, J. C. (2009) Reified Bayesian modeling and inference for physical systems, *J. Statistical Planning & Inference*, 139, 1121–1239
3. Hastings, W. K. (1970) Monte Carlo sampling methods using Markopv chains and their applications, *Biometrika*, 57, 97–109
4. Higdon, D., Gattiker, J. R. & Williams, B. J. (2008) Computer model calibration using high dimensional output, *J. Amer. Stat. Assn.*, 103, 570–583
5. Kaipio, J. & Somersalom E. (2007), Statistical inverse problems: Discretization, model reduction and inverse crimes, *J. Computational & Applied Mathematics*, 198, 493–504
6. Kalman, R. E., A new approach to linear filtering and prediction problems, *J. Basic Engineering*, 82, 35–45
7. Kennedy, M. & O'Hagan, A. (2001) Bayesian calibration of computer models (with discussion), *J. Roy. Stat. Soc. B*, 68, 425–464
8. Linkletter, C., Bingham, D., Hengartner, N. Higdon, D. & Ye, K. (2006), Variable selection for Gaussian process models in computer experiments, *Technometrics*, 48, 478–490
9. Metropolis, N., Rosenbluth, A., Rosenbluth, M., Teller, A. & Teller, E. (1953) Equations of state calculations by fast computing machines, *J. Chem. Phys.*, 21, 1087–1091
10. Oakley, J. & O'Hagan, A. (2004), Probabilistic sensitivity analysis of complex models, *J. Roy. Stat. Soc. B*, 66, 751–769
11. Oliver, D.S. & Chen, Y. (2010) Recent progress on reservoir history matching: a review, *Computational Geosciences*, 1–37
12. Ramsey, J. O. & Silverman, B. W. (1997) *Functional Data Analysis*, Springer
13. Sacks, J., Welch, W. J., Mitchell, T. J. & Wynn. H. P. (1989) Design and analysis of computer experiments (with discussion), *Statistical Science*, 4, 409–423
14. Santner, T. J., Williams, B. J. & Notz, W. J. (2003), *Design and Analysis of Computer Experiments*, Springer
15. Skjervheim, J. & Evensen, G. (2011), *An Ensemble Smoother for Assisted History Matching*, SPE Reservoir Simulation Symposium
16. Tegmark, M., et al. (2004) The three-dimensional power spectrum of galaxies from the Sloan Digital Sky Survey, *Astron. J.*, 606, 702–740

Chapter 5
Commentary: Simulation-Aided Inference in Cosmology

Carlo Graziani

Abstract Higdon's use of Gaussian Process (GP) emulation to analyze SDSS data using simulated power spectra from N-body simulations supplies a textbook case study of a set of techniques that are likely to become a standard part of the astrostatistics toolbox. The problems addressed by these techniques models based on expensive computer simulations that run on high-performance computing (HPC) platforms, which can only sparsely sample a large-dimensional input parameter space are likely to be of interest to a growing community of computational astrophysicists wishing to compare models to data, as this style of computing becomes "democratized" by the increasing availability of HPC platforms in University research settings. We comment here on the computational challenges of Gaussian Process modeling, the fidelity of model hierarchies, and strategies for the adaptive design of numerical experiments.

5.1 Gaussian Process Emulation

The relation of a computer model's output to its input has a term of art: the *Response Surface*, essentially the function that maps the input parameter manifold to the output space (usually a vector space). The input parameter space is often high-dimensional.

The fact that the dense probing of input parameter space is unaffordable creates a new situation with respect to statistical inference. In effect, the response surface must be interpolated to general parameter values based on a limited sampling of the parameter space corresponding to a limited number of simulations. This means that a new source of uncertainty, separate from instrumental noise (AKA "statistical

C. Graziani (✉)
Department of Astronomy, Flash Center For Computational Physics, University of Chicago, 5747 S. Ellis Avenue, Jones 314, Chicago, IL 60637, USA
e-mail: carlo@oddjob.uchicago.edu

E.D. Feigelson and G.J. Babu (eds.), *Statistical Challenges in Modern Astronomy V*, Lecture Notes in Statistics 209, DOI 10.1007/978-1-4614-3520-4_5,

error") and model inadequacy (AKA "systematic error") must be factored into the error budget: the uncertainty introduced by the interpolation. This uncertainty is represented by building an *emulator*—a stochastic representation of the simulator "trained" using the available model evaluations [2–5].

The stochastic nature of the response surface representation is frequently implemented using a Gaussian Process (GP) model [1]. Briefly, this is a methodology whereby a prior Gaussian distribution is specified on a space of functions describing the response surface, and then updated using data to produce a posterior summary of what is known about the surface. The result is an interpolation of the response surface to arbitrary points not sampled by simulations, with the interpolation uncertainty encoded as a Gaussian covariance. One benefit of this style of emulation is that if the simulations are to be compared with measurement data with Gaussian measurement uncertainties, those uncertainties may be naturally convolved with the GP interpolation uncertainty in a simple analytic manner [3, 6, 7].

In effect, the interpolation performed by the emulator allows us to transition to a new view of the problem: we regard the model output as *data* from a family of models (the code, at all possible parameter settings). The comparison of the computer model to measurement data is carried out by *joint model fitting to the computer data and the measurement data to simultaneously estimate the full response surface and the "true" model parameters.*

5.2 Computational Challenges

A difficulty that must be overcome in GP emulation (as in most GP modeling of large systems) is that the evaluation of likelihoods requires the inversion of large, symmetric, positive-definite covariance matrices (or rather, the solution of their associated linear problem), and the computation of the determinants of those matrices. In GP emulation, the dimension N of the space in which the covariance matrix operates is $N = N_{sim} \times N_{output}$, where N_{sim} is the number of simulations and N_{output} is the dimensionality of the output space. Since the computational cost of inversion scales as $\mathcal{O}(N^3)$, direct approaches (such as Cholesky factorization) rapidly lose their usefulness.

Higdon partly abates this problem through a data-reduction strategy, using a Principal Components Analysis (PCA) on the model output to create a manageable representation of the simulation output. By keeping only $N_{components}$ of the singular values (the largest ones, representing the "most active" components), and placing a GP model on each parameter weight function in the resulting decomposition, Higdon reduces the problem to one with a computational cost that scales as $\mathcal{O}(N_{components} \times N_{sim}^3)$.

This is a substantial savings, but it leaves in place an important $\mathcal{O}(N_{sim}^3)$ scaling, which has the extremely galling consequence that the very act of performing more simulations to improve our knowledge of the response surface can quickly result in an infeasible computational cost. For problems with high input parameter space dimensionality, and with complex response surface structure, there is simply no alternative to growing N_{sim} to the point where the structure can be resolved, at least in parameter space regions corresponding to high posterior density. It is therefore necessary to consider approximation schemes that control the cost of GP emulation.

Gibbs and MacKay [8], adapt methods due to Skilling [9] to exhibit approximations to linear problem solutions that involve only matrix-vector multiplications, and which scale as N^2. These methods are related to approaches that note the equivalence of the required inversion problem to a quadratic form minimization, and adopt conjugate gradient minimization as the minimization strategy. By terminating the minimization at an adequate level of accuracy, but well before formal exact convergence, such methods achieve N^2 complexity cost [10].

In addition, it is possible to adopt covariance models based on kernels of compact support—that is to say, covariances that vanish when the distance between points exceeds a certain limit. Such kernels give rise naturally to sparse covariance matrices, which can then be handled at costs approaching N for operations such as matrix-vector multiplication in the case of expanding domain asymptotic regime. For a discussion of a family of such kernels, see p. 88 of Rasmussen and Williams [1]. Compact-support kernels may also be combined in Schur products with more general kernels, a technique called "tapering" [11], which can provide the benefits of sparse matrices with the more complex covariance structure of non-compact kernels.

5.3 Model Fidelity Hierarchies

Even if one has abated the curse of dimensionality problem by some approximation scheme, one often still confronts a computational cost issue associated with running the simulations themselves. Complex, high-fidelity, multi-physics, multi-scale simulations may require so much computational time on an HPC platform that they may simply not be available in the required abundance for an adequate resolution of the response surface.

This circumstance may be addressable by supplementing the highest-fidelity simulations with cheaper—and more abundant—lower-fidelity simulations, at the cost of some inaccuracy which we may hope to cross-calibrate against the high-fidelity simulations. Examples include simulations of lower spatial resolution, or including approximate physics, or excluding computationally-expensive physics, or using spatial symmetry assumptions (such as cylindrical, planar, or spherical symmetry) to reduce the dimensionality of the problem.

It is noteworthy that it is not necessarily the case that the quantitative accuracies of the available types of simulations fall into a natural hierarchical rank-ordering. It may be the case that some simulations are more accurate than others in some input parameter regimes, but less so in others. In addition, it may occur that some types of approximations leading to faster simulations in some parameter regimes may simply *fail* in some parameter regimes—the code may crash, or numerical instabilities may develop, or the approximation may simply break down, leading to results bearing no relation to the true physical situation. In such a circumstance, levels of the fidelity hierarchy may simply go missing in certain parameter regimes. Therefore, while it is usually clear that there exists a maximum-fidelity level of simulation corresponding to the highest computational cost, in general the remaining levels of the hierarchy may not be strictly ordered by accuracy.

The research efforts that I am aware of to fit such multi-fidelity level simulations into a GP emulator scheme [12–14] make some relatively strong assumptions about the nature of the relationship between the levels of the hierarchy. These are spelled out in [12], and include a strictly-ordered hierarchy of fidelity levels, a Markov-like assumption on the relative informativeness of neighboring levels, and stationarity (i.e. translational invariance) of the underlying GP over the parameter space. In view of the considerations above, and of the desirability of generalizing GP emulation away from stationary models, it seems worth exploring somewhat more agnostic schemes for connecting the simulation fidelity levels.

5.4 Adaptive Numerical Experiment Design

At what parameter values are we to run the simulations? This is the issue of numerical experimental design. When potentially expensive computations are invoked to probe a response surface over a potentially high-dimensional input parameter space, it is urgent that simulations not be wasted on parameter space regions that neither illuminate interesting structure of the response surface nor reside in neighborhoods where the surface closely resembles the measurement data. It seems hopeless to accomplish this sort of optimization efficiently with ab initio designs such as Latin Hypercubes. Existing information from analysis of the data and the response surface using the current design must be used to guess parameter choices for future runs that are, in some sense, optimal.

The two objectives of globally characterizing response surface structure and of using what is already known about the response function for specific inference goals (e.g., modeling measurement data) are in a tension that is known from the global optimization and adaptive learning literatures by a term of art: the "Exploration-Exploitation Tradeoff", wherein (in the current instance) the "exploration" imperative to understand the response surface everywhere competes with the "exploitation" necessity of focusing on regions appearing to resemble the experimental situation under study. Both activities are essential, and their reconciliation is necessarily an important objective of adaptive experimental design theory.

The efforts that have been dedicated to adaptive numerical experiment design have been largely focused on the exploration aspect of the tension [13, 15, 16]. There is more to be learned about the full tension from the literature of *physical* experiment design. In particular, Loredo [17] exhibited a Bayesian experimental design scheme wherein observations currently "in the can" can be used to calculate the expected information gain—negative Shannon entropy—from a future observation with selected experimental parameters, and to choose those parameters so as to maximize that information.

This scheme is generalizable to numerical experimental design. Suppose we have measurement data y corresponding to an unknown true parameter setting θ_T. Suppose also that the existing design of N_{sim} simulations at parameter settings $\Theta = (\theta_1, \ldots, \theta_{N_{sim}})$ with outputs $Y = (y_1, \ldots, y_{N_{sim}})$ is to be augmented by a proposed simulation with parameters θ_+. Let the GP posterior predictive of the augmented design be $\pi(y_+|\theta_+, \Theta, Y)$ and have Shannon entropy $H(\theta_+, \Theta)$. Also, let the GP posterior predictive of the augmented design conditioned on the data and on the true parameter values be $\pi'(y_+|\theta_+, \Theta, Y, \theta_T, y)$, with Shannon entropy $H'(\theta_+, \Theta, \theta_T)$. Then it can be shown [18] that the expected information gain from the proposed new simulation is

$$EI(\theta_+) = H(\theta_+, \Theta) - \int d\theta_T \, P(\theta_T|y, Y, \Theta) H'(\theta_+, \Theta, \theta_T). \tag{5.1}$$

The first term in (5.1) embodies exploration (by itself, it yields Maxent sampling). The second term embodies exploitation, rewarding smaller predictive uncertainty near best-fit parameter point. The expected information gain may thus be used to drive a "Simulation-Inference-Design" cycle analogous to the "Observation-Inference-Design" cycle described in [17].

References

1. C. Rasmussen, C. Williams: *Gaussian Processes for Machine Learning*, (MIT Press, 2006)
2. J. Sacks, W. J. Welch, T. J. Mitchell, H. P. Wynn: Design and analysis of computer experiments. Statistical Science **4** (4), 409–423 (1989)
3. M. Kennedy, A. O'Hagan: Bayesian calibration of complex computer models. Journal of the Royal Statistical Society Series B **63**, 425–464 (2001)
4. T. Santner, B. Williams, W. Notz: The Design and Analysis of Computer Experiments. (Springer, 2003)
5. A. O'Hagan: Bayesian Analysis of Computer Code Outputs: A Tutorial. Reliability Engineering and System Safety **91** (10–11), 1290–1300 (2006)
6. D. Higdon, M. Kennedy, J. Cavendish, J. Cafeo, R. Ryne: Combining field data and computer simulations for calibration and prediction. SIAM Journal on Scientific Computing **26** (2), 448–466 (2004)
7. K. Heitmann, D. Higdon, M. White, S. Habib, B. J. Williams, E. Lawrence, C. Wagner: The Coyote Universe. II. Cosmological Models and Precision Emulation of the Nonlinear Matter Power Spectrum. The Astrophysical Journal **705**, 156–174 (2009)
8. M. Gibbs, D. MacKay: Efficient implementation of Gaussian processes. Cavendish Lab., Cambridge, UK, Tech. Rep. (1997)
9. J. Skilling: Bayesian numerical analysis. In: Physics & Probability: Essays in honor of Edwin T. Jaynes, 207–221 (1993)
10. G. Wahba, D. Johnson, F. Gao, J. Gong: Adaptive tuning of numerical weather prediction models: Randomized GCV in three- and four-dimensional data assimilation. Monthly Weather Review **123**, 3358–3369 (1995) (also available by anonymous ftp from ftp.stat.wisc.edu in pub/wahba)
11. R. Furrer, M. Genton, D. Nychka: Covariance tapering for interpolation of large spatial datasets. Journal of Computational and Graphical Statistics **15** (3), 502–523 (2006)

12. M. Kennedy, A. O'Hagan: Predicting the output from a complex computer code when fast approximations are available. Biometrika **87** (1), 1–13 (2000)
13. J. Cumming, M. Goldstein: Small Sample Bayesian Designs for Complex High-Dimensional Models Based on Information Gained Using Fast Approximations. Technometrics **51** (4), 377–388 (2009)
14. P. Qian, C. Wu: Bayesian hierarchical modeling for integrating low-accuracy and high-accuracy experiments. Technometrics **50** (2), 192–204 (2008)
15. R. Gramacy, H. Lee: Adaptive design and analysis of supercomputer experiments. Technometrics **51** (2), 130–145 (2009)
16. R. B. Gramacy, H. K. H. Lee, W. Macready: Adaptive exploration of computer experiment parameter spaces. Tech. rep., Bulletin of the International Society for Bayesian Analysis (ISBA) (December 2004)
17. T. Loredo, D. Chernoff: Bayesian adaptive exploration. In: AIP Conference Proceedings, Vol. 707, pp. 330–346 (2004)
18. C. Graziani, T. J. Loredo, M. Anitescu: Adaptive Design of Computer Experiments and Simulation Fidelity Hierarchies. In preparation (2012)

Chapter 6
The Matter Spectral Density from Lensed Cosmic Microwave Background Observations

Ethan Anderes and Alexander van Engelen

Abstract We use local likelihood estimates of gravitational shear and convergence from lensed cosmic microwave background observations to estimate the projected mass spectral density. Typically there is an additive bias when using a plug-in estimate of the spectral density from a noisy estimate of the random field. We explore the possibility of adjusting this bias by subtracting an approximate power spectrum of the noise in the reconstruction using unlensed simulations. We demonstrate some empirical results that suggest the remaining biases complement those seen in the quadratic estimate developed by Hu and Okamoto (ApJ 557:L79–L83, 2001; ApJ 574:566–574, 2002; Phys Rev D 67:083002, 2003). We finish the paper with a discussion regarding the potential scientific applications and the challenges associated with estimating the noise spectrum from simulations.

6.1 Introduction

Over the past decade the cosmic microwave background (CMB) has emerged as a fundamental probe of cosmology and astrophysics. In addition to the primary fluctuations of the early Universe, the CMB contains signatures of the gravitational bending of CMB photon trajectories due to matter, called gravitational lensing. Mapping this gravitational lensing is important for a number of reasons including, but not limited to, understanding cosmic structure, constraining cosmological parameters [10, 16] and detecting gravitational waves [11, 12, 15]. In this paper

E. Anderes (✉)
University of California, Davis, CA 95616, USA
e-mail: anderes@stat.ucdavis.edu

A. van Engelen
McGill University, Montréal H3A 2T8, QC, Canada
e-mail: engelen@physics.mcgill.ca

E.D. Feigelson and G.J. Babu (eds.), *Statistical Challenges in Modern Astronomy V*, Lecture Notes in Statistics 209, DOI 10.1007/978-1-4614-3520-4_6,

we investigate the possibility of using simulations to correct a bias when using a plug-in estimate of the matter spectral density from local likelihood estimates of gravitational lensing.

Two estimates have emerged for reconstructing the gravitational potential: the quadratic estimator (developed in [8, 9, 14]) and a global maximum likelihood estimate (developed in [6, 7]). The quadratic estimator, which is arguably the most popular, uses a first order Taylor approximation to establish mode coupling in the Fourier domain which can be estimated to recover the gravitational potential (real space analogs to these estimators can be found in [2, 3]). The maximum likelihood estimate, on the other hand, uses likelihood approximations to find an MLE for estimating the lensing potential. A new estimate developed in [1] uses a local Bayesian approach that avoids the computational difficulties associated with a full scale likelihood approach. This approach estimates the local curvature of the gravitational potential on sliding local neighborhoods of the observed CMB temperature and polarization fields. A low pass filter of the true gravitational potential is then constructed by stitching together local curvature estimates. The local analysis allows one to avoid using the typical first order Taylor expansion for the quadratic estimator and avoids the likelihood approximations used in global estimates. Moreover, the likelihood is computed in position space and therefore can easily deal with point source foregrounds, masking, nonstationary noise and nonstationary beams.

In [1] the local Bayesian method is shown to accurately reconstruct the gravitational potential under nearly ideal experimental conditions when observing both the temperature and the polarization field. In this paper, we consider the temperature fluctuations only. For more realistic experimental conditions the estimated projected mass can be noisy, especially at high frequency. However, using the isotropic assumption one can radially average the squared modulus of the Fourier transform of the estimate to approximate the spectral density. In doing so, one potentially gets accurate estimates of the mass spectral density even with small signal-to-noise ratios at each individual frequency of the mapping estimate.

There are two difficulties that arise when using locally estimated maps to estimate the spectral density. First, the observational noise weakens the amount of local information for gravitational shear and convergence. This has the impact of shrinking the local Bayes estimates toward the prior mean (at zero). The alternative, a local MLE estimate, is not as regularized and can have large estimation noise in the presence of weak local information. Using either of these estimates for estimating the spectral density yields significant biases: high bias for local MLE and low bias for local Bayes. In Fig. 6.1 we show the plug-in estimates of spectral density using the local MLE and Bayes estimates from one simulation of a lensed temperature field on a $10^\circ \times 10^\circ$ patch of the flat sky observed on 1 arcmin pixels with 2-μK noise and beam FWHM of 4 arcmin. The dashed line with stars shows the plug-in estimate from the local Bayes technique, which is clearly shrunk toward zero. The dashed line with triangles shows the local MLE technique, which has a high bias from the estimation error.

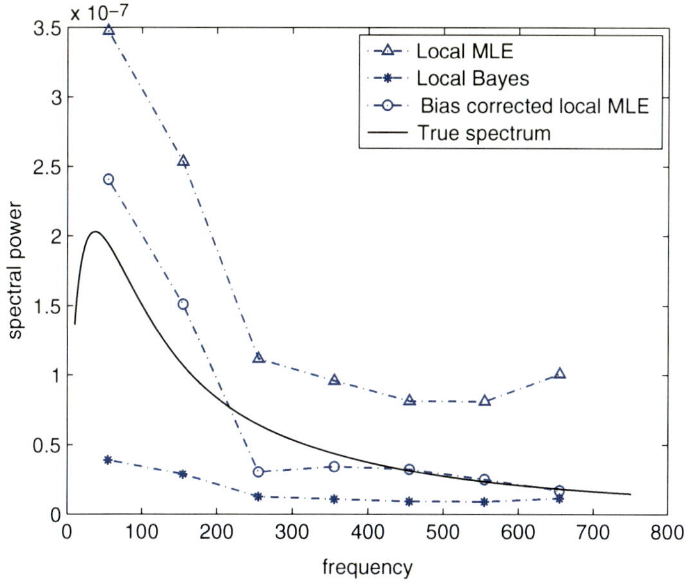

Fig. 6.1 *Solid line* shows the input theoretical spectrum; *triangles* show a local MLE estimate of the spectrum; *stars* show the local Bayes estimate of the spectrum; *circles* show the bias corrected local MLE estimate of spectral density. The simulation is on a $10^\circ \times 10^\circ$ patch of the flat sky observed on 1 arcmin pixels with 2-μK noise and beam FWHM of 4 arcmin

In an attempt to mitigate these biases we work with the overly noisy MLE estimate but correct the resulting bias in the plug-in spectral density estimate using simulations. The dashed line with circles in Fig. 6.1 shows this new estimate. It is clear that this technique has significantly less bias than either the local MLE or the Bayes estimate. However, to make this new technique scientifically useful one needs a theoretical understanding of the behavior of the local MLE estimate in both the lensed and unlensed case (since unlensed simulations are used to correct the bias). There are two main difficulties in deriving such an understanding. First, the estimates are implicitly defined as a maximizer of the local likelihood and, as such, there is no closed form. Secondly, the typical asymptotic arguments used for MLE estimates hold as the signal-to-noise ratio approaches infinity. Since the signal-to-noise ratio is very low on each local neighborhood one might expect the estimates to behave differently than their asymptotic cousins.

The remainder of the paper is organized as follows. In Sect. 6.2 we give a detailed account of the local MLE and Bayesian estimates. Then in Sect. 6.3 we discuss how estimation error propagates to biases in plug-in estimates of spectral density and how to estimate the bias with simulations. We present numerical evidence that one can subtract this estimated bias to produce estimates of spectral density that are comparable to the quadratic estimator found in the current literature. Finally, in Sect. 6.5, we discuss the challenges associated with local estimates of lensing and the resulting estimates of spectral density. We emphasize that the goal of this paper is to partly give some hints at the success of a new method but primarily to illuminate the challenges associated with local likelihood estimates in general.

6.2 Local Estimates of Shear and Convergence

The CMB radiation measures temperature fluctuations of the early Universe some 400,000 years after the big bang. Let $T(\boldsymbol{x})$ denote these fluctuations (measured in units μK) on the observable sky. In this paper we work with the small angle limit and use a flat sky approximation so that $\boldsymbol{x} \in \mathbb{R}^2$. Instead of directly observing T we observe a remapping of the CMB due to the gravitational effect of intervening matter. This lensed CMB can be written $T(\boldsymbol{x} + \nabla\phi(\boldsymbol{x}))$ where ϕ denotes the gravitational potential (see [4], for example).

To describe the local estimate of ϕ from the lensed CMB, developed in [1], first consider a small circular observation patch with diameter δ in the flat sky centered at some point $\boldsymbol{x}_0$, denoted $\mathscr{N}_\delta(\boldsymbol{x}_0) \subset \mathbb{R}^2$. Over this small region we decompose ϕ into an overall local quadratic q^ϕ and error term ε so that

$$\phi = q^\phi + \varepsilon.$$

The global estimate of ϕ is based on stitching together local estimates of q^ϕ, denoted $\hat{q}^\phi$, from the lensed CMB observed on $\mathscr{N}_\delta(\boldsymbol{x}_0)$. Notice that as $\delta \to 0$ the expected magnitude of the error ε approaches zero. This has the effect of improving the following Taylor approximation

$$T(\boldsymbol{x} + \nabla\phi(\boldsymbol{x})) = T(\tilde{\boldsymbol{x}}) + \nabla\varepsilon(\boldsymbol{x}) \cdot \nabla T(\tilde{\boldsymbol{x}}) + \cdots \tag{6.1}$$

for $\boldsymbol{x} \in \mathscr{N}_\delta(\boldsymbol{x}_0)$, where we use the notation $\tilde{\boldsymbol{x}} \equiv \boldsymbol{x} + \nabla q^\phi(\boldsymbol{x})$. Notice that $\tilde{\boldsymbol{x}}$ depends not only on $\boldsymbol{x}$ but also the unknown coefficients of the quadratic term q^ϕ. Now when δ is sufficiently small we can truncate the expansion in (6.1) to get

$$T(\boldsymbol{x} + \nabla\phi(\boldsymbol{x})) \approx T(\boldsymbol{x} + \nabla q^\phi(\boldsymbol{x})) \tag{6.2}$$

on the local neighborhood $\mathscr{N}_\delta(\boldsymbol{x}_0)$. By regarding q^ϕ as unknown we can use the right hand side of (6.2) to develop a likelihood for estimating the coefficients of q^ϕ. Nominally q^ϕ has six unknown coefficients for which to estimate. However, we can ignore the linear terms in q^ϕ since the CMB temperature and the polarization are statistically invariant under the resulting translation in ∇q^ϕ. Therefore, one can write q^ϕ as $c_1(x - x_0)^2/2 + c_2(x - x_0)(y - y_0) + c_3(y - y_0)^2/2$ for unknown coefficients $c_1 = q^\phi_{xx}, c_2 = q^\phi_{xy}, c_3 = q^\phi_{yy}$.

6.2.1 The Local Likelihood

Using the Gaussian approximation of the CMB along with the quadratic potential approximation given by (6.2) one can construct the likelihood as a function of the unknown quadratic coefficients in q^ϕ. Let $\boldsymbol{x}_1,\dots,\boldsymbol{x}_n$ denote the observation locations of the CMB within the local neighborhood $\mathcal{N}_\delta(\boldsymbol{x}_0)$ centered at $\boldsymbol{x}_0$. Using approximation (6.2), the CMB observables in this local neighborhood are well modeled by white noise corruption of a convolved (by the beam) lensed intensity field T. Let $\boldsymbol{t}$ denote the n-vector of observed CMB values at the corresponding pixel locations in $\mathcal{N}_\delta(\boldsymbol{x}_0)$ for the intensity T. Let φ denote the instrumental beam so that the kth entry of $\boldsymbol{t}$ is modeled as

$$t_k \approx \int_{\mathbb{R}^2} d^2\boldsymbol{x}\, \varphi(\boldsymbol{x}) T(\tilde{\boldsymbol{x}}_k - \tilde{\boldsymbol{x}}) + \sigma_T n_k \tag{6.3}$$

where the n_k's are independent standard Gaussian random variables, $\tilde{\boldsymbol{x}}_k = \boldsymbol{x}_k + \nabla q^\phi(\boldsymbol{x}_k)$ and $\tilde{\boldsymbol{x}} = \boldsymbol{x} + \nabla q^\phi(\boldsymbol{x})$. Note that this is an approximate model for t_k based on (6.2). In actuality, the kth temperature measurement is $\int_{\mathbb{R}^2} d^2\boldsymbol{x}\, \varphi(\boldsymbol{x}) T(\boldsymbol{x}_k - \boldsymbol{x} + \nabla\phi(\boldsymbol{x}_k - \boldsymbol{x})) + \sigma_T n_k$, but the linearity of ∇q^ϕ allows us to write $\boldsymbol{x}_k - \boldsymbol{x} + \nabla\phi(\boldsymbol{x}_k - \boldsymbol{x}) \approx \textit{constant} + \tilde{\boldsymbol{x}}_k - \tilde{\boldsymbol{x}}$ on the small neighborhood $\mathcal{N}_\delta(\boldsymbol{x}_0)$. Since one can write $\boldsymbol{x} + \nabla q^\phi(\boldsymbol{x}) = M\boldsymbol{x}$ where the M is a 2×2 real matrix, the sheared temperature $T(\tilde{\boldsymbol{x}})$ is a stationary random field with spectral density given by $C^{TT}_{M^{-1}\boldsymbol{\ell}} \det M^{-1}$. After adjusting for the beam (which is applied after lensing) the covariance between the observations in $\boldsymbol{t}$ can be written

$$\left\langle t_k t_j \right\rangle_{\mathrm{T}} \approx \sigma_T^2 \delta_{ij} + \int_{\mathbb{R}^2} \frac{d^2\boldsymbol{\ell}}{(2\pi)^2} e^{i\boldsymbol{\ell}\cdot(\boldsymbol{x}_k - \boldsymbol{x}_j)} |\varphi(\boldsymbol{\ell})|^2 \frac{C^{TT}_{M^{-1}\boldsymbol{\ell}}}{\det M}. \tag{6.4}$$

Remark. We use the notation $\langle\cdot\rangle_{\mathrm{T}}$ to denote expectation, or ensemble average, with respect to both the CMB temperature field $T(\boldsymbol{x})$ and the observational noise n_k. Conversely, we use the notation $\langle\cdot\rangle_\phi$ to denote expectation with respect to the large scale structure ϕ and for brevity we write $\langle\cdot\rangle \equiv \langle\langle\cdot\rangle_{\mathrm{T}}\rangle_\phi$ where the expectations are done under the assumption that T and ϕ are independent.

Now, using Gaussianity of the full vector of CMB observables the log likelihood (up to a constant), as a function of the quadratic fit q^ϕ, can be written

$$\mathscr{L}(q^\phi|\boldsymbol{t}) = -\frac{1}{2}\boldsymbol{t}^\dagger \left(\Sigma_{q^\phi} + \sigma_T^2 I\right)^{-1} \boldsymbol{t} - \frac{1}{2}\ln\det\left(\Sigma_{q^\phi} + \sigma_T^2 I\right) \tag{6.5}$$

where $\Sigma_{q^\phi} + \sigma_T^2 I$ is the covariance matrix of the observation vector $\boldsymbol{t}$ containing the covariances $\left\langle t_k t_j \right\rangle_{\mathrm{T}}$ given in (6.4) and $\sigma_T^2 I$ is the noise covariance structure where I is the $n\times n$ identity matrix. Notice that the noise structure does not depend on the unknown quadratic q^ϕ. In addition, one can utilize a single FFT to quickly compute the integral (6.4) for sufficient resolution in the argument $\boldsymbol{x}_k - \boldsymbol{x}_j$ to recover $\left\langle t_k t_j \right\rangle_{\mathrm{T}}$ for all pairs k, j.

6.2.2 The Local Posterior

In [1] it is argued that the local estimates of q^ϕ are modeled by a lowpass filter of the true gravitational potential. In particular, the quadratic function q^ϕ can be modeled by

$$q^\phi(\boldsymbol{x}) \approx \int \frac{d^2\boldsymbol{\ell}}{2\pi} e^{i\boldsymbol{x}\cdot\boldsymbol{\ell}} \phi^{\mathrm{lp}}(\boldsymbol{\ell})$$

over $\boldsymbol{x} \in \mathcal{N}_\delta(\boldsymbol{x}_0)$, where $\phi^{\mathrm{lp}}(\boldsymbol{\ell}) \equiv \varphi_\delta(\boldsymbol{\ell})\phi(\boldsymbol{\ell})$, with low-pass filter defined by

$$\varphi_\delta(\boldsymbol{\ell}) \approx \min\Big\{1, \Big[2 - \frac{\delta}{\pi}|\boldsymbol{\ell}|\Big]^+\Big\}. \tag{6.6}$$

Therefore a natural candidate for the prior on the coefficients of q^ϕ is the distribution of the random variables $\frac{\partial^2\phi^{\mathrm{lp}}(0)}{\partial x_k \partial x_j}$. These are mean zero and Gaussian with variances obtained by the corresponding spectral moments of ϕ^{lp}. Letting this prior be denoted by $\pi(q^\phi)$ the posterior distribution on q^ϕ, which we maximize to estimate q^ϕ in the local Bayesian case, is

$$p(q^\phi|\boldsymbol{z}) \propto e^{\mathcal{L}(q^\phi|\boldsymbol{z})}\pi(q^\phi). \tag{6.7}$$

6.2.3 Stitching Together the Local Curvatures

The local MLE estimates of q^ϕ are found by maximizing (6.5) whereas the local Bayesian estimates are found by maximizing (6.7). These estimates give local quadratic fits to the true potential ϕ, i.e. local curvature estimates: $\hat{\phi}^{\mathrm{lp}}_{xx}, \hat{\phi}^{\mathrm{lp}}_{xy}, \hat{\phi}^{\mathrm{lp}}_{yy}$. The global estimate of ϕ^{lp} is found by stitching together these local the estimates. This is done in [1] by performing a gradient fit to $(\hat{\phi}^{\mathrm{lp}}_{xx}, \hat{\phi}^{\mathrm{lp}}_{xy})$ which gives $\hat{\phi}^{\mathrm{lp}}_x$ and a gradient fit to $(\phi^{\mathrm{lp}}_{xy}, \phi^{\mathrm{lp}}_{yy})$ which gives an estimate $\hat{\phi}^{\mathrm{lp}}_y$. A final gradient fit is then fit to the vector field $(\hat{\phi}^{\mathrm{lp}}_x, \hat{\phi}^{\mathrm{lp}}_y)$ to obtain an estimate $\hat{\phi}^{\mathrm{lp}}$. The result of this iterated gradient fit is shown in Fig. 6.2 for a simulated lensed temperature field.

6.3 Spectral Density Estimates of Projected Mass

In this section we discuss the plug-in estimate of spectral density and show how estimation error propagates to biases in the spectral density estimate. We discuss this in the context of estimating the projected mass spectral mass density $C_\ell^{\kappa\kappa}$ where κ denotes the convergence field (which is a tracer for mass fluctuations) and is defined by

$$\kappa \equiv -(\phi_{xx} + \phi_{yy})/2$$

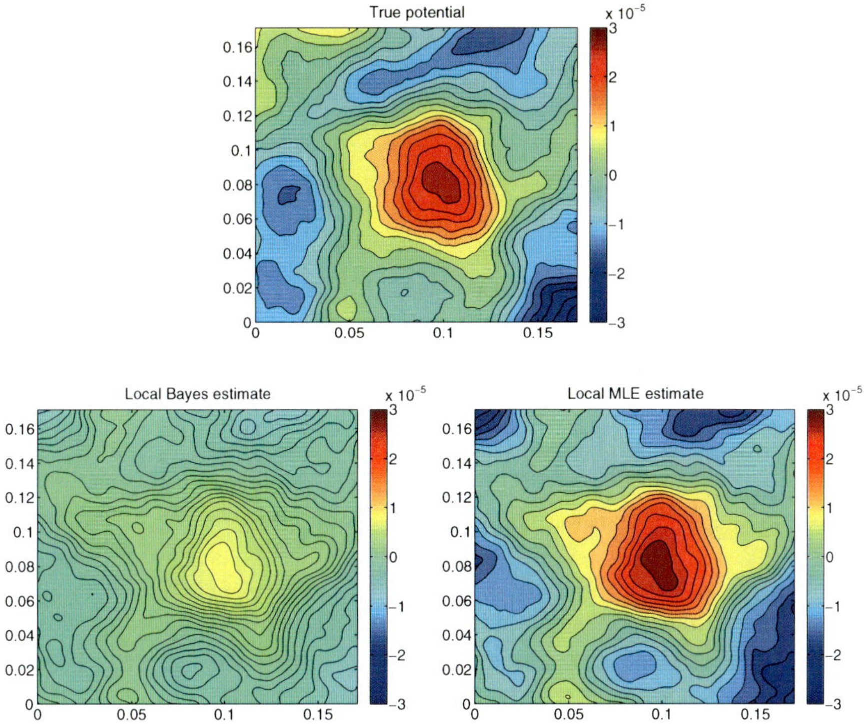

Fig. 6.2 Estimated gravitational potential from simulated lensed CMB: input gravitational potential (*top*); local Bayes estimate (*bottom left*); local MLE estimate (*bottom right*). The lensed temperature simulation was observed on a $10^\circ \times 10^\circ$ patch of the flat sky with 1 arcmin pixels, 2-μK noise and a beam FWHM of 4 arcmin

using the shear notation given in [17]. The spectral density $C_\ell^{\kappa\kappa}$ is defined as the Fourier transform of the autocovariance function:

$$C_\ell^{\kappa\kappa} = \int d^2\boldsymbol{x}\, e^{-i\boldsymbol{\ell}\cdot\boldsymbol{x}} \langle \kappa(\boldsymbol{x})\kappa(\boldsymbol{0})\rangle_\phi.$$

Notice that $\langle \kappa(\boldsymbol{x})\kappa(\boldsymbol{0})\rangle_\phi$ gives the autocovariance since $\langle \kappa(\boldsymbol{x})\rangle_\phi = 0$.

To develop the estimate of $C_\ell^{\kappa\kappa}$ we need the following identity when κ:

$$\langle \kappa(\boldsymbol{\ell})\kappa(\boldsymbol{\ell}')^*\rangle_\phi = \delta_{\boldsymbol{\ell}-\boldsymbol{\ell}'} C_\ell^{\kappa\kappa}$$

This follows directly from the definition of spectral density, the assumption that $\kappa(\boldsymbol{x})$ is isotropic and the definition $\delta_{\boldsymbol{\ell}} \equiv \int \frac{d^2\boldsymbol{x}}{(2\pi)^2} e^{i\boldsymbol{x}\cdot\boldsymbol{\ell}}$. Notice that $\kappa(\boldsymbol{\ell})$ is technically a generalized process which behaves like $\sqrt{C_\ell^{\kappa\kappa}}W(\boldsymbol{\ell})$ where $W(\boldsymbol{\ell})$ is white noise. Therefore when working with finite sky observations of $\kappa(\boldsymbol{x})$ one can produce a discrete version of $\kappa(\boldsymbol{\ell})$ (using discrete Fourier transform) which satisfies $\left\langle|\kappa(\boldsymbol{\ell})|^2\right\rangle_\phi \approx \frac{C_\ell^{\kappa\kappa}}{\Delta\boldsymbol{\ell}}$ where $\Delta\boldsymbol{\ell} \equiv \Delta\ell_1\Delta\ell_2$ is the grid area in Fourier space. For the remainder of this paper we work with this discrete version of κ.

If one knew the convergence field κ then one can estimate $C_\ell^{\kappa\kappa}$ by

$$\widehat{C_{\ell_0}^{\kappa\kappa}} = \frac{\Delta\boldsymbol{\ell}}{\#A_{\ell_0}} \sum_{\boldsymbol{\ell}\in A_{\ell_0}} \left|\kappa(\boldsymbol{\ell})\right|^2 \tag{6.8}$$

where $\Delta\boldsymbol{\ell}$ denotes the area of the observation grid in $\boldsymbol{\ell}$; A_{ℓ_0} denotes a gridded annulus with radius ℓ_0; $\#A_{\ell_0}$ denotes the number of grid points in A_{ℓ_0}. Notice that this estimate is unbiased: $\langle\widehat{C_\ell^{\kappa\kappa}}\rangle_\phi = C_\ell^{\kappa\kappa}$.

In the case of the local MLE or Bayes estimates one has—$\hat{\phi}_{xx}^{\rm lp}$, $\hat{\phi}_{yy}^{\rm lp}$ and $\hat{\phi}_{xy}^{\rm lp}$—the estimates of the mixed partial derivative as a function of local neighborhood midpoint. This leads to an estimate of κ as

$$\hat{\kappa}(\boldsymbol{\ell}) \equiv -(\hat{\phi}_{xx}^{\rm lp}(\boldsymbol{\ell}) + \hat{\phi}_{yy}^{\rm lp}(\boldsymbol{\ell}))/(2\varphi_\delta(\boldsymbol{\ell}))$$

where φ_δ is the band pass filter, defined in (6.6), which approximates the local neighborhood effect discussed in [1]. One can then use $\hat{\kappa}$ to construct a plug-in estimate of $C_\ell^{\kappa\kappa}$ defined as

$$\widehat{C_{\ell_0}^{\hat{\kappa}\hat{\kappa}}} \equiv \text{“plug-in estimate”} = \frac{\Delta\boldsymbol{\ell}}{\#A_{\ell_0}} \sum_{\boldsymbol{\ell}\in A_{\ell_0}} \left|\hat{\kappa}(\boldsymbol{\ell})\right|^2. \tag{6.9}$$

The main problem with the plug-in estimate (6.9) is that estimation error from $\hat{\kappa}$ propagates to biases in $\widehat{C_\ell^{\hat{\kappa}\hat{\kappa}}}$. In the local Bayesian case, the estimation error results in a multiplicative shrinking bias as is seen in Fig. 6.1. Conversely there is a large additive bias for the local MLE plug-in estimate shown in Fig. 6.1. This bias has a simple explanation. If one lets N denote the κ estimation error (so that $\hat{\kappa} = \kappa + N$) then by assuming isotropy $\Delta\boldsymbol{\ell}\langle|\hat{\kappa}(\boldsymbol{\ell})|^2\rangle \approx C_\ell^{\kappa\kappa} + C_\ell^{\kappa N} + C_\ell^{N\kappa} + C_\ell^{NN}$ so that

$$\left\langle\widehat{C_\ell^{\hat{\kappa}\hat{\kappa}}}\right\rangle \approx C_\ell^{\kappa\kappa} + \underbrace{\left(C_\ell^{\kappa N} + C_\ell^{N\kappa} + C_\ell^{NN}\right)}_{\text{additive bias}} \tag{6.10}$$

where C_ℓ^{NN} is the spectrum for N (assuming isotropy) and $C_\ell^{\kappa N}$ is the cross spectrum between κ and N so that $\langle\kappa(\boldsymbol{\ell})N(\boldsymbol{\ell}')^*\rangle \equiv C_\ell^{\kappa N}\delta_{\boldsymbol{\ell}-\boldsymbol{\ell}'}$. For the local Bayes estimate the dominant source of bias is from the first two terms $C_\ell^{\kappa N} + C_\ell^{N\kappa}$ which is from the multiplicative shrinkage bias. Conversely, it seems the dominant source of bias for the local MLE estimate is from the last term, C_ℓ^{NN}, which causes the upward bias seen in the dashed line with triangles in Fig. 6.1.

6.3.1 *Bias Correcting Local MLE Spectral Estimates with Simulations*

In the previous section we used $\hat{\phi}^{\rm lp}_{xx}$, $\hat{\phi}^{\rm lp}_{yy}$ and $\hat{\phi}^{\rm lp}_{xy}$ to approximate the convergence κ and construct the plug-in estimate $\widehat{C^{\hat{\kappa}\hat{\kappa}}_\ell}$. We argued that estimation error results in spectral density estimation bias which is quantified by $C^{\kappa N}_\ell + C^{N\kappa}_\ell + C^{NN}_\ell$. The two terms $C^{\kappa N}_\ell + C^{N\kappa}_\ell$ dominate the bias when using a local Bayesian estimate. Conversely, we will see that the dominant source of bias when using a local MLE estimate is from C^{NN}_ℓ. The advantage of this scenario is that C^{NN}_ℓ has potential to be estimated using unlensed simulations (i.e. where $\kappa = 0$) whereas one must simulate κ under some fiducial model to approximate $C^{\kappa N}_\ell + C^{N\kappa}_\ell$. It is for this reason that we choose to use the noisy local MLE estimates, but correct the resulting bias in the plug-in spectral density estimate by approximating the noise spectrum C^{NN}_ℓ from unlensed simulations. In particular, we use the following bias-adjusted local MLE estimate of $C^{\kappa\kappa}_\ell$

$$\widetilde{C^{\kappa\kappa}_\ell} \equiv \widehat{C^{\hat{\kappa}\hat{\kappa}}_\ell} - \widehat{C^{NN}_\ell} \tag{6.11}$$

where $\hat{\kappa}$ is the local MLE estimate of κ and $\widehat{C^{NN}_\ell}$ is approximated using simulations. To construct $\widehat{C^{NN}_\ell}$ we use the local MLE estimation procedure for $\hat{\kappa}$ and run it on multiple realizations of unlensed CMB (with noise and beam) on the same pixel configuration of the observations. Since these simulations are done with $\kappa = 0$, the result is pure noise N. A spectral density estimate, based on N, is computed for each realization, which are then averaged over multiple realizations to construct $\widehat{C^{NN}_\ell}$.

Remark. In this paper we assume the noise spectrum is radially symmetric so that $\widehat{C^{NN}_\ell}$ is estimated by the same radial averaging as done in (6.9). If the beam or noise is asymmetric this assumption is unlikely to be true. However, one can still estimate the noise spectrum from simulations and subtract the resulting bias in $\widehat{C^{\hat{\kappa}\hat{\kappa}}_\ell}$.

6.4 Simulation

We use four types of simulations in this section, each summarized in Table 6.1. The lensed simulations (with additional noise and beam) are used to generate estimates of κ, using both the local MLE and quadratic estimates, which are then used to construct the plug-in estimates $\widehat{C^{\hat{\kappa}\hat{\kappa}}_\ell}$ given in Sect. 6.3. The unlensed simulations (also with additional noise and beam) are used to estimate the error spectrum, $\widehat{C^{NN}_\ell}$, derived in Sect. 6.3.1 for both the quadratic estimates and the local MLE estimates. We use periodic boundary conditions for the quadratic estimates to avoid complicated appodization issues inherent in the quadratic estimate based on

Table 6.1 The four types of simulations used to compare the bias adjusted local MLE and quadratic estimates of $C_\ell^{\kappa\kappa}$

Simulation	Boundary type	Useage	Number of simulations
$T(\boldsymbol{x}+\nabla\phi(\boldsymbol{x}))$	Periodic	$\widehat{C_\ell^{\hat\kappa\hat\kappa}}$ (using the quadratic estimate)	100
$T(\boldsymbol{x})$	Periodic	$\widehat{C_\ell^{NN}}$ (using the quadratic estimate)	100
$T(\boldsymbol{x}+\nabla\phi(\boldsymbol{x}))$	Non-periodic	$\widehat{C_\ell^{\hat\kappa\hat\kappa}}$ (using local MLEs)	35
$T(\boldsymbol{x})$	Non-periodic	$\widehat{C_\ell^{NN}}$ (using local MLEs)	35

non-periodic sky cuts. Due to computational time constraints only 35 simulations were made for the local MLE estimates (verses 100 simulations for the quadratic estimate).

The non-periodic lensed CMB fields are simulated[1] by generating a high resolution simulation of $T(\boldsymbol{x})$ and the gravitational potential $\phi(\boldsymbol{x})$ on a periodic $17^\circ \times 17^\circ$ patch of the flat sky with 0.25 arcmin pixels . The lensing operation is performed by taking the numerical gradient of ϕ, then using linear interpolation to obtain the lensed field $T(\boldsymbol{x}+\nabla\phi(\boldsymbol{x}))$. We down-sample the lensed field, every 4th pixel, and restrict to a $10^\circ \times 10^\circ$ patch to obtain the desired arcmin pixel resolution for the simulation output. A Gaussian beam with a FWHM of 4 arcmin is applied in Fourier space using FFT of the lensed fields. Finally white noise is added in pixel space with a standard deviation of 2 μK-arcmin. A similar procedure is performed for the periodic lensed CMB fields, except the initial high resolution simulation of $T(\boldsymbol{x})$ and $\phi(\boldsymbol{x})$ are done on a periodic $10^\circ \times 10^\circ$ patch of the flat sky with 0.25 arcmin pixels.

The top plot of Fig. 6.3 summarizes the results using the local MLE estimates with a non-periodic cut sky. The bottom plot of Fig. 6.3 summarizes the corresponding results using the quadratic estimate on a periodic cut sky. Both show the ensemble average of the bias adjusted spectral density estimates $\widetilde{C_\ell^{\kappa\kappa}}$ (blue) compared to the true spectral density $C_\ell^{\kappa\kappa}$ (black) and the ensemble averaged spectrum $\widehat{C_\ell^{\kappa\kappa}}$ one would obtain if one had access to the true κ field for each simulation (red). The bars denote standard deviation error bars. The reason we include $\widehat{C_\ell^{\kappa\kappa}}$ is to show the pixelization and appodization bias which is present irrespective of estimation procedure for κ.

Both estimates of $\widetilde{C_\ell^{\kappa\kappa}}$ based on the quadratic estimate and the local MLE estimate do a good job of tracking the true spectral density. It appears there is more variability in the local MLE estimate, especially at low ℓ. However, at low ℓ the local MLE estimate looks nearly unbiased. The observed power suppression bias

[1]The fiducial cosmology used in our simulations is based on a flat, power law ΛCDM cosmological model, with baryon density $\Omega_b = 0.044$; cold dark matter density $\Omega_{\mathrm{cdm}} = 0.21$; cosmological constant density $\Omega_\Lambda = 0.74$; Hubble parameter $h = 0.71$ in units of $100\,\mathrm{km\,s^{-1}\,Mpc^{-1}}$; primordial scalar fluctuation amplitude $A_s(k = 0.002\,\mathrm{Mpc}^{-1}) = 2.45 \times 10^{-9}$; scalar spectral index $n_s(k = 0.002\,\mathrm{Mpc}^{-1}) = 0.96$; primordial helium abundance $Y_P = 0.24$; and reionization optical depth $\tau_r = 0.088$. The CAMB code is used to generate the theoretical power spectra [13].

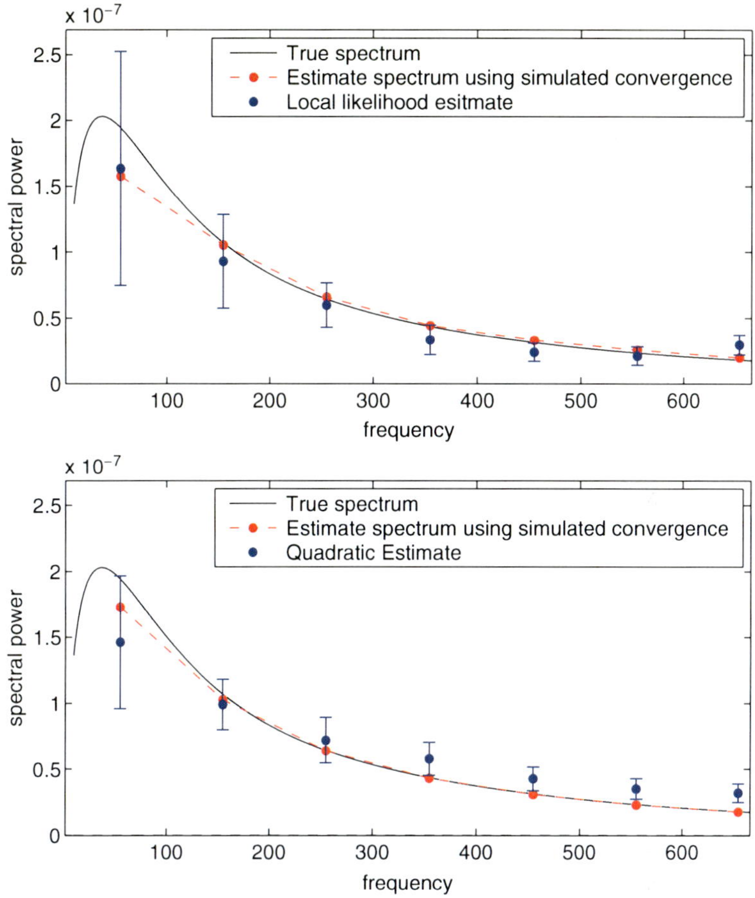

Fig. 6.3 *Top:* The simulation results for the bias adjusted local MLE estimate of $C_\ell^{\kappa\kappa}$. The *blue dots* show the ensemble mean of the estimates (1σ error bars). The *red dots* shows the mean of the estimates $\widehat{C_\ell^{\kappa\kappa}}$ if one had access to the true κ. *Bottom:* The corresponding results for the bias adjusted (with simulations) quadratic estimate of $C_\ell^{\kappa\kappa}$ (See Sect. 6.4 for the simulation details)

at low ℓ and power amplification bias at high ℓ, for the quadratic estimator, is well documented in [5, 11]. It is interesting that the power amplification bias at high ell is opposite to the bias in the local MLE estimate. This may be due to a different Taylor truncation error used to derive the two different estimates. Irrespective of where the bias comes from, it is potentially scientifically useful that the biases are complementary.

Note: The ensemble averaged spectrum $\widehat{C_\ell^{\kappa\kappa}}$ based on the true κ (red) is different at low ℓ in the top plot versus the bottom plot in Fig. 6.3. This is presumably do to the appodization effect which is present in the local MLE simulations, since we are using non-Periodic sky cuts, but not present for the period sky simulations used for the quadratic estimator.

6.5 Challenges

The main challenge for the local MLE procedure is the difficulty in deriving global properties of the noise structure. Since the local MLE estimate is based on a different Taylor truncation it may provide an important complement to the quadratic estimator. Indeed, the spectral density bias in the frequency range 300–600 seems entirely complimentary to the bias in the quadratic estimator. Moreover, the estimation variability also seems comparable in this range. However, the bias and variance most likely depends on the true, but unknown, spectrum $C_\ell^{\kappa\kappa}$. It remains to be seen if the bias remains complementary under alternative models for $C_\ell^{\kappa\kappa}$. Therefore, before it can be used in conjunction with the quadratic estimator, one must get some theoretical quantification of the nature of bias and variance.

It is clear from the results given in Sect. 6.4 that simulations can provide a partial answer to the quantification of bias and variance of the spectral density estimation. Unfortunately, the local MLE estimate is somewhat computationally expense. Each local estimate on a small neighborhood can be done quickly. However, these local estimates are required at a sufficient resolutions to get adequate coverage in Fourier space. Therefore a complete understanding of bias and variance seems unattainable through simulation.

One potential advantage of the local MLE estimate is the apparent unbiasedness at low ℓ. This contrasts with the situation for the quadratic estimate, where the bias at low ℓ has been quantified by Hanson et. al. [5]. They present a method for correcting the low ℓ bias in the quadratic estimator. However, this method depends on a fiducial model for $C_\ell^{\kappa\kappa}$. Indeed, this problem persists when the quadratic estimator is applied to non-periodic sky cuts where quantification of the appodization effect is usually done with simulations under a fiducial model for $C_\ell^{\kappa\kappa}$. The advantage of the local MLE estimates, in this case, is that it does not require a fiducial model for first order bias correction or appodization. The cost of this unbiasedness, it seems, is the apparent increase in variability at low ℓ.

Acknowledgements We thank Lloyd Knox for numerous helpful discussions.

References

1. Anderes, E., Knox, L. & van Engelen, A., Phys, Rev D 83, 043523 (2011)
2. Bucher, M., Carvalho, C. S., Moodley, K., Remazeilles, M., arXiv:1004.3285 (2010)
3. Carvalho, C. S., Moodley, K., Phys. Rev. D 81, 123010 (2010)
4. Dodelson, S., *Modern cosmology*, Academic Press (2003)
5. Hanson, D., Challinor, A., Efstathiou, G., Bielewicz, P., Phys, Rev D 83, 043005 (2011)
6. Hirata, C., & Seljak, U., Phys. Rev. D 67, 043001 (2003a)
7. Hirata, C., & Seljak, U., Phys. Rev. D 68, 083002 (2003b)
8. Hu, W., ApJ 557: L79-L83 (2001)
9. Hu, W., & Okamoto, T., ApJ 574: 566–574 (2002)
10. Kaplinghat, M., Knox, L., Song, Y., Phys. Rev. Lett. 91, 241301 (2003)

11. Kesden, M., Cooray, A., Kamionkowski, M., Phys. Rev. Lett. 89, 011304 (2002)
12. Knox, L., Song, Y., Phys. Rev. Lett. 89, 011303 (2002)
13. Lewis, A. and Challinor, A. and Lasenby, A., ApJ, 538: 473–476 (2000)
14. Okamoto, T., & Hu, W., Phys. Rev. D 67, 083002 (2003)
15. Seljak, U, & Hirata, C., Phys. Rev. D 69, 043005 (2004)
16. Smith, K., Hu W., Manoj, K., Phys. Rev. D 74, 123002 (2006)
17. Zaldarriaga, M., & Seljak, U., Phys. Rev. D 59, 123507 (1999)

Chapter 7
Commentary: 'The Matter Spectral Density from Lensed Cosmic Microwave Background Observations'

Alan Heavens

Abstract Weak gravitational lensing of the Cosmic Microwave Background (CMB) by the intervening clumpy Universe is an important effect which affects parameter estimation in cosmology if not correctly accounted for, and which limits our ability to measure primordial gravitational waves from inflation. Quantifying its effects is an important task, and one which is challenging in practice. In this commentary, I give some physical context and describe the statistical properties of the CMB and lensing fields, and argue that in principle it is an ideal topic for statisticians to get involved with. In practice, there are several challenges which make detailed study quite challenging, and the accompanying paper addresses one of these with a novel approach. This is the effect of non-uniform sky coverage, due to regions of the sky being masked by, for example, point sources. The paper by Drs. Anderes and van Engelen addresses this with a new idea—a local maximum likelihood estimator of the lensing potential, stitching together the estimates to give a global lensing map. It discusses the challenges inherent in the approach, and offers some possibilities to meet the challenges.

7.1 The Cosmological and Statistical Appeal of the CMB

Ethan Anderes and Alexander van Engelen address one of the most important topics in cosmology today. They analyse the effects of gravitational lensing—the bending of light by gravity—on the CMB radiation. The CMB photons give a snapshot of the Universe at recombination, about 300,000 years after the Big Bang; the photons have travelled largely unimpeded since then, being subject to relatively few physical processes, one of which is the deflection due to the gravitational influence of the

A. Heavens (✉)
SUPA, Institute for Astronomy, University of Edinburgh, Blackford Hill, Edinburgh EH9 3HJ, UK
e-mail: afh@roe.ac.uk

E.D. Feigelson and G.J. Babu (eds.), *Statistical Challenges in Modern Astronomy V*, Lecture Notes in Statistics 209, DOI 10.1007/978-1-4614-3520-4_7,

increasingly clumpy intervening Universe. There are a number of observational probes of cosmology, but it is fair to say that the CMB is the prime source of information about the Universe, for two main reasons. The first is that the detailed statistical properties of the fluctuating radiation field depend quite sensitively on the key parameters of the Universe, such as its expansion rate, matter content and so on, as well as giving us a window into the early Universe, where a period of rapid accelerating expansion, known as inflation, is thought to have provided the seeds for subsequent structure formation. These early fluctuations can be detected in the temperature field of the Universe, and also yield a small polarisation due to Thomson scattering of the radiation from free electrons at the recombination era. The second reason is that the physics of the CMB is rather well understood, as at the time of emission (in the standard cosmological model, which is a very successful description of the Universe), the Cosmos was an almost uniform mixture of photons, ordinary matter and dark matter, with a small component of dark energy. This is a simple system to analyse, so the confrontation of observation with theory is very robust, and firm conclusions can be drawn with high confidence. From a statistical point-of-view, the CMB is also a very appealing hunting ground, as its statistical properties are simple, principally as a result of the central limit theorem, so we know pretty much what we are dealing with. In practice, there are important complications, and subtle effects which may indicate new physics. Lensing is one known physical effect, and as it changes the power spectrum of the observed CMB, including its effects is important to get accurate estimation of cosmological parameters. Apart from using the statistical properties of the temperature field to determine cosmological parameters, a very exciting future opportunity is to look for rather direct evidence for inflation, the process in the early Universe which is thought to be responsible for the present expansion of the Universe. This manifests itself through polarisation signals in the CMB as a result of gravitational waves generated during inflation. These give rise to so-called B-mode perturbations, at a level which depends on the energy scale of inflation, manifested in the tensor-to-scalar ratio r. Measurement of primordial B-modes is enormously challenging as the expected level is very low, and furthermore, gravitational lensing of the polarised emission caused by Thomson scattering (which produces E modes) causes a B-mode polarisation signal which dominates the power spectrum on scales less than about a degree. For these reasons, understanding the effect of lensing on the CMB is of vital importance.

7.2 Gravitational Lensing of the CMB

Gravitational lensing of the CMB causes deflection of the photon direction, whilst preserving surface brightness (through Liouville's theorem). The deflection can be described in terms of a lensing potential ϕ, which is an integral along the line-of-side of the gravitational potential, weighted with a lensing kernel. The CMB map is therefore distorted in a way which depends on the distribution of matter along the line-of-sight. The deflections are typically rather small, of the order of a few arcminutes, but the distortions of the map are correlated over scales of around 10°. An excellent review of the subject appears in [1]. From a statistical point-of-view, this is an (almost) ideal situation, since the CMB is (almost) a random gaussian field, and the lensing potential is (almost) another gaussian random field, the exception being on the smallest scales, where nonlinear collapse makes the field slightly nongaussian. Thus in principle it is open to a full Bayesian treatment, a necessary ingredient of which is that one can predict the probability of the data vector given the parameters of the model. Unfortunately in practice this is computationally too demanding. This provides additional motivation for the accompanying paper.

The effect of lensing is as follows: the deflection means that at angular position $\mathbf{x}$ the temperature is given by the unlensed temperature at $\mathbf{x} + \nabla\phi(\mathbf{x})$, which is normally expanded as:

$$T(\mathbf{x} + \nabla\phi(\mathbf{x})) = T(\mathbf{x}) + \nabla^{\mu}\phi\nabla_{\mu}T(\mathbf{x}) + \frac{1}{2}\nabla^{\mu}\phi\nabla^{\nu}\phi\nabla_{\mu}\nabla_{\nu}T(\mathbf{x}) + \dots \tag{7.1}$$

It is known that truncation of this series as shown is a good approximation, but is inaccurate at the level of about 10% on arcminute scales. Analysis of this expansion allows a quadratic estimator for the lensing potential to be written down, via its Fourier transform:

$$\phi_{\mathbf{l}} \propto \int \frac{d^2\mathbf{l}'}{2\pi} T_{\mathbf{l}'} T_{\mathbf{l}-\mathbf{l}'} g(\mathbf{l},\mathbf{l}') \tag{7.2}$$

where g is a known function. The details can be found in [1] or in the original references contained there, but the point is here that this is evaluated in Fourier space, so this is effective if we have all-sky coverage (actually flat-sky is assumed here), but this becomes problematic when the CMB has holes due to bright point sources, or if the noise in the map is non stationary. Both of these are normal, so in practice this is rather difficult to do.

7.3 Anderes and Van Engelen's Method

The method proposed by Anderes and van Engelen reconstructs the lensing potential *locally*, not using the Fourier analysis which represents a global method. For a sufficiently small patch, they approximate the lensing potential as a quadratic function of the coordinates, $q^{\phi}(\mathbf{x})$, and estimate the quadratic coefficients using a local maximum-likelihood method. This has the obvious and attractive advantage that it is immune to the effects of holes elsewhere in the map, and one can approximate the noise as stationary across each patch. The estimation of the lensing potential can be done with Bayesian methods (which may drive the solution to zero because of noise), or MLE estimators, for which the noise bias has a different form. In the accompanying work, correction of the MLE bias is effected using simulations.

There is no doubt that this is a rather challenging problem, and this work is to a certain extent exploratory, highlighting the issues which will need to be addressed very well, but not yet providing a full solution. The main innovation is to approximate the lensing potential as a quadratic function of the coordinates; this must be a good approximation on sufficiently small scales; the important question of course is whether it is adequate on the minimum scales on which it can be applied. The lensed temperature field is then approximated by

$$T(\mathbf{x}+\nabla\phi(\mathbf{x})) \simeq T(\mathbf{x}+\nabla q^{\phi}(\mathbf{x})). \tag{7.3}$$

By simplifying the lensing potential, it is possible to use standard MLE techniques to estimate the three independent coefficients of the quadratic form, exploiting the gaussianity in the problem. Notice that the lensing potential is slightly nongaussian on the very smallest scales, due to nonlinear evolution of the matter density field along the line-of-sight, but the effects are at the percent level on arcminute scales; they are probably unimportant, but this would need checking. Two approaches are taken at this point, either using a MLE or computing the mode of the posterior in a Bayesian analysis. The prior in the Bayesian treatment tends to bias the potential towards zero, so the authors concentrate on the MLE, which seems to have the advantage of an additive, noise-dominated bias which is correctable, compared with a lensing-dependent multiplicative bias which is harder to deal with. There is an immediate issue to contend with, and that is how to patch together the different MLE of the potential in different areas of sky. A variety of gradient fits is employed, to give a potential reconstruction which is visually good, but naturally one would want to know to what extent this patching introduces artefacts in the reconstruction. Perhaps in order to assess how good the constructions are, the authors analyse the power spectrum of the recovered convergence field and compare in simulations with the input. This seems natural, although if this was the main goal, then the reconstruction step may be unnecessary and there may be more direct ways to work only with statistical quantities. The bias corrections are rather large, but the authors show that after correction both the MLE and Bayesian methods yield reasonable estimates of the convergence power spectrum. Given that at some level the bias correction may depend on the true convergence power spectrum, and the authors have yet to test the sensitivity of the correction to this.

With some analyses of lensing of the CMB, computational expense is an issue, and this is no exception, but on the positive side the problem of large-scale biases in the power spectrum which besets other methods appears to be absent. In summary the novel approach which is presented here is an interesting addition to the list of techniques which can be applied to the challenging problem of accurate analysis of the lensed CMB. It is a work in progress, and it will be interesting to see whether the challenges which are identified in this work can be met in practice. If so, the scientific gains are very worthwhile, so one hopes so.

Reference

1. Lewis A., Challinor A.: Weak Gravitational Lensing of the CMB. Physics Reports, **429**, 1–65 (2006)

Chapter 8
Needlets Estimation in Cosmology and Astrophysics

Domenico Marinucci

Abstract Needlets are a form of spherical wavelets which has recently drawn a lot of interest in the cosmological and astrophysical literature. We shall briefly recall the most important features of the needlets construction, and explain why their properties make possible a succesful application to several issues of interest in the analysis of Cosmic Microwave Background data. Many of these possibilities have been exploited already, and we review some results. We shall then explore the role of needlets in adaptive estimation, with a focus on cosmic rays experiments and future weak gravitational lensing and polarization observations on spin random fields.

Dedicated to the memory of Daryl Geller. Much of what is presented here is based upon the contributions of Daryl Geller (1951–2011). In particular, besides developing Mexican needlets (with A.Mayeli), he is to be credited for most of the work on spin needlets/mixed needlets: very little of the mathematical theory behind these developments would exist without him.

8.1 Introduction

Over the last decade, wavelet techniques have become a well-established tool for the analysis of cosmological and astrophysical data, see for instance [51] and the references therein. In particular, a growing interest has been devoted in the last 5 years to the application in a cosmological environment of a new form of spherical

D. Marinucci (✉)
Department of Mathematics, University of Rome Tor Vergata, Via della Ricerca Scientifica 1, Roma, Italy,
e-mail: marinucc@mat.uniroma2.it

E.D. Feigelson and G.J. Babu (eds.), *Statistical Challenges in Modern Astronomy V*, Lecture Notes in Statistics 209, DOI 10.1007/978-1-4614-3520-4_8,

wavelets, called needlets. Needlets were introduced in the mathematical literature by Narcowich et al. [41, 42], see also [20–22] for extensions and generalizations. The investigation of the stochastic properties of needlets when implemented on spherical random fields is due to [2, 3, 35, 36, 40], where applications to several statistical procedures are also considered. Several applications to experimental data have already been implemented: for instance [44] have focussed on estimation of cross-power spectrum from CMB and large scale structure data provided by the NVSS catalogue; [38] have given an overview of the method and various possible applications to CMB; [45] considered search for asymmetries and local estimators of the angular power spectrum; [6, 35, 46, 48, 49] have focussed on the analysis of the needlets bispectrum, non-Gaussianities, estimation of the nonlinearity parameter f_{nl} and its directional variations; [8, 12, 23, 24] discussed the numerical properties of the needlets and exploited them for map-making and angular power spectrum estimation; [14, 15] considered the search for bubble as a test of eternal inflation.

More recently, a few papers have focussed on the use of needlets to develop estimators within the thresholding paradigm, in the framework of directional data. Thresholding estimates were introduced in the statistical literature by Donoho et al. in [10], where it was proved that nonlinear wavelet estimators based on thresholding techniques achieve nearly optimal minimax rates (up to logarithmic terms) for a wide class of nonparametric estimation of unknown density and regression functions. The theory has been enormously developed ever since—we refer to [25] for a textbook reference. In an astrophysical context, needlet-based thresholding algorithms are discussed by Baldi et al. [4] and Kerkyacharian et al. [29, 30]; applications to cosmic rays data analysis are provided for instance by Faÿ et al. [13] and Iuppa et al. [27, 28]. Earlier results on minimax estimators for spherical data, outside the needlets approach, are due to Kim and coauthors (see [31, 32, 34]). Another very active area involves the use of needlets for the analysis of spin data, i.e. those arising when considering the polarization of CMB data and/or future weak gravitational lensing experiment such as the projected mission Euclid [5, 33]). Some results in this area have been provided by Geller and Marinucci [18] and Gelleret al. [16], with further developments discussed by Geller et al. [17], Geller and Marinucci [19], and Durastanti et al. [11].

In this presentation, we shall first review briefly the main features of the needlet construction, and explain how its properties make it a suitable tool for data analysis in many area of cosmological interest. After reviewing briefly applications to CMB, we shall discuss adaptive properties and their importance in the framework of gamma rays, weak gravitational lensing and polarization of the CMB.

8.2 Needlets Construction and Main Properties

Consider any function f defined on the sphere S^2, and such that $f \in L^2(S^2)$, that is to say $\int_{S^2} f^2(x)dx < \infty$. It is well known that the following spectral representation holds:

$$f(x) = \sum_{lm} a_{lm} Y_{lm}(x), \tag{8.1}$$

$$a_{lm} = \int_{S^2} f(x) \overline{Y}_{lm}(x) dx, \tag{8.2}$$

where the set $\{Y_{lm}\}$ represents the array of so-called spherical harmonics on the sphere, defined to be the eigenfunctions of the spherical Laplacian

$$\Delta_{S^2} Y_{lm} = -l(l+1) Y_{lm} \, , \Delta_{S^2} = \frac{1}{\sin \vartheta} \frac{\partial}{\partial \vartheta} \left(\sin \vartheta \frac{\partial}{\partial \vartheta} \right) + \frac{1}{\sin^2 \vartheta} \frac{\partial^2}{\partial \varphi^2} \, .$$

When the function $f(x)$ is random, as in the case of the CMB temperature data (which is assumed to be the realized of an isotropic, finite variance random field) we have also that

$$E a_{lm} = 0 \, , E a_{lm} \overline{a}_{l'm'} = C_l \delta_l^{l'} \delta_m^{m'} \, ,$$

where the bar denotes complex conjugation and C_l the so-called angular power spectrum of the random field.

In the presence of a partially observed sky (as happens for CMB, where some regions are masked by the presence of the Milky Way and other foreground contaminants), the evaluation of the inverse Fourier transform (8.2) becomes unfeasible. Moreover, localization in both the real and the harmonic space is indeed necessary when searching for localized features, such as for instance the highly debated *Cold Spot* [7]. In view of these considerations, the double localization properties of spherical wavelets become most valuable. Among spherical wavelets, we shall be concerned with needlets, whose construction we review as follows.

Let $b(.)$ be a weight function satisfying three conditions, namely

- *Compact support*: $b(t)$ is strictly larger than zero only for $t \in [B^{-1}, B]$, some $B > 1$
- *Smoothness:* $b(t)$ is C^∞
- *Partition of unity:* for all $l = 1, 2, \ldots$ we have

$$\sum_{j=0}^{\infty} b^2 \left(\frac{l}{B^j} \right) = 1 \, .$$

Recipes to construct a function $b(.)$ that satisfy these conditions are easy to find and are provided for instance by Marinucci et al. [38] and Marinucci and Peccati [39].

Next step in the construction is the introduction of a set of *cubature points and weights,* namely a grid of points $\{\xi_{jk}\}$ on the sphere and a grid of weights λ_{jk} such that

$$\sum_{jk} \lambda_{jk} Y_{l_1 m_1}(\xi_{jk}) \overline{Y}_{l_2 m_2}(\xi_{jk}) = \int_{S^2} Y_{l_1 m_1}(x) \overline{Y}_{l_2 m_2}(x) dx \, , \text{ for } B^{j-1} \leq l_1, l_2 \leq B^{j+1}.$$

In other words, cubature points provide a grid of pixels such that the integrals of spherical harmonics are equal to the corresponding Riemann sums on this grid. In practice, cubature points can be identified with the pixel centres of a standard package such as HealPix, and cubature weights can be taken to be equal to the pixel areas, with a very minor numerical approximation.

We have now all the background material to introduce the needlet system, which is defined by

$$\psi_{jk}(x) = \sqrt{\lambda_{jk}} \sum_{l=B^{j-1}}^{B^{j+1}} \sum_{m=-l}^{l} b\left(\frac{l}{B^j}\right) Y_{lm}(x)\overline{Y}_{lm}(\xi_{jk}) ,$$

with the corresponding needlet coefficients provided by

$$\beta_{jk} = \int_{S^2} f(x)\psi_{jk}(x)dx = \sqrt{\lambda_{jk}} \sum_{l=B^{j-1}}^{B^{j+1}} \sum_{m=-l}^{l} b\left(\frac{l}{B^j}\right) a_{lm}Y_{lm}(\xi_{jk}) . \tag{8.3}$$

The coefficients $\{\lambda_{jk}\}$ are such that $cB^{-2j} \leq \lambda_{jk} \leq CB^{-2j}$, with $c, C \in R$, and $N_j = card\{\xi_{jk}\} \approx B^{2j}$, see for instance [3] for more details.

It is now well-known that needlets enjoy quite a few important properties that make them very suitable for spherical data analysis (see for instance [38]). Indeed,

1. *Numerical implementation*: Needlets have important numerical advantages: they do not rely on any tangent plane approximation, but they are naturally embedded in the manifold structure of the sphere and perfectly adapted to standard packages, such as HealPix.
2. *Localization*: Needlets are compactly supported in the harmonic space, i.e. at each scale j needlets are supported on a finite number of multipoles which are perfectly controlled by the data analyst. As far as real space is concerned, for every $M = 1,2,3\ldots$, there exist some constant c_M such that

$$|\psi_{jk}(x)| \leq \frac{c_M B^{2j}}{\{1+B^j d(x,\xi_{jk})\}^M} , \text{ for all } x \in S^2,$$

   where $d(.,.)$ denotes the standard geodesic distance on the sphere. In other words, for any fixed angular distance the tail of the needlets decay faster than any polynomial, i.e. quasi-exponentially as the frequency increases.
3. *Reconstruction property*: As established by Narcowich, Petrushev and Ward, needlets make up a *tight frame system,* meaning that for any (random or deterministic) function $f \in L^2(S^2)$ we have

$$\int_{S^2} f^2(x)dx = \sum_l \frac{2l+1}{4\pi} C_l = \sum_{jk} \beta_{jk}^2 ,$$

a sort of conservation of energy condition. This property yields many important consequences: the first and most important is the following *reconstruction property,* again in the L^2 sense:

$$f(x) = \sum_{jk} \beta_{jk} \psi_{jk}(x) \,, \tag{8.4}$$

so that the pair (8.3)–(8.4) makes a sort of analogue of standard results in Fourier analysis such as (8.1)–(8.2).

4. *Asymptotic uncorrelation*: In the random case, the needlet coefficients β_{jk} are random variables, with correlation such that for all $M = 1, 2, \ldots$, there exist $\widetilde{c}_M$ ensuring that

$$Corr(\beta_{jk}, \beta_{jk'}) \leq \frac{\widetilde{c}_M}{\left\{1 + B^j d(x, \xi_{jk})\right\}^M} \,, \text{ for all } x \in S^2.$$

5. *Flexible implementation*: as discussed by Scodeller et al. [50], needlets can be adapted to specific problems by suitable tuning in the choice of the weight function $b(.)$ and the bandwidth parameter B.

More recently, the needlet idea has been extended by Geller and Mayeli with the construction of so called Mexican needlets, see [20–22] for the definition and discussion of their properties and [50] for numerical analysis and implementation in a cosmological framework. Loosely speaking, the idea is to replace the compactly supported kernel $b(\frac{l}{B^j})$ by a smooth function of the form

$$b\left(\frac{l}{B^j}\right) = \left(\frac{l}{B^j}\right)^{2p} exp\left(-\frac{l^2}{B^{2j}}\right),$$

for some integer parameter p. Because the function $b(.)$ is not compactly supported, an exact reconstruction function cannot hold; Geller and Mayeli show, however, that the corresponding error can be made arbitrary small by a suitable choice of (approximate) cubature points and weights. Apart from that, Mexican needlets undeniably enjoy some very interesting properties: in particular, they have extremely good localization properties in real space, they allow for flexible and numerical convenient implementation, and for $p = 1$ they provide at high frequencies a good approximation to the so-called Spherical Mexican Hat Wavelet construction. Their statistical properties are also encouraging: although the uncorrelation property does not hold for arbitrary angular power spectra, it does hold for the parameter range of interest in the analysis of CMB data, and indeed in these circumstances the numerical evidence in [50] suggest that they may even outperform standard needlets. Most of the analysis we report below in a CMB-related environment have been duplicated with Mexican needlets, with very positive results.

In the following Section, we review briefly some applications of needlets to CMB data analysis where these properties have been fully exploited.

8.3 Applications to CMB Data Analysis

Uncorrelation was first established by Baldi et al. [2], and then used to derive statistical properties of several estimators of interest for CMB data analysis. For instance, consider the statistic

$$\widehat{\Gamma}_j := \sum_k \beta_{jk}^2 ;$$

it is immediate to see that $\widehat{\Gamma}_j$ provides an unbiased estimator for (a binned form of) the angular power spectrum,

$$E\widehat{\Gamma}_j = \Gamma_j = \sum_l b^2\left(\frac{l}{B^j}\right)\frac{2l+1}{4\pi}C_l ;$$

moreover $\widehat{\Gamma}_j$ is also consistent, that is $\widehat{\Gamma}_j/\Gamma_j$ converges in probability to one as the frequency diverges, and asymptotically Gaussian, i.e. it is possible to construct standard confidence intervals. These properties were established by Baldi et al. [2], and then used by Pietrobon et al. [44] to supply an estimator for cross-spectra between background radiation and large scale structure data, when investigating the Integrated Sachs-Wolfe effect. By the same approach it is possible to search for asymmetries in the power spectra of CMB data, for instance between the Northern and Southern hemisphere, an idea introduced by Baldi et al. [3] and then implemented by Pietrobon et al. [45], or to supply estimators of the angular power spectra in CMB temperature data, see [12].

Several other applications focussed for instance on using analogous arguments to construct needlet-based estimators of the bispectrum and/or the nonlinearity parameter f_{nl}, see for instance [46–49]. Here the idea can be summarized as follows. It is well-known that, under isotropy and Gaussianity, the angular power spectrum provides full information on the dependence structure of a random field. To search for non-Gaussianity, it is necessary to consider higher order statistics, for instance the so-called bispectrum, defined by

$$Ea_{l_1m_1}a_{l_2m_2}a_{l_3m_3} = B_{l_1m_1l_2m_2l_3m_3} = \begin{pmatrix} l_1 & l_2 & l_3 \\ m_1 & m_2 & m_3 \end{pmatrix} b_{l_1l_2l_3} . \tag{8.5}$$

The second equality in (8.5) is a crucial consequence of isotropy, entailing that the physical information is concentrated in the *reduced bispectrum* $b_{l_1l_2l_3}$, while isotropy is enforced by the appearance of the Wigner-s 3j symbols on the left, see [26,37,39] for more discussion and details. A natural question to ask is then how to estimate $b_{l_1l_2l_3}$, especially in the presence of missing data. The following needlet bispectrum estimator was introduced by Lan and Marinucci [35] and applied on real data by Pietrobon et al. [46,47] and Rudjord et al. [48,49]:

$$I_{j_1j_2j_3} = \sum_{k_1k_2k_3} \widehat{\beta}_{j_1k_1}\widehat{\beta}_{j_2k_2}\widehat{\beta}_{j_3k_3}h_{j_1j_2j_2}$$

$$EI_{j_1 j_2 j_3} = \sum_{l_1 l_2 l_3} b\left(\frac{l_1}{B^{j_1}}\right) b\left(\frac{l_2}{B^{j_2}}\right) b\left(\frac{l_3}{B^{j_3}}\right) b_{l_1 l_2 l_3} \ ,$$

where $h_{j_1 j_2 j_2}$ is a normalizing factor; we refer to the previous references for more details. In short, needlet coefficients provide a computationally very convenient estimator for a binned form of the bispectrum, with the added bonus of localization in real space. The latter property makes it possible to investigate not only possible non-Gaussianities, but also their variation over the CMB sky, a task carried over by Pietrobon et al. [47] and Rudjord et al. [49]. We refer to these and the previous references for results on applications of these procedures to WMAP data, in particular estimated values of the fundamental nonlinearity parameter f_{nl}.

In a CMB related framework, needlets have become popular for other applications as well. For brevity's sake, we avoid to report each of them here—we simply recall, for instance, various approaches to map-making (the so-called Needlet Internal Linear Combination method is now the standard procedure in the Planck pipeline, see [1,8,23,54]), and very recently the use of needlets to search for bubbles as a test of eternal inflation [14, 15]. Rather than going further into these issues, we prefer to consider more recent developments, such as those concerning polarization and/or weak gravitational lensing data, and those related to directional data and cosmic rays.

8.4 Directional Data

We shall now consider the analysis of directional data, i.e. those emerging from large area surveys for cosmic rays detections. Examples of this setting include the search for ultra high energy cosmic rays considered by experiments such as *AUGER*, gamma rays as investigated by satellites *AGILE* and *Fermi-LAT*, and ground-based observatories such as *ARGO-YBJ*. Many other examples could also be considered.

In each of these cases, the statistical problem can be formulated as observing independent directions $\{X_1, \dots X_n\}$, each $X_i \in S^2$ representing an incoming direction on the sky, possibly observed with error. We shall consider the case where we are interested to reconstruct the density of observed data.

The idea which we shall discuss here follows from classical approaches to wavelet-based density estimation, as discusses on the real line by Donoho et al. [10] and Hardle et al. [25] and many following references. Let $f(x)$ denote the population density of incoming cosmic rays; we have the expansion

$$f(x) = \sum_{jk} \beta_{jk} \psi_{jk}(x) \ , \ \beta_{jk} = \int_{S^2} f(x) \psi_{jk}(x) dx \ .$$

Consider the needlet coefficient estimator

$$\widehat{\beta}_{jk} = \frac{1}{n} \sum_{i=1}^{n} \psi_{jk}(X_i) \ ; \tag{8.6}$$

we have easily

$$E\widehat{\beta}_{jk}=\frac{1}{n}\sum_{i=1}^{n}E\psi_{jk}(X_i)=\frac{1}{n}\sum_{i=1}^{n}\int_{S^2}f(x)\psi_{jk}(x)dx=\beta_{jk}\ .$$

An immediate idea to estimate cosmic rays density is then to implement the so-called linear wavelet estimator [25]

$$\widehat{f}_n(x)=\sum\widehat{\beta}_{jk}\psi_{jk}(x)\ . \tag{8.7}$$

A more refined approach is to rely instead on needlet based *thresholding* estimates, as discussed for instance by Baldi et al. [4], and then extended and generalized by Kerkyacharian et al. [29, 30] and Faÿ et al. [13]. The idea of thresholding is now classical in statistics (see [10, 25]) and can be intuitively explained as follows. Start from the linear estimate (8.7); the smallest coefficients are expected to be dominated by noise, and hence can be dropped, keeping just those coefficients which are above a given threshold.

More precisely, we can consider the nonlinear estimate

$$\widehat{f}_n^*(x)=\sum\widehat{\beta}_{jk}^*\psi_{jk}(x)\ ,\ \widehat{\beta}_{jk}^*=\widehat{\beta}_{jk}I(|\widehat{\beta}_{jk}|>ct_n)\ ,$$

where t_n is a threshold level and $I(A)$ denotes the indicator function of the event A, taking value 1 if A is verified, 0 otherwise. Such estimates can be shown to be nearly optimal (in the *minimax sense*) over a wide class of density functions (described by *Besov spaces*) and different loss functions, i.e. norms by which to measure when the estimate is "close" to the density to be estimated. We refer to the above mentioned papers for discussion and technical details; results on data collected by the ARGO-YBJ collaboration will be released soon.

8.5 Spin Nonparametric Regression

8.5.1 Background

Another generalization of the needlet approach has been recently advocated by Geller and Marinucci [18]; applications to statistics can be found in [17]. In particular, we recall that the CMB satellite missions WMAP and Planck are currently collecting data also on the so-called polarization of CMB. The latter can be loosely described as observations on random ellipses living on the tangent planes for each location on the celestial sphere. Mathematically, this can be expressed by defining random sections of so-called spin fiber bundles, a generalization of the notion of scalar random fields (see [17–19] and below for much more details and discussion). Quite interestingly, exactly the same mathematical framework describes the so-called weak gravitational lensing induced on the observed shape of distant galaxies by clusters of matter (see for instance [5, 33] and the references therein). Huge amount of observational data are expected in the next decade, by means of satellite missions in preparations such as *Euclid*.

The applications of spin needlets to CMB polarization data is discussed in [16]; in [11] spin nonparametric regression was introduced, with a view to applications to polarization and weak lensing data. We refer also to [52] and the references therein for alternative approaches to wavelets analysis in this framework.

We shall then be concerned with the regression model:

$$Y_{i;s} = F_s(X_i) + \varepsilon_{i;s} \ , \tag{8.8}$$

where $F_s(\cdot)$ is an spin function, to be discussed below; for instance, for $s = 2$ F_s can be taken to represent the geometric effect of the gravitational shear. On the other hand, we assume the $\varepsilon_{i;s}$ are i.i.d. spin random variables, which can be viewed as an observational error (to be interpreted, for instance, as the intrinsic shape of the galaxy).

The concept of a spin function was introduced in the 1960s by Newman and Penrose in [43], while working on gravitational radiation. Loosely speaking, a function F represents a spin s quantity if, whenever a tangent vector at point $x \in S^2$ is rotated by an angle ψ under a coordinate change, F transforms as $F' = e^{is\psi}F$ (see [18] for mathematical formalization). Note that for $s = 0$ we are back to the usual scalar functions.

It is also possible to introduce the system of *spin spherical harmonics* $Y_{lm;s}$ as the eigenfunctions of a second-order differential operator which generalizes the spherical Laplacian (refer again to [18, 53] for more details).

The spin spherical harmonics are themselves an orthonormal system, i.e. they satisfy

$$\int_{S^2} Y_{lm;s}\overline{Y}_{lm;s}dx = \int_0^{2\pi}\int_0^{\pi} Y_{lm;s}(\vartheta,\varphi)\overline{Y}_{lm;s}(\vartheta,\varphi)\sin\vartheta d\vartheta d\varphi = \delta_l^{l'}\delta_m^{m'} \ .$$

As for the scalar case, the following representation holds

$$F_s(x) = \sum_l \sum_m a_{lm;s} Y_{lm;s}(x) \ .$$

Here, the spherical harmonics coefficients $a_{lm;s} := \int_{S^2} F_s \overline{Y}_{lm} dx$ are such that

$$a_{lm;s} = a_{lm;E} + i a_{lm;M} \ ,$$

where $\{a_{lm;E}\}, \{a_{lm;M}\}$ are the coefficients of two standard (scalar-valued) spherical functions, which in the physical literature are labelled the electric and magnetic components of the spin function F_s, see again [18, 19] for more discussion.

8.5.2 Spin and Mixed Needlets

The construction of spin needlets (as provided by Geller and Marinucci [18]) is formally similar to the scalar case, although as we discuss below it entails deep differences in terms of the spaces involved. Indeed, spin needlets are defined as follows:

$$\psi_{jk;s}(x) = \sqrt{\lambda_{jk}} \sum_{l} b\left(\frac{\sqrt{e_{l,s}}}{B^j}\right) \sum_{m=-l}^{l} \overline{Y}_{lm;s}(\xi_{jk}) Y_{lm;s}(x), \tag{8.9}$$

where $\{\lambda_{jk}, \xi_{jk}\}$ are, as before, cubature weights and cubature points, $b(\cdot) \in C^{\infty}$ is nonnegative, it is compactly supported in $[1/B, B]$ and satisfies the partition of unity property. Note, however, that the mathematical meaning of (8.9) is rather different from the scalar case; indeed $\psi_{jk;s}(x)$ is to be viewed as a spin s function with respect to rotations of the tangent plane $\mathbb{T}_x$, and a spin $-s$ function with respect to rotations of the tangent plane $\mathbb{T}_{\xi_{jk}}$. The spin needlet operators acts on spin s functions to produce spin s coefficients

$$\int_{S^2} F_s(x)\overline{\psi}_{jk;s}(x)\,dx = \sqrt{\lambda_{jk}} \sum_{lm} b\left(\frac{\sqrt{e_{l,s}}}{B^j}\right) a_{lm;s} Y_{lm;s}(\xi_{jk}) =: \beta_{jk;s}\,. \tag{8.10}$$

We report some important properties for spin needlets, very similar to those in scalar case (see [41,42]). The following *reconstruction formula* holds:

$$F_s(x) = \sum_j \sum_k \beta_{jk;s} \psi_{jk;s}(x).$$

Also, from the previous discussion it follows easily that $\left|\psi_{jk;s}\right|^2$ is a well-defined scalar quantity (It is simple to check that also the squared coefficients $\left|\beta_{jk;s}\right|^2$ are scalar). The following localization property is hence well-defined (see [18]): for any $M \in \mathbb{N}$, there exists a constant $c_M > 0$ such that for every $x \in S^2$:

$$\left|\psi_{jk;s}(x)\right| \leq \frac{c_M B^j}{\left(1 + B^j \arccos\left(\langle \xi_{jk}, x\rangle\right)\right)^M}\,.$$

As an alternative construction, [19] have considered so-called mixed needlets, defined as

$$\psi_{jk;s,\mathcal{M}}(x) = \sqrt{\lambda_{jk}} \sum_{l \geq |s|} b\left(\frac{\sqrt{e_{ls}}}{B^j}\right) \sum_m Y_{lm;s}(x) \overline{Y}_{lm}(\xi_{jk})\,.$$

The construction is similar to the one discussed earlier, the main difference being the fact that the resulting needlet coefficients $\beta_{jk;s,\mathcal{M}}$ are scalar, rather than spin, quantities. We refrain from a full comparison here for brevity's sake; it suffices to say that the procedures we shall discuss below can be implemented with both kind of needlets.

Spin and mixed needlets can actually be used for polarization data analysis much the same way as we have seen for the scalar case. In particular, they can be used to derive angular power spectrum estimators for the so-called E and B modes of polarization, they can be implemented to test non-Gaussianity, they can be exploited to search for asymmetries and local features. In the section below, however, we shall discuss a different application, i.e. their exploitation to obtain adaptive estimation for fields observed with errors.

8.5.3 Nonparametric Regression on Spin Runctions

We start by recalling the regression formula (8.8):

$$Y_{i;s} = F_s(X_i) + \varepsilon_{i;s} \ .$$

As discussed earlier, we envisage a situation where it is possible to collect data which can be viewed as measurements on a spin field, i.e. for instance the polarization of the Cosmic Microwave Background (see [9]), or the Weak Gravitational Lensing effect on the images of distant Galaxies (see [5]).

The procedure we are going to investigate can be viewed again as a form of needlet thresholding in the spin fiber bundles case. Our approach could be implemented for both mixed and spin needlets. We start by defining, as usual, an unbiased estimator for needlet coefficients. More precisely, we define

$$\widehat{\beta}_{jk;s} := \frac{1}{n}\sum_{i=1}^{n} Y_i \overline{\psi}_{jk;s}(X_i) \ , \ i = 1,2,\ldots,n \ .$$

We have immediately:

$$E\left(\widehat{\beta}_{jk;s}\right) = \int_{S^2} \overline{\psi}_{jk;s}(X_i) F_s(X_i) = \beta_{jk;s} \ . \tag{8.11}$$

The thresholding estimator is then defined, as usually, (see for instance [10, 25])

$$F_s^*(x) = \sum_{j=1}^{J_n}\sum_{k=1}^{N_j} \beta_{jk;s}^* \psi_{jk;s}(x). \tag{8.12}$$

In (8.12), J_n represents a cut-off frequency, which we shall fix at $B^{J_n} = \sqrt{\frac{n}{\log n}}$, whereas N_j is the cardinality of the cubature point set at frequency j; it is known (see for instance [3]) that there exist positive constants c_1, c_2 such that $c_1 B^{2j} \leq N_j \leq c_2 B^{2j}$ (written $N^j \approx B^{2j}$). It is then possible to show (see [11]) that thresholding estimates achieve 'nearly optimal' (up to logarithmic factors) rates with respect to general loss functions.

Theorem 8.1. *Let $F_s \in \mathscr{B}^r_{\pi q;s}(G)$, the "Besov ball" such that $\|F_s\|_{\mathscr{B}^r_{\pi q;s}} \leq G < \infty$, $r - \frac{2}{\pi} > 0$, and consider F^*_s defined by (8.12). For $1 \leq p < \infty$, there exist $\kappa > 0$ such that we have*

$$\sup_{F_s \in \mathscr{B}^r_{\pi q;s}} E\left\|F^*_s - F_s\right\|^p_{L^p_s} \leq C_p \{\log n\}^p \left[\frac{n}{\log n}\right]^{-\alpha(r,\pi,p)},$$

$$\alpha(r,\pi,p) = \begin{cases} \frac{rp}{2r+2} \text{ for } \pi \geq \frac{2p}{2r+2} \text{ (regular zone)} \\ \frac{p(r-2(\frac{1}{\pi}-\frac{1}{p}))}{2(r-2(\frac{1}{\pi}-\frac{1}{2}))} \text{ for } \pi \leq \frac{2p}{2r+2} \text{ (sparse zone)} \end{cases}.$$

Also, for $p = \infty$

$$\sup_{F_s \in \mathscr{B}^r_{\pi q;s}} E\left\|F^*_s - F_s\right\|_{L^\infty_s} \leq C_\infty \left[\frac{n}{\log n}\right]^{-\alpha(r,\pi,\infty)}, \quad \alpha(r,\pi,\infty) = \frac{(r-\frac{2}{\pi})}{2(r-2(\frac{1}{\pi}-\frac{1}{2}))}.$$

Remark 8.1. The definitions of "regular" and "sparse" zones are classical, and so are the rates obtained, which indeed correspond (for instance) to those presented by Baldi et al. [4] for density estimation. The results are basically saying that over a broad class of functions thresholding estimates converge (up to logarithmic factors) as fast as any other possible estimator, even without prior knowledge on the regularity of the (spin) function to be estimated. This is exactly the sort of robustness property we were looking for. Of course $\alpha(r,\pi,p) < \frac{1}{2}$, $\lim_{r\to\infty} \alpha(r,\pi,p) = \frac{1}{2}$. This is to say that for "very regular" functions, thresholding estimates converge as fast as the pure parametric case.

Remark 8.2. For $s = 0$, the previous results cover adaptive nonparametric regression for complex-valued, scalar functions. Again, the rates correspond to the usual nearly minimax bounds.

Acknowledgements The material in this survey covers research which has been developed in collaboration with several other mathematicians, among which we mention P. Baldi, D. Geller, G. Kerkyacharian, D. Picard and the PhD students X. Lan and C. Durastanti. On the physical side, we mention especially Davide Pietrobon and Frode Hansen for applications of needlets to CMB data analysis, and Roberto Iuppa and Rinaldo Santonico for gamma rays applications.

References

1. Ade, P.A.R. and the Planck LFI Core Team (2011) Planck Early Results: The Low Frequency Instrument Data Processing, arXiv 1101.2048
2. Baldi, P.; Kerkyacharian, G.; Marinucci, D.; Picard, D. (2009) Asymptotics for Spherical Needlets, *Annals of Statistics*, Vol. 37, No. 3, 1150–1171, arXiv: math.st/0606599
3. Baldi, P.; Kerkyacharian, G.; Marinucci, D.; Picard, D. (2009) Subsampling Needlet Coefficients on the Sphere, *Bernoulli*, Vol. 15, 438–463, arXiv: 0706.4169

4. Baldi, P.; Kerkyacharian, G.; Marinucci, D.; Picard, D. (2009) Adaptive Density Estimation for Directional Data Using Needlets, *Annals of Statistics,* Vol. 37, No. 6A, 3362–3395, arXiv: 0807.5059
5. Bridles, S. et al. (2009) Handbook for the GREAT08 Challenge: an Image Analysis Competition for Gravitational Lensing, *Annals of Applied Statistics*, Vol. 2, pp.6–37
6. Cabella, P.; Pietrobon, D.; Veneziani, M.; Balbi, A.; Crittenden, R.; de Gasperis, G.; Quercellini, C.; Vittorio, N. (2010) Foreground influence on primordial non-Gaussianity estimates: needlet analysis of WMAP 5-year data, arXiv: 0910.4362, *Monthly Notices of the Royal Astronomical Society*, Volume 405, Issue 2, pp. 961–968
7. Cruz,M.; Cayon, L.; Martinez-Gonzalez, E.; Vielva, P. ; Jin, J. (2007) The non-Gaussian Cold Spot in the 3-year WMAP data, *Astrophys.J.* 655:11–20
8. Delabrouille, J., Cardoso, J.-F., Le Jeune, M. , Betoule, M., Fay, G., Guilloux, F. (2008) A Full Sky, Low Foreground, High Resolution CMB Map from WMAP, *Astronomy and Astrophysics*, Volume 493, Issue 3, 2009, pp.835–857, arXiv 0807.0773
9. Dodelson, S. (2003) *Modern Cosmology*, Academic Press
10. Donoho, D.; Johnstone, I.; Kerkyacharian, G.; Picard, D. (1996) Density estimation by wavelet thresholding, *Annals of Statistics*, 24, 508–539
11. Durastanti, C., Geller, D. and Marinucci, D. (2012) Adaptive Nonparametric Regression on Spin Fiber Bundles, *Journal of Multivariate Analysis*, Vol. 104, 1, pp. 16–38, arXiv:1009.4345
12. Faÿ, G.; Guilloux, F.; Betoule, M.; Cardoso, J.-F. ; Delabrouille, J.; Le Jeune, M. (2008) CMB Power Spectrum Estimation Using Wavelets, *Physical Review D,* D78:083013, arxiv:0807.1113
13. Faÿ, G.; Delabrouille, J.; Kerkyacharian, G.; Picard, D.(2011) Testing the isotropy of high energy cosmic rays using spherical needlets, arXiv:1107.5658
14. Feeney, S.M.; Johnson, M.C.; Mortlock, D.J.; Peiris, H.V. (2010) First Observational Tests of Eternal Inflation, preprint, arXiv:1012.1995
15. Feeney, S.M.; Johnson, M.C.; Mortlock, D.J.; Peiris, H.V. (2010) First Observational Tests of Eternal Inflation: Analysis Methods and WMAP 7-Year Results, preprint, arXiv:1012.3667
16. Geller, D.; Hansen, F.K.; Marinucci, D.; Kerkyacharian, G.; Picard, D. (2008), Spin Needlets for Cosmic Microwave Background Polarization Data Analysis, *Physical Review D*, D78:123533, arXiv:0811.2881
17. Geller, D.; Lan, X.; Marinucci, D. (2009) Spin Needlets Spectral Estimation, *Electronic Journal of Statistics*, Vol. 3, 1497–1530, arXiv:0907.3369
18. Geller, D.; Marinucci, D. (2010) Spin Wavelets on the Sphere, *Journal of Fourier Analysis and its Applications*, Vol. 16, pp.840–844, arXiv: 0811.2835
19. Geller, D.; Marinucci, D. (2011) Mixed Needlets, *Journal of Mathematical Analysis and Applications*, Vol.375, pp.610–630, arXiv: 1006.3835
20. Geller, D.; Mayeli, A. (2009) Continuous Wavelets on Manifolds, *Math. Z.,* Vol. 262, pp. 895–927, arXiv: math/0602201
21. Geller, D.; Mayeli, A. (2009) Nearly Tight Frames and Space-Frequency Analysis on Compact Manifolds, *Math. Z.,* Vol, 263 (2009), pp. 235–264, arXiv: 0706.3642
22. Geller, D.; Mayeli, A. (2009) Besov Spaces and Frames on Compact Manifolds, *Indiana Univ. Math. J.,* Vol. 58, pp. 2003–2042, arXiv:0709.2452.
23. Ghosh, T.; Delabrouille, J.; Remazeilles, M.; Cardoso, J.-F.; Souradeep, T. (2010) Foreground Maps in WMAP Frequency Bands, arxiv: 1006.0916
24. Guilloux, F.; Fay, G.; Cardoso, J.-F. (2009) Practical Wavelet Design on the Sphere, *Applied and Computational Harmonic Analysis,* Vol. 26, pp.143–160, arxiv 0706.2598
25. Hardle, W.; Kerkyacharian, G.; Picard, D.; Tsybakov, A. (1997) *Wavelets, Approximations and statistical application.* Springer, Berlin
26. Hu, W. (2001) The Angular Trispectrum of the CMB, *Physical Review D*, 64 (2001) 083005
27. Iuppa, R. ; Di Sciascio, G.; Hansen, F.K.; Marinucci, D.; Santonico, R. and the ARGO collaboration (2011) A needlet-based approach to the shower-mode data analysis in the ARGO-YBJ experiment, *Proceedings of the 32nd ICRC, Beijing*, - conf. ID 0518.

28. Iuppa, R. ; Di Sciascio, T.; Hansen, F.K.; Marinucci, D.; Santonico, R. and the ARGO collaboration (2011) Spherical needlets for data analysis in wide field of view experiments, in preparation.
29. Kerkyacharian, G.; Nickl, R.; Picard, D. (2011) Concentration Inequalities and Confidence Bands for Needlet Density Estimators on Compact Homogeneous Manifolds, *Probability Theory and Related Fields*, to appear, arXiv: 1102.2450
30. Kerkyacharian,G.; Pham Ngoc, T.M. ; Picard, D. (2011) Localized Deconvolution on the Sphere, *Annals of Statistics*, to appear.
31. Kim, P.T.; Koo, J.-Y. (2002) Optimal Spherical Deconvolution, *Journal of Multivariate Analysis*, 80, 21–42
32. Kim, P.T.; Koo, J.-Y., Luo, Z.-M. (2009) Weyl Eigenvalue Asymptotics and Sharp Adaptation on Vector Bundles, *Journal of Multivariate Analysis*, 100, 1962–1978
33. Kitching, T. et al. (2010) Gravitational Lensing Accuracy Testing 2010 (GREAT10) Challenge Handbook, preprint, arXiv: 1009.0779
34. Koo, J.-Y.; Kim, P.T. (2008) Sharp Adaptation for Spherical Inverse Problems with Applications to Medical Imaging, *Journal of Multivariate Analysis*, 99, 165–190
35. Lan, X.; Marinucci, D. (2008) The Needlets Bispectrum, *Electronic Journal of Statistics*, Vol. 2, pp.332–367, arXiv:0802.4020
36. Lan, X.; Marinucci, D. (2009) On the Dependence Structure of Wavelet Coefficients for Spherical Random Fields, *Stochastic Processes and their Applications*, 119, 3749–3766, arXiv:0805.4154
37. Marinucci, D. (2006) , High-Resolution Asymptotics for the Angular Bispectrum of Spherical Random Fields, *The Annals of Statistics*, Vol. 34, pp. 1–41, arXiv: math/0502434
38. Marinucci, D.; Pietrobon, D.; Balbi, A.; Baldi, P.; Cabella, P.; Kerkyacharian, G.; Natoli, P.; Picard, D.; Vittorio, N. (2008) Spherical Needlets for CMB Data Analysis, *Monthly Notices of the Royal Astronomical Society*, Volume 383, Issue 2, pp. 539–545, January 2008, arXiv: 0707.0844
39. Marinucci, D.; Peccati, G. (2011) *Random Fields on the Sphere: Representations, Limit Theorems and Cosmological Applications*, London Mathematical Society Lecture Notes, Cambridge University Press
40. Mayeli, A. (2010), Asymptotic Uncorrelation for Mexican Needlets, *Journal of Mathematical Analysis and Applications*, Vol. 363, Issue 1, pp. 336–344, arXiv: 0806.3009
41. Narcowich, F.J.; Petrushev, P.; Ward, J.D. (2006) Localized Tight Frames on Spheres, *SIAM Journal of Mathematical Analysis* Vol. 38, pp. 574–594
42. Narcowich, F.J.; Petrushev, P.; Ward, J.D. (2006) Decomposition of Besov and Triebel-Lizorkin Spaces on the Sphere, *Journal of Functional Analysis*, Vol. 238, 2, 530–564
43. Newman, E.T.; Penrose, R. (1966) Note on the Bondi-Metzner-Sachs Group, *Journal of Mathematical Physics*, Vol.7 No.5, 863–870
44. Pietrobon, D.; Balbi, A.; Marinucci, D. (2006) Integrated Sachs-Wolfe Effect from the Cross Correlation of WMAP3 Year and the NRAO VLA Sky Survey Data: New Results and Constraints on Dark Energy, *Physical Review D*, id. D:74, 043524
45. Pietrobon, D.; Amblard, A.; Balbi, A.; Cabella, P.; Cooray, A.; Marinucci, D. (2008) Needlet Detection of Features in WMAP CMB Sky and the Impact on Anisotropies and Hemispherical Asymmetries, *Physical Review D*, D78 103504, arXiv: 0809.0010
46. Pietrobon, D.; Cabella, P.; Balbi, A.; de Gasperis, G.; Vittorio, N. (2009) Constraints on Primordial non-Gaussianity from a Needlet Analysis of the WMAP-5 Data, arXiv: 0812.2478, *Monthly Notices of the Royal Astronomical Society*, Volume 396, Issue 3, pp. 1682–1688
47. Pietrobon, D.; Balbi, A.; Cabella, P.; Gorski, K.M. (2009) Needatool: A Needlet Analysis Tool for Cosmological Data Processing, *Astrophysical Journal*, Vol. 723, N.1 , arXiv: 1010.1371
48. Rudjord, O.; Hansen, F.K.; Lan, X.; Liguori, M.; Marinucci, D.; Matarrese, S. (2009) An Estimate of the Primordial Non-Gaussianity Parameter f_{NL} Using the Needlet Bispectrum from WMAP, *Astrophysical Journal*, 701, 369–376, arXiv: 0901.3154

49. Rudjord, O.; Hansen, F.K.; Lan, X.; Liguori, M.; Marinucci, D.; Matarrese, S. (2010) Directional Variations of the Non-Gaussianity Parameter f_{NL}, *Astrophysical Journal*, Volume 708, Issue 2, pp. 1321–1325, arXiv: 0906.3232
50. Scodeller, S.; Rudjord, O.; Hansen, F.K.; Marinucci, D.; Geller, D.; Mayeli, A. (2011) Introducing Mexican Needlets for CMB Analysis: Issues for Practical Applications and Comparison with Standard Needlets, *The Astrophysical Journal*, Volume 733, Issue 2, article id. 121, arXiv: 1004.5576
51. Starck, J.-L.; Bobin, J. (2010) Astronomical Data Analysis and Sparsity: from Wavelets to Compressed Sensing, *Proceedings of the IEEE Special Issue on: Applications of Sparse Representation and Compressive Sensing*, Vol. 98, No. 6, pp 1021–1030, arXiv: 0903.3383
52. Starck, J.-L.; Moudden, Y.; Bobin, J. (2009) Polarized wavelets and curvelets on the sphere, *Astronomy and Astrophysics*, 497, pp 931–943, arXiv: 0902.0574
53. Wiaux, Y.; Jacques, L.; Vandergheynst, P. (2007) Fast Spin ± 2 Spherical Harmonics and Applications in Cosmology, *Journal of Computational Physics*, Vol. 226, p. 2359–2371
54. Zacchei A. and the Planck LFI Core Team (2011) Planck Early Results: The Low Frequency Instrument Data Processing, arXiv 1101.2040

Part II
Bayesian Analysis Across Astronomy

Chapter 9
Parameter Estimation and Model Selection in Extragalactic Astronomy

Martin D. Weinberg

Abstract Astronomy is rife with multi-instrument, multiple wave band data sets and complex physical theories. An astronomer, therefore, needs to (1) infer the parameters of models from multiple hypotheses; (2) inter-compare hypotheses; and (3) test that the data is sufficiently well explained by the models. Most often, all three needs are inseparably linked. The Bayesian approach allows these to be addressed simultaneously and consistently. Although Bayesian inference is well-suited to problems of inference in astronomical science, the most commonly used tools best treat idealized or specialized models. Here, I describe our experience based on two such problems in extragalactic science—testing models based on galaxy images and exploring recipes galaxy evolution using *semi-analytic* models—using the UMass Bayesian Inference Engine (BIE), a parallel-optimized software package for parameter inference and model selection. The BIE is designed as a collaborative platform for Bayesian methodology for astronomical problems.

9.1 Introduction

Inference is fundamental to the scientific process. We may broadly identify two categories of inference problems: (1) *estimation*—finding the parameter of a theory or model from data; and (2) *hypothesis testing*—determining which theory, indeed if any, is supported by the data. These are computationally difficult problems in practice. The different data characteristics for each survey and engenders varied selection effects and inhomogeneous error models. Moreover, the information content of large survey databases can in principle determine models with many parameters but exhaustive exploration of parameter space is infeasible. In brief,

M.D. Weinberg (✉)
Department of Astronomy, University of Massachusetts, Lederle Graduate Research Center, Amherst, MA 01003, USA
e-mail: weinberg@astro.umass.edu

E.D. Feigelson and G.J. Babu (eds.), *Statistical Challenges in Modern Astronomy V*, Lecture Notes in Statistics 209, DOI 10.1007/978-1-4614-3520-4_9,

astronomers increasingly rely on numerical data analysis, but most cannot take full advantage of the power afforded by present-day computational statistics for attacking the inference problem owing to a lack of tools.

These classes of estimation problems are readily posed by Bayesian inference (BI), which determines model parameters while allowing for straightforward incorporation of heterogeneous selection biases. Combined with the modern theory of probability and Monte Carlo computation, Bayes theorem provides a rich framework for the quantitative investigation of a wide variety of inference problems, such as classification and cluster analysis, that broadly extends and unifies the two categories described above. Unfortunately, the computational complexity of BI grows quickly with the number of model parameters and becomes intractable before the volume of currently available large data sets is reached. Beginning in 2000, a multi-disciplinary investigator team from the Departments of Astronomy and Computer Science at UMass designed and implemented the Bayesian Inference Engine,[1] a Markov chain Monte Carlo (MCMC) parallel software platform for performing statistical inference over very large data sets. This presentation is a research 'travel log', describing our experience in applying BI to complex problems with current algorithms using the BIE. Please see our companion posters for additional details on two of our projects.

9.2 What do Astronomers Want and Need?

9.2.1 *Parameter Estimation*

Many astronomical data analysis problems are posed as parameter estimates. For example: one observes the flux profile of a disk galaxy and would like to estimate its scale length. In this problem, we are asserting that the underlying model is true and testing the hypothesis that the parameter, a scale length, has a particular value. BI approaches these problems with the following steps, reflecting standard practice of scientific method: (1) numerically quantify a prior belief in the hypothesis; (2) collect data that will either be consistent or inconsistent with the hypothesis; (3) compute the new confidence in the hypothesis given the new data. These steps may be repeated to achieve the desired degree of confidence. A clever observer will design campaigns that refine confidence efficiently (i.e., that makes the confidence high or low).

A prime motivation for the BIE project is our thesis that the power of expensive and large survey data sets is underutilized by targeting parameter estimation through maximum likelihood (ML) as the goal. For example, let us consider our

[1]Previously funded by NASA's Applied Information and System Research (AISR) Program. For further information see http://www.astro.umass.edu/bie.

example above. We determine the posterior probability distribution for scale lengths for some subset of survey images. Alongside scale length, we determine other parameters such as luminosity, axis ratios or inclination, and possibly higher-moments such as asymmetry of the disk. We use the scale length with maximum probability as the *best* estimate. We then attempt to correlate the scale lengths with some other parameter of interest, luminosity or asymmetry, say. Then, if a correlation exists, we attempt to interpret the correlation in the context of theories of galaxy formation and evolution. Observe, that in the first step, we are throwing out much of the information implicit in the posterior distribution. In particular, the luminosity estimate is most likely correlated with the scale-length estimate. If we were to plot the posterior distribution in these two parameters, we might find that the distribution is elongated in the scale-length–asymmetry plane, possible in the same sense as the putative correlation! In other words, the confidence in the hypothesis of a correlation should include the full posterior distribution of parameter estimates, not just the maximum probability estimate. See Sect. 9.4.2 and Fig. 9.2 for an example.

9.2.2 Which Model Is Right?

This leads naturally to the following question: which model is right one? This question is a critical piece of the scientific method. Astronomers typically do not address it quantitatively but *need* to do so. I will separate the general question "which model is right?" into two: (1) "does the model explain the data?", the *goodness-of-fit* problem; and (2) "which of two (or more) models better explains the data?", the *model selection* problem. Let us begin with (1) and describe (2) in the next section.

Suppose we have performed a parameter estimation and determined the parameter region(s) containing a large fraction of the probability. We may compute the predicted models for each parameter vector in the region and assess the fit to the data. This is often done by eye. But, *model checking*, or assessing the fit of a model, is a crucial part of any statistical analysis. Before making any conclusions from the application of a statistical model to a data set, an investigator should assess the fit of the model to make sure that the model can explain adequately the important aspects of the data set. Serious misfit (failure of the model to explain important aspects of the data that are of practical interest) should result in the replacement or extension of the model. Even if a model has been assumed to be final, it is important to assess its fit to be aware of its limitations before making any inferences.

The posterior predictive check (PPC) is a commonly-used Bayesian model evaluation method. It is simple and has a clear theoretical basis. To apply the method, one first defines a set discrepancy measures, $T(\mathbf{D}, \boldsymbol{\theta})$. A discrepancy measure, like a classical test statistic, measures the difference between an aspect of the observed data set and the theoretically predicted or *replicated* data set. That is, the predicted distribution of some future data $\mathbf{D}^{rep}$ after having observed the data $\mathbf{D}$ is

$$p(\mathbf{D}^{rep}|\mathbf{D}) = \int p(\mathbf{D}^{rep}, \boldsymbol{\theta}|\mathbf{D})\, d\boldsymbol{\theta} = \int p(\mathbf{D}^{rep}|\boldsymbol{\theta}, \mathbf{D}) p(\boldsymbol{\theta}|\mathbf{D})\, d\boldsymbol{\theta} \tag{9.1}$$

(e.g. [1]). Practically, a number of replicated data sets are generated from $P(\mathbf{D}|\boldsymbol{\theta}, \mathscr{M})$ with $\boldsymbol{\theta}$ selected from posterior distribution. Any systematic differences between the observed data set and the replicated data sets indicate potential failure of the model to explain the data. For example, one may use the distribution of a discrepancy measure conditional of the replicated data set to generate estimate a Bayesian p-value for the true data:

$$\begin{aligned} p_B &= P[T(\mathbf{D}^{rep}, \boldsymbol{\theta}) \geq T(\mathbf{D}, \boldsymbol{\theta})|\mathbf{D}] \\ &= \int\int I_{T(\mathbf{D}^{rep}, \boldsymbol{\theta}) \geq T(\mathbf{D}, \boldsymbol{\theta})} p(\mathbf{D}^{rep}|\boldsymbol{\theta}) p(\boldsymbol{\theta}|\mathbf{D}) d\mathbf{D}^{rep} d\boldsymbol{\theta} \end{aligned} \tag{9.2}$$

where I_q is the indication function for the condition q. This incorporates both the variance of the observations $\mathbf{D}$ *and* the distribution of parameter values $\boldsymbol{\theta}$. If p_B is in the tails of the distribution for the DM, one rejects the fit. The critical region has the usual meaning here. If probability is too small, the analysis model is regarded as invalid for the given statistic.

Another approach attempts to fits a non-parametric model to the data. If the non-parametric model better explains the data than the fiducial model, we reject the fiducial model as a good fit. A naive implementation of this idea is difficult, requiring a second high-dimensional MCMC simulation to infer the posterior distribution for the non-parametric model and a careful specification of the prior distribution. A clever scheme for doing this, described in [19], is based on the following observation: if a cumulative distribution function is strictly increasing and continuous, the inverse $F^{-1}(u)$ for $u \in [0,1]$ is the unique real number θ such that $F(\theta) = u$. In the multivariate case, the inverse will not be unique generally, but, instead, we may define

$$F^{-1}(u) = \inf_{\theta \in \mathbb{R}^d} \{F(\theta) \geq u\} \tag{9.3}$$

for a parameter vector θ of rank d. Then, rather than defining a general class of densities in $\mathbb{R}^d$ to propose the alternative, Verdinelli and Wasserman consider a functional perturbation to F, $G(F(\theta))$ say, such that G maps the unit interval onto itself. The identity, $G(u) = u$, is the unperturbed probability distribution. Then, the test evaluates the uniformity of the distribution of probabilities under the hypothesis.

Intuitively, this development is closely related to the probabilistic interpretation of the marginal likelihood. To see this, consider the one-dimensional case for simplicity: let $f(\mathbf{D}|\theta) = P(\theta)L(\mathbf{D}|\theta)$ and $F_0(\theta) = \int_{-\infty}^{\theta} d\theta\, f(\mathbf{D}|\theta)$ and therefore $P(\mathbf{D}) = F_0(\infty)$. If the distribution of $F_0(\theta_i)$ for $\{\theta_i\}$ is not uniform in $[0,\ 1]$, we can perturb $f(\mathbf{D}|\theta)$ by moving some density from a region of under sampling to a region of over sampling and, thereby, increase $P(\mathbf{D})$.

9.2.3 Model Selection and Bayes Factors

We often have doubts about our parametric models, even those that appear to fit the data. This is especially true when the models are phenomenological rather than the results of *first principle* theories. Therefore, we need to estimate which competing model better represents the data. Astronomers are becoming better versed in the more traditional statistical *rejection* tests but astronomers often really want *acceptance* tests. Bayes factors provide this: one can straightforwardly evaluate the evidence *in favor* of the null hypothesis rather than only test evidence for rejecting it. Rather than using the posterior extremum as in ML, one marginalizes over the parameter space to get the marginal probability of the data under each model or hypothesis, and their ratio that provides evidence in favor of one model specification over another [15]. Bayes factors are very flexible, allowing multiple hypotheses to be compared simultaneously or sequentially. The posterior probability for competing models can be evaluated over an ensemble of data and used to decide whether or not a particular family of models should be preferred. Similarly, common parameters can be evaluated over a field of competing models with appropriate posterior model probabilities assigned to each. A tutorial illustrating this may be found in the BIE documentation.

Given all of these advantages, why are Bayes factors not more commonly used? There are two main difficulties. First, multidimensional integrals are difficult to compute. To compute the factor, we need the marginal likelihood integral: $P(\mathbf{D}) = \int \pi(\boldsymbol{\theta})P(\mathbf{D}|\boldsymbol{\theta})d\boldsymbol{\theta}$. For a real world model, the dimensionality of $\boldsymbol{\theta}$ is likely to be >10. Such a quadrature is infeasible using standard techniques. On the other hand, a typical MCMC calculation has generated a large number of evaluations of the integrand at considerable expense. Can we use the posterior sample to evaluate the integral?

Raftery [14] suggests a "Laplace-Metropolis" estimator which uses the MCMC posterior simulation to approximate the marginal density of the data using Laplace's approximation (see Raftery op. cit. for details). As part of the BIE development, [20] describes two new approaches evaluating the marginal likelihood from the MCMC-generated posterior sample and both of these are implemented in the BIE as secondary analysis routines (see Sect. 9.3.3). In short, we believe that BIE together with recent advances for computing marginal likelihood makes the wholesale computation of Bayes factors feasible, at least in some cases of interest.

A second well-known difficulty is the sensitivity of Bayes factors to the choice of prior. This may be tested through direct sensitivity analyses, such as resimulation with chains at different resolutions and approximate priors. We believe that the BIE project currently provides a useful platform for investigating the use of Bayesian model comparison and hypothesis testing and we hope it helps pave the way for new applications. In some cases, computing the Bayes factor will be infeasible. For these, the BIE includes an MCMC algorithm that selects between models as part of the posterior simulation. This is described in Sect. 9.3.4.

9.2.4 Observational Requirements

The likelihood function—the probability of the data given the parameter vector and model $P(D|\theta,M)$—is fundamental to any inference, Bayesian or otherwise. The more direct the construction of the likelihood given a parameter vector from the physical theory, that is, the smaller the information loss, the easier the calculation and the higher the quality of the result. In other words, the more the data is 'reduced' through summary statistics and 'cleaned' by applying complex matched filters, the less information remains, the more cumbersome the error model, and the greater the affect of difficult-to-model correlations. This is somewhat contrary to standard practice in astronomy.

Moreover, astronomers often quote their error models in the form of uncorrelated pseudo-standard errors. The *cultural* interpretation is that the data, typically a data bin or pixel, should be within the range specified by the error bar most of the time. Quoted error bars are often inflated to make this condition obtain. This leads to a number of fundamental flaws that makes the error model (and therefore the data) unsuitable for BI:

1. Binned and pixelated data are nearly always correlated. For example, a flat-field correction correlates the pixels of an image over its entire scale. Sky brightness removal has similar effects. There are many additional sources of indirect correlations. Parameter estimations are often sensitive to these correlated excursion in the data values and ignoring strongly correlations will lead to erroneous inferences. *Data archivers can enable accurate inference by providing correlation matrices for all error models.*
2. *Selection effects must be modeled in the likelihood function, and therefore these effects must be well specified by the archivist and chosen to be straightforwardly computable whenever possible.* For example, consider a multiband flux-limited source catalog. A color-magnitude or Hess diagram in two flux bands will have a non-rectangular boundary owing to the flux limit. Although this is a simple example, selection effects may be terribly difficult to model; consider spatial variation in source completeness to the diffraction spikes from bright stars.
3. *Astronomers tend to use traditional summary data representations that inadvertently complicate computation of* $P(D|\theta,M)$. For a simple example, the magnitude-magnitude diagram contains the same information as the color-magnitude diagram but the selection effects lie along data coordinate boundaries. For a more complicated example, consider the Tully-Fisher diagram. The input data set may contain flux limits, morphology selection, image inclination cuts, redshift range limits, to name a few.

By way of example, [9] describes the parameter inference for a semi-analytic model of galaxy formation conditioned on a galaxy mass function with both correlated and uncorrelated data bins. The differences in the posterior distributions for these two cases is dramatic (see Sect. 9.4.1). When the error model is in doubt, the sensitivity of the inference to the error model may be investigated in the Bayesian paradigm using hierarchical models. Although this is expensive and rarely done, we should consider performing such sensitivity analyses regularly.

9.3 Solutions Provided by the Bayesian Inference Engine

Our experience suggest that there is no single *best* MCMC algorithm for all applications. Rather, each choice represents a set of trade offs: more elaborate algorithms with multiple chains, augmented spaces, etc., are more expensive to run but may be the only solution for a complex posterior distribution. Moreover, combinations of MCMC algorithms in multiple-chain schemes are often useful. For example, we have found DE (Sect. 9.3.2) to be very helpful because of its adaptive tuning of the proposal function, but DE does not efficiently explore the parameter space. The exploration happens early on in the simulation and depends on the number of chains and the prior distribution of the state particles. However, a DE chain ensemble may be combined with parallel tempering (see Sect. 9.3.1) to circumvent this limitation. This application has been explored in [9].

The BIE provides extensible support for convergence testing. For multiple chains, the work horse is the commonly used Gelman and Rubin [2] statistic. For single-chain algorithms, we have had good success with a diagnostic method that assesses the convergence of both marginal and joint posterior densities following [4].

9.3.1 Simulated and Parallel Tempering

Metropolis-Hastings is one of the fundamental MCMC algorithms [6, 10] and is stated as follows. Let $P(\boldsymbol{\theta})$ be the desired distribution to be sampled and $q(\boldsymbol{\theta},\boldsymbol{\theta}')$ be a known easy-to-compute transition probability between two states. Let $a(\boldsymbol{\theta},\boldsymbol{\theta}')$ be the probability of accepting state $\boldsymbol{\theta}'$ given the current state $\boldsymbol{\theta}$. One can show that if the detailed balance condition holds then the chain will sample $P(\boldsymbol{\theta})$ and it is straightforward to show that $a(\boldsymbol{\theta},\boldsymbol{\theta}') = \min\{1,[P(\boldsymbol{\theta}')q(\boldsymbol{\theta}',\boldsymbol{\theta})/P(\boldsymbol{\theta})q(\boldsymbol{\theta},\boldsymbol{\theta}')]\}$ solves this equation (see [8] for additional discussion).

However, for complex and high-dimensional posteriors, the state easily gets trapped in isolated modes, between which the Markov chain moves only rarely. There are a number of techniques for mitigating the mixing problem. For the BIE, we adopted a synthesis of Metropolis-coupled Markov chains [3] and a simulated tempering method proposed by [11] called *tempered transitions*. To sample from a distribution $P_0(\boldsymbol{\theta})$ with isolated modes, one defines a series of n auxiliary distributions, $P_1(\boldsymbol{\theta}),\ldots,P_n(\boldsymbol{\theta})$, with P_k being easier to sample than P_{k-1}. For example, one may choose $P_k(\boldsymbol{\theta}) \propto P_0^{\beta_k}(\boldsymbol{\theta})$ with $1 = \beta_0 > \beta_1 > \cdots > \beta_{n-1} > \beta_n > 0$. Then, the method defines a pair of base transitions for each k, $\hat{T}_k$ and $\check{T}_k$, which both have P_k as an invariant distribution and satisfy the following mutual reversibility condition for all $\boldsymbol{\theta}$ and $\boldsymbol{\theta}'$: $P_k(\boldsymbol{\theta})\hat{T}_k(\boldsymbol{\theta},\boldsymbol{\theta}') = \check{T}_k(\boldsymbol{\theta}',\boldsymbol{\theta})P_k(\boldsymbol{\theta}')$. A tempered transition first finds a candidate state by applying the base transitions in the sequence $\hat{T}_1 \cdots \hat{T}_n$. After each upward transition, new states are sampled from a broader distribution. In most cases, this liberates the candidate state from confinement by the mode of the initial state. This is then followed by a series of downward transitions $\check{T}_n \cdots \check{T}_1$. This candidate state is then accepted or rejected based on ratios of probabilities involving intermediate states.

The *parallel tempering* algorithm inverts the order of the previous solution: it simultaneous simulates n chains, each with target distribution P_j and proposes to swap states between adjacent members of the sequence at predefined intervals. After each interval, a pair of adjacent simulations in the series is chosen at random and a proposal made to swap their parameter states. The swap is accepted with a Metropolis-Hastings criterion. Final results are based on samples from the $\beta_0 = 1$ chain. As in the previous algorithm, the high-temperature states will mix between modes more efficiently and subsequent swapping with lower-temperature chains will promote their mixing. We have found that some tempering is an essential ingredient for many of our problems.

9.3.2 Differential Evolution

Real-world high-dimensional likelihood functions often have complex topologies with strong anisotropies about their maxima (see Sect. 9.4.1, Fig. 9.1). Difficulty in tuning the Metropolis-Hastings proposal function to achieve a good acceptance rate and good mixing plagues high-dimensional MCMC simulations of the posterior probability. This affects all of algorithms discussed up to this point. Recently [18] introduced an approach based on a genetic algorithm called Differential Evolution (DE) [13, 16, 17]. The DE uses an ensemble of chains, run in parallel, to adaptively compute the Metropolis-Hastings proposal function. In addition, Ter Braak suggests a combination of proposals that facility exploring within and jumping between modes.

Assume that our ensemble has n chains to start, for example, initialized from the prior probability. The DE algorithm [13] proposes to update member i as follows: $\pi_p = \pi_{R0} + \gamma(\pi_{R1} - \pi_{R2})$ where π_{R0}, π_{R1} and π_{R2} are randomly selected without replacement from the population without π_i. The proposal vector replaces the chosen one if the fitness of π_p is higher than the fitness of π_i. Ter Braak [18] shows that with minor modifications the proposal function and acceptance condition for DE obeys detailed balance. The new algorithm takes the form $\pi_p = \pi_i + \gamma(\pi_{R1} - \pi_{R2}) + \varepsilon$ where ε is drawn from a symmetric distribution with a small variance compared to that of the target, but with unbounded support, e.g. $\varepsilon \sim N^d(0,\sigma^2)$ for very small variance σ^2. The random variate ε is demanded by the recurrence condition: the domain for non-zero values of the posterior P must be reached infinitely often for an infinite length chain.

9.3.3 Computation of Bayes Factors and Marginal Likelihoods

Weinberg [20] presents two computationally-modest families of quadrature algorithms that use the full generality sample posterior but without the instability of the harmonic mean approximation [12] or the specificity of the Laplace approximation [7]. The Laplace approximation works well when the posterior distribution appears as a multivariate normal distribution, but this is a rare occurrence in my experience.

The first algorithm begins with the normalized Bayes theorem: $Z \times P(\boldsymbol{\theta}|\mathbf{D}) = \pi(\boldsymbol{\theta})L(\mathbf{D}|\boldsymbol{\theta})$. Dividing by $L(\mathbf{D}|\boldsymbol{\theta})$ and integrating over $\boldsymbol{\theta}$ we have

$$Z \times \int_{\Omega} d\boldsymbol{\theta} \frac{P(\boldsymbol{\theta}|\mathbf{D})}{L(\mathbf{D}|\boldsymbol{\theta})} = \int_{\Omega} d\boldsymbol{\theta}\, \pi(\boldsymbol{\theta}) \tag{9.4}$$

where Ω is the domain of $\boldsymbol{\theta}$, often $\Omega \subset \mathbb{R}^k$. Since the Markov-chain samples the posterior, $P(\boldsymbol{\theta}|\mathbf{D})$, the computation of the integral on the left from the chain appears as an inverse weighting with respect to the likelihood. This is poorly conditioned due to the inevitable small values of $L(\mathbf{D}|\boldsymbol{\theta})$. The second approach is a direct numerical integration of

$$Z = \int_{\Omega} d\boldsymbol{\theta}\, P(\boldsymbol{\theta}|\mathbf{D}) = \int_{\Omega} d\boldsymbol{\theta}\, \pi(\boldsymbol{\theta})L(\mathbf{D}|\boldsymbol{\theta}). \tag{9.5}$$

For weakly informative prior distributions, the entire domain of support is not sampled by the Markov chain. However, if the integrals in (9.4) are dominated by the domain sampled by the chain, the integrals may be approximated by quadrature over a truncated domain, Ω_s, that eliminates a small number of values $L(\mathbf{D}|\boldsymbol{\theta})$ in the sample.

To evaluate the r.h.s. of (9.4), we may use the sampled posterior distribution itself to tessellate the sampled volume in $\Omega_s \subset \Omega$. This may be done straightforwardly using a space-partitioning structure. A computationally efficient structure is a binary space partition (BSP) tree, which divides a region of parameter space into two exclusive sub regions at each node. The most easily implemented tree of this type for arbitrary dimension is the kd-tree (short for k-dimensional tree). The kd-tree algorithms split $\mathbb{R}^k$ on planes perpendicular to one of the coordinate system axes. The implementation provided for the BIE uses the median value along one of axes (a *balanced* kd-tree). We have also implemented a hyper-octree. The hyper-octree generalizes the octree by splitting each n-dimensional parent node into 2^n hypercubic children. Unlike the kd-tree, the hyper-octree does not split on point location and the size of the cells is not strictly coupled to the number of points in the sample. This helps provide a better representation of the volume containing sample points. In addition, the cells in the kd-tree maybe have extreme axis ratios. See [20] for additional details, tests, and discussion.

In summary, the choice between the various algorithms depends on the problem at hand. The Laplace approximation will be a good choice for posterior distributions that are unimodal with light tails. I continue to investigate performance of algorithms in [20] for high-dimensional distributions. For dimension $n < 10$, the direct volume tessellation gives higher-accuracy smaller-variance results. However, as n increases, the sample-size requirement also increases exponentially with n. A paper describing additional strategies is in preparation.

9.3.4 Reversible Jump Simulation

When large sample sizes are impractical, one may aggregate a number of models with an additional indicator variable to designate the active model. This results in a general state space consisting of the discrete range of the indicator and continuous ranges for each parameters for each model. Green [5] showed that the detailed balance equation can be formulated in such a general state space. This allows one to propose models of different dimensionality and thereby incorporate model selection into the probabilistic simulation itself. This requires a transition probability to and from each subspace. For model comparison, we are interested in the posterior probabilities of different models k to draw some conditional or marginal conclusions about different models. Such estimates follow directly from the simulated posterior. For example, an estimate of the marginal probability for each model follows directly from the occupation frequency in each subspace. We have found that this approach is very hard to tune.

9.4 Case Studies

9.4.1 Semi-analytic Galaxy Formation Models: BIE-SAM

Many of the physical processes parametrized in semi-analytical models of galaxy formation remain poorly understood and under specified. This has two critically important consequences for inferring constraints on the physical parameters: (1) prior assumptions about the size of the domain and the shape of the parameter distribution will strongly affect resulting inference; and (2) a very large parameter space must be fully explored to obtain an accurate inference. Moreover, both *must* be done together. Both of these issues are naturally tackled with a Bayesian approach that allows one to constrain theory by data in a probabilistically rigorous way. We have presented [9] a semi-analytic model (SAM) of galaxy formation in the framework of BI and illustrated its performance on a test problem using BIE; we call the combined approach BIE-SAM. Our 16-parameter semi-analytic model incorporates the most commonly used parametrizations of important physical processes from existing SAMs including star formation, SN feedback, galaxy mergers, and AGN feedback.

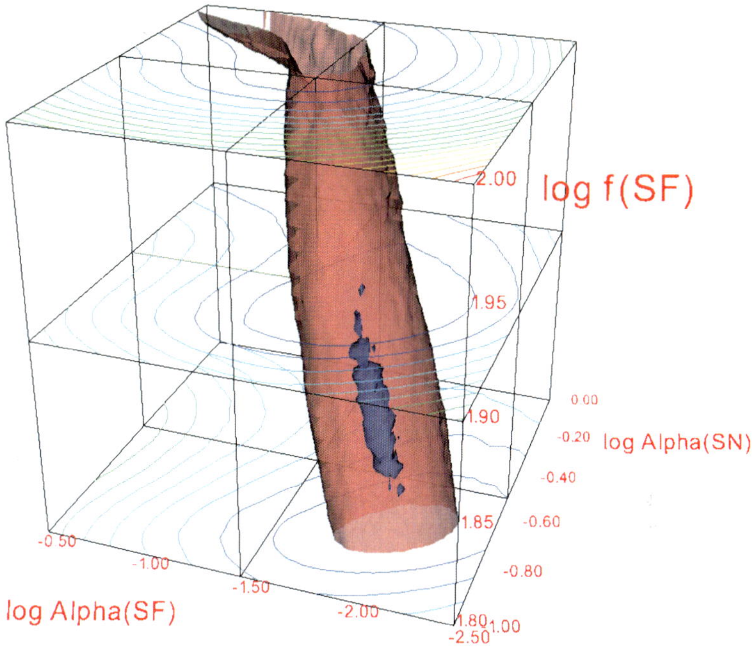

Fig. 9.1 Marginal posterior distribution for 3 out of 13 parameters in the BIE-SAM: the star-formation threshold surface density f_{SF}, the star-formation efficiency power-law index α_{SF}, and the supernova feedback energy fraction α_{SN}. The blue (*red*) surfaces enclose approximately 10% (67%) of the density (See [9] for additional detail)

To demonstrate the power of this approach, we used the observationally derived stellar mass function of galaxies to constrain a number of important model parameters. We find that the posterior distribution has very complex structure and topology, indicating that finding the best fit by tweaking model parameters is improbable. As an example, Fig. 9.1 describes isosurfaces of the marginalized posterior distribution. The surfaces have complex geometry and are strongly inhomogeneous in parameter direction. Moreover, the posterior clearly shows that many model parameters are strongly covariant, and therefore the inferred value of a particular parameter can be significantly affected by the priors used for other parameters. As a consequence, one *may not* tune a small subset of model parameters while keeping other parameters fixed and expect a valid result. With the use of synthetic data to mimic systematic uncertainties in the reduced data, we have shown that resulting model parameter inferences can be significantly affected by the use of an incorrect error model. This fact will be obvious to the statisticians but is not well-appreciated by the astronomers.

The method developed here can be straightforwardly applied to other data sets and to multiple data sets simultaneously. In addition, the Bayesian approach explicitly builds on previous results by incorporating the constraints from previous inferences into new data sets; the BIE is designed to do this automatically. For many processes in galaxy formation, competing models have been proposed but not quantitatively compared. Bayes factor analyses (see Sect. 9.3.3) or explicit model comparison techniques such as the reversible jump algorithm ([5], see also Sect. 9.3.4) provide a quantitative comparison of different models for given data.

9.4.2 *Galphat*

A recent paper [21] describes Galphat (GALaxy PHotometric ATtributes), a Bayesian galaxy image analysis package built for the BIE, designed to efficiently and reliably generate the posterior probability distribution of model parameters from an image. Using the various tempering algorithms available in the BIE, our tests have demonstrated that we can achieve a steady-state distribution and the simulated posterior will include nearly all possible multiple modes consistent with the prior distribution. Given the posterior distribution, we may then consistently estimate the confidence levels. We show that the surface-brightness model will often have correlated parameters. Any hypothesis testing that uses the ensemble of posterior information will be affected by these correlations. The full posterior distributions from Galphat identify these correlations and incorporate them in subsequent inferences.

As an example, Fig. 9.2 shows the size–Sérsic index relation inferred from a synthetically-generated sample of elliptical galaxy images. The left-hand panel shows the traditional scatter diagram of maximum posterior parameter values while the right-hand panel shows the inferred distribution based on the full posterior distribution of the ensemble. The left-hand panel *incorrectly* suggests that smaller galaxies are less concentrated while the right-hand panel *correctly* reveals that the size and concentration are uncorrelated.

9.5 BIE: Technical Overview

At the core is a software library of inter-operable components necessary for performing Bayesian computation. The BIE classes are available as both C++ libraries and as a stand-alone system with integrated command-line interface. The command-line interface is well tested and is favored by most users so far. A user does not need to be an expert or even an MPI programmer to use the system; the simple user interface is similar to MatLab or gnuplot. In addition to the engine itself, the BIE package includes a number of stand-alone programs for viewing and analyzing output from the BIE and for testing the components. Although source code is available, we recommend using one of the pre-compiled

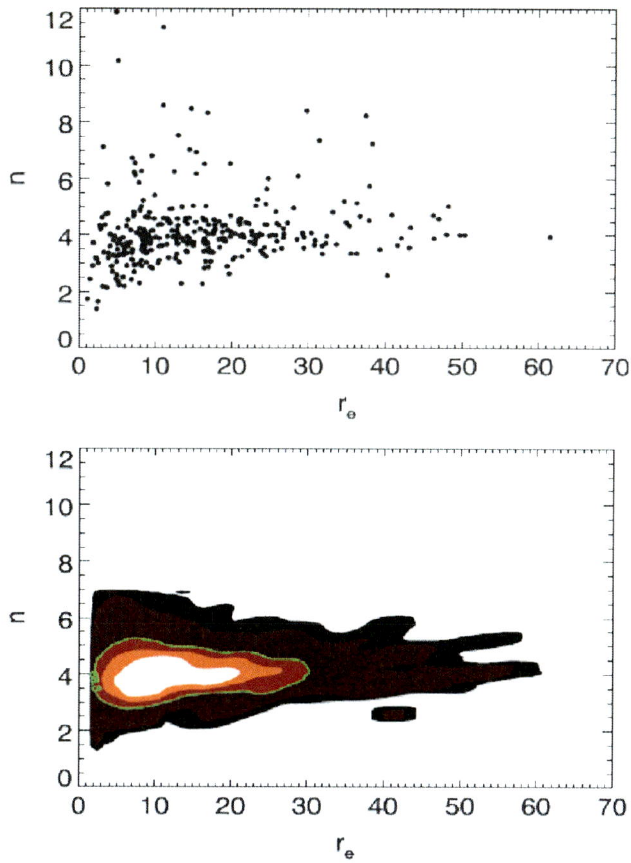

Fig. 9.2 *Top*: Scatter plot using the best-fit parameters. *Bottom*: marginal posterior density for the same parameters

Debian or Ubuntu packages,[2] if possible, to avoid library version dependencies. We use SVN version management (autoconf, automake), GNU coding standards, and DejaGNU regression testing.

The researcher needs to be able to stop, restart, and possibly refocus inferential computations for both technical and scientific reasons. The BIE was designed with these scenarios in mind. The BIE's persistence system is built on top of the BOOST[3] serialization library. The BIE classes inherit from a base serialization class that provides the key serialization members and a simple mnemonic scheme to mark persistent data. The most common use of BIE persistence to date is

[2] http://www.astro.umass.edu/bie.

[3] http://www.boost.org.

checkpointing and recovery, Checkpointing guards against loss of computation by saving intermediate data to support recovery in the middle of long-running computational steps; and it allows one to "freeze" or "shelve" a computation and pick it up later. It also provides the basic support needed to interrupt a computation, do some reconfiguring, and resume, as when machines need to be added to or removed from a cluster, etc.

9.6 Discussion and Summary

This presentation reports our experience in applying MCMC methods to observational and theoretical Bayesian inference problems in astronomy using the UMass Bayesian Inference Engine (BIE).

We began by outlining our motivation. Most researchers are well-versed in the identifying the "best" parameters for a particular model for some data using the maximum likelihood method (ML). For example, consider the fit of a surface brightness model to galaxy images. Parameters from the ML solutions are typically plotted in a scatter diagram and trends are interpreted physically. Section 9.2.1 describes the pitfalls of this approach. Rather, this is a complex *hypothesis test*: the astronomer wants to test the hypothesis that the data is correlated with a coefficient larger than some predetermined value α. However, without incorporating the correlations imposed by both the theoretical model, the error model and selection process, the significance of the apparent correlation is uncertain. Moving on, combining plotting scatter diagrams from multiple data sources inadvertently mixes error models and selection effects which further complicates a quantitative interpretation. Similarly, the astronomer needs methods of assessing whether a posited model is correct. I have divided these needs into two: goodness-of-fit tests (Sect. 9.2.2) and model Sect. 9.2.3. As an example of the former, the astronomer may have found the best parameters using ML, but does the model fully explain all of the features of the data? If it does not, one must either modify or reject the model before moving on the next step. As an example of the latter, suppose an inference results in two parameter regions or multiple models that explain the data. Which model *best* explains the data?

All of these wants and needs—combining data from multiple sources, estimating the probability of model parameters, assessing goodness of fit, and selecting between competing models—are naturally addressed in a single probabilistic framework known as *Bayesian inference* (BI). In particular, BI provides a data-first discipline that demands the error model and selection effects are specified by the probability distribution for the data given the model $\mathcal{M}$, $P(\mathbf{D}|\boldsymbol{\theta},\mathcal{M})$, colloquially known as the likelihood function $L(\mathbf{D}|\boldsymbol{\theta},\mathcal{M})$. Prior results including quantified expert opinion are specified in the prior probability function $P(\theta|\mathcal{M})$. The inferential computation may be incremental: the data may be added in steps and new or additional observations may be motivated at each step, true to the scientific method. In the end, this approach may be generalized to locating the most likely models in the generalized space of models; this leads to goodness-of-fit and model comparison tests.

With these advantages comes a major disadvantage: BI is expensive! Nonetheless, the elegance and promise of BI motivated us to attempt a computational solution and this became the BIE project. The algorithms and techniques described here are available in the BIE have proved useful in address the complications found in research problems. In short, the BIE fills a gap between tools developed for small-scale problems or those designed to test new algorithms and a computational platform designed for production-scale inference problems typical of present-day astronomical survey science. Its primary product is a sample from a posterior distribution to be used form parameter estimation and model selection. Other Bayesian applications, such as non-parametric inference and clustering, should be possible with little modification, but have not been investigated so far. The BIE is designed to run on HPC-class clusters, although it will also run on workstations and laptops. The open object-oriented architecture allows for cross-fertilization between researchers and groups with both mathematical and scientific interests, for example, those both developing new algorithms and astronomical models for different applications.

Acknowledgements This work was supported in part by NSF IIS Program through award 0611948 and by NASA AISR Program through award NNG06GF25G. I thank Neal Katz and Eliot Moss for comments on an early draft of this manuscript.

References

1. Gelman, A.: A bayesian formulation of exploratory data analysis and goodness-of-fit testing. International Statistical Review **71**(2), 369–382 (2003)
2. Gelman, A., Rubin, D.B.: Inference from iterative simulation using multiple sequences. Stat. Sci. **7**, 457–511 (1992)
3. Geyer, C.J.: Markov Chain Monte Carlo maximum likelihood. In: Keramidas (ed.) Computing Science and Statistics: Proceedings of the 23rd Symposium on the Inferface, pp. 156–163. Interface Foundation (1991)
4. Giakoumatos, S., Vrontos, I., Dellaportas, P., Politis, D.: An MCMC convergence diagnostic using subsampling. J. Comput. Graph. Statistics **8**(3), 431–451 (1999)
5. Green, P.J.: Reversible jump Markov Chain Monte Carlo computation and Bayesian model determination. Biometrica **82**, 711–732 (1995)
6. Hastings, W.K.: Monte Carlo sampling methods using Markov chains and their applications. Biometrika **57**, 97–109 (1970)
7. Lewis, S.M., Raftery, A.E.: Estimating Bayes factors via posterior simulation with the Laplace-Metropolis estimator. J. Am. Stat. Assoc. **440**, 648 (1997)
8. Liu, J.S.: Monte Carlo Strategies in Scientific Computing. Springer Series in Statistics. Springer (2004)
9. Lu, Y., Mo, H.J., Weinberg, M.D., Katz, N.S.: A Bayesian approach to the semi-analytic model of galaxy formation: The methodology. MNRAS(2010). Submitted

10. Metropolis, N., Rosenbluth, A., Rosenbluth, M., Teller, A., Teller, E.: Equations of state calculations by fast computing machines. Journal of Chemical Physics **21**, 1087–1091 (1953)
11. Neal, R.M.: Sampling from multimodal distributions using tempered transitions. Statistics and Computing **6**, 353–366 (1996)
12. Newton, M.A., Raftery, A.E.: Approximate Bayesian inference by the weighted likelihood bootstrap. Journal of the Royal Statistical Society, Ser. B **56**, 3–48 (1994)
13. Price, K.: Differential evolution. Dr. Dobbs Journal **264**, 18–24 (1997)
14. Raftery, A.E.: Hypothesis testing and model selection with posterior simulation. In: Markov Chain Monte Carlo in Practice (1995)
15. Robert, C.P.: The Bayesian Choice. Springer (2007)
16. Storn, K.: An introduction to differential evolution. In: D. Corne, M. Dorigo, F. Glover (eds.) New Ideas in Optimization, pp. 79–108. McGraw-Hill, London (1999)
17. Storn, R., Price, K.: Differential evolution—a simple and effective heuristic for global optimization over continuous spaces. Journal of Global Optimization **11**, 341–359 (1997)
18. Ter Braak, C.J.F.: A Markov Chain Monte Carlo version of the genetic algorithm differential evolution: easy Bayesian computing for real parameter spaces. Stat. Comput. **16**, 239–249 (2006)
19. Verdinelli, I., Wasserman, L.: Bayesian goodness-of-fit testing using infinite-dimensional exponential families. Ann. Statist. **26**, 1215–1241 (1998)
20. Weinberg, M.D.: Computing the Bayes factor from a Markov Chain Monte Carlo simulation of the posterior distribution. Bayesian Analysis (2009). Submitted
21. Yoon, I., Weinberg, M.D., Katz, N.S.: New insight on galaxy structure from GALPHAT: Motivation, methodology, and benchmarks. MNRAS(2010). Submitted

Chapter 10
Commentary: Bayesian Model Selection and Parameter Estimation

Philip C. Gregory

Abstract Bayesian model selection and parameter estimation is attracting a lot of interest in the astronomical community because of its power and logical consistency. Markov chain Monte Carlo provides the computational power for Bayesian parameter estimation problems in large parameter spaces but needs to be supported with other numerical techniques for efficient exploration of multi-modal probability distributions. Bayesian model selection is easy in concept but remains a difficult challenge for large parameter spaces. My comments here on the paper by Roberto Trotta are based on lessons learned from developing a controlled statistical fusion approach to some of these issues.

10.1 Introduction

Martin Weinberg reports on using the UMass Bayesian Inference Engine (BIE) package for model selection and parameter estimation in extragalactic astronomy. I have independently developed a Bayesian approach to accomplish similar goals with particular emphasis in the arena of exoplanets. This conference provides an opportunity to exchange lessons learned. Not surprisingly, because of the large number of model parameters involved, both groups employ a Markov chain Monte Carlo (MCMC) integration engine. The BIE philosophy is that there is no single best MCMC algorithm and develop a variety of MCMC algorithms augmented by different tools like parallel tempering, simulated annealing and differential evolution depending on the complexity of the problem. My approach has been to attempt

---

P.C. Gregory (✉)
Physics and Astronomy, University of British Columbia, 6224 Agricultural Rd, Vancouver, B.C. V6T 1Z1, Canada
e-mail: gregory@phas.ubc.ca

E.D. Feigelson and G.J. Babu (eds.), *Statistical Challenges in Modern Astronomy V*, Lecture Notes in Statistics 209, DOI 10.1007/978-1-4614-3520-4_10,

to fuse together the advantages of all of the above tools together with a genetic crossover operation in a single MCMC algorithm to facilitate the detection of a global minimum in χ^2.

My latest algorithm is called fusion MCMC [10]. This fusion has only been possible through the development of a unique adaptive control system to automate the choice of an efficient set of MCMC proposal distributions even if the parameters are highly correlated. The control system also supervises the operation of the different components. Figure 10.1 shows two schematics on the operation of an eight parallel chain fusion MCMC and the control system. In applications to real precision radial velocity data the algorithm has proved highly effective [5–8, 10]. The *Mathematica* based parallel code is run on a 8 core PC and requires 10 h for a 6 planet model with 37 parameters and one million iterations. The execution time scales with the number of planets.

10.2 Some Useful Lessons

10.2.1 Highly Correlated Parameters

For some models the data is such that the resulting estimates of the model parameters are highly correlated and the MCMC exploration of the parameter space can be very inefficient. One solution to this problem is Differential Evolution Markov Chain (DE-MC) [2]. DE-MC is a population MCMC algorithm, in which multiple chains are run in parallel, typically from 15 to 40, although Weiner's experience suggests that 64 chain would be the bare minimum. DE-MC solves an important problem in MCMC, namely that of choosing an appropriate scale and orientation for the jumping distribution.

For the fusion MCMC algorithm, I developed and tested a new method [9], in the spirit of DE, that automatically achieves efficient MCMC sampling in highly correlated parameter spaces without the need for additional chains. The block in the lower left panel of Fig. 10.1 automates the selection of efficient proposal distributions when working with model parameters that are independent or transformed to new independent parameters. New parameter values are jointly proposed based on independent Gaussian proposal distributions ('I' scheme), one for each parameter. Initially, only this 'I' proposal system is used and it is clear that if there are strong correlations between any parameters the σ values of the independent Gaussian proposals will need to be very small for any proposal to be accepted and consequently convergence will be very slow. However, the accepted 'I' proposals will generally cluster along the correlation path. In the optional third stage of the control system (see right panel of Fig. 10.1) every second accepted 'I' proposal is appended to a correlated sample buffer. There is a separate buffer for each parallel tempering level. Only the 300 most recent additions to the buffer are retained. A 'C' proposal is generated from the difference between a pair of randomly selected

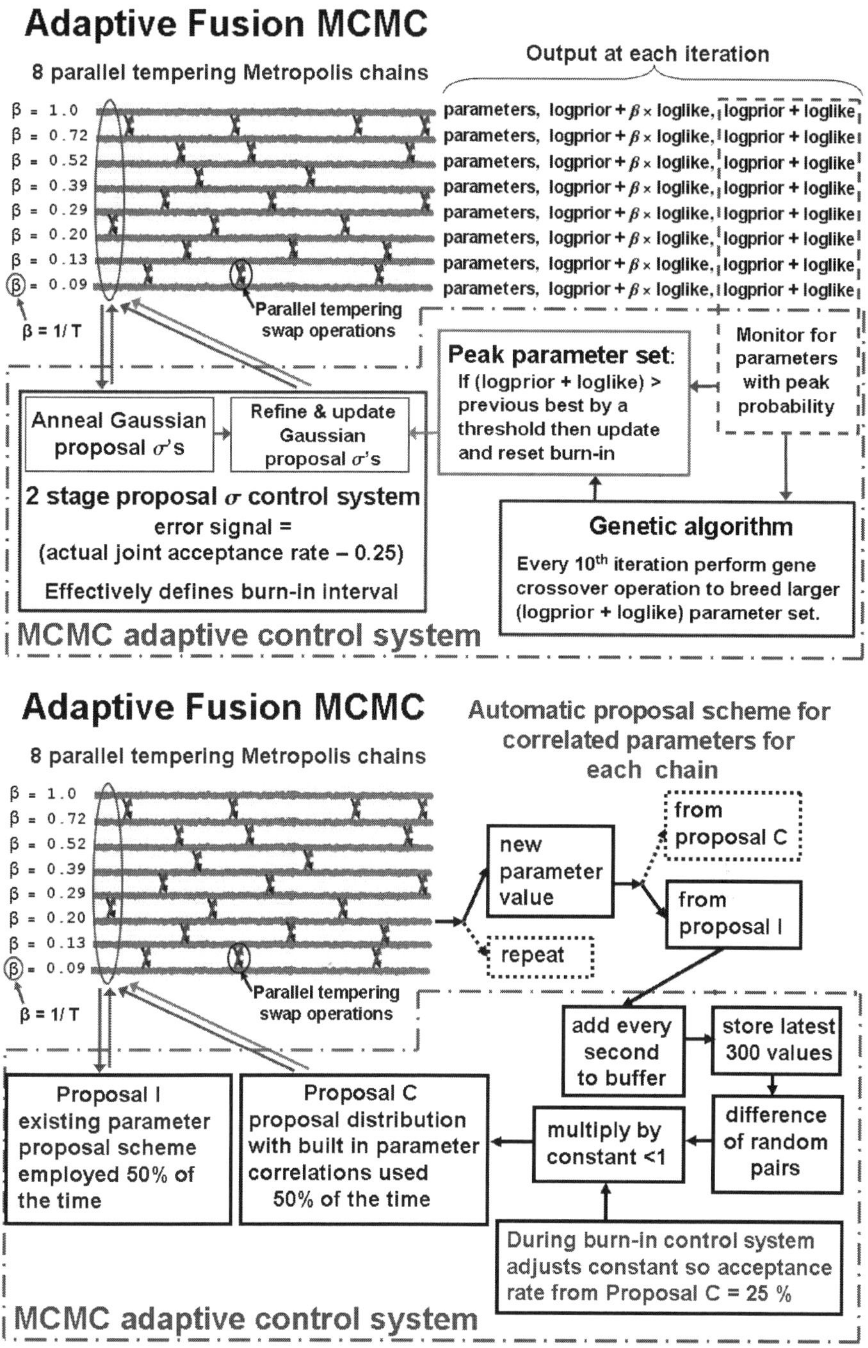

Fig. 10.1 Two schematics on the operation of the adaptive fusion MCMC algorithm. The *right panel* illustrates the automatic proposal scheme for handling correlated parameters

samples drawn from the correlated sample buffer for that tempering level, after multiplication by a constant. The value of this constant (for each tempering level) is computed automatically [9] by another control system module which ensures that the 'C' proposal acceptance rate is close to 25%. With very little computational overhead, the 'C' proposals provide the scale and direction for efficient jumps in a correlated parameter space.

The final proposal distribution is a random selection of 'I' and 'C' proposals such that each is employed 50% of the time. The combination ensures that the whole parameter space can be reached and that the FMCMC chain is aperiodic. The parallel tempering feature operates as before to avoid becoming trapped in a local probability maximum.

Because the 'C' proposals reflect the parameter correlations, large jumps are possible allowing for much more efficient movement in parameter space than can be achieved by the 'I' proposals alone. Once the first two stages of the control system have been turned off, the third stage continues until a minimum of an additional 300 accepted 'I' proposals have been added to the buffer and the 'C' proposal acceptance rate is within the range ≥ 0.22 and ≤ 0.28. At this point further additions to the buffer are terminated and this sets a lower bound on the burn-in period.

Figure 10.2 shows the autocorrelation functions of post burn-in MCMC samples for two highly correlated parameters χ and ω. The solid black trace corresponds to a search in χ and ω using only 'I' proposals. The light gray trace corresponds to a search in χ and ω with 'C' proposals turned on. The dashed trace corresponds to a search in the transformed orthogonal coordinates $\psi = 2\pi\chi + \omega$ and $\phi = 2\pi\chi - \omega$ using only 'I' proposals. It is clear that a search in χ and ω with 'C' proposals turned on achieves the same excellent results as a search in the transformed orthogonal coordinates ψ and ϕ using only 'I' proposals.

10.2.2 Noise Model

Based on their results, Weinberg concludes that a data-model comparison without an accurate error model is likely to be erroneous. I have found it very useful to incorporate an extra noise parameter, s, that can allow for any additional noise beyond the known measurement uncertainties.[1] We assume the noise variance is finite and adopt a Gaussian distribution with a variance s^2. Thus, the combination of the known errors and extra noise has a Gaussian distribution with variance $= \sigma_i^2 + s^2$, where σ_i is the standard deviation of the known noise for ith data point.

[1] In the absence of detailed knowledge of the sampling distribution for the extra noise, we pick an independent Gaussian model because for any given finite noise variance it is the distribution with the largest uncertainty as measured by the entropy, i.e., the maximum entropy distribution [1, 11].

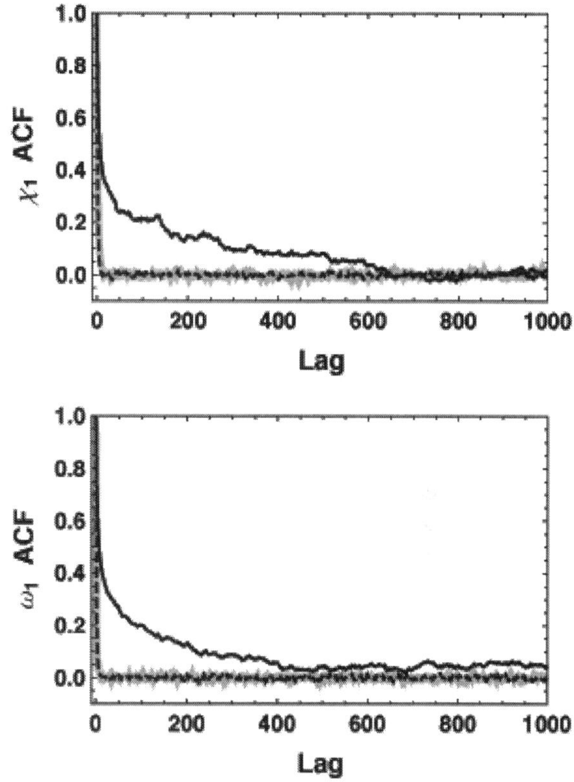

Fig. 10.2 The two panels show the MCMC autocorrelation functions for two highly correlated parameters χ and ω. The *solid black* trace corresponds to a search in χ and ω using only 'I' proposals. The *light gray* trace corresponds to a search in χ and ω with 'C' proposals turned on. The *dashed trace* corresponds to a search in the transformed orthogonal coordinates $\psi = 2\pi\chi + \omega$ and $\phi = 2\pi\chi - \omega$ using only 'I' proposals

In general, nature is more complicated than our model and known noise terms. Marginalizing s has the desirable effect of treating anything in the data that can't be explained by the model and known measurement errors as noise, leading to more conservative estimates of the parameters. See Sects. 9.2.3 and 9.2.4 of [1] for a tutorial demonstration of this point. If there is no extra noise then the posterior probability distribution for s will peak at $s = 0$.

Incorporating an extra noise parameter also results in an automatic annealing operation whenever the Markov chain is started from a location in parameter space that is far from the best fit values. When the χ^2 of the fit is very large, the Bayesian Markov chain automatically inflates s to include anything in the data that cannot be accounted for by the model with the current set of parameters and the known measurement errors. This results in a smoothing out of the detailed structure in the χ^2 surface and, as pointed out by [3], allows the Markov chain to explore the large scale structure in parameter space more quickly. The chain begins to decrease the value of the extra noise as it settles in near the best-fit parameters. An example of this is shown in Fig. 10.3. This is similar to simulated annealing, but does not require choosing a cooling scheme.

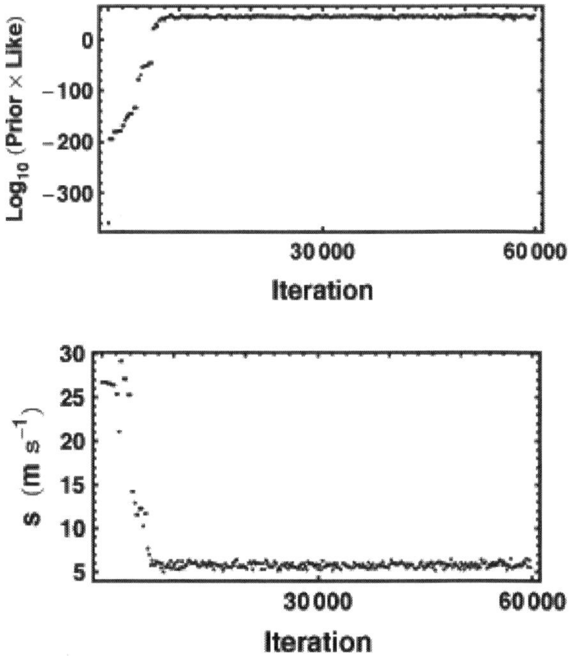

Fig. 10.3 The *top panel* is a plot of the Log_{10}[Prior × Likelihood] versus MCMC iteration. The *bottom panel* is a similar plot for the extra noise term s. Initially s is inflated and then rapidly decays to a much lower level as the best fit parameter values are approached

10.2.3 Model Selection

One of the great strengths of Bayesian analysis is the built-in Occam's razor. More complicated models contain larger numbers of parameters and thus incur a larger Occam penalty, which is automatically incorporated in a Bayesian model selection analysis in a quantitative fashion (see [1] for example, p. 45). Bayesian model selection relies on the ratio of marginal likelihoods where the marginal likelihood is the weighted average of the conditional likelihood, weighted by the prior probability distribution of the model parameters and any unknown additional noise parameter. At the last SCMA conference Clyde et al. [2] reviewed the state of techniques for model selection from a statistics perspective and Ford and Gregory [4] evaluated the performance of a variety of marginal likelihood estimators in the exoplanet context. The bottom line is that Bayesian model selection is easy in concept but becomes progressively more difficult to compute as the number of model parameters increase. Here we compare recent results obtained from two different methods: (1) nested restrictive Monte Carlo (NRMC), and (2) the ratio estimator (RE).

Nested restrictive Monte Carlo (NRMC) is a recent improvement [8, 10] on the RMC method. In RMC [4], the volume of parameter space sampled is restricted to a region delineated by the outer borders (e.g., 99% credible region) of the MCMC marginal parameter distributions for the dominant mode. In principle, the contribution from a secondary mode can be computed in a like fashion.

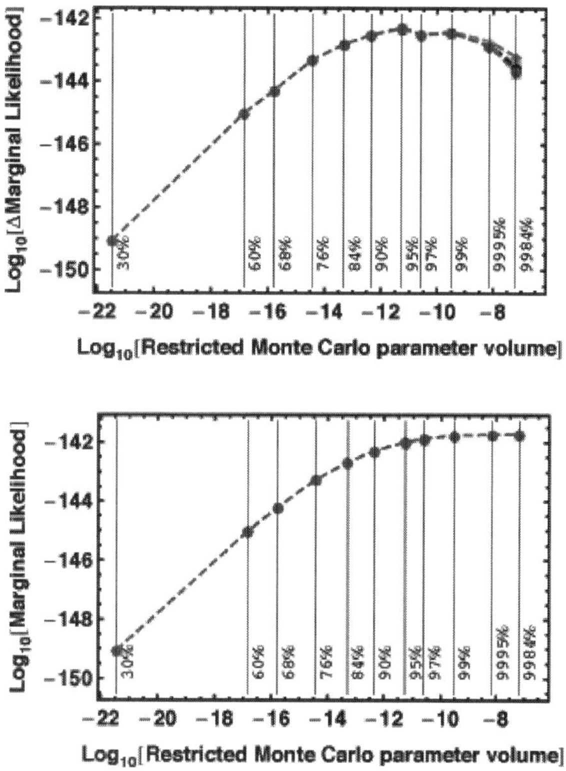

Fig. 10.4 The *top panel* shows the contribution of the individual nested intervals to the NRMC marginal likelihood for the three planet model (17 parameters) for five repeats. The *bottom panel* shows the integral of these contributions versus the parameter volume of the credible region

In NRMC integration, multiple boundaries are constructed based on credible regions ranging from 30% to ≥99%, as needed. The contribution to the total integral from each nested interval is computed. For example, for the interval between the 30% and 60% credible regions, we generate random parameter samples within the 60% region and reject any sample that falls within the 30% region. Using the remaining samples we can compute the contribution to the NRMC integral from that interval.

The left panel of Fig. 10.4 shows the contributions from the individual intervals for five repeats of the NRMC evaluation for a three planet model fit to the Gliese 581 [10] exoplanet system. The right panel shows the summation of the individual contributions versus the volume of the credible region. The credible region listed as 9995% is defined as follows. Let X_{U99} and X_{L99} correspond to the upper and lower boundaries of the 99% credible region, respectively, for any of the parameters, with X_{U95} and X_{L95} similarly defined. Then $X_{U9995} = X_{U99} + (X_{U99} - X_{U95})$ and $X_{L9995} = X_{L99} + (X_{L99} - X_{L95})$. Similarly, $X_{U9984} = X_{U99} + (X_{U99} - X_{U84})$.

Table 10.1 shows a comparison of the NRMC method to a second marginal likelihood estimator called the Ratio Estimator [4] (RE), for three planet (17 parameters) and four planet (22 parameters) exoplanet models for three different stars HD 11964, 47 UMa, and Gliese 581. The RE method employed a mixture

Table 10.1 The ratio of the NRMC and RE marginal likelihoods estimates for three planet (17 parameters) and four planet (22 parameters) exoplanet models

Star	# planets	NRMC estimator / RE estimator (improved version)
HD 11964	3	0.9
47 UMa	3	0.75
Gliese 581	3	1.01
Gliese 581	4	0.016

of 150 multivariate Normals [7] to approximate the MCMC samples. The latest version improves the handling of wrap around angular parameters in the calculation of the covariance matrix of each multivariate Normal. For the three planet models the NRMC and RE methods agree within 25%. In the case of HD11964, one of the three signals is a suspected artifact but this is of no consequence for the present comparison of marginal likelihood estimators. At sufficiently high dimensions, the NRMC method is expected to underestimate the marginal likelihood and the factor by which it underestimates is expected to grow with increasing dimension. Thus NMRC estimated Bayes factor should not falsely support a more complicated model and in this sense the NRMC method is expected to fail in a conservative fashion. On the other hand, the RE method has the potential to pay too much attention to the mode as each integrand in the ratio involves the square of the posterior density and is expected to overestimate the marginal likelihood at sufficiently high dimensions. As the table indicates, by the time we reach a four planet model (22 parameters) one or both of these methods is failing.

Acknowledgements The author would like to thank Wolfram Research for providing a complementary license for gridMathematica.

References

1. Gregory, P. C.: Bayesian Logical Data Analysis for the Physical Sciences: A Comparative Approach with *Mathematica* Support, Cambridge University Press (2005)
2. Clyde, M. A., Berger, J. O., Bullard, F., Ford, E. B., Jeffreys, W. H., Luo, R., Paulo, R., Loredo, T.: Current Challenges in Bayesian Model Choice. In 'Statistical Challenges in Modern Astronomy IV,' G. J. Babu and E. D. Feigelson (eds.), ASP Conf. Ser., 371, 224–240 (2007)
3. Ford, E. B.: Improving the Efficiency of Markov Chain Monte Carlo for Analzing the Orbits of Extrasolar Planets. ApJ, 620, 481 (2006)
4. Ford, E. B., & Gregory, P. C.:Bayesian Model Selection and Extrasolar Planet Detection. In 'Statistical Challenges in Modern Astronomy IV,' G. J. Babu and E. D. Feigelson (eds.), ASP Conf. Ser., 371, 189–204 (2007)
5. Gregory, P. C.: A Bayesian Analysis of Extrasol Planet Data for HD 73526. ApJ, 631, 1198–1214 (2005)
6. Gregory, P. C.: A Bayesian Kepler Periodogram Detects a Second Planet in HD 208487. MNRAS, 374, 1321–1333 (2007)
7. Gregory, P. C.: A Bayesian Periodogram Finds Evidence for Three Planets in HD 11964. MNRAS, 381, 1607–1619 (2007)

8. Gregory, P. C., and Fischer, D. A.: A Bayesian Periodogram Finds Evidence for Three Planets in 47 Ursae Majoris. MNRAS, 403, 731–747, (2010)
9. Gregory, P. C.: Bayesian Exoplanet Tests of a New Method for MCMC Sampling in Highly Correlated Parameter Spaces. MNRAS, 410, 94–110 (2011)
10. Gregory, P. C.: Bayesian Re-analysis of the Gliese 581 Exoplanet System. MNRAS, in press (2011)
11. Jaynes, E. T., 1957, Stanford University Microwave Laboratory Report 421, Reprinted in 'Maximum Entropy and Bayesian Methods in Science and Engineering', G. J. Erickson and C. R. Smith, eds, Dordrecht: Kluwer Academic Press, p.1 (1988)

Chapter 11
Cosmological Bayesian Model Selection: Recent Advances and Open Challenges

Roberto Trotta

Abstract The cosmology community has been increasingly focusing on Bayesian model selection as a tool to discriminate between competing theories to explain a large amount of data about our Universe. In this paper, I summarize the conceptual underpinnings and the algorithmic implementations of Bayesian model comparison. I then discuss two representative applications of Bayesian model comparison to cosmological problems: determining whether the Universe is infinite and selecting the "best" model of inflation. I conclude by offering some reflections about open challenges and interpretational issues. Help and suggestions from the statistics community would be appreciated in further developing the field.

11.1 Introduction

In the last decade, cosmology has been revolutionized by a large amount of highly accurate data, which have allowed physicists to test in unprecedented ways our current concordance cosmological model. As we enter the second decade of the twenty-first century, our vanilla cosmological model is remarcably simple, and in excellent agreement with most data sets. This model is called "the ΛCDM model", as it contains both a cosmological constant (usually represented by the symbol Λ) and cold dark matter (CDM) particles, both of which have yet to be discovered in laboratory experiments. The ΛCDM model rests on two fundamental assumptions: (a) that the expansion of the Universe is described by Einstein's theory of General Relativity and (b) that the Universe obeys the cosmological principle, i.e., that on sufficiently large scales it is statistically homogeneous and isotropic. The simplest

R. Trotta (✉)
Astrophysics Group, South Kensington Campus, Imperial College London,
London, SW7 2AZ UK
e-mail: r.trotta@imperial.ac.uk

E.D. Feigelson and G.J. Babu (eds.), *Statistical Challenges in Modern Astronomy V*,
Lecture Notes in Statistics 209, DOI 10.1007/978-1-4614-3520-4_11,

version of the model contains six free parameters, which will be denoted collectively by Θ. The values of those quantities are not predicted by the theory but have to be constrained from observation:

- Parameters describing the matter-energy content of the Universe (density of baryonic—i.e., "normal"—matter, density of cold dark matter, cosmological constant).
- Parameters describing the spatial distribution of primordial density fluctuations emerging from the Big Bang.
- One parameter describing the effect of ionizing radiation being injected into the Universe at a later time (e.g., by a first generation of stars).

The data d being used to infer the value of Θ span a wide range of scales, both spatially and temporally, in the Universe. Broadly speaking, they can be classified as observations of the primordial fluctuations in the cosmic microwave background (CMB), the relic radiation from the Big Bang; observations of the growth of structures in the Universe (galaxies and clusters); observations of standard candles (e.g., supernovae type Ia) and/or standard rulers (e.g., baryonic acoustic oscillations); observations of the gravitational bending of light (strong and weak gravitational lensing). All of those phenomena can be predicted (often to very high accuracy, as is the case for the CMB) from the ΛCDM model, as a function of Θ.

Bayes Theorem is ubiquitously used in cosmology (see e.g. [11]) to obtain the posterior pdf $p(\Theta|\mathbf{D},\mathscr{M})$ for Θ:

$$p(\Theta|\mathbf{D},\mathscr{M}) = \frac{p(\mathbf{D}|\Theta,\mathscr{M})p(\Theta|\mathscr{M})}{p(\mathbf{D}|\mathscr{M})}. \tag{11.1}$$

Here, $p(\Theta|\mathscr{M})$ is the prior, $p(\mathbf{D}|\Theta,\mathscr{M})$ the likelihood and $p(\mathbf{D}|\mathscr{M})$ the Bayesian evidence (or model likelihood). We have made explicit the conditioning on a specific cosmological model, $\mathscr{M}$, for example the ΛCDM model sketched above. Posterior constraints on Θ are nowadays typically at the percent or sub-percent level, and the prior influence is often minimal for parameter inference. A recent example of cosmological data from CMB observations is shown in Fig. 11.1. The ensuing constraints on some of the cosmological parameters Θ from a combination of data sets are displayed in Fig. 11.2.

Having largely solved the parameter inference problem, in recent years the focus of the cosmological community has been increasingly shifting towards model selection questions [17, 19, 35]. A typical problem is to decide whether there is evidence in the data for extra parameters in the ΛCDM model, beyond the vanilla ones. There are however also alternative models (e.g., Bianchi models or modified gravity models) which are *not* extension of ΛCDM, and whose parameter space is largely or completely disjoint. This question is being addressed by the use of Bayesian model selection techniques, to which we now turn our attention.

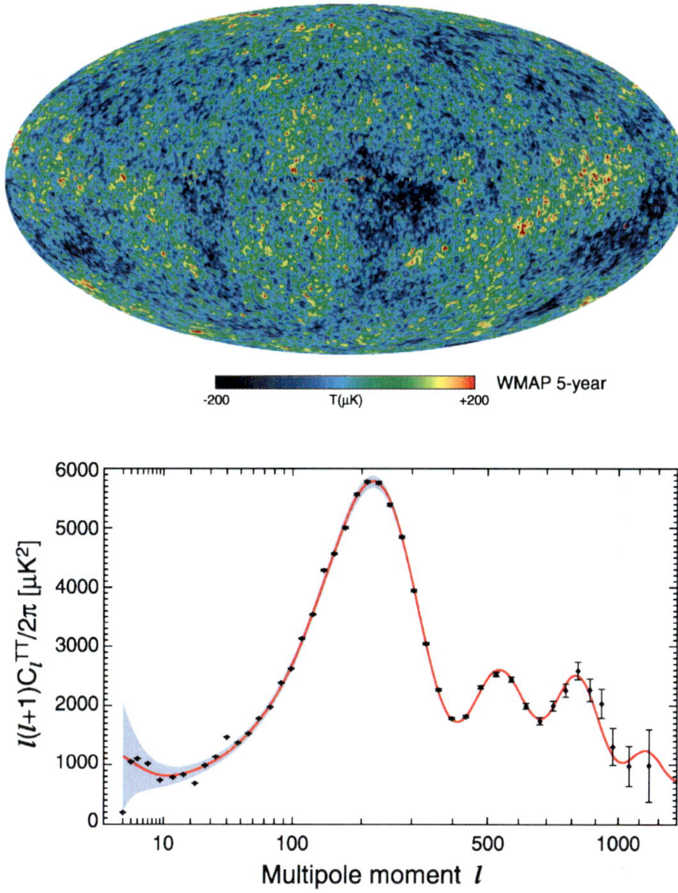

Fig. 11.1 Example of cosmological data sets. *Top*: WMAP 5-year all sky map of the cosmic microwave background temperature fluctuations (from [10]). *Bottom*: temperature power spectrum measurements (i.e., the harmonic transform of the 2-point correlation function) extracted from the map (from [18]) (notice that data points are correlated); along the x-axis, the multipole moment is inversely proportional to the angular separation on the sky. *Black* errorbars give statistical errors, while the shaded *gray* band gives the sampling error (cosmic variance). The *red line* is the (remarkably good) best-fit from the ΛCDM model after the cosmological parameters Θ have been fitted to the data

11.2 Cosmological Model Selection

11.2.1 Shaving Theories with Occam's Razor

When there are several competing theoretical models, Bayesian model comparison provides a formal way of evaluating their relative probabilities in light of the data and any prior information available. The "best" model is then the one which strikes

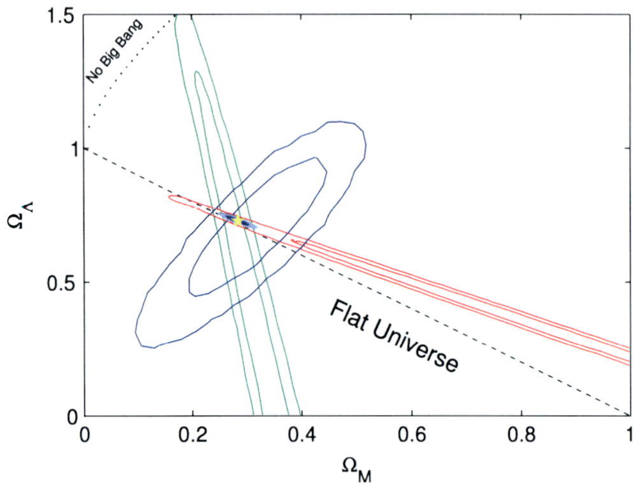

Fig. 11.2 Example of constraints on cosmological parameters Ω_m (describing the fractional energy density of matter) and Ω_Λ (the fractional energy density in a cosmological constant). The *red* contours are 68%, 95% marginalized posterior constraints from cosmic microwave background data (such as the ones displayed in Fig. 11.1), the *green* contours from Baryonic Acoustic Oscillations data, the *blue* contours from supernovae type Ia data. The combined constraints from all three data sets are given by the filled contours. The *star* denotes the best-fit parameters value (From [25])

an optimum balance between quality of fit and predictivity. In fact, it is obvious that a model with more free parameters will always fit the data better (or at least as good as) a model with less parameters. However, more free parameters also mean a more "complex" model (a precise definition of "model complexity" can be found in [16]). Such an added complexity ought to be avoided whenever a simpler model provides an adequate description of the observations. This guiding principle of simplicity and economy of an explanation is known as *Occam's razor*—the simplest theory compatible with the available evidence ought to be preferred.

An important feature is that an alternative model must be specified against which the comparison is made. In contrast with frequentist goodness-of-fit tests, Bayesian model comparison maintains that it is pointless to reject a theory unless an alternative explanation is available that fits the observed facts better (for more details about the difference in approach with frequentist hypothesis testing, see e.g. [22]). In other words, unless the observations are totally impossible within a model, finding that the data are improbable given a theory does not say anything about the probability of the theory itself *unless we can compare it with an alternative*. A consequence of this is that the probability of a theory that makes a correct prediction can increase if the prediction is confirmed by observations, provided competitor theories do not make the same prediction.

In the context of model comparison it is appropriate to think of a model as a specification of a set of parameters Θ *and* of their prior distribution, $p(\Theta|\mathcal{M})$. It is the number of free parameters *and* their prior range that control the strength of the Occam's razor effect in Bayesian model comparison: models that have many parameters that can take on a wide range of values but that are not needed in the light of the data are penalized for their unwarranted complexity. Therefore, *the prior choice ought to reflect the available parameter space under the model* $\mathscr{M}$, *independently of experimental constraints we might already be aware of.* This is because we are trying to assess the economy (or simplicity) of the model itself, and hence the prior should be based on theoretical or physical constraints on the model under consideration. Often these will take the form of a range of values that are deemed "intuitively" plausible, or "natural". Thus the prior specification is inherent in the model comparison approach.

11.2.2 The Bayesian Evidence

The evaluation of a model's performance in the light of the data is based on the *Bayesian evidence* (as it is usually called in the cosmology and astrophysics community), which in the statistical literature is often called *marginal likelihood* or *model likelihood*. The evidence is the normalization integral on the right-hand-side of Bayes' theorem, (11.1):

$$p(\mathbf{D}|\mathscr{M}) \equiv \int p(\mathbf{D}|\Theta,\mathscr{M})p(\Theta|\mathscr{M})\mathrm{d}\Theta. \tag{11.2}$$

Thus the Bayesian evidence is the average of the likelihood under the prior for a specific model choice. From the evidence, the model posterior probability given the data is obtained by using Bayes' Theorem to invert the order of conditioning:

$$p(\mathscr{M}|\mathbf{D}) \propto p(\mathscr{M})p(\mathbf{D}|\mathscr{M}), \tag{11.3}$$

where $p(\mathscr{M})$ is the prior probability assigned to the model itself. Usually this is taken to be non-committal and equal to $1/N_m$ if one considers N_m different models. When comparing two models, $\mathscr{M}_0$ versus $\mathscr{M}_1$, one is interested in the ratio of the posterior probabilities, or *posterior odds*, given by

$$\frac{p(\mathscr{M}_0|\mathbf{D})}{p(\mathscr{M}_1|\mathbf{D})} = B_{01}\frac{p(\mathscr{M}_0)}{p(\mathscr{M}_1)} \tag{11.4}$$

and the Bayes factor B_{01} is the ratio of the models' evidences:

$$B_{01} \equiv \frac{p(\mathbf{D}|\mathscr{M}_0)}{p(\mathbf{D}|\mathscr{M}_1)} \quad \text{(Bayes factor)}. \tag{11.5}$$

A value $B_{01} > (<)\ 1$ represents an increase (decrease) of the support in favour of model 0 versus model 1 given the observed data. From (11.4) it follows that the Bayes factor gives the factor by which the relative odds between the two models

Table 11.1 Empirical scale for evaluating the strength of evidence when comparing two models, $\mathcal{M}_0$ versus $\mathcal{M}_1$ (so-called "Jeffreys' scale"). Threshold values are empirically set, and they occur for values of the logarithm of the Bayes factor of $|\ln B_{01}| = 1.0$, 2.5 and 5.0. The right-most column gives our convention for denoting the different levels of evidence above these thresholds. The probability column refers to the posterior probability of the favoured model, assuming non-committal priors on the two competing models, i.e. $p(\mathcal{M}_0) = p(\mathcal{M}_1) = 1/2$ and that the two models exhaust the model space, $p(\mathcal{M}_0|\mathbf{D}) + p(\mathcal{M}_1|\mathbf{D}) = 1$

$\|\ln B_{01}\|$	Odds	Probability	Strength of evidence
< 1.0	$\lesssim$3:1	< 0.750	Inconclusive
1.0	~3:1	0.750	Weak evidence
2.5	~12:1	0.923	Moderate evidence
5.0	~150:1	0.993	Strong evidence

have changed after the arrival of the data, regardless of what we thought of the relative plausibility of the models before the data, given by the ratio of the prior models' probabilities.

Bayes factors are usually interpreted against the Jeffreys' scale [13] for the strength of evidence, given in Table 11.1. This is an empirically calibrated scale, with thresholds at values of the odds of about 3 : 1, 12 : 1 and 150 : 1, representing weak, moderate and strong evidence, respectively.

Bayesian model comparison *does not* replace the parameter inference step (which is performed within each of the models separately). Instead, model comparison *extends* the assessment of hypotheses in the light of the available data to the space of theoretical models, as evident from (11.4).

11.3 Numerical Evaluation of the Evidence

The computation of the Bayesian evidence, (11.2), is in general a numerically challenging task, as it involves a multi-dimensional integration over the whole of parameter space. Fortunately, several methods are now available, each with its own strengths and domains of applicability. Some of them have been developed by astronomers/cosmologists and are rapidly finding applications in other domains.

1. The numerical method of choice until recently has been thermodynamic integration, whose computational cost can however be fairly large. In typical cosmological applications [2,3,33], thermodynamic integration can require up to $\sim 10^7$ likelihood evaluations, two orders of magnitude more than MCMC-based parameter estimation. Recently, population Monte Carlo algorithms have been used successfully to compute the evidence [14].
2. Skilling [30, 32] has put forward an elegant algorithm called "nested sampling", which has been implemented in the cosmological context by Bassett et al. [1],

Mukherjee [27], Shaw [31], and Feroz and Hobson [7] (for a theoretical discussion of the algorithmic properties, see [4]). The gist of nested sampling is that the multi-dimensional evidence integral is recast into a one-dimensional integral that is easy to evaluate numerically. This technique allows to reduce the computational burden to about $\sim 10^5$ likelihood evaluations. Recently, the development of what is called "multi-modal nested sampling" has allowed to increase significantly the efficiency of the method [7, 37], reducing the number of likelihood evaluations by another order of magnitude.

3. Useful approximations to the Bayes factor, (11.5), are available for situations in which the models being compared are *nested* into each other, i.e. the more complex model ($\mathscr{M}_1$) reduces to the original model ($\mathscr{M}_0$) for specific values of the new parameters. This is a fairly common scenario in cosmology, where one wishes to evaluate whether the inclusion of the new parameters is supported by the data. For example, we might want to assess whether we need isocurvature contributions to the initial conditions for cosmological perturbations, or whether a curvature term in Einstein's equation is needed, or whether a non-scale invariant distribution of the primordial fluctuation is preferred. Writing for the extended model parameters $\Theta = (\alpha, \beta)$, where the simpler model $\mathscr{M}_0$ is obtained by setting $\beta = 0$, and assuming further that the prior is separable (which is usually the case in cosmology), i.e. that

$$p(\alpha, \beta | \mathscr{M}_1) = p(\beta | \mathscr{M}_1) p(\alpha | \mathscr{M}_0), \tag{11.6}$$

the Bayes factor can be written in all generality as

$$B_{01} = \left. \frac{p(\beta | \mathbf{D}, \mathscr{M}_1)}{p(\beta | \mathscr{M}_1)} \right|_{\beta=0}. \tag{11.7}$$

This expression is known as the Savage–Dickey density ratio (SDDR, see [35, 40]). The numerator is simply the marginal posterior under the more complex model evaluated at the simpler model's parameter value, while the denominator is the prior density of the more complex model evaluated at the same point. This technique is particularly useful when testing for one extra parameter at the time, because then the marginal posterior $p(\beta | \mathbf{D}, \mathscr{M}_1)$ is a 1-dimensional function and normalizing it to unity probability content only requires a 1-dimensional integral, which is simple to do using for example the trapezoidal rule.

4. An instructive approximation to the Bayesian evidence can be obtained when the likelihood function is unimodal and approximately Gaussian in the parameters [9]. Expanding the likelihood around its peak to second order one obtains the Laplace approximation

$$p(\mathbf{D} | \Theta, \mathscr{M}) \approx \mathscr{L}_{\max} \exp\left[-\frac{1}{2} (\Theta - \Theta_{\rm ML})^t L (\Theta - \Theta_{\rm ML}) \right], \tag{11.8}$$

where Θ_{ML} is the maximum-likelihood point, $\mathscr{L}_{\max}$ the maximum likelihood value and L the likelihood Fisher matrix (which is the inverse of the covariance matrix for the parameters). Assuming as a prior a multinormal Gaussian distribution with zero mean and Fisher information matrix P one obtains for the evidence, (11.2)

$$p(\mathbf{D}|\mathscr{M}) = \mathscr{L}_{\max} \frac{|F|^{-1/2}}{|P|^{-1/2}} \exp\left[-\frac{1}{2}({\Theta_{\mathrm{ML}}}^t L \Theta_{\mathrm{ML}} - \overline{\Theta}^t F \overline{\Theta})\right], \tag{11.9}$$

where the posterior Fisher matrix is $F = L + P$ and the posterior mean is given by $\overline{\Theta} = F^{-1} L \Theta_{\mathrm{ML}}$.

From (11.9) we can deduce a few qualitatively relevant properties of the evidence. First, the quality of fit of the model is expressed by $\mathscr{L}_{\max}$, the best-fit likelihood. Thus a model which fits the data better will be favoured by this term. The term involving the determinants of P and F is a volume factor, encoding the Occam's razor effect. As $|P| \leq |F|$, it penalizes models with a large volume of wasted parameter space, i.e. those for which the parameter space volume $|F|^{-1/2}$ which survives after arrival of the data is much smaller than the initially available parameter space under the model prior, $|P|^{-1/2}$. Finally, the exponential term suppresses the likelihood of models for which the parameters values which maximise the likelihood, Θ_{ML}, differ appreciably from the expectation value under the posterior, $\overline{\Theta}$. Therefore when we consider a model with an increased number of parameters we see that *its evidence will be larger only if the quality-of-fit increases enough to offset the penalizing effect of the Occam's factor.*

On the other hand, it is important to notice that the Bayesian evidence does *not* penalize models with parameters that are unconstrained by the data. It is easy to see that unmeasured parameters (i.e., parameters whose posterior is equal to the prior) do not contribute to the evidence integral, and hence model comparison does not act against them, awaiting better data.

11.3.1 Cosmological Applications

There is a rapidly growing literature in cosmology applying the above ideas to cosmological model selection, some of which is surveyed in [36]. Here we present but two recent examples, as this will serve to highlight some of the open questions in the next section.

11.3.1.1 Is the Universe Infinite?

One of the key cosmological parameters Θ is a quantity, usually denoted by Ω_κ, characterizing the spatial curvature of the Universe. A Universe with $\Omega_\kappa = 0$ is spatially flat (i.e., its geometry is Euclidean, and parallel lines meet at infinity) and infinite in extent; for $\Omega_\kappa > 0$ the Universe is finite and closed (its geometry is the 3D analogous of a sphere and parallel lines converge), while for $\Omega_\kappa < 0$ the geometry is hyperbolic (so-called "open" Universe, where parallel lines diverge from each other) and the Universe is infinite. Models predicting the curvature of the Universe are rooted in fairly well understood physics, a feature which helps in setting physically motivated priors on Ω_κ. For example, the possibility of a flat, $\Omega_\kappa \sim 0$ Universe has long been favoured by theoretical prejudice, as a flat or close-to-flat Universe is a generic prediction of the inflationary scenario, which is in good agreement with observations of the CMB. Parameter inference on the value of Ω_κ delivers posterior constraints of the order $\Omega_\kappa = -0.0057^{+0.0066}_{-0.0068}$ (68 % region) [15].

As there are only three possibilities (i.e., models) for the curvature in a Universe obeying the cosmological principle, the question of whether the Universe is finite (closed) or infinite (open or flat) is well suited to be tackled with Bayesian model selection techniques. A detailed analysis can be found in [38]. Here we just summarize the main results. Starting from a non-committal prior on the three models under consideration, $p(\mathscr{M}_i) = 1/3$ ($i = 1,2,3$), the posterior probability for the Universe being infinite is evaluated using (11.3), for two different choices for the prior on Ω_κ (as the flat model is nested within the non-flat ones, the only relevant prior for the model comparison is the one on the extra parameter of the more complex models, namely Ω_κ, as can be seen from (11.7)). The prior selection is motivated by different physical considerations: the "Astronomer's prior" (a uniform prior in the range $-1 \leq \Omega_\kappa \leq 1$) is motivated by basic consistency with observable properties of the Universe, such as the age of the oldest objects, while the "Curvature scale prior" (a uniform prior in the range $-5 \leq \log|\Omega_\kappa| \leq 0$) is based on theoretical considerations of the inflationary scenario.

The resulting posterior probability for an infinite Universe is (for the most constraining data combination and the simplest parameterization of the dark energy sector) is 98% from the Astronomer's prior, but only 69% for the Curvature scale prior. This reflects the stronger Occam's razor effect implied by the Astronomer's prior. Although in both cases the posterior probability for an infinite Universe has increased from the a prior proabability of $\sim$67%, it is clear that the amount by which this scenario is preferred by the data is strongly dependent on the theoretical prior assumptions one makes.

11.3.1.2 Inflationary Model Comparison

The second example I would like to discuss is the inflationary model comparison carried out in Ref. [26]. Although the technical details are fairly involved, the underlying idea can be sketched as follows.

The term "inflation" describes a period of exponential expansion of the Universe in the very first instants of its life, some 10^{-32} s after the Big Bang, during which the size of the Universe increased by at least 25 orders of magnitude. This huge and extremely fast expansion is required to explain the observed isotropy of the cosmic microwave background on large scales. It is believed that inflation was powered by one or more "scalar fields". The behaviour of the scalar field during inflation is determined by the shape of its potential, which is a real-valued function $V(\phi)$ (where ϕ denotes the value of the scalar field). The detailed shape of $V(\phi)$ controls the duration of inflation, but also the spatial distribution of inhomogeneities (perturbations) in the distribution of matter and radiation which emerge from inflation. It is from those perturbations that galaxies and cluster form out of gravitational collapse. Hence the shape of the scalar field can be constrained by observations of the large scale structures of the Universe and of the CMB anistropies.

Theories of physics beyond the Standard Model motivate certain functional forms of $V(\phi)$, which however typically have a number of free parameters, Ψ. The fundamental model selection question is to use cosmological observations to discriminate between alternative models for $V(\phi)$ (and hence alternative fundamental theories). The major obstacle to this programme is that very little if anything at all is known a priori about the free parameters Ψ describing the inflationary potential. What is worse, such parameters can assume values across several orders of magnitude, according to the theory. Hence the Occam's razor effect of Bayesian model comparison can vary in a very significant way depending on the prior choices for Ψ. Furthermore, a non-linear reparameterization of the problem (which leaves the physics invariant) does in general change the Occam's razor factor, and hence the model comparison result.

In Ref. [26] a first attempt was made to tackle inflationary model selection from a principled point of view. The main result of the analysis is shown in Fig. 11.3, which presents the Bayes factors between models (suitably normalized w.r.t. a reference model, here the so-called LFI_2 model). Two classes of models for $V(\phi)$ have been considered, namely so-called Small Field Inflation (SFI) models and Large Field Inflation (LFI) models. The two classes of model differ in the parameterized form of $V(\phi)$, and have different sets of parameters, differing in dimensionality, as well. Within each class of models, sub-classes are defined (denoted by subscripts in Fig. 11.3) based on theoretical considerations, e.g. by fixing some of the parameters to certain values. The priors on the models' parameters have been chosen based on theoretical considerations of possible values achievable under each class of models. Typical priors are uniform on the log of the parameter (to reflect indifference w.r.t. the characteristic scale of the quantity), within a range chosen as a reflection of physical model building. The models' priors are chosen in such a way to lead to non-committal priors for the two classes as a whole, i.e. $p(\mathrm{SFI}) = p(\mathrm{LFI}) = 1/2$.

Figure 11.3 shows that some models in the LFI class are fairly strongly disfavoured by the data (e.g., LFI_3 and LFI_4), while the model comparison is inconclusive in most other cases. One finds that the posterior probability for the SFI model class evaluates to $p(\mathrm{SFI}|d) \approx 0.77$. Therefore, the probability of the SFI class has increased from 50% in the prior to about 77% in the posterior, signalling a weak preference for this type of models in the light of the data.

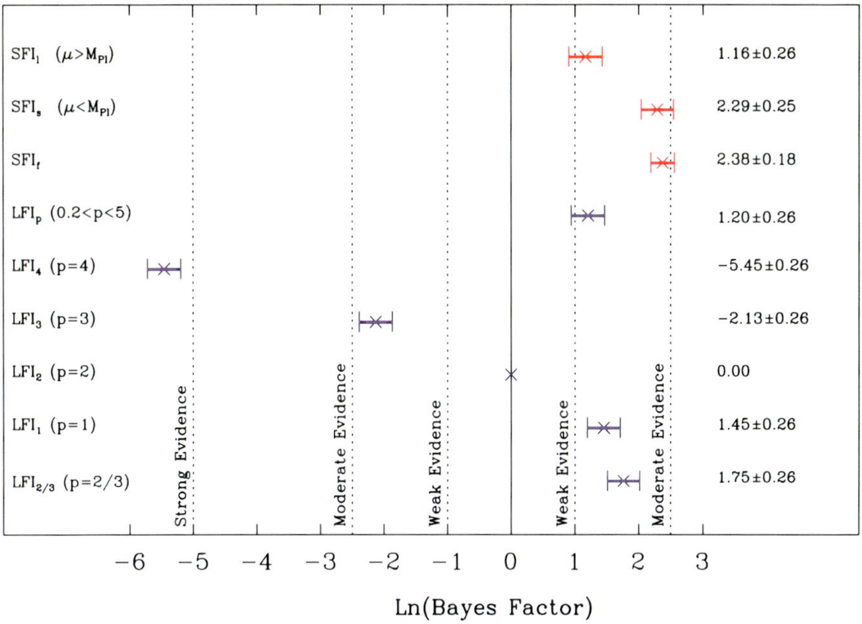

Fig. 11.3 Results of Bayesian model comparison between nine inflationary models (*vertical axis*), subdivided in two categories (*SFI* models and *LFI* models), from Ref. [26]. Errorbars reflect the 68% uncertainty on the value of the Bayes factor from the numerical evaluation

11.4 Interpretational Challenges and Open Questions

I conclude by listing what I think are some of the open questions and outstanding challenges in the application of Bayesian model selection to cosmological model building.

- **Is Bayesian model selection always applicable?** The Bayesian model comparison approach as applied to cosmological and particle physics problems has been strongly criticized by some authors. E.g., George Efstathiou [6] and Bob Cousins [5] pointed out (in different contexts) that often insufficient attention is given to the selection of models and of priors, and that this might lead to posterior model probabilities which are largely a function of one's unjustified assumptions. This draws attention to the difficult question of how to choose priors on phenomenological parameters, for which theoretical reasoning offers poor or no guidance (as in the inflationary model comparison example above). In the statistics literature, several approaches are available, e.g. nonsubjective, intrinsic and fractional Bayes factors [8]. It would be interesting to learn about real-data experience in using such methods, and to investigate whether they can be useful in the cosmological context. Also, a thorough investigation of approximate criteria for model selection (BIC, AIC, DIC, etc. [12, 19]) in cosmology would be desirable.

- **How do we deal with Lindley's paradox?** It is simple to construct examples of situations where Bayesian model comparison and classical hypothesis testing disagree (Lindley's paradox [21]). This is not surprising, as frequentist hypothesis testing and Bayesian model selection really ask different questions of the data [29]. As Louis Lyons aptly put it: "Bayesians address the question everyone is interested in by using assumptions no-one believes, while frequentists use impeccable logic to deal with an issue of no interest to anyone" [23]. However, such a disagreement is likely to occur in situations where the signal is weak, which are precisely the kind of "frontier science" cases which are the most interesting ones (e.g., discovery claims). Is there a way to evaluate e.g. the loss function from making the "wrong" decision about rejecting/accepting a model?
- **How do we assess the completness of the set of known models?** Bayesian model selection always returns a best model among the ones being compared, even though that model might be a poor explanation for the available data. Is there a principled way of constructing an *absolute* scale for model performance in a Bayesian context? Recently, the notion of Bayesian doubt, introduced in [24], has been used to extend the power of Bayesian model selection to the space of unknown models in order to test our paradigm of a ΛCDM cosmological model. It would be useful to have feedback from the statistics community about the validity of such an approach, and whether similar tools have already been developed in other contexts.
- **Is Bayesian model averaging useful?** Bayesian model averaging can be used to obtain final inferences on parameters which take into account the residual model uncertainty (examples of applications in cosmology can be found in [20, 28, 39]). However, it also propagates the model selection problems discussed above to the level of model-averaged parameter constraints. Is it useful to produce model-average parameter constraints, or should this task be left to the user, by providing model-specific posteriors and Bayes factors instead?
- **Is there such a thing as a "correct" prior?** In fundamental physics, models and parameters (and their priors) are supposed to represent (albeit in an idealized way) the real world, i.e., they are not simply useful representation of the data (as they are in other statistical problems, e.g. as applied to social sciences). In this sense, one could imagine that there exist a "correct" prior for e.g. the parameters Θ of our cosmological model, which could in principle be derived from fundamental theories such as string theory (e.g., the distribution of values of cosmological parameters across the landscape of string theory [34]). This raises interesting statistical questions about the relationship between physics, reality and probability.

11.5 Conclusions

I have briefly surveyed the status and recent advances in the application of Bayesian model selection in cosmology. I am sure that the input of the statistics community will be invaluable in further advancing the topic. A discussion forum such as the SCMA conference is an extremely useful way of promoting cross-disciplinary dialogue between the two communities, and as such should be taken as a blueprint for future initiatives in Astrostatistics.

Acknowledgements The author would like to thank the organizers of SCMA V Conference for the invitation to present this work and for partial travel support. I am grateful to Andrew Jaffe, Martin Kunk, Andrew Liddle, Mike Hobson, Tom Loredo, Louis Lyons, David van Dyke, Daniel Mortlock, Ofer Lahav, Ben Wandelt for many stimulating discussions. The use of WMAP images from LAMBDA is acknowledged.

References

1. B. A. Bassett, P. S. Corasaniti and M. Kunz, Astrophys. J. **617** L1–L4 (2004).
2. M. Beltran, J. Garcia-Bellido, J. Lesgourgues, A. R. Liddle, *et al.*, Phys. Rev. **D71** 063532 (2005).
3. M. Bridges, A. N. Lasenby and M. P. Hobson, Mon. Not. Roy. Astron. Soc. **369** 1123–1130 (2006).
4. N. Chopin and C. P. Robert, Contemplating Evidence: properties, extensions of, and alternatives to Nested Sampling.. (Available as preprint from: http://www.crest.fr/pageperso/Nicolas.Chopin/Nicolas.Chopin.htm. Accessed Jan 2008.).
5. R. D. Cousins, Phys. Rev. Lett. **101**, 029101 (2008).
6. G. Efstathiou, preprint: arXiv:0802.3185.
7. F. Feroz and M. P. Hobson, Mon. Not. Roy. Astron. Soc., 384, 2, 449–463 (2008).
8. J. K. Ghosh, M. Delampady and T. Samanta, *An Introduction to Bayesian analysis*, Springer (2006).
9. A. F. Heavens, T. D. Kitching and L. Verde, Mon. Not. Roy. Astron. Soc. **380** 1029–1035 (2007).
10. G. Hinshaw *et al.* [WMAP Collaboration], Astrophys. J. Suppl. **180**, 225–245 (2009).
11. *Bayesian Methods in Cosmology*, M. Hobson *et al.*, (Eds), Cambridge University Press (2010).
12. G. Claeskens and N. L. Hjort, *Model selection and model averaging*, Cambridge University Press (2008).
13. H. Jeffreys, *Theory of probability*, 3rd edn , Oxford Classics series (reprinted 1998) (Oxford University Press, Oxford, UK, 1961).
14. M. Kilbinger *et al.*, preprint: arXiv:0912.1614.
15. Komatsu E., et al., Astrophys. J. Supp., 192, 18 (2011).
16. M. Kunz, R. Trotta and D. Parkinson, Phys. Rev. **D74** 023503 (2006).
17. A. H. Jaffe, Astrophys. J. **471** 24 (1996).
18. D. Larson, J. Dunkley, G. Hinshaw, E. Komatsu, M. R. Nolta, C. L. Bennett, B. Gold, M. Halpern *et al.*, Astrophys. J. Suppl. **192**, 16 (2011).
19. A. R. Liddle, Mon. Not. Roy. Astron. Soc. **351** L49–L53 (2004).
20. A. R. Liddle, P. Mukherjee, D. Parkinson, Y. Wang, Phys. Rev. **D74**, 123506 (2006).
21. D. Lindley, Biometrika **44** 187–192 (1957).
22. T. J. Loredo, From Laplace to Supernova SN 1987A: Bayesian Inference in Astrophysics, in T. Fougere (Editor) *Maximum-Entropy and Bayesian Methods*, Available from: http://bayes.wustl.edu/gregory/articles.pdf (accessed Jan 15 2008) (Kluwer Academic Publishers, Dordrecht, The Netherlands, 1990), pp. 81–142.
23. L. Lyons, A particle physicist's perspective on astrostatistics, in *Proceedings of the Statistical Challenges in Modern Astronomy IV Conference*, 371, Pennsylvania State University, Penn-

sylvania, USA, 12–15 June 2006 (Astronomical Society of the Pacific, San Francisco, 2007), pp. 361–372.

24. M. C. March, G. D. Starkman, R. Trotta, P. M. Vaudrevange, Mon. Not. Roy. Astron. Soc. **410**, 2488–2496 (2011).
25. M. C. March, R. Trotta, P. Berkes, G. D. Starkman, P. M. Vaudrevange, Mon. Not. Roy. Astron. Soc. in press, preprint: arXiv:1102.3237.
26. J. Martin, C. Ringeval, R. Trotta, Phys. Rev. D83, 063524 (2011), arXiv:1009.4157.
27. P. Mukherjee, D. Parkinson and A. R. Liddle, Astrophys. J. **638** L51–L54 (2006).
28. D. Parkinson, A. R. Liddle, Phys. Rev. **D82**, 103533 (2010).
29. T. Sellke, M. Bayarri and J. O. Berger, American Statistician **55** 62–71 (2001).
30. J. Skilling, Nested sampling, in R. Fischer, R. Preuss and U. von Toussaint (Eds) *Bayesian Inference and Maximum Entropy Methods in Science and Engineering*, 735 (Amer. Inst. Phys. conf. proc. 2004), pp. 395–405.
31. R. Shaw, M. Bridges and M. P. Hobson, Mon. Not. Roy. Astron. Soc. **378** 1365–1370 (2007).
32. J. Skilling, Bayesian Analysis **1** 833–861 (2006).
33. A. Slosar *et al.*, Mon. Not. Roy. Astron. Soc. **341** L29 (2003).
34. M. Tegmark, JCAP **0504**, 001 (2005)
35. R. Trotta, Mon. Not. Roy. Astron. Soc. **378** 72–82 (2007).
36. R. Trotta, Contemp. Phys. **49**, 71 (2008)
37. R. Trotta, F. Feroz, M. P. Hobson, L. Roszkowski and R. Ruiz de Austri, JHEP **0812**, 024 (2008)
38. M. Vardanyan, R. Trotta and J. Silk, Mon. Not. Roy. Astron. Soc. **397**, 431 (2009)
39. M. Vardanyan, R. Trotta, J. Silk, Mon. Not. Roy. Astron. Soc. in print, preprint: arXiv:1101.5476.
40. I. Verdinelli and L. Wasserman, J. Amer. Stat. Assoc. **90** 614–618 (1995).

Chapter 12
Commentary: Cosmological Bayesian Model Selection

David A. van Dyk

Abstract Model selection methodology is an active field of discussion among statisticians, particularly for disjoint, non-nested models. Roberto Trotta has reviewed the issue in the context of model selection within the context of ΛCDM cosmological models. I briefly discuss the issue from both frequentist and Bayesian perspectives, expressing cautions about use of priors, Bayes factors, and p-values. There are no silver bullets, but Bayes factors seem most promising.

12.1 Introduction

Doctor Trotta is to be congratulated for his lucid summary of recent advances in Bayesian fitting of cosmological models and of the outstanding challenges in the more difficult problem of model selection. This situation is not unique to cosmology. Differences among statistical paradigms such as frequency-based or Bayesian methods are generally much more pronounced in model checking and selection than in fitting. Indeed no consensus exists even among Bayesians or among frequentists as to the best way forward in model selection. As such this remains an active area of statistical research where the experience of cosmologists may lead to insight with impact on more general statistical methodology. It is also a subtle area where one must be wary of all-purpose solutions. As Doctor Trotta points out, model selection in cosmology is not confined to nested models (e.g., adding "extra parameters in the ΛCDM beyond the vanilla ones") but includes the more technically challenging case of comparing non-nested models "whose parameter spaces are largely or completely disjoint". Such seemingly innocuous differences may be highly consequential and lead to subtle technical issues.

D.A. van Dyk (✉)
Statistics Section, Department of Mathematics, Imperial College London,
London SW7 2AZ, UK
e-mail: dvandyk@imperial.ac.uk

E.D. Feigelson and G.J. Babu (eds.), *Statistical Challenges in Modern Astronomy V*,
Lecture Notes in Statistics 209, DOI 10.1007/978-1-4614-3520-4_12,

I hope to illustrate some of the subtleties involved and the advantages of a mixed approach that considers and compares various methods in the context of a specific model selection problem.

12.2 Methods for Model Selection and Checking

Model checking problems often begin with a default or presumed model,

Null Hypothesis: E.g., the Universe is "Flat".

The scientist asks whether the model is consistent with the data or if it is plausible that the data were generated under the model. If not, we aim to characterize the inconsistency, improve the model, and recheck the improved model. In principle this cycle of model improvement can be iterated, perhaps with the acquisition of new data, until a satisfactory model is obtained.

We may also have a model that we suspect or hope is better than the null model,

Alternative Hypothesis: E.g., the Universe is "Hyperbolic".

With a competing model in hand, we typically aim to decide between or weigh the evidence for the two (or more) models. These procedures are known as *model selection* and *model comparison*. In some situations we may wish to assume the null hypothesis until we have substantial evidence it is implausible. This is analogous to assuming a defendant is innocent, until proven guilty in a court of law. Similarly we may not wish to overturn a long standing standard model without truly solid evidence. In other situations we may not have any particular reason to favor one model over another and may wish to simply weigh the relative evidence for each.

These are surprisingly subtle problems and despite decades of research, discussion, and sometimes heated arguments, little consensus exists among statisticians as how to best tackle them. This is especially concerning because competing methods may lead to very different conclusions. Part of the difficulty is that model selection is somewhat ill-posed. Statisticians view models as parsimonious mathematical summaries of complex phenomena. They are not meant to capture the full complexity of that which they summarize. As such different models can be viewed as approximations with various tradeoffs between simplicity and detail, no one of which may be 'true' or even better than the others; they are simply different. Nonetheless we may wish to investigate how a particular model differs from reality (i.e., model checking) or which of a set of models better approximates a particular aspect of reality (i.e., model comparison). Remembering that models are not meant to be perfect, however, it is no surprise that there is no completely general theory for model selection nor is there always a clear cut answer to model selection problems. Model checking, comparison, and selection are nuanced endeavors into the shades of grey.

Frequency-Based Methods. The standard frequency based method begins with a statement of a null and an alternative hypothesis,

H_0: E.g., the Universe is Flat: $\Omega_\kappa = 0$, and
H_1: E.g., the Universe is not Flat: $\Omega_\kappa \neq 0$,

and computes a test statistic, T, with known distribution under H_0. A threshold, $T^\star$ is then computed as, e.g., the smallest value such that $\text{Prob}(T > T^\star | \Omega_\kappa = 0, \text{other parameters}) \leq \alpha$, where α is the *significance level* of the test. Under the assumption that H_0 holds, T is greater than $T^\star$ with probability less than α. This is an infrequent occurrence if α is small. Thus, we typically choose a small value of α and if we observe $T > T^\star$ conclude that there is sufficient evidence to declare H_0 implausible. In this example, we would conclude that the Universe is not flat.

This paradigm is generally advocated on the basis of its control of the probability of *false positive*. That is, we will wrongly conclude that H_0 is implausible with probability less than α, when H_0 actually holds. On the other hand the method offers no characterization of the strength of evidence, a task left to the notorious p-value, see below. Another important sticking point lies in the derivation of a test statistic with known distribution under H_0. This can be a difficult if not impossible task in complex models that have numerous unknown parameters.

Bayesian methods. Because Bayesian methods treat parameters as random quantities there is no problem in principle with unknown parameters under either H_0 or H_A. In particular the *prior predictive distribution*, given in Trotta's equation (2) specifies how likely the data, d, is under model $i \in \{0,1\}$. In a Bayesian paradigm the model consists of a specification of both the likelihood and the prior distribution and both are compared together. The typical method for comparing two models involves the *Bayes Factor*, or the *posterior probability of* H_0. Unlike standard frequency-based methods, both the Bayes Factor and $p(H_0|\mathcal{M})$ treat H_0 and H_1 essentially symmetrically. There is no need to treat H_0 as the default or a priori assumed model.

A typical criticism of Bayesian methods in general is their requirement that one specifies a prior distribution. Of course, when informative prior information is available, Bayesian methods offer a principled method of combining this information with the current data. In many situations, these concerns are of little practical importance because the posterior distribution, parameter estimates, error bars, and interval estimates are quite insensitive to the choice of prior distribution. Unfortunately, the same is not true of prior predictive distributions and Bayes factors which can be quite sensitive to the choice of prior distribution. As an example, suppose we observe a Gaussian variable with mean μ and variance one, use a Gaussian prior distribution on μ with mean zero and variance τ^2, and are interested in testing $H_0 : \mu = 0$ against $H_1 : \mu \neq 0$. Using (2) we can compute the prior predictive distribution of d which is a Gaussian distribution with mean zero and variance $1 + \tau^2$ and is plotted in the lefthand panel of Fig. 12.1 for several values

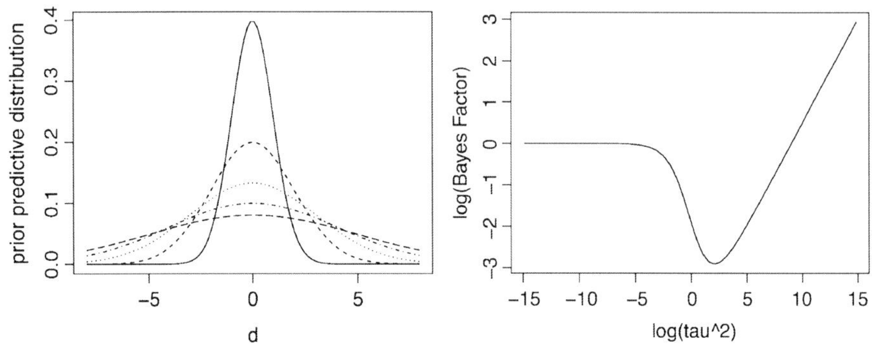

Fig. 12.1 The dependence of the prior predictive distribution and the Bayes factor on the choice of prior distribution. The *lefthand* panel plots the prior predictive distribution for the Gaussian example describe in the text with five choices of the prior distribution. The *righthand* plot shows the effect of the prior variance, τ^2, on the Bayes factor. Results are highly dependent on the prior distribution and the prior must be chosen with care to accurately represent available prior information

of τ^2. The prior predictive distribution is highly dependent on the prior distribution and $p(d|\mathscr{M})$ can be made arbitrarily small for any value of d by choosing τ^2 large enough. The righthand panel of Fig. 12.1 illustrates the effect on the log Bayes factor, which varies from indifference between H_0 and H_1 to strong support for H_1 to strong support for H_0 as τ^2 increases.

Reflecting on Fig. 12.1, it is clear that we must think carefully about our choice of prior distribution and it is critical that the prior distribution accurately summarizes available prior information. The typical strategy of using "non-informative" prior distributions with large variances clearly effects the Bayes factor. In fact "improper" prior distributions (e.g., with infinite variance) result in improper prior predictive distributions and undefined Bayes factors. There is no simple default prior distribution available when computing Bayes Factors. This is especially problematic when the parameter space is large and in particular when the H_A and H_0 are either not nested or the dimension of the parameter space under H_A is much larger than that under H_0. Specifying subjective prior distributions in large multivariate spaces involves careful consideration of the correlations and likely relationships among the parameters. In model selection problems, the hypothesized models may be rather speculative and little prior information about the values of the parameters may be forthcoming. Thus, we may have little information for a the choice of prior and the prior may heavily influence results. In such situations, it is absolutely critical that the choice of prior distribution be reported along with the Bayes factor.

I worry about the application of Bayes factors in cosmology, just as I generally worry about their use by scientists and statistician alike. Doctor Trotta mentions the "Astronomer's Prior" ($\Omega_\kappa \sim \text{Unif}(-1,1)$) and the "Curvature Scale Prior" ($\log|\Omega_\kappa| \sim \text{Unif}(-5,0)$). In the inflationary model he notes that "little if anything is known a priori about the free parameter Ψ..." and that "non-linear transformations

... in general change ... the model comparison results." Understandably convenient prior distributions are used in the absence of well quantified substantive prior knowledge. Unfortunately, Bayes factors based on such priors lead to questionable results.

P-values. In the context of frequency based hypothesis testing, the *p-value* is often reported to quantify the degree of evidence, p-value $=\text{prob}(T > T^{\star} | \Omega_{\kappa}=0, otherparameters)$. Although they are endemic in data analysis, there is a large literature on the difficulties of interpreting p-values, especially when testing precise null hypotheses (e.g., [2]). When compared to Bayes factors and the posterior probability of H_0, p-values *vastly overstate the evidence for* H_1, even when compared to Bayesian methods that use the prior most favorable to H_1 from a large class of priors. This is because p-values are computed given data *as extreme or more extreme* than d. This is *much stronger evidence* for H_1 than d. (In some cases p-values agree with Bayesian measures computed with "as extreme or more extreme' data [4]). P-values cannot be simply recalibrated to agree with Bayesian measures because the magnitude of the discrepancy depends on the sample size, the model, and the precision of H_0. In short p-values should be avoided because they are difficult to interpret, have questionable frequency properties, and bias inference in the direction of false discovery.

12.3 The Bottom Line

There are other statistical paradigms and hybrid methods that aim to evaluate models and decide between them, e.g., posterior-predictive-p-values [6], conditional error probabilities [1], decision theory (e.g., [5, 7]), etc. Still there are no silver bullets. Most statisticians agree that model selection should be rephrased into model fitting problems whenever possible. In the case of nested models, it is often possible to fit the larger model and report interval for the parameters that are free in the larger but not the smaller model. The added value of the larger model can be assessed by examining the likely values of these parameters. This avoids the problem of model selection, but may not adequately address the scientific question. In such cases, I agree with Doctor Trotta that Bayesian methods are most promising. Despite their dependence on the choice of prior distribution, Bayes factors represent a principled probability-based assessment of the relative evidence for H_0 and H_1. Unlike p-vales, they aim to answer the right questions and like other Bayesian methods, they have no problem with nuisance parameters. Various strategies exist for mitigating their dependence on the prior distribution. For example, Berger and Delampady [2] recommends optimizing the Bayes factor over a class of priors and Berger and Pericchi [3] review methods that use a subset of the data to construct an informative prior distribution and the remainder to compute the Bayes factor. Overall, I view Bayes factors as the most promising method for model selection. Clearly care must be taken when selecting prior distributions and sensitivity analyses must be conducted. But at a fundamental level Bayes factors answer the questions of most interest to scientists.

References

1. Berger, J. O., Boukai, B., and Wang, Y. (1997). Unified frequentist and Bayesian testing of a precise hypothesis (with discussion). *Statistical Science* **12**, 133–160.
2. Berger, J. O. and Delampady, M. (1987). Testing precise hypotheses (with discusion). *Statistical Science* **2**, 317–352.
3. Berger, J. O. and Pericchi, L. R. (2001). Objective Bayesian methods for model selection: Introduction and comparision (with discussion). In *Model Selection* (Editor: P. Lahiri), 135–207. IMS, Beachwood, Ohio.
4. Berger, J. O. and Sellke, T. (1987). Testing a point null hypothesis: The irreconcilability of P-values and evidence (with discussion). *Journal of the American Statistical Association* **82**, 112–139.
5. Casella, G. and Berger, R. L. (1990). *Statistical Inference.* Wadsworth & Brooks/Cole, Pacific Grove, CA.
6. Gelman, A., Meng, X.-L., and Stern, H. (1996). Posterior predictive assessment of model fitness (with discussion). *Statistica Sinica* **6**, 733–807.
7. van Dyk, D. A. (2011). Setting Limits, Computing Intervals, and Detiction. In *Phystat 2011 Proc.* (Eds: H. Prosper and L. Lyons), in press. CERN Yellow Report.

Chapter 13
Measurement Error Models in Astronomy

Brandon C. Kelly

Abstract I discuss the effects of measurement error on regression and density estimation. I review the statistical methods that have been developed to correct for measurement error that are most popular in astronomical data analysis, discussing their advantages and disadvantages. I describe functional models for accounting for measurement error in regression, with emphasis on the methods of moments approach and the modified loss function approach. I then describe structural models for accounting for measurement error in regression and density estimation, with emphasis on maximum-likelihood and Bayesian methods. As an example of a Bayesian application, I analyze an astronomical data set subject to large measurement errors and a non-linear dependence between the response and covariate. I conclude with some directions for future research.

13.1 Introduction

Measurement error is ubiquitous in astronomy. Astronomical data consists of passive observations of objects, whereby astronomers are able to directly measure the flux of an object as a function of wavelength, its location on the sky, and the time of the observation. Because the number of photons detected from an astronomical object follows a Poisson process, this makes the measurement of a source's intensity intrinsically subject to measurement error, even if none is introduced from the detector. Therefore, the very nature of astronomical data makes measurement error unavoidable. Moreoever, quantities that are derived from an object's observed emission, either by fitting a model to the spectral energy distribution (SED) or by employing scaling relationships, are also 'measured' (derived) with error. Examples

B.C. Kelly (✉)
Department of Physics, University of California,
Santa Barbara CA, USA
e-mail: bckelly@cfa.harvard.edu

E.D. Feigelson and G.J. Babu (eds.), *Statistical Challenges in Modern Astronomy V*, Lecture Notes in Statistics 209, DOI 10.1007/978-1-4614-3520-4_13,

include mass, metallicity, and distance. Often the measurement error on the derived quantities is significant. This is unfortunate as inference on the derived quantities is often the goal of astronomical data analysis. Therefore, there has been considerable interest in how to perform statistical inference in the presence of measurement error.

Measurement error is a problem that affects, at various levels, all scientific research. Because of this, numerous methods for handling measurement errors have been developed ([6, 7, 9] are good references). In this contribution, I will present a survey of methods for handling measurement error that have been developed and used in astronomical data analysis. Because astronomical measurement errors are, in general, heteroskedastic (having different variances), I will limit my discussion to methods developed for heteroskedasticity. I will focus on situations where a deterministic relationship is not assumed between the variables, but where all variables of interest are random and are measured with error. Because of this, I will ignore situations where the measurement error is the only source of randomness in one's data. An example of this type of situation is fitting a model to an observed spectrum, where the measurement error is the only source of randomness; i.e., in the absence of measurement error a deterministic relationship is assumed between, say, flux density and wavelength. Methods for handling measurement error in this case are relatively well-established, and typically one simply minimizes the usual χ^2 statistic (e.g., [3]). However, it is worth pointing out that many complications may still exist, and more sophisticated methods may be needed, especially when dealing with low-count X-ray and γ-ray data (e.g., [24]) or when incorporating calibration unceratinties [14]. Instead, I will focus on methods for analyzing data from astronomical samples, where the variables are a random sample from an underlying distribution. Within the context of regression, this implies that intrinsic scatter (referred to as equation error in the statistics literature) exists in the relationship among the variables, and thus a deterministic relationship is not assumed between the variables even without the presence of measurement error.

Most of the techniques I will discuss focus on accounting for measurement error in regression. The goal of regression is often to understand how one variable changes with another. For example, how does the mass of a black hole change as a function of the stellar velocity dispersion of the host galaxy's bulge? Typically one simply estimates how the average value and dispersion of one variable depends on another. Measurement error statistical models are typically divided into two types: 'functional' and 'structural' models. In functional modeling, one assumes that the unknown true values of the variables are fixed, whereas in structural modeling the unknown true values of the variables have their own intrinsic distribution. As a result, in structural modeling one must parameterically model the distribution of the true values of the variables, whereas in functional modeling one does not. Density estimation is another important technique in astronomical data analysis, being the foundation for luminosity and mass function estimation. The methods I will discuss for handling measurement error in structural models are also applicable to density estimation, as in this case regression and density estimation are based on the same formalism. When discussing regression methods, I will refer to the 'dependent' variable as the response, and the 'independent variables' as the covariates.

13.2 Notation and Error Model

Throughout this work I will denote the measured response for the ith data point as y_i, and the measured covariate for the ith data point as x_i. I will denote the true values as η_i and ξ_i, respectively. If there are $p > 1$ covariates, then I will use the vectors $\mathbf{x_i}$ and $\xi_\mathbf{i}$. I will use $\mathbf{y}$ to denote the set of values of y_i for each of the n data points, $\mathbf{y} = [y_1, \ldots, y_n]$. To denote the set of x_i, I will use $\mathbf{x} = x_1, \ldots, x_n$ if there is one covariate, and the $n \times p$ matrix $X = [\mathbf{x_1}, \ldots, \mathbf{x_n}]$ if there are multiple covariates. I assume the classical additive error models throughout this review, unless otherwise specified:

$$\eta_i = f(\xi_\mathbf{i}, \theta) + \varepsilon_i \tag{13.1}$$

$$\mathbf{x_i} = \xi_\mathbf{i} + \varepsilon_{\mathbf{x},\mathbf{i}} \tag{13.2}$$

$$y_i = \eta_i + \varepsilon_{y,i}. \tag{13.3}$$

The function $f(\xi, \theta)$ describes how the mean value of η depends on ξ as a function of the parameters, θ. For example, for linear regression $f(\xi, \theta) = \alpha + \beta^T \xi$ with $\theta = (\alpha, \beta)$ denoting the slopes and intercept. The terms $\varepsilon_i, \varepsilon_{\mathbf{x},\mathbf{i}}$, and $\varepsilon_{y,i}$ are random variables denoting the intrinsic scatter in η at fixed ξ (i.e., the equation error), the measurement error in $\mathbf{x_i}$, and the measurement error in y_i, respectively. The random variables $\varepsilon_i, \varepsilon_{\mathbf{x},\mathbf{i}}$, and $\varepsilon_{y,i}$ are assumed to have zero mean and variances $Var(\varepsilon_i) = \sigma^2, Var(\varepsilon_{y,i}) = \sigma^2_{y,i}$, and $Var(\varepsilon_{\mathbf{x},\mathbf{i}}) = \Sigma_{\mathbf{x},i}$. As is typical in astronomy, the parameter σ^2 is assumed to be unknown and a free parameter in the model, while the variances in the measurement errors, $\sigma^2_{y,i}$ and $\Sigma_{\mathbf{x},i}$, are assumed known. The measurement errors are assumed to be independent of ε_i. In addition, for simplicity I also assume that the measurement errors in y_i and $\mathbf{x_i}$ are independent, unless otherwise specified. However, this is not always true, and many methods are able to handle correlated measurement errors, see the references for individual techniques for further details.

Following Gelman et al. [10], I will also typically use the notation $p(\cdot)$ to denote the probability density of the argument. For example, $p(x)$ denotes the marginal probability density of x, $p(y|x)$ denotes the conditional probability density of y given x, and $p(y,x)$ denotes the joint probability density of y and x. It should be understood that $p(\cdot)$ will not always have the same functional form, and that this must be inferred from context, i.e., it is not necessarily true that $p(x) = p(y)$ even if $x = y$. When this may be confusing, I use different symbols to denote different probability densities.

13.3 Effects of Measurement Error

Measurement error has the effect of blurring and broadening the distribution of quantities, similar to the blurring of astronomical images by a point spread function.

This makes statistical inference based on the measured values biased, and smears out any trends in the data. The distribution of the measured quantities is obtained as

$$p(y,\mathbf{x}) = \int\int p(y,\mathbf{x}|\eta,\xi)p(\eta,\xi)\, d\eta\, d\xi. \tag{13.4}$$

Under the additive error model of Sect. 13.2, (13.4) simplifies to

$$p(y,\mathbf{x}) = \int f(y-\eta)\int g(\mathbf{x}-\xi)p(\eta,\xi)\, d\xi\, d\eta, \tag{13.5}$$

where $f(\cdot)$ and $g(\cdot)$ denote the probability distributions of the measurement errors ε_y and ε_x, respectively. Equation 13.5 shows that under additive measurement error, the observed distribution of a set of quantities is the convolution of the intrinsic distribution with the measurement error distribution. Convolution has the effect of broadening distributions, which biases density estimation and masks trends.

Some of the effects of measurement error are illustrated in Fig. 13.1. Here, I simulated a sample of covariates from a bimodal distribution, and simulated the response assuming a nonlinear relationship between η and ξ. I then added large measurement error to both η and ξ. As can be seen, measurement error has blurred out many of the features in the data set, and broadened the distributions.

To further see how measurement error biases statistical inference for regression, consider the additive error model for linear regression, assuming one covariate. In addition, for simplicity assume that the measurement errors are homoskedastic (having the same variance) for both the response and covariate. If one were to ignore measurement error and proceed through the usual ordinary least-squares (OLS) analysis, then one would obtain the following estimates for the slope, variance in the intrinsic scatter, and uncertainty in the estimated slope (assume the intercept, α, is known):

$$\hat{\beta}_{OLS} = \frac{Cov(x,y)}{Var(x)} = \frac{Cov(\xi,\eta)}{Var(\xi)+\sigma_x^2} \tag{13.6}$$

$$\begin{aligned}\hat{\sigma}^2_{OLS} &= Var(y-\alpha-\hat{\beta}_{OLS}x)\\ &= (\beta^2-\hat{\beta}^2_{OLS})Var(\xi)+\hat{\beta}^2_{OLS}\sigma_x^2+\sigma_y^2+\sigma^2\end{aligned} \tag{13.7}$$

$$Var(\hat{\beta}_{ols}) = \frac{\hat{\sigma}^2_{OLS}}{Var(x)} = \frac{\hat{\sigma}^2_{OLS}}{Var(\xi)+\sigma_x^2}, \tag{13.8}$$

where β and σ^2 are the true values of the slope and variance in intrinsic scatter. From (13.6) to (13.8) we can deduce the following:

- Equation 13.6 shows that measurement error in the covariate attenuates the regression slope, biasing it toward zero. Therefore, trends between the response and the covariate will appear weaker than they really are. If the measurement error in the covariate is negligible, then there is no bias in the slope even if the measurement errors in the response are large.

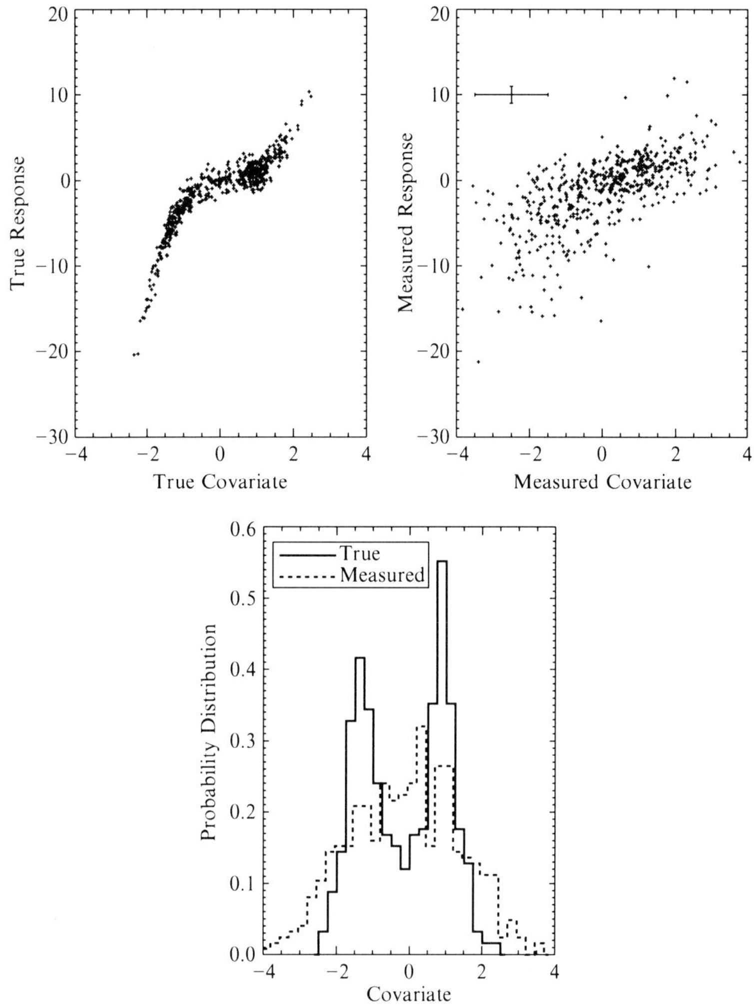

Fig. 13.1 Illustration of the effect of measurement error on regression and density estimation, using a simulated sample. The true distribution of the response and covariate (*upper left*), compared with the measured distribution (*upper right*). The error bars in the *center plot* denote the standard deviation of the Gaussian measurement errors. The measurement errors have effectively washed out any visual evidence for a tight non-linear relationship between the response and covariate. The *lower plot* shows the distribution of the true and measured values of the covariate. The measurement errors have washed out any evidence for bimodality in the distribution, and significantly broadened it

- Equation 13.7 shows that measurement error in both the response and covariate bias the estimate of σ^2 upward. Therefore, the variance in the response about the regression line will appear larger than it really is.

- Equation 13.8 show that measurement error in the covariate causes one to underestimate the error in the estimated slope. Thus, if the covariate is significantly contaminated by measurement error, then one would incorrectly conclude that the slope is precisely estimated to be ≈ 0, and therefore conclude that there is no relationship between the response and covariate!

Clearly measurement error can have a significant effect on one's data analysis, and ignoring it can lead to erroneous conclusions. Luckily, a number of statistical methods have been developed for handling measurement errors.

13.4 Functional Methods for Accounting for Measurement Error in Regression

A variety of functional models have been proposed for handling measurement errors in regression, and here I summarize the methods that are commonly used in the astronomical literature. Since heteroskedastic measurement errors are the norm in astronomy, I only discuss methods that allow the variances in the measurement error to vary among the observations. Moreover, as discussed earlier, I focus on methods that incorporate intrinsic scatter in the relationship between the response and covariate. The reader is referred to Carroll et al. [6] for a more thorough and general discussion of methods developed for handling measurement error.

13.4.1 Method of Moments Approach for Linear Regression

In linear regression the least-squares estimates of the intercept, slope, and intrinsic dispersion are obtained from the moments of the data. In the previous section I showed that the moments of the observed data are biased estimates of the moments of the intrinsic distribution when the data are measured with error. Therefore a simple method of accounting for measurement error in linear regression is to estimate the moments of the true values of the data, and then use these estimated moments to estimate the regression parameters. This is the idea behind the method of moments (MM) estimators, where the moments of the observed data are 'debiased' by removing the contribution from the measurement errors.

Akritas and Bershady [1] describe a methods of moments approach for linear regression that handles heteroskedastic measurement error in both the response and covariate, intrinsic scatter, and correlation between the response and covariate measurement error. Akritas and Bershady used their method to characterize the color-luminosity and Tully-Fisher relationships for galaxies. Their estimators, as is typical for the method of moments, assume the additive error model of Sect. 13.2 with the mean value of η depending linearly on ξ: $f(\xi, \theta) = \alpha + \beta\xi$. They do not assume a particular distribution for the measurement errors, the covariate, or the intrinsic scatter. However, their approach does assume that the variance in the measurement errors and correlation between the measurement errors are known. They call their estimator the BCES estimator, for bivariate correlated errors and intrinsic scatter.

Denote the covariance between the measurement errors in the response and covariate as $Cov(\varepsilon_{y,i}, \varepsilon_{x,i}) = \sigma_{yx,i}$. Also, denote the sample average for x as $\bar{X}$, the sample average for y as $\bar{Y}$, the sample variance for x as V_x, the sample variance for y as V_y, and the sample covariance between x and y as V_{xy}. Then, the methods of moments estimators are

$$\hat{\beta}_{MM} = \frac{V_{xy} - \bar{\sigma}_{yx}}{V_x - \bar{\sigma}_x^2} \tag{13.9}$$

$$\hat{\alpha}_{MM} = \bar{Y} - \hat{\beta}_{MM}\bar{X} \tag{13.10}$$

where $\bar{\sigma}_{yx} = \sum_{i=1}^{n} \sigma_{yx,i}/n$ and $\bar{\sigma}_x^2 = \sum_{i=1}^{n} \sigma_{x,i}^2/n$. Akritas and Bershady [1] show that the MM estimators are asymptotically unbiased, that the sampling distribution of the MM estimators is asymptotically normal, and describe how to estimate the asymptotic covariance matrix of $\hat{\alpha}_{MM}$ and $\hat{\beta}_{MM}$. Patriota and Bolfarine [18] derive the asymptotic covariance matrix of the MM estimators under the additional assumption that the measurement errors are normally distributed, creating more powerful hypothesis testing when this is true. In addition, Cheng and Riu [8] give the MM estimator for the variance in the intrinsic scatter:

$$\hat{\sigma}_{MM}^2 = V_y - \hat{\beta}_{MM}(V_{xy} - \bar{\sigma}_{yx}) - \bar{\sigma}_y^2, \tag{13.11}$$

where $\bar{\sigma}_y^2$ is the sample average of $\sigma_{y,i}^2$.

The main advantage of the MM estimators are that they do not make any assumptions about the distribution of the measurements errors, about the distribution of the covariate, nor about the distribution of the intrinsic scatter. This is attractive, is it makes the MM estimators robust. One of the disadvantages of the MM estimators is that they are not as precise as some other methods, such as structural models, when the distributions of $\varepsilon_x, \varepsilon_y, \varepsilon$, and ξ are known, or at least when they can be accurately modeled parameterically, as the MM estimators do not impose prior assumptions about the distributions. Another disadvantage is that the MM estimators tend to be highly variable when the sample size is small, and/or the measurement errors are large. This is on account of the term $V_x - \bar{\sigma}_x^2$ in the denominator of the equation for $\hat{\beta}_{MM}$. When the sample size is small, then V_x is more variable, and it is possible that $V_x \sim \bar{\sigma}_x^2$. This is also possible when measurement errors are large, as the variance in x becomes dominated by the measurement errors. When this occurs, the estimate for the slope can become very large, or change sign. Similarly, if the measurement errors in y are large, then the MM estimator for the intrinsic dispersion can become negative, which is impossible. Therefore, despite the robustness of the MM estimators, more stable estimators should be used when the sample size is small, or when the measurement errors make up a significant component to the variance in the data.

13.4.2 Modified Loss Function Approach

Modified loss function methods modify the figure of merit function (i.e., the 'loss' function), to incorporate measurement error. The weighted squared error loss function is the most common loss function used in astronomy. A weighted least squares (WLS) estimator for linear regression was proposed by Sprent [22] to minimize the following loss function for the special case of no intrinsic scatter:

$$Q_{WLS}(\alpha,\beta) = \sum_{i=1}^{n} \frac{(y_i - \alpha - \beta x_i)^2}{\sigma_{y,i}^2 + \beta^2 \sigma_{x,i}^2}. \tag{13.12}$$

The weights in (13.12) reflect the contribution of the measurement errors to the squared error. Here I have used the notation $Q_{WLS}(\alpha,\beta)$ instead of the more commonly used χ^2 to emphasize the fact that (13.12) is a loss (or figure of merit) function, and will not necessarily follow a χ^2 distribution even if the errors are Gaussian (although one can still use (13.12) regardless of the distribution of the measurement errors). Note that this implies that one cannot derive uncertainties in the parameters by looking for regions of constant $\Delta Q_{WLS}(\alpha,\beta)$. As with the method of moments estimators, the WLS estimators do not make any assumptions about the distribution of the measurement errors, covariate, or intrinsic scatter.

The loss function defined by (13.12) assumes that there is no intrinsic scatter in the relationship between the response and covariate. How then to modify (13.12) to include the intrinsic scatter? Motivated by their work on characterizing the M_{BH}–σ_* relationship, Tremaine et al. [23] suggested using the following modified WLS loss function:

$$\tilde{Q}_{WLS}(\alpha,\beta,\sigma^2) = \sum_{i=1}^{n} \frac{(y_i - \alpha - \beta x_i)^2}{\sigma^2 + \sigma_{y,i}^2 + \beta^2 \sigma_{x,i}^2}. \tag{13.13}$$

While the addition of σ^2 to the denominator of (13.13) is intuitive, as it reweights the loss function to incorporate the intrinsic scatter, the unknown value of σ^2 creates difficulties for the WLS estimator based on $\tilde{Q}_{WLS}(\alpha,\beta,\sigma^2)$. As discussed in Kelly [13], (13.13) can only be minimized with respect to α and β at fixed σ^2, as the minimum of (13.13) occurs at $\sigma^2 \to \infty$ for any value of α and β. Clearly, one cannot estimate the regression parameters by minimizing $\tilde{Q}_{WLS}(\alpha,\beta,\sigma^2)$. Instead, the most common approach (as suggested by Tremaine et al. [23]) is to initially use $\sigma^2 = 0$, and then find the values of α and β which minimize (13.12). Then, using these best-fit values for α and β, σ^2 is estimated by finding the value such that $\tilde{Q}_{WLS}(\alpha,\beta,\sigma^2)/(n-2) = 1$. Unfortunately, the properties of the WLS estimator based on this procedure, such as its bias and asymptotic distribution, are unknown. Kelly [13] performed simulations to study the behavior of the WLS estimator based on (13.13) when the data are contaminated by large measurement error, and compared with the MM estimator and a maximum-likelihood estimator (see Sect. 13.5.1). In general, the WLS estimator gave biased values for the slope, while the MM estimator for the slope was approximately unbiased except in the limit of extreme measurement error, and the maximum-likelihood estimator was approximately unbiased except in the limit of a small sample with extreme measurement error. Therefore, based on the problems associated with the WLS estimator based on (13.13), I do not recommend its use.

While the modification to the least squares loss function by (13.13) exhibits some problems, it is still possible to derive consistent estimators for the regression parameters by modifying the least squares loss function. Instead, consider the following modified loss function:

$$Q(\alpha,\beta,\sigma^2) = \frac{1}{\sigma^2}\sum_{i=1}^{n}\left[(y_i-\alpha-\beta x_i)^2-\sigma_{y,i}^2-\beta^2\sigma_{x,i}^2\right]. \tag{13.14}$$

Equation 13.14 corrects the usual least-squares loss function by subtracting off the contribution to the squared error from the measurement errors, and is therefore an estimate of the loss function that would have been obtained if there was no measurement error. Minimization of (13.14) with respect to (α,β,σ^2) results in the MM estimators given by (13.9)–(13.11) [7]. Therefore, the method of moments estimators can be understood as resulting from a corrected least squares loss function.

Thus far I have focused on linear regression. However, there are cases where a non-linear relationship may exist between the average value of the response and the covariate, and one desires to use a functional model. Patriota and Bolfarine [17] describe a corrected score method for polynomial regression under the heteroskedastic additive error model (Sect. 13.2), which they applied to an astronomical data set. The reader is referred to their work for further details.

13.5 Structural Methods for Regression and Density Estimation

Structural models for regression are those that make assumptions about the distribution of the covariate. As such, they are also applicable to density estimation. I will focus on structural models that rely on the construction of a likelihood function,[1] therefore requiring one to specify a parameteric model for the distributions of the measurement errors, intrinsic scatter, and covariates. These methods include both maximum-likelihood estimators and Bayesian methods. Likelihood-based techniques have the advantage that they are flexible and may be applied to a variety of problems, including those requiring non-linear forms for $f(\xi,\theta)$,

[1]The likelihood function is the probability of observing the data, given some parameters. It requires assuming a parameteric form for the sampling distribution of the data.

variance in intrinsic scatter that depends on the covariate, and data sets that include censoring[2] and truncation. However, they have the disadvantages that they are computationaly expensive, and that one must assume a parameteric form for all distributions involved, decreasing their robustness. That being said, it is possible to use highly flexible parameteric forms, increasing the robustness of likelihood based methods [11]. Moreover, the additional assumptions involved in the parameteric modeling typically buys one an increase in efficiency, providing smaller standard errors for the maximum-likelihood and Bayesian estimators when the parameteric statistical model is a good description of the data.

13.5.1 Constructing the Likelihood Function

The basic idea behind likelihood-based methods is to treat the measurement errors as a missing data problem. Little and Rubin [15] describe methods for handling missing data, while Gelman et al. [10] describe Bayesian approaches to the missing data problem. First, one formulates the likelihood function for the complete data, i.e., the likelihood function for both the measured and true values of the data. In general, for regression we have the following hierarchical model:

$$\xi_i \sim p(\xi|\psi) \tag{13.15}$$

$$\eta_i|\xi_i \sim p(\eta|\xi,\theta) \tag{13.16}$$

$$y_i,\mathbf{x}_i|\eta_i,\xi_i \sim p(y,\mathbf{x}|\eta,\xi). \tag{13.17}$$

The notation $z \sim p(z)$ means that the random variable z is drawn from the probability distribution $p(z)$. The distributions $p(\xi|\psi), p(\eta|\xi,\theta)$, and $p(y,\mathbf{x}|\eta,\xi)$ are the distributions for the covariates, the response given the covariate, and the measured data, respectively. The distribution for the covariate is parameterized by ψ, while the distribution for η at a given ξ is parameterized by θ; note that here I have absorbed the parameter describing the variance in the intrinsic scatter into θ, whereas in the previous sections I have kept σ^2 seperate from θ. For simplicity, I assume that the distribution of the measurement errors is considered known, as is typically the case in astronomy. If additional parameters are needed to describe the distribution of the measured data, e.g., if the variance in the measurement errors is unknown, then these should be included in (13.17). Most of the interest in regression lies in inference on θ, which describes how the response depends on the covariates. If, instead of regression we are interested in density estimation, then there is no response variable and only (13.15) and (13.17) are used.

[2]Data are said to be censored when only an upper or lower limit is available.

Under the statistical model given by (13.15)–(13.17), the complete data likelihood function for the ith data point is

$$p(y_i, \mathbf{x}_i, \eta_i, \xi_i | \theta, \psi) = p(y_i, \mathbf{x}_i | \eta_i, \xi_i) p(\eta_i | \xi_i, \theta) p(\xi_i | \psi). \tag{13.18}$$

In order to calculate the observed data likelihood function for the ith data point, we integrate out the missing (and thus unknown) data from the complete data likelihood function:

$$p(y_i, \mathbf{x}_i | \theta, \psi) = \int\int p(y_i, \mathbf{x}_i | \eta, \xi) p(\eta | \xi, \theta) p(\xi | \psi)\, d\eta\, d\xi \tag{13.19}$$

When the data points are statistically independent, as is almost always the case, the observed data likelihood function for the entire data set is the product of (13.19) over the $i = 1, \ldots, n$ data points. Further details on this procedure can be found in Carroll et al. [6]. Once one has chosen parameteric forms for the distributions involved in (13.15)–(13.17), one can use (13.19) to compute the maximum-likelihood estimate for the parameters (θ, ψ) and use the likelihood ratio to estimate confidence regions for the parameters. That's it! Of course, in practice this is not so simple, as computing the integrations involved in (13.19) and performing the optimization of (13.19) can be numerically difficult. The Expectation-Maximization (EM) algorithm is often helpful, and additional numerical techniques are described in, for example, Press et al. [19] and Robert and Casella [20].

As an example of the likelihood approach, consider the following simple model. Assume the measurement errors to be normally distribution with zero mean and known variances, as described in Sect. 13.2. For the regression model, assume that the response (η) at fixed covariate (ξ) is normally distributed with mean $f(\xi, \theta) = \alpha + \beta^T \xi$ and variance σ^2; this is the usual linear regression model with Gaussian intrinsic scatter. The distribution of the covariates is assumed to be a p-dimensional multivariate normal density with mean μ and covariance matrix T. Under this model, the parameters are $\theta = (\alpha, \beta, \sigma^2)$ and $\psi = (\mu, T)$. For this model, the integrals in (13.19) can be done analytically. Denoting $\mathbf{z}_i = (y_i, \mathbf{x}_i)$, the measured data likelihood is

$$p(\mathbf{y}, X | \theta, \psi) = \prod_{i=1}^{n} \frac{1}{(2\pi)^{(p+1)/2} |V_i|^{1/2}} \exp\left\{ -\frac{1}{2} (\mathbf{z}_i - \zeta)^T V_i^{-1} (\mathbf{z}_i - \zeta) \right\} \tag{13.20}$$

$$\zeta = (\alpha + \beta^T \mu, \mu) \tag{13.21}$$

$$V_i = \begin{pmatrix} \beta^T T \beta + \sigma^2 + \sigma_{y,i}^2 & \beta^T T \\ T\beta & T + \Sigma_{x,i} \end{pmatrix}. \tag{13.22}$$

The Gaussian likelihood model described here is commonly used, but it is not robust and can be subject to considerable systematic error due to model mispecification (e.g., [11]). Motivated by this, several authors have proposed using a mixture of Gaussian functions as a model for the distribution of the covariates (e.g., [5,13,21]). Bovy et al. [4] describe a mixture of Gaussian functions model for density estimation when some of the measurements are missing at random. Kelly et al. (2008) describe a mixture of Gaussian functions model for density estimation of a truncated sample, with emphasis on luminosity function estimation. The mixture of Gaussian functions model inherits much of the mathematical simplicity of the Gaussian model, enabling an analytic calculation of the observed data likelihood, while still being flexible enough to model most realistic astrophysical distributions. In addition, Andreon [2] describe a model for incorporating contamination from a background distribution, and model the distribution of the covariates as a mixture of Schechter functions.[3]

13.5.2 Bayesian Methods and an Example

Bayesian methods build on the likelihood methods described in Sect. 13.5.1 and compute the probability distribution of the parameters, given the observed data; this is called the 'posterior' distribution. This is done by first assuming a 'prior' distribution on the parameters, $p(\theta, \psi)$, where the prior distribution quantifies our information on the parameters θ and ψ before we take any of the data. The posterior distribution is then related to the prior and the likelihood by

$$p(\theta, \psi | \mathbf{y}, X) = p(\theta, \psi) p(\mathbf{y}, X | \theta, \psi). \tag{13.23}$$

For example, for the Gaussian model described by (13.20)–(13.22), and assuming a uniform prior on the parameters ($p(\alpha, \beta, \sigma^2, \mu, T) \propto 1$[4]), the posterior distribution for $(\alpha, \beta, \sigma^2, \mu, T)$ is proportional to (13.20) as a function of these parameters. Bayesian methods differ from the frequentist likelihood methods, such as maximum-likelihood, in that the inclusion of the prior distribution enables one to calculate the probability of the parameters, given the observed data. This implies that, in theory, the posterior distribution is exact, and therefore uncertainties on the parameters are reliable and easy to interpret regardless of the sample size and complexity of the statistical model. In contrast, the maximum-likelihood methods compute a point estimate of the parameters, and then use various methods (e.g., the likelihood ratio or bootstrap) to estimate the sampling distribution of the parameters, from which confidence regions are derived. The maximum-likelihood methods are useful, but it can become difficult to estimate the sampling distribution when the sample size is small, or for highly complex and difficult models.

---

[3]The Schechter function is an unnormalized Gamma distribution. It is commonly used in astronomy as a model for the number density of galaxies in the universe as a function of their luminosity.

[4]Technically this is uniform subject to the conditions that $\sigma^2 > 0$ and $|T| > 0$.

Bayesian methods have become increasingly popular in astronomy, as well as in other scientific disciplines. The primary driver of this increase in popularity has been the advancements in statistical computing that have enabled Bayesian inference, namely the use of Markov Chain Monte Carlo (MCMC) methods. Details of MCMC methods may be found in Gelman et al. [10] and Liu [16], and for an example of an MCMC algorithm under linear regression and heteroskedastic measurement errors, see Kelly [13]. One of the primary advantages of MCMC methods is that they are modular, and we can divide the computational problem up into smaller computational problems that are easier to solve. Because the true values of the data are not known, they are treated as additional parameters, and thus can also be updated via MCMC. We can also incorporate upper and lower limits in a straightforward manner through this approach by treating their true values as missing data [13], although the definition of upper limit in astronomy is not always straightforward [12]. These properties of MCMC samplers are a significant advantage of the Bayesian approach, as we avoid the integration over the true values of the data required in (13.19) for the maximum-likelihood approach, and we obtain improved estimates for the true values of the data. In fact, often it is easier to program a MCMC sampler and perform Bayesian inference than it is to do the optimization and numerical integration required for maximum-likelihood.

As an illustration of the Bayesian approach, I consider a data set from Constantin et al. (2011, in prep) comparing the X-ray photon index, Γ_X, with the luminosity relative to the Eddington limit (i.e., the Eddington ratio, L/L_{Edd}) for a sample of Active Galactic Nuclei (AGN).[5] The measured data are shown in Fig. 13.2a. The X-ray photon index provides a measure of how much energy is being released through soft X-rays as opposed to hard X-rays, and the Eddington Luminosity is the luminosity at which outward radiation and inward gravitational pressure balance for a spherical geometry. This data set provides a good illustration of the power of the Bayesian approach, as the average value of the response exhibits a non-linear and non-monotonic dependence on the covariate, and the measurement errors are very large in both the response and covariate. The values of the Eddington ratio (i.e., the covariate) where the X-ray photon index (i.e., the response) changes its dependence on L/L_{Edd} are of particular interest, as models of black hole accretion flows suggest that the accretion flow geometry changes at certain critical values of the Eddington ratio. Because of this, and the non-linear appearance in the data, I have chosen to model the data using a segmented line with two knots, where the slope of the line changes at the knots. I modeled the intrinsic distribution of $\log L_X/L_{Edd}$ as a mixture of three Gaussian distributions. To make the model robust against outliers, I assume that the both measurement errors and the intrinsic scatter follow a Student's t-distribution with eight and four degrees of freedom, respectively. I used the MCMC algorithms described in Chap. 9 of Carroll et al. [6] and Kelly [13] as the basis for my MCMC sampler under this model, and include

[5] AGN are believed to be supermassive black holes that are accreting gas and are located in the center of a galaxy.

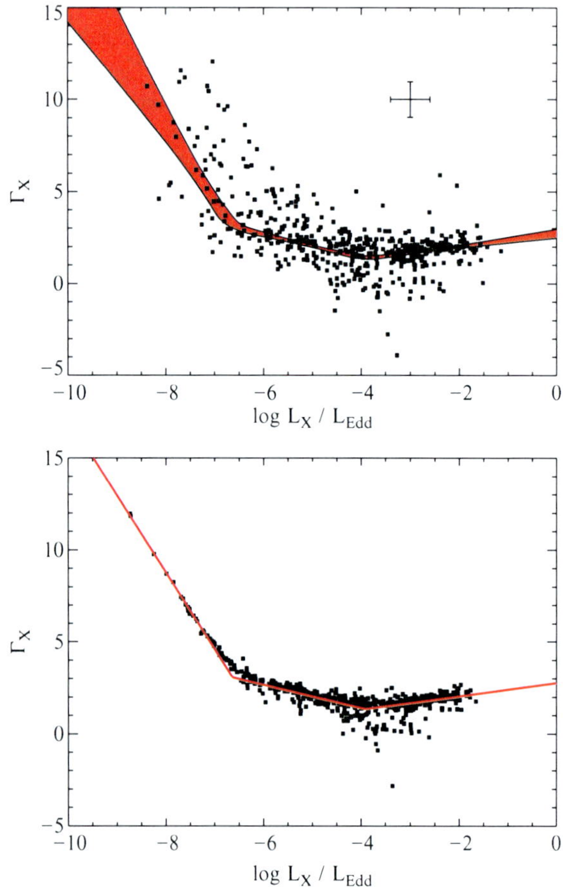

Fig. 13.2 The measured values of Γ_X and $\log L_X/L_{Edd}$ compared with the region containing 68% of the posterior probability for the mean value of Γ_X at fixed L_X/L_{Edd} (*left*). The data point with error bars is not real and only used to illustrate the typical size of the error bars. Also shown are the posterior mean values for the true values of Γ_X and $\log L_X/L_{Edd}$, compared with the best-fitting *segmented line* (*right*). A non-linear trend is apparent in both the *segmented line* model and in the estimated distribution of Γ_X and $\log L_X/L_{Edd}$ using the *segmented line* as a prior

an ancillarity-sufficiency interweaving strategy for increased efficiency [26]. This MCMC algorithm produces both random draws of the parameters for the segmented line model from their posterior distribution, but also random draws of the true values of the Eddington ratio and photon index from their posterior distribution.

The region containing 68% of the posterior probability on the mean value of Γ_X as a function of L_X/L_{Edd} is also shown in Fig. 13.2a. The location of the knots are estimated to be $\log L_X/L_{Edd} = -6.65 \pm 0.25$ and -3.91 ± 0.21, respectively. The segmented line model of Γ_X at fixed L_X/L_{Edd} is preferred over a simple line model, illustrating the complex dependence of Γ_X on L_X/L_{Edd}. In Fig. 13.2b I show the posterior mean values of Γ_X and $\log L_X/L_{Edd}$, as well as the segmented line computed from the posterior mean for its parameters. The posterior mean estimates for the true (i.e., not measured) values of Γ_X and $\log L_X/L_{Edd}$ represent a more model-independent estimate of the dependence of the photon index on L_X/L_{Edd}. This represents a real advantage of the Bayesian approach, as not only are we able to estimate the probability distribution of the parameters of interest, but we can also estimate the probability distribution of the true values of the data as well, conditional on our assumed statistical model, the measured values of the data, and the amplitude of the measurement errors. The non-linear trend is also apparent from the values of Γ_X and L_X/L_{Edd} estimated from the Bayesian method. The knot at $L_X/L_{Edd} \sim 2 \times 10^{-7}$ may represent the increasing prevalence of additional astrophysical components to the X-ray spectrum as the AGN becomes faintier, such as hot gas not associated with the AGN, while the knot at $L_X/L_{Edd} \sim 10^{-4}$ may represent a change in the accretion flow geometry. Figure 13.2b suggest that the scatter in Γ_X at fixed L_X/L_{Edd} increases near the knot at $L_X/L_{Edd} \sim 10^{-4}$, which may be indicative of instabilities when the accretion flow changes geometry, or of uncorrected intrinsic absorption. Further analysis of this data set will be discussed in Constantin et al. (2011, in prep).

13.6 Outstanding Issues in Measurement Error Models for Astronomical Data: Directions for Future Research

I will conclude by listing a couple of unsolved problems in dealing with measurement errors in astronomical data analysis, which I hope will lead to further research in this area.

- **Data subject to large, non-Gaussian measurement errors.** Non-gaussian errors are common in astronomical data, especially when one is analyzing a set of derived quantities. Often, the most physically-interesting quantities are those derived by fitting an astrophysical model to the measured flux values at various wavelengths. Often the unertainties in these derived quantities are large, skewed, or exhibit multiple modes. There is currently no well-established method for handling the measurement errors in this case, although Bayesian hierarchical models such as that proposed by van Dyk et al. [25] hold promise.
- **Handling measurement errors in massive astronomical data sets.** Current and planned astronomical surveys will provide an explosion of data, allowing one to construct data sets with millions to billions of objects, each with multiple quantities measured. Many powerful methods developed for data mining will be applied to these data, potentially providing a powerful route to knowledge discovery. Unfortunately, all of the quantities obtained from these data sets will be measured with error, and most methods developed for data mining of massive data sets do not incorporate measurement error. This is especially a problem when dealing with derived quantities, which will likely require a more careful statistical analysis on account of their sometimes highly irregular error distributions. Currently, algorithms, such as MCMC, that allow one to perform reliable statistical inference on complicated statistical models do not scale well to massive data sets. If we want to perform inference on massive data sets subject to measurement error using more complicated and realistic statistical models, we will need advances on the computational side.

Acknowledgements I would like to thank Anca Constantin for sharing her data set with me before publication, and Aneta Siemiginowska, Xiaohui Fan, and Tommaso Treu for helpful comments on an earlier version of this manuscript. I acknowledges support by NASA through Hubble Fellowship grant #HF-51243.01 awarded by the Space Telescope Science Institute, which is operated by the Association of Universities for Research in Astronomy, Inc., for NASA, under contract NAS 5-26555.

References

1. Akritas, M. G., & Bershady, M. A. 1996, T. Astrophys. J., **470**, 706
2. Andreon, S. 2006, Monthly Notic. of the Royal Astron. Soc., **369**, 969
3. Bevington, P. R., & Robinson, D. K., *Data Reduction and Error Analysis for the Physical Sciences*, 3rd edn. (McGraw-Hill, New York, 2003)
4. Bovy, J., Hogg, D. W., & Roweis, S. T. 2009, arXiv:0905.2979
5. Carroll, R. J., Roeder, K., & Wasserman, L., 1999, Biometrics, **55**, 44
6. Carroll, R. J., Ruppert, D., Stefanski, L. A., Crainiceanu, C. M., *Measurement Error in Nonlinear Models: A Modern Perspective*, 2nd edn. (Chapman & Hall/CRC, Boca Raton, 2006)
7. Cheng, C-L., & Van Ness, J. W., *Statistical Regression with Measurement Error* (Arnold, London, 1999)
8. Cheng, C-L., & Riu, J. 2006, Technometrics, **48**, 511
9. Fuller, W. A., *Measurement Error Models* (John Wiley & Sons, New York, 1987)
10. Gelman, A., Carlin, J. B., Stern, H. S., & Rubin, D. B., *Bayesian Data Analysis*, 2nd edn. (Chapman & Hall/CRC, Boca Raton, 2004)
11. Huang, X., Stefanski, L. A., & Davidian, M. 2006, Biometrika, **93**, 53
12. Kashyap, V. L., van Dyk, D. A., Connors, A., Freeman, P. E., Siemiginowska, A., Xu, J., & Zezas, A. 2010, T. Astrophys. J., **719**, 900
13. Kelly, B.C., Fan, X. & Vestergaard, M. 2008, Astrophys. J., **682**, 874
14. Lee, H., et al. 2011, T. Astrophys. J., **731**, 126
15. Little, R. J. A., & Rubin, D. B. *Statistical Analysis with Missing Data*, 2nd ed. (John Wiley & Sons, Hoboken, 2002)
16. Liu, J.S., *Monte Carlo Strategies in Scientific Computing*, (Springer, New York, 2004)
17. Patriota, A. G., & Bolfarine, H. 2008, T. Indian J. of Stat., **70**, 267
18. Patriota, A. G., Bolfarine, H., & de Castro, M. 2009, Statist. Method., **6**, 408
19. Press, W. H., Teukolsky, S. A., Vetterling, W. T., & Flannery, B. P., *Numerical Recipes: The Art of Scientific Computing*, 3rd edn. (Cambridge Unv. Press, New York, 2007)
20. Robert, C. P., & Casella, G., *Monte Carlo Statistical Methods*, 2nd edn. (Springer, New York, 2004)
21. Roy, S., Banerjee, T., 2006, Ann. Instit. Statist. Math., **58**, 153
22. Sprent, P. 1966, J. Royal Stat. Soc. Ser. B, **28**, 278
23. Tremaine, S., et al. 2002, T. Astrophys. J., **574**, 740
24. van Dyk, D. A., Connors, A., Kashyap, V. L., & Siemiginowska, A. 2001, T. Astrophys. J., **548**, 224
25. Van Dyk, D. A., DeGennaro, S., Stein, N., Jefferys, W. H., & von Hipple, T. 2009, T. Ann. of App. Stat., **3**, 117
26. Yu, Y., & Meng, X-L. 2011, to appear, J. of Comput. & Graph. Stat.,

Chapter 14
Commentary: "Measurement Error Models in Astronomy" by Brandon C. Kelly

David Ruppert

Abstract Bayesian analysis offers a general approach to measurement error that has many advantages—it focuses attention on careful modeling, is widely applicable, and provides efficient estimators. Bayesian analysis is relatively easy using `WinBUGS` software. We discuss here the paper by Brandon Kelly, and present an example of fitting a quadratic regression model with `WinBUGS` called from `R`, with the `WinBUGS` and `R` code provided.

14.1 Introduction

Dr. Kelly has written an excellent introduction to measurement error, and I have no disagreements with anything in his paper. In these comments, I will expand upon what I believe are some key points.

There are many special-purpose methods for handling measurement error for particular sets of models. Some of these methods are suitable only for linear regression models, but of course many astrophysical models are nonlinear. Other methods such as regression calibration and SIMEX [2] are widely applicable but use approximations which, though often valid, are not guaranteed to produce accurate inference. SIMEX in particular can be inefficient for many models.

An astrostatistician who is not widely read in the measurement error literature would benefit from a single approach to measurement errors that is widely applicable, is not unduly complicated, needs no approximations, and is efficient. Such an approach exists: Bayesian analysis. Dr. Kelly's has one section on Bayesian methods, and the intent of these comments is to feature them more prominently.

D. Ruppert (✉)
School of Operations Research and Information Engineering, Department of Statistical Science, Cornell University, 1170 Comstock Hall, Ithaca, NY 14853, USA
e-mail: dr24@cornell.edu

E.D. Feigelson and G.J. Babu (eds.), *Statistical Challenges in Modern Astronomy V*, Lecture Notes in Statistics 209, DOI 10.1007/978-1-4614-3520-4_14,

14.2 Advantages of Bayesian Modeling

There are many advantages to a Bayesian approach to measurement errors or, in fact, to any statistical problem.

A Bayesian analysis focuses attention on careful modeling. Section 14.4 discusses an example of a non-Bayesian estimator, orthogonal regression, that is easily misapplied because it is simple to use, but practitioners do not always understand the restrictive conditions under which it is valid. This type of misapplication is less likely if one takes a Bayesian approach where one can focus on the model, since estimation is straightforward once a satisfactory model is found—one generates a Monte Carlo sample from the posterior, say by MCMC, and then computes the posterior mean or posterior quantiles. In contrast, a non-Bayesian approach requires one to develop an estimator which may then require a careful theoretical or Monte Carlo study to make sure that it is consistent and reasonably efficient. This concentration on estimation draws attention away from modeling.

In Bayesian measurement error modeling, the unknown true covariate values are just another set of unknowns and are treated in the same way as the parameters. To a Bayesian, anything unknown is random, one conditions on whatever is known, and then finds the conditional distribution of whatever is unknown. If MCMC is used, this means that the unknown true values of mismeasured covariates are multiply imputed. The analysis is conditional on the mismeasured values.

There are some strong theoretical reasons for using Bayesian methods. Under fairly general conditions, Bayesian estimators are competitive with the best frequentist estimators even if one takes a frequentist perspective. For example, Bayesian estimators are asymptotically efficient and they are optimal in a decision theoretic framework. In particular, under weak assumptions, all admissible estimators are Bayesian. This means that to avoid using a Bayesian estimator, one must use an inadmissible estimator, that is, one that is dominated by some other estimator.

A newcomer to Bayesian analysis may be daunted by the need for priors. However, it is usually easy to specify "non-informative" priors to cover situations where one has little prior information. In other cases, strong prior information does exist. For example, in astronomy it is often assumed that measurement error variances are known. However, it may be that they are only known up to a small amount of uncertainty. In such situations, the use of informative priors is natural and will account for the uncertainty about the variances. In contrast, a non-Bayesian analysis that assumes that the variances are known exactly will underestimate uncertainty.

A Bayesian analysis is applicable to virtually any parametric statistical problem and to many nonparametric problems. If one is confronted with a challenging astrostatistical problem, there may be no known frequentist estimator. The only generally applicable frequentist technique is maximum likelihood estimation. However, with a measurement error problem, computation of the likelihood can be difficult because the unknown covariate values must be integrated out of the likelihood. A Bayesian can perform this integration as part of a MCMC computation. Often the MCMC computations can be easily done using the `BUGS` software, e.g., with `WinBUGS`.

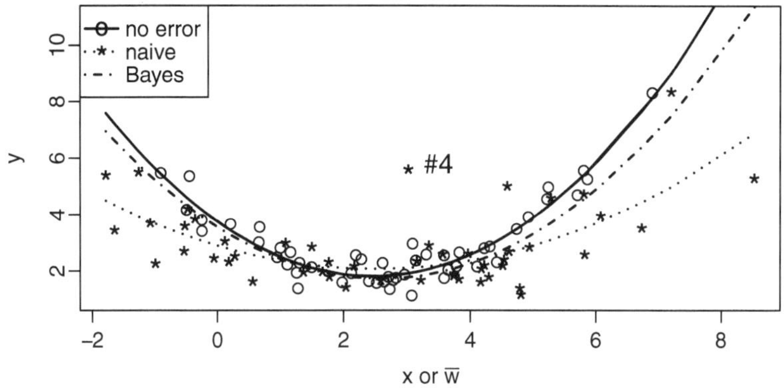

Fig. 14.1 A simulated sample from a quadratic regression model with measurement error. The *circles* are a scatterplot of the data without measurement error, that is, of (X_i, Y_i), and the *solid line* is a quadratic polynomial least-squares fit to these data. The *asterisks* are a scatterplot of observed data, $(\overline{W}_i, Y_i)$, and the *dotted line* is a quadratic least-squares fit to those data. The *dashed*-and-*dotted line* is the Bayes estimate using the observed data and is close to the least-squares fit using the correctly measured data. The fourth data point is of particular interest (see text) so it is marked

14.2.1 An Example: Quadratic Regression

The purpose of this example is to illustrate the components of a measurement error model and to show how such a model can be fit using the `WinBUGS` software and the `R2WinBUGS` package in `R`.

A measurement error model has three components: the regression model, the measurement model, and the model for the distribution of the true covariate values. The regression model specifies the conditional distribution of the response Y given the covariates X. In this example, the regression model is $Y_i = \alpha + \beta X_i + \gamma X_i^2 + \varepsilon_i$ where X_i is scalar and $\varepsilon_i \overset{\text{iid}}{\sim} N(0, \sigma_\varepsilon^2)$. The measurement model is $W_{ij} = X_i + U_{ij}$, $j = 1, 2$, where $U_{ij} \overset{\text{iid}}{\sim} N(0, \sigma_u^2)$ independently of $X_1 g, \ldots, X_n$, and σ_U^2 is unknown. Define $\overline{W}_i = (W_{i1} + W_{i2})/2$. The model for the distribution of the true covariate values is $X_i \overset{\text{iid}}{\sim} N(\mu_x, \sigma_x^2)$. Here "$\overset{\text{iid}}{\sim}$" means "independent and identically distributed as".

A random sample of size 50 was generated from this model and plotted in Fig. 14.1. With simulated data we can, of course, compare the estimates with and without measurement error. The least-squares fits using the data without error (solid) and the mismeasured data (dotted) are quite different. In particular, the estimate of γ is much smaller with the mismeasured data because of bias.

An interesting feature here is that $\overline{W}_4$ is approximately 3, but the plots suggest that X_4 is near either -1 or 6. In fact, $X_4 = 5.8$. The Bayes estimator is able to use the values of both $\overline{W}_4$ and Y_4 as well as the quadratic shape of the regression function to impute X_4. In contrast, the commonly used frequentist method of regression calibration [2] uses only $\overline{W}_4$ to impute X_4 and will be less accurate than the Bayes estimator. It is the ability of the Bayes estimator to use all information in the data and in the likelihood that is the basis for its efficiency.

The BUGS program for this model is:

```
1.  model{
2.  for(i in 1:N){
3.  w1[i] ~ dnorm(x[i],tauw)
4.  w2[i] ~ dnorm(x[i],tauw)
5.  x[i] ~ dnorm(mux,taux)
6.  y[i] ~ dnorm(muy[i],taue)
7.  muy[i] <- alpha + beta*x[i]+ gamma*x[i]*x[i]
8.  }
9.  mux ~ dnorm(0.0,1.0E-6)
10. alpha ~ dnorm(0.0,1.0E-6)
11. beta ~ dnorm(0.0,1.0E-6)
12. gamma ~ dnorm(0.0,1.0E-6)
13. tauw ~ dgamma(0.1,0.01)
14. taux ~ dgamma(0.1,0.01)
15. taue ~ dgamma(0.1,0.01)
16. }
```

Lines 3 and 4 specify the measurement error model, line 5 the model for the distribution of the true covariate values, and lines 6 and 7 the regression model. Lines 9–15 specify the prior. The symbol "~" means "is distributed as." `dnorm` is the normal distribution and its arguments are its mean and precision (reciprocal of the variance). Similarly, `dgamma` is the gamma distribution with arguments the shape and scale parameters. We see that μ, α, and β are given normal priors with mean 0 and variance 10^6. The variances σ_X^2, σ_U^2, and σ_ε^2 are given inverse-gamma priors with shape parameter 0.1 and scale parameter 0.01. These priors are intended to be noninformative, that is, they should have little influence on the posterior distribution.

One of the advantages of using `WinBUGS` is that it is easy to vary the model. In this example, the measurement error variances are equal but unknown. As Dr. Kelly mentions, in astronomy the measurement error variances are typically unequal but often treated as known. In that case, `tauw` would not be a scalar parameter as here but would instead be a data vector of known precisions.

The Bayesian analysis was done in `R` with the following program:

```
library(R2WinBUGS) # to call bugs
library(coda) # for output analysis
dat = read.csv("eiv.csv",header=TRUE)
attach(dat)
wbar = (w1+w2)/2
N = length(w1)
data=list("N","w1","w2","y") # data list for bugs
inits=function(){list(mux=2,tauw=1,taux=1,x=wbar,
     alpha=0,beta=0,gamma=1,taue=1)}
eiv.sim = bugs(data,inits,model.file="eiv.bug",
```

```
11. parameters=c("alpha","beta","gamma","x[4]"),
        n.chains = 5,
12. n.iter=35000,n.burnin=5000,n.thin=100,
        bugs.seed="8877",
13. bugs.directory="c:/Program Files/WinBUGS14/",
        codaPkg=T)
14. mcmcout = read.bugs(eiv.sim) # read.bugs()
        is in coda
15. options(digits=3)
16. summary(mcmcout) # gives summary of MCMC output
17. effectiveSize(mcmcout) # effectivesize() is
        in coda
18. postscript("traceDensity.ps",width=7,height=5)
19. par(mfrow=c(2,4))
20. plot(mcmcout,auto.layout=F) # trace and
        density plots
21. graphics.off()
```

Line 1 loads the `R2WinBUGS` package which contains the `bugs` command calling WinBUGS in lines 10–13. Line 2 loads the `coda` package used to analyze the MCMC output. Lines 3–6 prepare the data and line 7 creates the data list that is passed to WinBUGS on line 10. Line 11 creates a function that generates initial values. For simplicity, deterministic starting values are used but random starting values are an option and are recommended in practice. Lines 14–21 produce output (not shown) and the plots in Fig. 14.2. We see in line 12 that there are five chains, each of 35,000 iterations with the first 5,000 discarded as burn-in. The chains are thinned so that only every 100th iteration is saved, and therefore each saved chain has 300 iterations.

Figure 14.2 contains trace plots and kernel estimates of the marginal posterior densities of the deviance and of X_4. The trace plots are plots of the MCMC samples of the parameters versus iteration number, one curve for each chain. One can see that X_4 has a bimodal posterior distribution with modes at approximately -1 and 6, which agrees with what is seen in Fig. 14.1. The chains move between the two modes but only occasionally.

14.3 Structural Models

I prefer structural to functional modeling for a number of reasons. First, a Bayesian approach requires a structural model, since anything unknown is modeled as random. Practitioners rightly worry about model misspecification when a simple structural assumption is used, for example, that the true covariate values are normally distributed. However, the true covariate values will always have an empirical distribution, and the use of a structural model should be satisfactory provided the model includes distributions close to the empirical distribution. This is insured if one of the flexible structural models mentioned by Dr. Kelly is used.

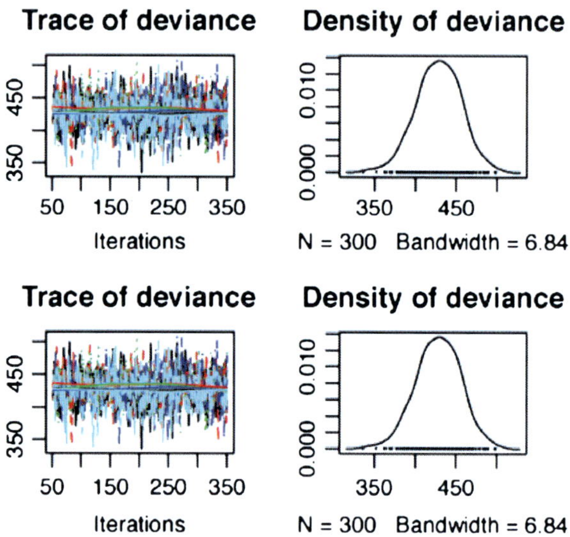

Fig. 14.2 Trace plots and kernel density estimates for the deviance (*left*) and X_4 (*right*)

14.4 The Need for Careful Modeling

If one is focused on estimators instead of modeling, then there is the danger of using an estimator whose underlying assumptions do not hold. Doing this can lead to serious biases. For example, Carroll and Ruppert [1] discuss how easy it is to misapply orthogonal regression (OR). The OR model is $y_{\text{true}} = \beta_0 + \beta_1 X$, $Y = y_{\text{true}} + \varepsilon$, and $W = X + U$. Here ε is the measurement error in Y, and U is the measurement error in X. It is assumed that there is no equation error. It is assumed further that we know, or at least have an estimate of, $\eta = \text{var}(Y|X)/\text{var}(W|X) = \sigma_\varepsilon^2/\sigma_U^2$. This assumption is reasonable if one knows the precision of the measurements so that both σ_ε^2 and σ_U^2 are known or if, instead, one knows that the measurements have equal precisions so that $\eta = 1$. The assumption of no equation error is crucial. Unfortunately, equation error is common and this create a trap for the unwary.

The OR estimator can be viewed as a functional estimator that treats $X_1,\dots,X_n$ as unknown parameters, so that $\beta_0,\beta_1,X_1,\dots,X_n$ are estimated by minimizing $\sum_{i=1}^n \{\eta^{-1}(Y_i - \beta_0 - \beta_1 X_i)^2 + (W_i - X_i)^2\}$ over $(\beta_0,\beta_1,X_1,\dots,X_n)$.

The danger is that it is easy to misapply OR in the presence of equation error. This leads to overcorrection if one uses $\eta = \sigma_\varepsilon^2/\sigma_U^2$ as if there were no equation error. Instead one should use

$$\eta_{\text{EE}} := \frac{\text{var}(Y|X)}{\text{var}(W|X)} = \frac{\sigma_Q^2 + \sigma_\varepsilon^2}{\sigma_U^2} \tag{14.1}$$

where σ_Q^2 is the equation error variance. Of course, this requires an estimate of σ_Q^2. There are techniques for estimating σ_Q^2; see Carroll and Ruppert [1] for references.

However, if one plugs an estimate of σ_Q^2 into (14.1) and treats it as known, then the uncertainty in the estimates of the other parameters will be underestimated. A Bayesian approach is recommended instead.

References

1. Carroll, R. J., and Ruppert, D.: The use and misuse of orthogonal regression estimation in linear errors-in-variables models, The American Statistician, **50**, 1–6 (1996)
2. Carroll, R.J., Ruppert, D., Stefanski, L., and Crainiceanu, C.M.: *Measurement Error in Nonlinear Models: A Modern Perspective*, 2nd edn. (Chapman and Hall, New York, 2003)

Chapter 15
Asteroseismology: Bayesian Analysis of Solar-Like Oscillators

Othman Benomar

Abstract Asteroseismology is gaining momentum nowadays. The unprecedent data quality obtained by the CoRoT and Kepler space-borne instruments allows us to probe the stellar physics with a precision never achieved before. Thanks to that, new discoveries have been raising new challenges requiring the use of robust new statistical approaches. F stars (i.e. hot solar-like stars) are among the most complicated solar-like oscillators to analyze. In this paper, we summarize the difficulties the asteroseismic community has faced with F stars and how Bayesian approaches help to solve the issues encountered.

15.1 Introduction

Space-borne instruments such MOST [12], CoRoT [3] and Kepler [7] allow us to observe stellar oscillations in stars all over the HR diagram. The study of these oscillations enable us to estimate stellar characteristics such the age, mass and the radius with a typical precision of 50%, 10% and 2% respectively. Historically, star's pulsations have been noticed since at least the seventeenth century on Mira and their study permitted significant discoveries. For example, measurements of solar pulsations in the 1980s showed that the core of the Sun was hotter than suggested by the neutrino flux. Many explanations were proposed including a revision of stellar physics models by introducing new Weakly Interacting Massive Particles (WIMPs), or the neutrino oscillations. Over time, as the amount of helioseismic data increased,

O. Benomar (✉)
Sydney Institute for Astronomy (SIfA), School of Physics, University of Sydney, Sydney, NSW 2006, Australia
e-mail: benomar@physics.usyd.edu.au

E.D. Feigelson and G.J. Babu (eds.), *Statistical Challenges in Modern Astronomy V*, Lecture Notes in Statistics 209, DOI 10.1007/978-1-4614-3520-4_15,

the neutrino problem stood unsolved. Finally in 1998, the Super-Kamiokande experiment was able to measure the flavor change between neutrinos and confirm the neutrino oscillation hypothesis, and ruling out the other possibilities.

Solar-like oscillators are among the most challenging to study because of the low amplitude of their oscillations [9]. Moreover, like it has been the case for the Sun decades ago, their study could unlock key concepts related, for example, to their evolution and their internal structure.

Pulsation modes in solar-like oscillators are known to be excited by the convection into their upper convective layer and correspond to pressure modes (p-modes). In the Sun these oscillations can be observed at frequencies in the range 2–4 mHz [8] in the power spectrum. In fact, a fine spectral structure of Lorentzian peaks is observed and identified as the eigenmodes of the Sun. Each eigenmode is characterized by three integer (n, ℓ, m) into a base of spherical harmonics, with n being the number of nodes along the radius of the star (radial order), ℓ the number nodal line at the surface (degree of the mode) and m, the number of nodal lines intersecting the equator (azimuthal order). Pulsation frequencies vary as a function of the age, mass and effective temperature of the star and depends on its internal structure.

15.2 The Difficult Case of F Stars

Continuously observed during 60 days, HD49933 was the first star observed by CoRoT. High quality photometric data[1] had never been achieved before for another star than the Sun, showing clear resolved p-modes oscillations. HD 49933 is an F star for which pulsations had been detected from ground based observations [11], but without being able to resolve the individual p-modes. Like in the Sun, HD 49933 has a signature of the surface convective motion at low frequency and a very rich spectrum of p-modes at higher frequency. But while in the Sun, one can clearly identify the low degree modes into the power spectrum (degree ℓ up to 3), in the case of HD 49933, we are not able to separate the odd pairs from the even pairs of modes (Fig. 15.1) and thus we are unable to identify the degrees of the p-modes by visual inspection of the power spectrum [2, 5]. Three main reasons explained this. First, the signal to noise is much lower than for the Sun, mainly because the granulation noise level is higher, the sign of an active star. As a consequence, the ℓ=3 is too weak to be extracted. Second, mode lifetimes are a function of the stellar temperature. F stars being hotter than G stars, modes are wider in HD 49933 than in the Sun (by a factor 4) and pairs of modes of same parities overlap (e.g. $\ell = 0$ and $\ell = 2$). Finally, the rotational splitting is about ten times higher than in the Sun (rotation in approximatively 3.4 days), thus increasing the overlapping. Almost all F stars suffer from these problems, the so-called ‘HD 49933 syndrome’.

[1] Luminosity variations at a level of few ppm are detected by CoRoT.

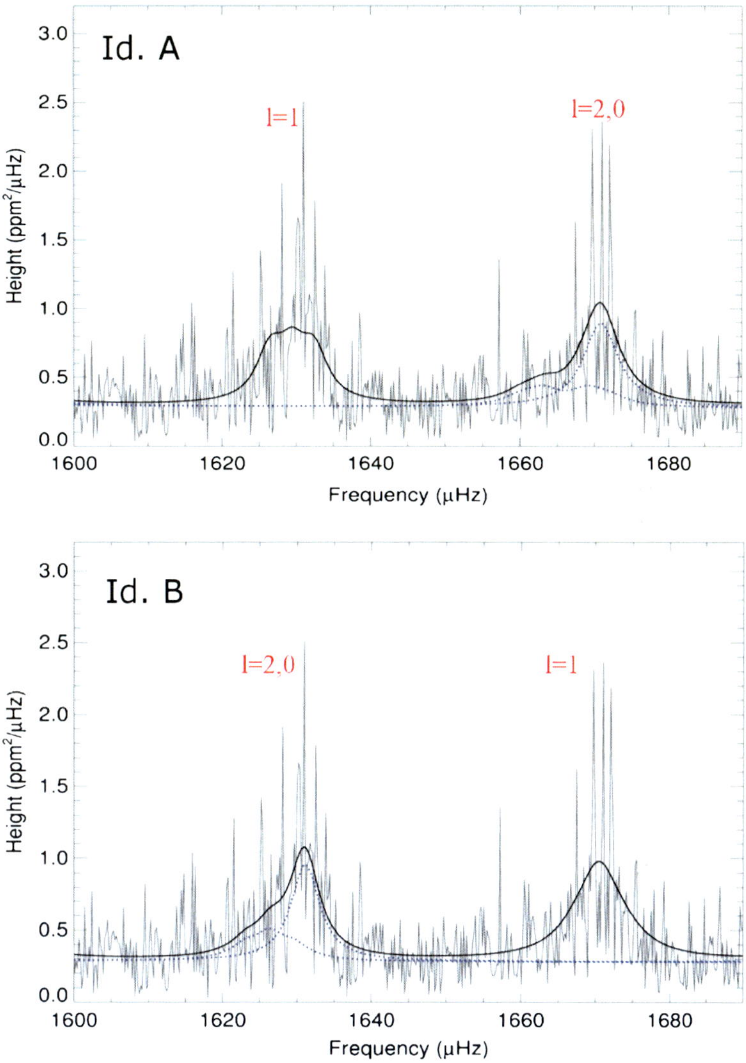

Fig. 15.1 A sample of the power spectrum of HD 49933. The two possible identifications of the modes are shown with a *black line*. *Dotted lines* represent the $l = 0$ and the $l = 2$ individual mode profiles

Faced with the HD 49933 syndrome and in order to extract the parameters of the p-mode pulsations (frequency, height and mode lifetime), the asteroseismology community involved in CoRoT used a recipe similar to the one successfully applied to the Sun. The initial approach presented in [2] consists of fitting of a sum of Lorentzian profiles to the power spectrum, based on a Maximum Likelihood

Estimation (MLE). If one assume that only the three lowest degrees (i.e. $\ell = 0$, $\ell = 1$ and $\ell = 2$) are significant in the power spectrum of HD 49933, then two mode identifications are possible, depending on how the power excesses exhibited by the power spectrum can be tagged. Reference [2] compared the significance of these mode identifications on the basis of the maximum likelihood ratio test. The most likely mode identification, hereafter called, identification A (the other one being identification B), was significant to more than 99.99%. Doubts about the degree to which we could trust such a result came up very quickly, mainly because such likelihood ratio tests had never been done by the asteroseismic community with such a low signal-to-noise ratio and with a relatively large number of parameters (around 80 parameters). In such conditions, the likelihood function is likely to possess several local maxima and convergence issues may arise, since it will become hard to find the absolute maximum of the likelihood function.

A more robust method had to be used in order to verify this mode identification and confirm the inferred values of the pulsation parameters, such as a Bayesian approach coupled with Markov Chain Monte Carlo algorithm [5, 6]. The Bayesian approach is well suited to extract as much as information as possible from a given data set, by including a priori knowledge from other data sets or theoretical assumptions. This approach becomes even more powerful if it utilizes all of the information contained in the posterior probability distribution function and not only on the determination of its maximum. To do so, one needs to sample the posterior probability distribution function and this can be achieved by using a Markov Chain Monte Carlo algorithm. A model comparison then relies on the computation of the so-called Bayes factor. Such an approach is far more reliable than MLE but requires intensive computation.

We used a Bayesian analysis, previously validated on simulated data [5], on HD 49933 that gave different results than [2]. While [2] have assumed that $\ell = 0$, $\ell = 1$ and $\ell = 2$ were present in the power spectrum, our approach was unable to show evidence of the presence of $\ell = 2$ modes. Moreover, no clear mode identification was possible.

In order to solve this critical problem of mode identifications, an additional observation of 137 days was carried out by CoRoT, and we proceeded to a combined analysis of the two data sets [6], in a similar fashion. The total observation duration is thus approximatively three times higher than the first observation. The new observation has substantially modified the odds and this time, $\ell = 2$ are clearly identified with odds strongly in favor of the identification B (opposite to [2]). Nowadays, several analysis of HD 49933 have been carried out and confirm the identification B [4, 10, 13]: the asteroseismic community finally has reached a consensus about the mode identification of this star and on the recipes to analyze stars affected by HD 49933 syndrome.

The difficult analysis of HD 49933 has paved the way for other F stars. Since its analysis, several Bayesian approaches have been carried out successfully on several tens of Kepler F stars [1]. Kepler was launched in March 7, 2009 and long duration observation[2] of the stars in the Kepler field render unlikely a wrong identification of the modes: odds ratio are always decisively in favor of one identification.

15.3 New Challenges

The mode identifications problem in F stars is now settled but the asteroseismic community is facing new challenges. The most problematic one concerns the number of observed stars. The asteroseismology is gaining momentum these last years and with the Kepler mission, we are overwhelmed by data. The present stellar signal analysis tools need human supervision in order to return a reliable result that render difficult the analysis of all the observed stars, in reasonable time. Moreover, in the forthcoming years, many asteroseismic programs allowing global sky-surveys (such as the SONG[3] network or the PLATO[4] mission) will probably be in place and we will need to find how to analyze several hundred thousands of power spectrum, while stars behaviors vary a lot and sometimes in an unexpected way.

References

1. Appourchaux, T., KASC community: . *Astro. Astrophys.*, **in prep** (2011)
2. Appourchaux, T., Michel, E., Auvergne, M., Baglin, A., Toutain, T., Baudin, F., Benomar, O., Chaplin, W.J., Deheuvels, S., Samadi, R., Verner, G.A., Boumier, P., García, R.A., Mosser, B., Hulot, J.C., Ballot, J., Barban, C., Elsworth, Y., Jiménez-Reyes, S.J., Kjeldsen, H., Régulo, C., Roxburgh, I.W.: CoRoT sounds the stars: p-mode parameters of Sun-like oscillations on HD 49933. *Astro. Astrophys.*, **488**, 705–714 (2008)
3. Baglin, A., Auvergne, M., Barge, P., Deleuil, M., Catala, C., Michel, E., Weiss, W., Team, T.C.: Scientific objective for a minisat: CoRoT. ESA Special Publication **1306**, 33 (2006)
4. Bedding, T.R., Kjeldsen, H.: Scaled oscillation frequencies and échelle diagrams as a tool for comparative asteroseismology. Communications in Asteroseismology **161**, 3–15 (2010)
5. Benomar, O., Appourchaux, T., Baudin, F.: The solar-like oscillations of HD 49933: a Bayesian approach. *Astro. Astrophys.*, **506**, 15–32 (2009).
6. Benomar, O., Baudin, F., Campante, T.L., Chaplin, W.J., García, R.A., Gaulme, P., Toutain, T., Verner, G.A., Appourchaux, T., Ballot, J., Barban, C., Elsworth, Y., Mathur, S., Mosser, B., Régulo, C., Roxburgh, I.W., Auvergne, M., Baglin, A., Catala, C., Michel, E., Samadi, R.: A fresh look at the seismic spectrum of HD49933: analysis of 180 days of CoRoT photometry. *Astro. Astrophys.*, **507**, L13–L16 (2009).

[2]Up to now, Among the 200,000 observed stars, several thousand have been observed almost continuously for more than 500 days. Many of them will be observed during the whole duration of the mission -nominally 3.5 years- and probably beyond.

[3]http://astro.phys.au.dk/SONG/

[4]http://sci.esa.int/plato

7. Borucki, W.J., Koch, D.G., Lissauer, J., Basri, G., Brown, T., Caldwell, D.A., Jenkins, J.M., Caldwell, J.J., Christensen-Dalsgaard, J., Cochran, W.D., Dunham, E.W., Gautier, T.N., Geary, J.C., Latham, D., Sasselov, D., Gilliland, R.L., Howell, S., Monet, D.G., Batalha, N.: KEPLER Mission Status. In: C. Afonso, D. Weldrake, & T. Henning (ed.) Transiting Extrapolar Planets Workshop, *Astronomical Society of the Pacific Conference Series*, vol. 366, p. 309 (2007)
8. Deubner, F.L., Noyes, R.W., Simon, G.W.: Observations of low wavenumber nonradial eigen modes of the sun. *Astro. Astrophys.*, p. 371 (1975)
9. Kjeldsen, H., Bedding, T.R.: Amplitudes of stellar oscillations: the implications for asteroseismology. *Astro. Astrophys.*, **293**, 87–106 (1995)
10. Mosser, B., Appourchaux, T.: On detecting the large separation in the autocorrelation of stellar oscillation times series. *Astro. Astrophys.*, **508**, 877–887 (2009).
11. Mosser, B., Bouchy, F., Catala, C., Michel, E., Samadi, R., Thévenin, F., Eggenberger, P., Sosnowska, D., Moutou, C., Baglin, A.: Seismology and activity of the F type star HD 49933. *Astro. Astrophys.*, **431**, L13–L16 (2005).
12. Walker, G., Matthews, J., Kuschnig, R., Johnson, R., Rucinski, S., Pazder, J., Burley, G., Walker, A., Skaret, K., Zee, R., Grocott, S., Carroll, K., Sinclair, P., Sturgeon, D., Harron, J.: The MOST Asteroseismology Mission: Ultraprecise Photometry from Space. *Pub. Astro. Soc. Pacific*, **115**, 1023–1035 (2003).
13. White , T., Bedding, T., Stello, D., Christensen-Dalsgaard, J., Huber, D., Kjeldsen, H.: Calculating asteroseismic diagrams for solar-like oscillations. *Astrophys. J.*, in press (2011)

Chapter 16
Semi-parametric Robust Event Detection for Massive Time-Domain Databases

Alexander W. Blocker and Pavlos Protopapas

Abstract The detection and analysis of events within massive collections of time-series has become an extremely important task for time-domain astronomy. In particular, many scientific investigations (e.g. the analysis of microlensing and other transients) begin with the detection of isolated events in irregularly-sampled series with both non-linear trends and non-Gaussian noise. We outline a semi-parametric, robust, parallel method for identifying variability and isolated events at multiple scales in the presence of the above complications. This approach harnesses the power of Bayesian modeling while maintaining much of the speed and scalability of more ad-hoc machine learning approaches. We also contrast this work with event detection methods from other fields, highlighting the unique challenges posed by astronomical surveys. Finally, we present results from the application of this method to 87.2 million EROS-2 sources, where we have obtained a greater than 100-fold reduction in candidates for certain types of phenomena while creating high-quality features for subsequent analyses.

16.1 Introduction

The analysis of massive time-domain astronomical surveys poses growing challenge within astrostatistics that demands both statistical rigor and computational efficiency. While such data provides a wide range of opportunities, the detection of isolated events is one ubiquitous problem that generally takes on a given outline:

A.W. Blocker (✉)
Department of Statistics, Harvard University, Cambridge, MA, USA
e-mail: ablocker@fas.harvard.edu

P. Protopapas
Harvard-Smithsonian Center for Astrophysics, Harvard University, 60 Garden Street, Cambridge, MA 02138, USA
e-mail: pprotopapas@cfa.harvard.edu

E.D. Feigelson and G.J. Babu (eds.), *Statistical Challenges in Modern Astronomy V*, Lecture Notes in Statistics 209, DOI 10.1007/978-1-4614-3520-4_16,

We are presented with a massive (10–100+ million) database of time series, possibly spanning multiple spectral bands. Our goal is to identify and classify time series containing events. How do we define an event? We are not interested in isolated outliers (as is the case in anomaly detection). Instead, we are looking for groups of observations that differ significantly from those nearby (i.e. "bumps" and "spikes"). In our applications of interest, such groups are differentiated from trends by their time scale—that is, they have structure at a higher frequency than we would consider a trend, but with a lower frequency than isolated outliers. Additionally, we would like to distinguish globally-variable light curves from isolated events, as they have very different scientific interpretations. This flavor of problem arises in many fields, but the case of astronomical time-domain surveys is particularly challenging.

There is an acute need for statistical methods that scale to these volumes of data throughout modern astronomy. This demands that we carefully manage the trade-off between statistical rigor and computational efficiency. In general, principled statistical methods yield better performance with messy, complex data, but scale poorly to massive datasets. In contrast, more ad-hoc machine learning methods handle clean data well, but often choke on issues we confront with complex astronomical data (outliers, nonlinear trends, irregular sampling, unusual dependence structures, etc.). Our approach is to inject probability modeling into our analysis in the right places, gaining much of the power of probability modeling without incurring its computational penalties.

We demonstrate the utility of this approach using a multi-stage technique for event detection. By combining a principled, flexible probability model with a discriminative classifier, we obtain excellent performance and computational efficiency analyzing the MACHO and EROS-2 surveys.

16.2 Previous Approaches and Unique Challenges

The astronomical literature contains a variety of approaches, among which scan statistics are prevalent. These have seen use in astronomical surveys [1, 2], but they often discard information by working with ranks and account for neither trends nor irregular sampling. Equivalent width methods (a scan statistic based upon local deviations) are also common in astrophysics. However, these typically rely upon Gaussian assumptions and relatively simple multiple testing corrections; the latter can unnecessarily decrease detection power. Numerous other approaches have been proposed in the literature, the vast majority of which rely upon Gaussian distributional assumptions, stationary, and/or regular sampling.

This problem also has a long history within the statistical community, often under the moniker of "change-point" or "regime-switching". Some recent examples include the work of Smyth and his collaborators [3, 4], who have used hidden Markov models to model deviations from learned baselines in sensor count data. There is a strong Bayesian lines of research on this topic; [5–7] are representative examples of this work. On the econometrics side, [8] and more recent work by Perron and collaborators [9, 10] are only a small part of the literature. Our setting is differs greatly from those seen in the vast majority of previous work.

Most preceding work has dealt with single, long time series which provide a high degree of internal replication. This allows methods to reliably ascertain what behavior is "typical" and find deviations from it with little outside information. In analyzing massive time-domain surveys, we have large sets of time series that are less informative individually. We must therefore rely on replication across series and prior scientific knowledge to find deviations from "typical" behavior. Furthermore, we must handle the additional complications of astronomical data.

These complications arise from both the measurement processes used in astronomical studies and the nature of the phenomena we study. The distribution of measurement errors from ground-based observations is typically fat-tailed (extreme outliers are prevalent). The resulting data requires more sophisticated noise models than the typical Gaussian. Non-linear, low-frequency trends are also common due to long-period variation in source intensity and/or calibration. Such trends render naïve, trend-free methods less effective; in particular, their specificity diminishes in this setting. The related but distinct problem of non-event light curves with variation at the time scale of interest also complicates our analysis and demands tools that can discriminate between these cases. Finally, irregular sampling is ubiquitous in astronomical surveys due to changes in the earth's orientation throughout the year and other factors. Irregular sampling can create artificial events in analyses that discard observation times; therefore, our method must take this information into account to maintain both high specificity and high sensitivity.

16.3 Models and Methods

Our analysis consists of two stages. First, we use a Bayesian probability model to detect of sources with variation at a time scale of interest (i.e. the time scale of events) and to reduce the dimensionality of our time series (using posterior summaries). Second, we employ a classifier based on these posterior summaries to discriminate among different types of variability. In the application described in Sects. 16.5 and 16.6, these types correspond to periodic and temporally-isolated (event-like) variability.

Formally, let V be the set of all time series with variation at a given time scale of interest (e.g., the range of lengths for isolated events), and let S be a subset of V corresponding to the time series of interest (events). For a given light curve Y_i, we want to estimate $P(Y_i \in S)$; that is, the probability that it is an event.

We decompose this probability as

$$P(Y_i \in S) = P(Y_i \in V \cap S) = P(Y_i \in V) \cdot P(Y_i \in S | Y_i \in V) \tag{16.1}$$

estimating or bounding each probability separately using the techniques described above. This decomposition allows us to employ generative techniques in the first stage while harnessing discriminative techniques in the second. We provide details of the models underlying these techniques below and cover the corresponding inference algorithms in Sect. 16.4.

16.3.1 Semi-parametric Model for Variable Light Curves

To flexibly model both non-linear trends and events at the time-scale of interest, we turn to wavelets. Their localization in both time and frequency allows us to separate event-like variation (characterized by a higher frequency) from trends (characterized by a lower frequency) while preserving local structure of our light curves.

We begin by specifying a linear model for each time series with a "split" incomplete wavelet basis:

$$y(t) = \beta_0 \phi_0(t) + \sum_{i=1}^{k_l} \beta_i \phi_i(t) + \sum_{j=k_l+1}^{M} \beta_j \phi_j(t) + \varepsilon(t) \qquad (16.2)$$

Here, $y(t)$ is the observed magnitude at time t. We define $(\phi_1, \dots, \phi_{k_l})$ as the k_l lowest-frequency components of a discrete-frequency wavelet basis, and $(\phi_{k_l+1}, \dots, \phi_M)$ as the higher-frequency components. The idea is for $(\phi_1, \dots, \phi_{k_l})$ to model structure due to trends, and $(\phi_{k_l+1}, \dots, \phi_M)$ to model structure at the scales of interest for events. We use an incomplete basis (excluding the highest frequencies) as we are not interested in modeling variation at time scales below those of interest for our events.

This basis formulation explicitly addresses irregular sampling as well. We simply evaluate the basis functions at the observation times to obtain a valid model for our light curve. This is simpler and more adaptable than, for example, using a continuous time autoregressive model.

To stabilize our inferences and regularize our estimates in under-sampled time periods (gaps), we impose a $N(0, \sigma^2 \cdot \tau)$ prior on $(\beta_1, \dots, \beta_M)$. This is conditionally conjugate to an augmented form of our model, which allows for efficient inference. The prior parameter τ is also readily interpretable: it is the number of artificial observations we are introducing for each coefficient. We set $\tau = \frac{1}{100}$ for our inference to reflect a diffuse prior; it is, however, sufficient to regularize our estimates in under-sampled periods.

To account for the extreme outliers observed in our light curves, we assume that our residuals $\varepsilon(t)$ are distributed as independent $t_\nu(0, \sigma^2)$ random variables. This allows our inference to ignore isolated outliers, focusing on variation with more structure. We fix ν for our model at 5; it is possible, although computationally expensive, to infer ν as well.

Selection of the wavelet basis ϕ is an important consideration for this method. It determines the trade-off between time and frequency localization for our inference, and it also constrains (due to incompleteness) the types of variation we can approximate well. In general, this choice depends upon the scientific context. We select the Symmlet 4 (a.k.a. Least Asymmetric Daubechies 4) wavelet basis for this work for its high degree of time localization, reasonable frequency localization, and quality of approximation for the phenomena of interest.

The final remaining choices are interval over which the basis is defined (to which our observation times are rescaled), the dimensionality of our basis M, and the number of "trend" components k_l. All three of these are interrelated and must be selected based on the time-scale of interest for events (as opposed to trends). We scale our basis to an interval of length 2,048 and set $M = 128$, $k_l = 8$. This provides enough resolution to capture events at the scale of interest while removing low-frequency trends and isolated outliers.

16.3.2 Screening for Variation at Frequencies of Interest

We screen light curves for further examination by testing $H_0 : \beta_{k_l+1} = \beta_{k_l+2} = \ldots = \beta_M = 0$ against the alternative that any of these coefficients differs from zero. This procedure will select many light curves that do not contain isolated events, but its primary purpose is to provide a high-quality set of candidate light curves of manageable size for further investigation and classification. These non-event light curves contain variation at the scale of interest, but this variation may be temporally diffuse. Our test statistic is $2(\hat{\ell}_1 - \hat{\ell}_0)$, where $\hat{\ell}_0$ is the log-likelihood of the null model evaluated at the MAP estimates; $\hat{\ell}_1$ is the analogous quantity for the alternative model. We use a χ^2 approximation for the reference distribution of this test statistic. Although this approximation is technically incorrect given the use of an informative prior, it provides a reasonable approximation that holds empirically. With this approximation, we employ a modified Benjamini-Hochberg FDR procedure with a maximum FDR of 10^{-4} to set the critical region for our test statistic [11, 12]. We present our validation for this technique in Sect. 16.6.2.

16.3.3 Classification Model for Isolated Variation

We engineered two features based on the model in Sect. 16.3.1 to discriminate between diffuse and isolated variability in the light curves selected by our screening procedure. Both are based on the normalized output of the preceding model, as this allows us to remove the nonlinear trends and isolated outliers. We thus obtain a high-quality, detrended and denoised representation of each light curve. We define for each light curve

$$\tilde{y}(t) = \sum_{j=k_l+1}^{M} \hat{\beta}_j \phi_j(t)\,; \quad z(t) = \frac{\tilde{y}(t) - \text{Mean}(\tilde{y}(t))}{\text{SD}(\tilde{y}(t))} \tag{16.3}$$

Our first feature is a monotonic transformation of a conventional CUSUM statistic, defined as $CUSUM$ via

$$S(t) = \sum_{s \le t} (z(s)^2 - 1)\,; \quad CUSUM = \log\left(1 + \frac{\max_t S(t) - \min_t S(t)}{\sqrt{n}}\right) \tag{16.4}$$

This captures the degree of temporal concentration for the variation in our fitted values—larger values will correspond to localized deviations from the baseline, while low values will correspond to deviations spread over a greater duration. It is maximized for a single spike with a flat baseline.

Our second feature is "directed variation". Our goal is for it to capture deviation from symmetric variation (as would be observed in periodic or quasi-periodic light curves). Letting z_{med} be the median of $z(t)$, we define:

$$DV = \frac{1}{\#\{t : z(t) > z_{\mathrm{med}}\}} \sum_{t:z(t)>z_{\mathrm{med}}} z(t)^2 - \frac{1}{\#\{t : z(t) < z_{\mathrm{med}}\}} \sum_{t:z(t)<z_{\mathrm{med}}} z(t)^2 \tag{16.5}$$

We tested a variety of classifiers including SVM (with linear and radial kernels), kNN, and LDA. However, in the end, we obtained our best performance from regularized logistic regression. We used a "weakly informative" prior as developed by Gelman et al. [13] to stabilize the estimates from this model. We describe its training and evaluation in Sect. 16.4.2.

16.4 Computation

Speed and scalability are the core goals of our computational strategy. We require a method that scales to databases of 200 million or more light curves (for the EROS-2 survey). As a result, our inference is optimization-based (as opposed to simulation) and highly-tuned for efficiency. We also manage the scale of training data where possible, preventing the computational cost of inference from scaling poorly with database size. We lay out the particulars of our algorithms below.

16.4.1 Efficient EM Inference for Semi-parametric Model

To obtain estimates of $\beta_0, \dots, \beta_M$ and σ^2 in our semi-parametric model, we first augment our model with a set of observation-specific variances. Let $z(t) \sim N(0,1)$ independent of $w(t) \sim \mathrm{InvGamma}(\frac{\nu}{2}, \frac{\nu}{2})$. Then, we can represent $\varepsilon(t)$ as $\varepsilon(t) \sim z(t) \cdot \sqrt{w(t)}$. This allows us to consider the set of $w(t)$ as missing data, opening our model to tools such as the EM algorithm [14].

Following this approach, we employ an EM algorithm with the optimal data augmentation scheme of [15] to obtain MAP estimates for the parameters of our semi-parametric model. Compared to a naïve EM implementation, we have found that this scheme offers a five to tenfold reduction in the iterations required for convergence.

We implemented this procedure in C with a direct interface to an optimized BLAS/LAPACK implementation. This allowed us to obtain an average time per complete estimation procedure (including EM estimation for both the null and alternative models, as specified in Sect. 16.3.2) of approximately 0.15–0.2 s, including file I/O, using a single processor on Harvard's Odyssey cluster. Memory usage was below 16 MB per light curve, and this algorithm is embarrassingly parallel across light curves. This combination allows our technique to scale to extremely large sets of time series.

16.4.2 Training the Classification Model via Simulation

We train our classification model on a combination of simulated data and curated, labeled light curves. Before descending into the details, we emphasize that this model must distinguish between local and global variation in light curves that have already passed the first-stage screen. Thus, our training data includes only such light curves.

The training data consisted of 12,365 labeled variable light curves from the MACHO dataset (periodic and quasi-periodic) and 9,170 simulated events (microlensing) that passed the given screening procedure. We obtained maximum a posteriori (MAP) estimates for this model via numerical maximization and performed tenfold cross-validation to assess its predictive ability. This validation showed excellent performance, with a mean cross-validated AUC of 0.991 on our training data.

16.5 Data

We used data from the MACHO survey for training and testing. The knowledge and information gained from this data was then used to analyze the EROS-2 survey.

The MACHO database consists of approximately 38 million LMC (Large Magellanic Cloud) sources, each observed in two spectral bands [16–18]. Data was collected from 1992 through 1999 on 50-inch telescope at Mount Stromlo Observatory, Australia on 94 43×43 fields in two bands, using eight $2,048 \times 2,048$ pixel CCDs. This data contains substantial gaps in observations due to seasonality and competing priorities for transient events.

The EROS-2 database consists of approximately 87.2 million sources, each observed in two spectral bands. Imaging was conducted with a 1 m telescope at ESO,

La Silla between 1996 and 2003, each camera consisting of mosaic of eight 2K $\times$ 2K LORAL CCDs. There are typically 800–1,000 observations per source, and irregular observations are prevalent (although less so than in the MACHO data).

16.6 Results

16.6.1 Semi-parametric Model: Empirical Properties

The semi-parametric model provided reasonable fits for both MACHO and EROS-2 data. It captured both non-linear trends (including changes in baseline between observing periods). We provide examples of fits for both the null and complete model on null and event light curves in Fig. 16.1.

16.6.2 Screening

To assess how well our LLR statistic and overall screening procedure performs on the data of immediate interest, we simulated $50,000$ events from a physics-based model (for microlensing) and 50,000 null time series based on the observed properties of the MACHO data. We obtain approximate power of 80% with an FDR of 10^{-4} based on this simulated data.

Running this on the EROS-2 data, we obtain a reduction of approximately 98% (from 87.2 million candidate light curves to approximately 1.5 million) from our screening procedure. This greatly eased the computational burden of subsequent analyses.

16.6.3 Classification of Isolated Events

Our classifier selected approximately 49,000 of the screened light curves as likely isolated events ($P \geq 0.5$). Of these, approximately 17,000 survived a final round of screening before further investigation. This final screen consisted of removing all fields with 20 or more identified events, as such clusters were not of scientific interest for the current investigation. One major example of this from EROS-2 is the supernova SN1987a, which affected light curves from the Large Magellanic Cloud. For other investigations, however, such screening may not be appropriate or necessary. We show the distribution of features for MACHO and EROS-2, with the estimated classification boundary, in Fig. 16.2.

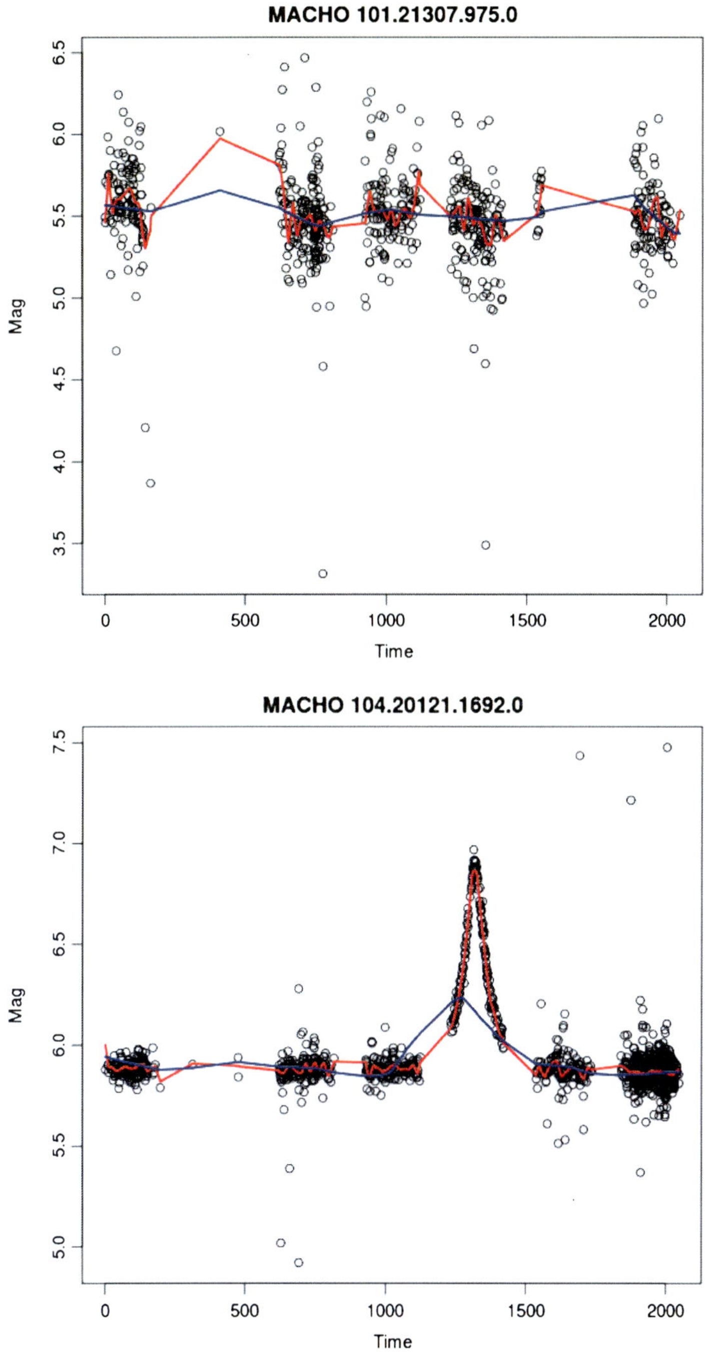

Fig. 16.1 Examples of fits for null and event MACHO light curves. Null model is in *blue*; complete model is in *red*

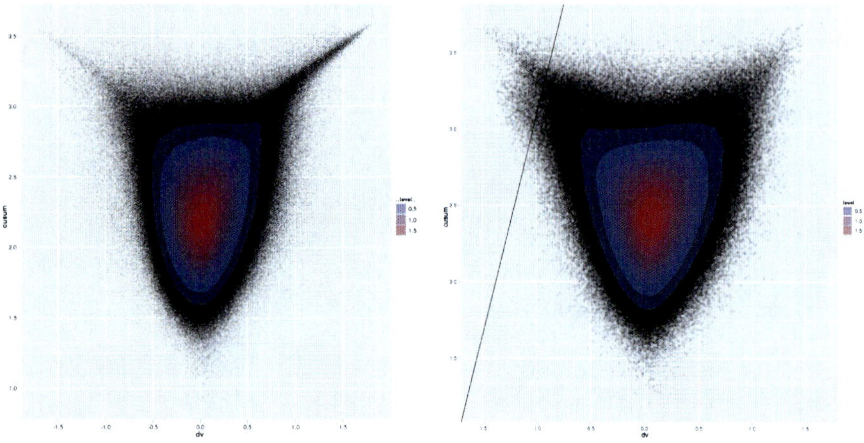

Fig. 16.2 Distribution of classification features for MACHO (*left*) and EROS-2 (*right*) databases. *DV* on the *horizontal axis*, *CUSUM* on the *vertical axis*

Within the events detected for EROS-2, we have found 68 known microlensing events, 42 known supernovas, and 25 known Cepheids with an (admittedly incomplete) database search (VizieR only). We have also identified several hundred previously unidentified transient phenomena that we are investigating further. These have been validated as previously unlabeled against a thorough database search (VizieR, Simbad, and VO).

16.7 Remarks

The method we have demonstrated combines the power of principled probability modeling with the speed and flexibility of more ad-hoc machine learning approaches. It scales to the analysis of massive astronomical time-domain surveys and can be adapted to detect a variety of temporally-isolated phenomena. It does not provide a final, scientific classification or analysis for light curves in these surveys; rather, we want to predict which time series are most likely to yield phenomena characterized by events (e.g. microlensing, blue stars, flares, etc.). Our technique is, at its core, a tool for rigorously-grounded discovery rather than approximate final analysis.

This, in turn, allows for the use of more complex, physically-motivated model on massive databases by pruning the set relevant data to a manageable size. We accomplish this while providing assessments of uncertainties at each stage of our screening and detection, and we provide a sufficiently rich framework to incorporate relevant domain knowledge.

We look forward to the application of this technique to more surveys and phenomena; in particular, we are currently investigating data from Pan-STARRS.

The approach demonstrated here can be applied to many other massive data challenges within astronomy and beyond, bringing the power of Bayesian probability modeling to massive data while maintaining computational tractability.

References

1. C.L. Liang, J.A. Rice, I.d. Pater, C. Alcock, T. Axelrod, A. Wang, S. Marshall, Statistical Science **19**(2), pp. 265 (2004). http://www.jstor.org/stable/4144411
2. D. Preston, P. Protopapas, C. Brodley, ArXiv e-prints. To appear in SIAM International Conference on Data Mining **arxiv:0901.3329v1 [astro-ph.IM]** (2009)
3. J. Hutchins, A. Ihler, P. Smyth, in *Proceedings of the Second International Workshop on Knowledge Discovery from Sensor Data (ACM SIGKDD Conference, KDD-08* (Citeseer, 2008)
4. A. Ihler, J. Hutchins, P. Smyth, ACM Transactions on Knowledge Discovery from Data (TKDD) **1**(3), 13 (2007)
5. A.F.M. SMITH, Biometrika **62**(2), 407 (1975). URL http://biomet.oxfordjournals.org/content/62/2/407.abstract
6. A.E. Raftery, V.E. Akman, Biometrika **73**(1), pp. 85 (1986). URL http://www.jstor.org/stable/2336274
7. B.P. Carlin, A.E. Gelfand, A.F.M. Smith, Journal of the Royal Statistical Society. Series C (Applied Statistics) **41**(2), pp. 389 (1992). URL http://www.jstor.org/stable/2347570
8. D.W.K. Andrews, Econometrica **61**(4), pp. 821 (1993). URL http://www.jstor.org/stable/2951764
9. J. Bai, P. Perron, Econometrica **66**(1), pp. 47 (1998). URL http://www.jstor.org/stable/2998540
10. P. Perron, Z. Qu, Journal of Econometrics **134**(2), 373 (2006). URL http://www.sciencedirect.com/science/article/B6VC0-4H21NGF-1/2/10d784957584e1f5ec0c8c4ff2b27f32
11. Y. Benjamini, Y. Hochberg, Journal of the Royal Statistical Society. Series B (Methodological) **57**(1), pp. 289 (1995). URL http://www.jstor.org/stable/2346101
12. Y. Benjamini, D. Yekutieli, The Annals of Statistics **29**(4), pp. 1165 (2001). URL http://www.jstor.org/stable/2674075
13. A. Gelman, A. Jakulin, M.G. Pittau, Y. Su, The Annals of Applied Statistics **2**(4), 1360 (2008). URL http://projecteuclid.org/euclid.aoas/1231424214
14. A.P. Dempster, N.M. Laird, D.B. Rubin, Journal of the Royal Statistical Society. Series B (Methodological) **39**(1), pp. 1 (1977). URL http://www.jstor.org/stable/2984875
15. X.L. Meng, D.v. Dyk, Journal of the Royal Statistical Society. Series B (Methodological) **59**(3), pp. 511 (1997). URL http://www.jstor.org/stable/2346009
16. C. Alcock, R.A. Allsman, T.S. Axelrod, D.P. Bennett, K.H. Cook, H.S. Park, S.L. Marshall, C.W. Stubbs, K. Griest, S. Perlmutter, W. Sutherland, K.C. Freeman, B.A. Peterson, P.J. Quinn, A.W. Rodgers, in *Sky Surveys. Protostars to Protogalaxies*, *Astronomical Society of the Pacific Conference Series*, vol. 43, ed. by B. T. Soifer (1993), *Astronomical Society of the Pacific Conference Series*, vol. 43, pp. 291–+
17. D.P. Bennett, C. Alcock, R.A. Allsman, T.S. Axelrod, K.B. Cook, K.C. Freeman, K. Griest, S.L. Marshall, B.A. Peterson, M.R. Pratt, P.J. Quinn, A.W. Rodgers, C.W. Stubbs, W. Sutherland, in *Clusters, Lensing, and the Future of the Universe*, *Astronomical Society of the Pacific Conference Series*, vol. 88, ed. by V. Trimble & A. Reisenegger (1996), *Astronomical Society of the Pacific Conference Series*, vol. 88, pp. 95–+
18. C. Alcock, R.A. Allsman, D.R. Alves, T.S. Axelrod, A.C. Becker, D.P. Bennett, K.H. Cook, A.J. Drake, K.C. Freeman, M. Geha, K. Griest, M.J. Lehner, S.L. Marshall, D. Minniti, C.A. Nelson, B.A. Peterson, P. Popowski, M.R. Pratt, P.J. Quinn, C.W. Stubbs, W. Sutherland, A.B. Tomaney, T. Vandehei, D. Welch, The Astrophysical Journal Supplement Series **136**, 439 (2001).

Chapter 17
Bayesian Analysis of Reverberation Mapping Data

Brendon J. Brewer

Abstract Reverberation mapping is a powerful technique for studying the broad line regions (BLR) and the masses of the central black holes in distant active galactic nuclei (AGN). By monitoring the temporal variations of the continuum emission and the broad emission lines, it is possible to measure the size of the broad line region, and combining this with velocity information from the line widths yields an estimator for the black hole mass. However, this estimator depends on an unknown dimensionless proportionality constant called the virial coefficient. Recently, we have developed an alternative, direct approach to analyzing reverberation mapping data that infers details of the astrophysical situation from the data, bypassing the need for a virial coefficient and providing information about the physical configuration of the BLR. In this contribution I will outline the method and discuss how it differs from traditional reverberation mapping analysis.

17.1 Introduction

Reverberation Mapping is an important technique for measuring the masses of the black holes that power Active Galactic Nuclei (AGN). The distribution of matter surrounding the black hole can also be studied, yielding constraints on AGN physics [3]. The measurement of black hole masses enables the study of the relations between supermassive black holes and their host galaxies [9]. The method makes use of the variability of the central continuum source [11], and the subsequent response of the broad lines, emitted by orbiting gas (Fig. 17.1). In the traditional method, the typical time delay, or lag, τ between the continuum variations and the broad line response is measured by cross correlating the continuum light curve

B.J. Brewer (✉)
Department of Physics, University of California, Santa Barbara, CA, USA
e-mail: brewer@physics.ucsb.edu

E.D. Feigelson and G.J. Babu (eds.), *Statistical Challenges in Modern Astronomy V*, Lecture Notes in Statistics 209, DOI 10.1007/978-1-4614-3520-4_17,

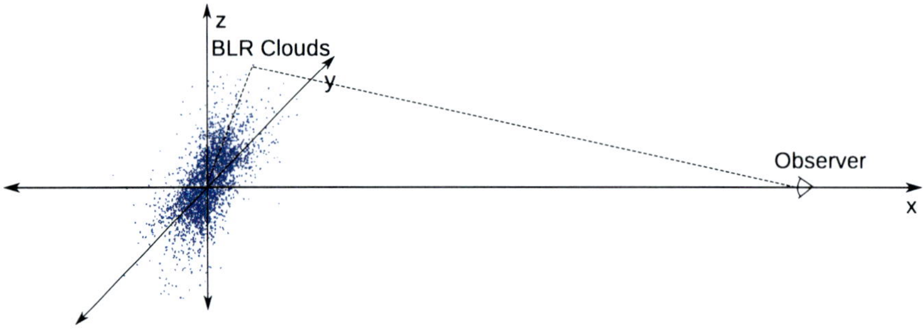

Fig. 17.1 The distribution of extra path lengths the light must travel from the central engine to a BLR cloud and then to the observer is the cause of the delayed response of the emission-line flux, and the variations in line shape. The distribution of BLR gas in this diagram corresponds to a probable configuration inferred from the Arp 151 data [5]

with the line flux light curve. The time delay τ measures the size of the broad line region (BLR), and the width of the broad lines, σ_l gives their typical orbital velocities. These measurements combined can give an estimate of the black hole mass, according to the formula [16]:

$$M_{BH} = f\frac{\sigma_l^2 c\tau}{G} \tag{17.1}$$

Here, the black hole mass is given in terms of physical constants and measurable quantities, but also depends on the dimensionless *virial coefficient* f. The virial coefficient is meant to encode the effect of the geometrical configuration of the BLR: for example, whether it is spherically symmetric, disky and face-on (this would imply a high value for f), disky and edge-on (implying low f), or whatever. However, if the value of f for any individual system is unknown, the black hole mass inherits this uncertainty. Typically, the distribution of f values for a population of AGN is used to indicate the uncertainty, implying that the uncertainty in an individual black hole mass is influenced by the diversity of f values across all systems, rather than on the data for that particular system.

The standard reverberation mapping procedure has been used with great success [1, 2, 7, 17], and has provided the basis for the calibration of less costly methods [8]. However, reverberation mapping data do not really arrive to us in the form of a value for τ and σ_l. These numbers are the results of procedures performed on the full data set, and there is no reason to think that they are sufficient statistics. The full data set consists of time series of the observed continuum flux, and spectral time series of the broad line response (i.e. the shape and flux of the chosen broad emission line, over time). To make the most of the data, we should perform an inference calculation based on the full data set (Fig. 17.2). For more details about our approach, please see [15] and [5].

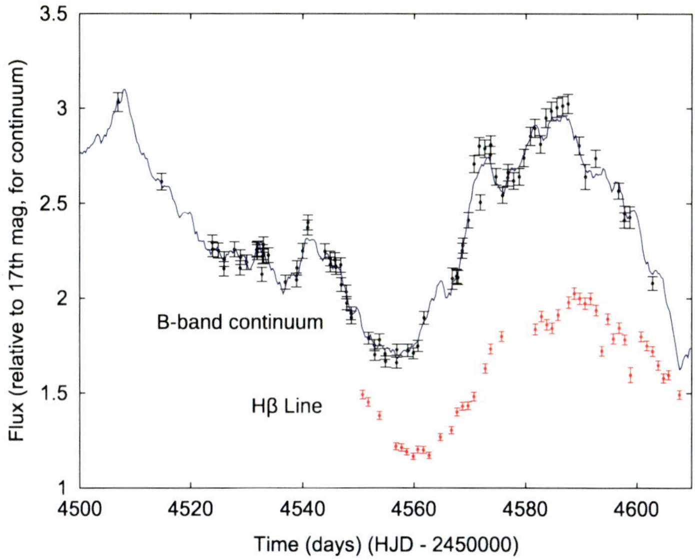

Fig. 17.2 Continuum flux time series for Arp 151, observed as part of the Lick AGN Monitoring Project (LAMP), and the corresponding Hβ flux time series. Note that the full Hβ data set actually consists of a spectrum at each time. i.e. there is also shape information, in addition to the flux plotted here

17.2 An Inference Approach

To best exploit the data, we should directly answer the question "what possible physical situations are plausible, in light of the data?". We begin by defining a hypothesis space and prior probabilities over that space, describing possible physical scenarios that might describe the physical system. By Bayes' rule, the posterior distribution for the parameters given the data is then given by:

$$p(\theta|D = D^*) \propto p(\theta)p(D|\theta)|_{D=D^*} \tag{17.2}$$

where D^* is the actual data set that was observed. The posterior distribution describes what is known about the parameters after taking into account the data, and suitable summaries can be derived from the full distribution (best estimates and error bars, credible intervals, etc.).

In our method, we numerically represent the geometry and kinematics of the BLR by a large number of point-like clouds (although we also have an implementation that describes the density function on a spatial grid). We do not aim to infer the position and velocity of every cloud. Instead we parameterize the distribution of the clouds by a small number of hyperparameters, and infer those hyperparameters. To generate a description of a 3D distribution of BLR clouds, we start by generating an axisymmetric distribution in the x-y plane, and then apply rotations to "puff up" the model into a 3D configuration. Finally, we weight the clouds by a non-axisymmetric illumination function to model non-axisymmetric distributions of gas. See [5] for details on how we parameterize the position and velocity distribution of the BLR gas.

Before we take into account a data set, we must describe our prior knowledge of the parameters. The parameter list is

$$\left\{M_{BH},\mu,F,\beta,\phi_{\text{open}},\phi_{\text{inc}},\kappa,q,\lambda,y_{\text{cont}}(t)\right\} \tag{17.3}$$

These are the black hole mass, the mean radius of the BLR, the fraction of the mean radius that is caused by a lower cutoff, the shape parameter of the radial BLR profile, the opening angle of the BLR, the inclination angle, the strength of the front/back asymmetry, the fraction of inflowing clouds, the departure from circularity of the orbits, and the continuum light curve at all times.

The prior distribution for $y_{\text{cont}}(t)$, before conditioning on the continuum data, is assigned to be a Gaussian process with mean y_0 and covariance function

$$C(\Delta t) = \sigma^2 \exp\left(-\left|\frac{\Delta t}{L}\right|^{\alpha}\right) \tag{17.4}$$

which introduces additional hyperparameters σ (the typical variation amplitude), L (the typical variation timescale) and $\alpha \in [1,2]$ (describing the smoothness of the variations). This Gaussian process model, in the case $\alpha = 1$, has been studied extensively in the context of AGN variability [11–13] and reverberation mapping [19]. This expands the full parameter list to

$$\left\{M_{BH},\mu,F,\beta,\phi_{\text{open}},\phi_{\text{inc}},\kappa,q,\lambda,y_{\text{cont}}(t),y_0,\sigma,L,\alpha,A\right\} \tag{17.5}$$

where we have also added a response coefficient A. The priors for all of these parameters are generic ignorance priors, uniform either in the parameter or its logarithm (where appropriate).

The sampling distributions $p(D|\theta)$ can be assigned by considering mock data. Specifically, if we knew all of the details of the physical situation, we would be able to predict mock noise-free spectra. The sampling distribution can then be assigned as a multivariate normal distribution, with mean values equal to the simulated noise-free data, and variances given by the supplied "error-bars" on the data. Of course, in reality, our model does not account for all effects apart from noise, but this common assumption provides a useful starting point.

With any complex Bayesian Inference problem in more than a few dimensions, the best way to summarise the posterior distribution is to generate random samples from it. To implement our parameter space exploration, we used Diffusive Nested Sampling [6] (DNS). DNS is an efficient MCMC-based version of Nested Sampling that works by exploring a mixture of the prior and a sequence of more constrained distributions that are created as the algorithm proceeds. For the purposes of exploring the parameter space efficiently, the main advantage of DNS is that it continually revisits the prior, "forgetting" its location along degeneracy curves.

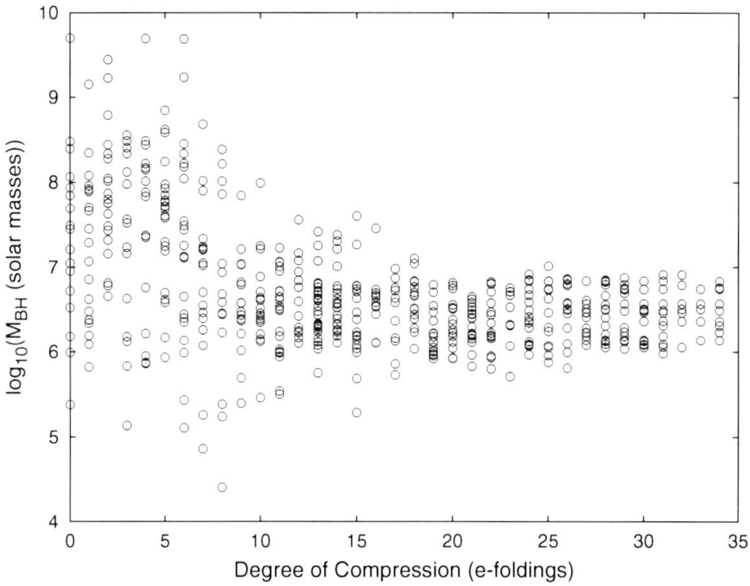

Fig. 17.3 The probability distribution for the black hole mass as a function of the compression of parameter space. Each step along the x-axis corresponds to selecting the best $1/e \approx 37\%$ of the remaining parameter space from the previous step, in terms of likelihood

17.2.1 Systematic Errors and Nested Sampling

In the application to the Arp 151 data [5], it was discovered that our BLR model could not fit the data to within the very small supplied error bars. Of course, the best solution to this is to develop more complex models with more freedom and more realistic physics, so that the data could be fit more exactly. However, the fit of this simple model to the data is still of some utility. In an attempt to obtain reliable inference of the black hole mass from an oversimplified model, we experimented with decreasing the constraining effect of the data, making the posterior distribution more conservative, i.e. more like the prior. Figure 17.3 shows the posterior distribution for the black hole mass as a function of compression. At $x = 0$ in this plot, the prior for the black hole mass is shown. As the parameter space is compressed by Nested Sampling (finding the best $1/e$ of the remaining prior mass), the x-value advances by 1. After about 20 compressions, the posterior inference on the black hole mass is remarkably insensitive to further compression. We discovered that all models above a compression of 20 reproduce the major qualitative aspects of the data, despite not fitting to within the error bars. However, most models below a compression of 20 do not resemble the data at all. Therefore, we selected models between a compression of 20 and 35 to form the posterior distribution. Note that the effect of this selection is very similar to the effect of raising the temperature of the likelihood function, or increasing the size of the error bars on the data. An important advantage of Nested Sampling over other sampling techniques is that all values for the temperature or compression are obtained in a single run, making it computationally trivial to consider the consequences of weakening the effect of the data.

17.3 Conclusions

The median and 68% credible interval for the black hole mass in Arp 151 was found to be $10^{6.51\pm0.28}$ $M_\odot$. This is lower than, but overlaps with, the value of $10^{6.85\pm0.07}$ $M_\odot$ obtained by [4] assuming $\log_{10} f = 0.74$ based on requiring active and inactive galaxies to obey the same correlation between M_{BH} and host-galaxy stellar velocity dispersion σ_* [14], and neglecting uncertainty in f. Recent measurements suggest that the intrinsic uncertainty in f from the standard method is at least 0.4 dex [10, 18], 33% higher than our uncertainty. Reversing the traditional argument, our measurement implies that $\log_{10} f = 0.40 \pm 0.28$, a low value, for this particular system. This low value agrees with the updated estimate of $\bar{f}$ from [9], although the low value may also just apply to this single system. Our method also allows for the inference of more structural parameters of the BLR, not just the mean radius and the black hole mass. Although the basic philosophy is sound, future work is needed to improve the realism of the physics and the flexibility of the BLR distribution in our model. This will result in more robust, and hopefully smaller, black hole mass uncertainties, and a more detailed picture of the physics of AGN.

Acknowledgements I would like to thank Tommaso Treu and Anna Pancoast for their substantial contributions to this work. I would also like to thank Aaron Barth for valuable discussions, and the entire LAMP team for the Arp 151 data and the encouragement. I acknowledge support by the NSF through CAREER award NSF-0642621.

References

1. Barth A. J., et al., 2011, ApJ, 732, 121
2. Bentz M. C., et al., 2008, ApJ, 689, L21
3. Bentz M. C., et al., 2010, ApJ, 720, L46
4. Bentz, M. C., et al. 2009, ApJ, 705, 199
5. Brewer B. J., et al., 2011, ApJ, 733, L33
6. Brewer, B., Partay, L., & Csanyi, G. 2010, Statistics and Computing, doi:10.1007/s11222- 010-9198-8
7. Denney K. D., et al., 2010, ApJ, 721, 715
8. Denney K. D., Peterson B. M., Dietrich M., Vestergaard M., Bentz M. C., 2009, ApJ, 692, 246
9. Graham A. W., Onken C. A., Athanassoula E., Combes F., 2011, MNRAS, 412, 2211
10. Greene, J. E., et al. 2010, ApJ, 721, 26
11. Kelly, B. C., Bechtold, J., & Siemiginowska, A. 2009, ApJ, 698, 895
12. Kozłowski, S., et al. 2010, ApJ, 708, 927
13. MacLeod, C. L., et al. 2010, ApJ, 721, 1014

14. Onken, C. A., Ferrarese, L., Merritt, D., Peterson, B. M., Pogge, R. W., Vestergaard, M., & Wandel, A. 2004, ApJ, 615, 645
15. Pancoast, A., Brewer, B. J., & Treu, T. 2011, ApJ, 730, 139
16. Peterson, B. M., et al. 2004, ApJ, 613, 682
17. Peterson B. M., Bentz M. C., 2006, NewAR, 50, 796
18. Woo, J., et al. 2010, ApJ, 716, 269
19. Zu, Y., Kochanek, C. S., & Peterson, B. M. 2010, ArXiv: 1008.0641

Chapter 18
Bayesian Mixture Models for Poisson Astronomical Images

Fabrizia Guglielmetti, Rainer Fischer, and Volker Dose

Abstract Astronomical images in the Poisson regime are typically characterized by a spatially varying cosmic background, large variety of source morphologies and intensities, data incompleteness, steep gradients in the data, and few photon counts per pixel. The Background-Source separation technique is developed with the aim to detect faint and extended sources in astronomical images characterized by Poisson statistics. The technique employes Bayesian mixture models to reliably detect the background as well as the sources with their respective uncertainties. Background estimation and source detection is achieved in a single algorithm. A large variety of source morphologies is revealed. The technique is applied in the X-ray part of the electromagnetic spectrum on *ROSAT* and *Chandra* data sets and it is under a feasibility study for the forthcoming *eROSITA* mission.

18.1 Introduction

One of the hot topics in X-ray (quantum energies > 0.1 keV) image analysis is the detection of faint sources. Both point-like and extended faint sources may provide important information about the Cosmos. For instance, a quantitative analysis of the abundance of galaxy clusters and groups as a function of redshift allows one to constrain cosmological parameters, to test the models for structure formation and

F. Guglielmetti (✉)
Max-Planck-Institut für extraterrestrische Physik, Giessenbachstrasse, D-85748 Garching, Germany
e-mail: fabrizia@mpe.mpg.de

R. Fischer • V. Dose
Max-Planck-Institut für Plasmaphysik, Boltzmannstrasse 2, D-85748 Garching, Germany
e-mail: Rainer.Fischer@ipp.mpg.de; vod@rzg.mpg.de

E.D. Feigelson and G.J. Babu (eds.), *Statistical Challenges in Modern Astronomy V*, Lecture Notes in Statistics 209, DOI 10.1007/978-1-4614-3520-4_18,

to provide the basis for follow-up studies of physical properties of these systems [1,3]. The detection and characterization of faint sources require advanced statistical methods.

18.1.1 The Data

X-ray images are characterized by few or no photon counts per pixel also for long exposures. The data consists of a diffuse background with superposed celestial objects, corrupted by Poisson noise and affected by instrumental complexities. Poisson noise dominates the signal especially at high frequencies of the electromagnetic spectrum. The instrumental complexities are, e.g., exposure variations, instrumental structures as detector ribs and charge-coupled device (CCD) gaps, smearing and vignetting effects, CCD failures and instrumental calibrations. An astronomical image is often a combination of several individual pointings, as for deep observations and mosaics of images, and the effects due to steep gradients in the data are cumbersome. Furthermore, the X-ray background is a composition of instrumental, particle and cosmic emissions. The cosmic background is not necessarily spatially constant. Celestial objects are characterized by a large variety of morphologies and apparent brightnesses. Sources, especially extended ones, can be superposed to both, smooth and highly, varying background.

18.1.2 Challenges in Image Analysis

The interpretation of observational data is a difficult task, especially when detecting faint sources and their (complex) morphologies. Several approaches have been developed so far. However, previous techniques do not jointly detect a large variety of source morphologies and describe large variations in the background.

An ideal source detection method should be capable to, preserve the statistics through the whole algorithm, detect faint sources, detect both point-like and extended sources, including complex morphologies, provide an accurate background estimation, include the exposure map in the background model, and provide uncertainties of estimates. Each of these desiderata entail a challenge in source detection and background estimation. In fact, the nature of the data of X-ray images is described by Poisson statistics and Poisson noise affects the data. Furthermore, joint background estimation and source detection is essential for a reliable detection of celestial objects and for a proper propagation of errors in background and source estimates. Conventional methods employ a threshold level for separating the sources from the background. Often, the threshold level is described in terms of the noise standard deviation, then translated into a probability (p-values). An ideal source detection method has to replace the threshold level by a measure of probability. In the same line of arguments, parameters entering the models need to be estimated from the data. In addition, the detection of extended sources is commonly achieved in several steps, e.g., reanalyzing the image after removing point sources from the image. Consequently, uncertainties in the data are not properly accounted for. In order to detect faint and extended sources, source features extending to the edge of the field of view and for providing good estimates in object photometry, a stable background model is essential. The estimation of a reliable background model and its uncertainties is a demanding task. Many techniques subtract an estimated background from the data, leading even to negative count rate values of the signal of interest: See, e.g., [4]. Moreover, the background model has to incorporate the exposure map. Exposure maps include also factors such as vignetting, defective pixels and instrumental structures, resulting in lack of data. The missing data must be handled consistently for the background estimation to prevent undesired artificial effects. Hence, the challenge is to preserve the statistics while taking into account the exposure map in the background model. The last demanding aspect for an ideal source detection method is the proper quantification of uncertainties of estimates.

Note that the knowledge of the instrumental point-spread-function (PSF) is not considered essential for source detection. A source detection algorithm designed for the detection of a large variety of source morphologies should be able to operate effectively without the PSF information. Source detection methods employing a PSF or its functional form are designed for the detection of point-like objects regardless of extended ones [5].

18.2 The Background-Source Separation Algorithm

The Background-Source separation (BSS) algorithm [2] is a probabilistic tool capable to satisfy the desiderata and tackle the challenges described in Sect. 18.1.2.

The BSS algorithm employs the single observed data set (photon image and exposure map) for source detection and background estimation. Bayesian probability theory (BPT) is the statistical tool used within the BSS technique, supplying a general and consistent frame for logical inference. Hence, the BSS algorithm takes advantage of all available information over a parameter set, which is described by a probability density over the corresponding parameter space. For each image pixel $\{ij\}$ two complementary hypotheses are considered $B_{\rm ij} : d_{\rm ij} = b_{\rm ij} + \varepsilon_{\rm ij}$ and $\overline{B}_{\rm ij} : d_{\rm ij} = b_{\rm ij} + s_{\rm ij} + \varepsilon_{\rm ij}$. Hypothesis $B_{\rm ij}$ specifies that the data $d_{\rm ij}$ consists only of background counts $b_{\rm ij}$ spoiled with noise $\varepsilon_{\rm ij}$, i.e., the (statistical) uncertainty associated with the measurement process. Hypothesis $\overline{B}_{\rm ij}$ specifies the case where additional source intensity $s_{\rm ij}$ contributes to the background. Two assumptions are taken into account: No negative values for $s_{\rm ij}$ and $b_{\rm ij}$ are allowed and $b_{\rm ij}$ is smoother than $s_{\rm ij}$. For modelling the structures arising in the background rate of the photon image, the thin-plate spline (TPS) is chosen. A weighted combination of TPSs centered about each supporting point gives the interpolation function that passes through the supporting points exactly while minimizing the bending energy. The TPS is not wavering between the supporting points, in opposite to cubic-splines, and is steady along the field, also where steep gradients in the data occurs, as at the field edge. The background amplitude, instead, consists of the TPS multiplied with the exposure map. Another important aspect of the technique is the likelihood for the mixture models, that arises applying BPT with the mixture model technique. The Bayesian mixture model is composed of two parts: a Poisson likelihood probability density function (pdf) $p(d_{\rm ij} \mid B_{\rm ij}, b_{\rm ij})$, i.e., $b_{\rm ij}$ contribution only, and a marginal Poisson likelihood pdf $p(d_{\rm ij} \mid \overline{B}_{\rm ij}, b_{\rm ij}, \gamma)$, i.e., $b_{\rm ij}$ plus $s_{\rm ij}$ components, where $s_{\rm ij}$ is marginalized and a parameter γ is introduced. According to BPT, prior pdfs have to be considered for both complementary hypotheses and for the source intensity parameter. The prior probability for the two complementary hypotheses, i.e., to have background only or additional source signal in a pixel or pixel cell,[1] is chosen to be β and $1-\beta$. Two prior pdfs over the source signal have been considered and tested: (1) an exponential prior pdf, (2) an inverse-Γ function prior pdf. Both prior pdfs of the source signal introduce the parameter γ. The likelihood for the mixture models results to be:

$$p(D \mid b, \beta, \gamma) = \prod_{ij} [\beta \cdot p(d_{\rm ij} \mid B_{\rm ij}, b_{\rm ij}) + (1-\beta) \cdot p(d_{\rm ij} \mid \overline{B}_{\rm ij}, b_{\rm ij}, \gamma)];$$

$$D = \{d_{\rm ij}\}, \quad b = \{b_{\rm ij}\}. \tag{18.1}$$

Equation 18.1 allows one to separate background and sources, considering all pixels for the background spline estimation, even those containing additional source contribution. It expresses our ignorance about the presence of background only or an additional source contribution in a certain pixel or pixel cell. This allows us to evaluate the posterior distribution over the background, $p(b|D)$, and the probability of having source contributions in pixels and pixel cells, $p(\overline{B}_{\rm ij}|d_{\rm ij})$.

$p(b|D)$ is the product of (18.1) and the prior pdf $p(b)$, that is chosen constant for positive values of b and null elsewhere. The maximum of the posterior pdf with respect to b gives an estimate of the background (amplitude) map. The background estimate is provided by the Gaussian approximation, where the Hessian matrix is used to extract the uncertainties of the background for each image pixel. Note that in (18.1) the model parameters γ and β appear. The values of γ and β and their uncertainties are estimated from the marginal posterior pdf $p(\beta, \gamma|D)$, under the assumption of the Laplace approximation.

The probability $p(\overline{B}_{\rm ij}|d_{\rm ij})$ for each pixel and pixel cells is approximated taking into account the optimal values of the background amplitude and the model parameters. $p(\overline{B}_{\rm ij}|d_{\rm ij})$ includes the Bayes factor. $p(\overline{B}_{\rm ij}|d_{\rm ij})$ estimated for varying correlation lengths of pixels give rise to the multiresolution analysis. The multiresolution analysis has the aim to analyze statistically source structures at multiple scales. The scales are the correlation lengths employed to create pixel cells. Source probability maps (SPMs) are created at different scales. SPMs allow one to separate

[1]We define pixels as the image finest resolution limited by instrumental design, while we define pixel cell a group of correlated neighboring pixels.

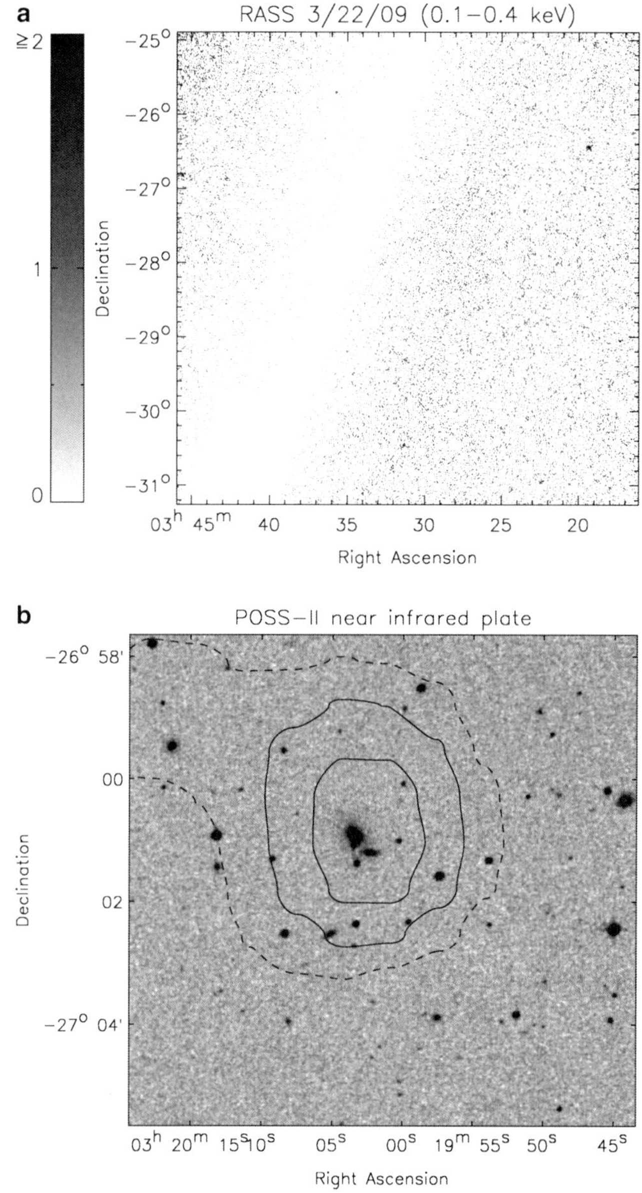

Fig. 18.1 Discovery of a cluster of galaxies and confirmed with optical sky surveys. **Panel a**: Soft band image of the *ROSAT* All-Sky survey (RASS) field RS932209n00. The image accounts for photon count/pixel in the range $0-9$. **Panel b**: POSS-II I plate with superposed X-ray contours from RASS field RS932209n00 (**panel a**) corresponding to 2, 3 and 4σ above the local X-ray background. This cluster of galaxies is known in the optical part of the electromagnetic spectrum as ACO S 340

point-like from extended sources. Note that background and sources are represented in the mixture components, ensuring no need of background subtraction for source detection. Furthermore, Poisson statistics is preserved also in the multiresolution analysis.

The BSS algorithm allows also for a multiband analysis. When multiband images are available, the information contained in each image can be statistically combined in order to extend the detection limit of the data. Conclusive posterior pdfs are provided for detected sources from combined energy bands.

The BSS algorithm is a general, powerful and flexible Bayesian technique for background and source separation. The technique is general since it is applicable to astronomical images coming from any count detector. The aim of providing more reliable results, with respect to previous techniques, for faint and extended sources is achieved: See, as example, in Fig. 18.1 the BSS detection of a new X-ray cluster of galaxies. The technique is flexible, because it can easily be extended to other statistics and astronomical problems in image analysis.

Acknowledgements The first author would like to thank H. Böhringer, H. Brunner and K. Dennerl (Max-Planck-Institut für extraterrestrische Physik, Germany), V. Mainieri and P. Rosati (European Southern Observatory, Germany) and P. Tozzi (INAF Osservatorio Astronomico di Trieste, Italy) for contributing to this work.

References

1. Böhringer, H., Pratt, G. W., Arnaud, M. et al.: Substructure of the galaxy clusters in the REXCESS sample: observed statistics and comparison to numerical simulations. A&A **514**, A32 (2010)
2. Guglielmetti, F., Fischer, R., Dose, V.: Background-source separation in astronomical images with Bayesian probability theory - I. The method. MNRAS **396**, 165–190 (2009)
3. Rosati, P., Borgani, S., Norman, C.: The Evolution of X-ray Clusters of Galaxies. ARA&A **40**, 539–577 (2002)
4. Śliwa, W., Soltan, A. M., Freyberg, M. J.: The harmonic power spectrum of the soft X-ray background. I. The data analysis. A&A **380**, 397–408 (2001)
5. Starck, J.-L., Pierre, M.: Structure detection in low intensity X-ray imagesA&AS, **128**, 397–407 (1998)

Chapter 19
Systematic Errors in High-Energy Astrophysics

Vinay Kashyap

Abstract Systematic errors are a crucial component of astronomical inference. In high-energy astrophysics, a great deal of effort is spent to minimize their effect, and analysis methods have matured over the years to automatically include high-quality calibration. However, calibration products are generally used as perfect representations of the instruments, and inherent uncertainties in their generation, both statistical and systematic, are ignored. We have developed a methodology by which such errors can be incorporated into analyses, via a modification of the MCMC process. Here we describe some recent developments by the Chandra calibration team to define, construct, and communicate the magnitude and characteristics of systematic calibration uncertainties. Our procedure can be generalized to incorporate different methods of defining the uncertainties.

19.1 Introduction

Astronomers have fully grasped the necessity of providing statistical error bars to parameter estimates. However, systematic errors are usually ignored while estimating the error budget. Consider the general observational model, which describes the translation from a known source model $S(E,\mathbf{x},t;\theta)$ to its expected signal in a detector,

$$M(E^*,\mathbf{x}^*,t;\theta) = \int dE\,d\mathbf{x}\; S(E,\mathbf{x},t;\theta)\; A(E,\mathbf{x}^*;\mathbf{x},t)\, R(E,E^*;\mathbf{x}^*,t)\; P(\mathbf{x},\mathbf{x}^*;E,t)\,, \tag{19.1}$$

where E is the intrinsic energy of the incoming photons and E^* is the energy measured in the detector, $\mathbf{x}$ is the location in the sky and $\mathbf{x}^*$ is the detector

V. Kashyap (✉)
Smithsonian Astrophysical Observatory, 60 Garden St. Cambridge, MA 02138, USA
e-mail: vkashyap@cfa.harvard.edu

E.D. Feigelson and G.J. Babu (eds.), *Statistical Challenges in Modern Astronomy V*, Lecture Notes in Statistics 209, DOI 10.1007/978-1-4614-3520-4_19,

location, t is the time, $A(.)$ is the effective area (aka ARF), $R(.)$ is the energy redistribution matrix (aka RMF), and $P(.)$ is the spatial redistribution matrix (aka PSF). The observed signal is usually Poisson(M), though there can be some more processing applied depending on the detector used. Calibration is defined by the values of the ARF, the RMF, and the PSF, and are usually determined through laboratory or in-flight measurements of objects with well-understood behavior. None of these are perfectly known, but they are used in analyses as though they are. Thus, when the data are analyzed to determine θ, given the nominal calibration products, the resulting uncertainty on it is invariably underestimated. In the following, we will consider how the errors on θ are affected and how they may be accounted for. In the process, we will describe a method that has the potential to take into account generally all types of systematic errors.

19.2 What is Systematic Error?

There are two forms of systematic error: *First,* it is the systemic, case-dependent kind where the uncertainty does not decrease when sample size is increased, unlike statistical error. This is manifested, for instance, in persistent deviations in fit residuals. *Second,* it is the kind that is introduced due to choosing a single realization of an underlying response function when in truth the response itself is subject to statistical error.

The typical tactic in dealing with systematic uncertainty is to square-add the errors, in analogy with error propagation in the Gaussian regime. But this is sub-optimal for many reasons: the error bars may be asymmetrical, and there is no way to incorporate biases and true systematics, and it is at best useful only in a strict Gaussian regime. An excellent example of the systematic errors that are present in the effective areas of current high-energy missions is shown by discrepancies in the simultaneously measured fluxes of the same object in different instruments (see, e.g., [4] in the case of the SNR G21.5, and [3] in the case of the SNR E0102).

19.3 Uncertainties in ACIS Effective Areas

A detector is a complex instrument, consisting of many subsystems. Each of the subsystems are separately calibrated, and they interact in highly non-linear ways. Analytical treatments of such systems are impossible to develop, and even a complete numerical modeling is beyond our computational ability at this stage.

Drake et al. [1] devised a Monte Carlo scheme where the uncertainty in each subsystem of the Chandra/ACIS-S detector was used to draw values that were then folded together to generate a plausible effective area curve, A_i. A number of such realizations of the effective area can be made, producing a sample of effective areas, $\mathscr{A} = \{A_i, i = 1..M\}$. From a statistical viewpoint, $\mathscr{A}$ represents a black-box sampling from the prior distribution on the effective area, $p(A)$. The effective areas thus generated demonstrate the complex nature of the calibration uncertainties. First, that there is a bias between the mean of the sample and the nominal effective area A_0 that is generated from the nominal values adopted for all the subsystems. Second, even though the values at a given energy are distributed in a manner that appears to be Gaussian, there are correlations across energy that preclude us from simply adopting independent error estimates at each energy.

19.4 Incorporating Calibration Errors

Once the calibration sample, $\mathscr{A}$, has been generated, the question becomes how to use it within data analysis. Here we describe a modified Markov-Chain Monte Carlo (MCMC) method that is flexible, robust, and fast (see [2]).

Consider an MCMC sample obtained by sampling the model parameters θ given the data, Y, and the calibration product, $A = A_0$,

$$\theta^{(k)} \sim p(\theta | Y, A_0),$$

where $\theta^{(k)}$ are the values of the parameters at iteration k. The set of parameter values thus obtained is used to estimate the best-fit values and the error bars. When calibration uncertainty is included, it is no longer possible to condition on A_0. Instead we add a new step that updates A according to the calibration uncertainties. In particular, $\theta^{(k)}$ is updated using the same iterative algorithm as above, with an additional step at each iteration that updates A. Suppose at iteration k, $A^{(k)}$ is the realization of the calibration product. Then the new algorithm consists of the following two steps:

$$A^{(k)} \text{ is sampled from } p(A|Y) \text{ and}$$
$$\theta^{(k)} \text{ is sampled from } p(\theta|Y, A^{(k)}).$$

If we assume that the data are not informative towards determining the values of the calibration, which is a good assumption for the vast majority of the cases, we can simplify the above by replacing $p(A|Y)$ with $p(A)$:

$$A^{(k)} \text{ is sampled from } p(A) \text{ and} \tag{19.2}$$
$$\theta^{(k)} \text{ is sampled from } p(\theta|Y, A^{(k)}). \tag{19.3}$$

This independence assumption gives us the freedom not to estimate the posterior distribution $p(A|Y)$ and simplifies the structure of the algorithm. It effectively separates the complex problem of model fitting in the presence of calibration uncertainties into two simpler problems: (1) fitting a model with known calibration and (2) the quantification of calibration uncertainties independent of the current data Y.

19.5 Generalization

The Bayesian method we have developed to deal with systematic calibration errors in the effective area (Sect. 19.4) can be generalized in two ways: first, to allow for the calibration sample to be modified by the data, and second, to extend the methods to other calibration products and other types of problems. These efforts are still in progress.

Fully Bayesian Method: We had originally assumed that the calibration sample $\mathscr{A}$ was invariant and that the data cannot be used to select a subspace in it that is more probable than other subspaces. This assumption is valid when the data quality is not as high as the data that was used to derive the calibration products in the first place. There is also an implicit assumption made that the calibration products are close to being correct. However, we can employ a fully Bayesian approach that bases inference on the full posterior distribution $p(\theta, A|Y)$. To accomplish this, we set up a two-step Gibbs sampler,

STEP 1: Sample $A^{(k+1)} \sim p(A|\theta^{(k)}, Y)$.
STEP 2: Sample $\theta^{(k+1)} \sim \mathscr{K}(\theta|\theta^{(k)}; Y, A^{(k+1)})$.

where $\mathscr{K}$ is the MCMC sampling kernel and k is the iteration step. Notice that STEP 1 requires that A be updated given the current data and parameter value. This is a computationally challenging step, and work is in progress to make the problem tractable.

Other Calibration Products: As noted in (19.1), calibration is defined using the ARF, RMF, and the PSF. Of these, the ARF is generally 1D, and the RMF and PSF are at least 2D. Carrying or generating samples of these quantities for every analysis can be prohibitive in both computational and storage costs. We have developed a PCA-based method to compress the information in ARFs so that the sample size can be reduced from O(10^3) to O(10). Preliminary analysis shows that a similar approach may work for the RMF and PSF. However, current PC decompositions of 2D calibration products are not physically interpretable, thus making it difficult to choose the number of components to compress the calibration sample. Nevertheless, our analysis of the effective area calibration sample suggests that a mode of writing the calibration sample that makes it feasible to generalize to any product,

$$\begin{aligned}\text{Replicate Calibration Product} = {} & \text{Mean} \\ & + \text{Offset} \\ & + \text{Explained Variability} \\ & + \text{Residual Variability}\,, \end{aligned} \tag{19.4}$$

where the Mean is the mean of the calibration sample, the Offset is the shift imposed on the center of the distribution of calibration uncertainty that accounts for biases, the Explained Variability is the portion of the variability that summarize in parametric and/or systematic way (e.g., using PCA), and the Residual Variability is the portion of the variability left unexplained by the systematic summary. This formulation allows a flexible presentation of the uncertainty:

1. When a large calibration sample is available, the random component may be set by selecting an index chosen randomly at each iteration, with the calibration product corresponding to that index used in that iteration. This process preserves the weights of the initial calibration sample, and in this case the residual component is identically zero.
2. In some cases, the calibration uncertainty is characterized by a multiplicative polynomial or spline factor that modifies the source term in (19.1). In this case, both the source model parameters θ and the modifying function parameters θ_{cal} are fit to the data that are used to determine the calibration products. Then, the Explained Variance component can be generated by sampling from the posterior distribution for θ_{cal}, with the offset and residual terms identically zero.
3. Often, only a vague estimate of a bias over a small passband is available to characterize the calibration uncertainty. This can be accounted for as a randomized additive constant term.

19.6 Summary

We have developed a new method for describing and incorporating systematic errors due to calibration uncertainties in high-energy data analysis. Our goal has been to obtain realistic error bars on astrophysical source model parameters that include both statistical and systematic errors. This work holds promise for generalizing the treatment of instrumental uncertainties to high-dimensional analyses.

Acknowledgements This work was supported by NASA AISRP grant NNG06GF17G, CXC NASA contract NAS8-39073, NSF grants DMS 04-06085, DMS 09-07522, DMS-0405953, and DMS-0907185.

References

1. Drake, J.J., et al., 2006, in Parameter Estimation Studies, Proc. SPIE, 6270, 49
2. Lee, H., et al., 2011, ApJ, 731, 126
3. Plucinsky, P., et al., 2010, *Update on the Thermal SNR WG*, in IACHEC, http://web.mit.edu/iachec/meetings/2010/Presentations/Plucinsky_thermalSNRs.pdf
4. Tsujimoto, M., et al. 2011, A&A, 525, 25

Chapter 20
Hierarchical Bayesian Models for Type Ia Supernova Inference

Kaisey S. Mandel

Abstract Type Ia supernovae (SN Ia) are the most precise cosmological distance indicators and are important for measuring the acceleration of the Universe and the properties of dark energy. Current cosmological analyses use rest-frame optical SN Ia light curves to estimate distances, whose accuracy is limited by the confounding effects of host galaxy dust extinction. The combination of broadband optical and near-infrared (NIR) light curves and spectroscopic data has the potential to improve inference in supernova cosmology. I describe a principled, hierarchical Bayesian framework to coherently model the multiple random and uncertain effects underlying the observed data, including measurement error, intrinsic supernova covariances, host galaxy dust extinction and reddening, peculiar velocities and distances. Using a new MCMC code, BAYESN, to compute probabilistic inferences for individual SN Ia and the population, I applied these hierarchical models to the joint analysis of the optical, near-infrared (NIR), and spectroscopic data from a large sample of nearby SN Ia. The combination of optical and NIR data better constrains estimates of the dust effects and approximately doubles the precision of cross-validated SN Ia distance predictions compared to using optical data alone. The hierarchical model is extended to include spectroscopic data to estimate significant correlations between the intrinsic optical colors and ejecta velocities. These applications demonstrate the power and flexibility of multi-level modeling in the analysis of SN Ia data.

20.1 Introduction

Although Type Ia supernovae (SN Ia) are generally thought to arise from the thermonuclear explosions of degenerate carbon–oxygen white dwarf stars, the exact nature of the progenitors and the mechanisms for the explosions are still uncertain.

K.S. Mandel (✉)
Harvard-Smithsonian Center for Astrophysics, 60 Garden St., Cambridge, MA 02138, USA
e-mail: kmandel@cfa.harvard.edu

E.D. Feigelson and G.J. Babu (eds.), *Statistical Challenges in Modern Astronomy V*, Lecture Notes in Statistics 209, DOI 10.1007/978-1-4614-3520-4_20,

Despite the uncertainty in their astrophysical nature, SN Ia rest-frame optical light curves have been of great utility for inferring the distances to galaxies and measuring fundamental quantities of the universe. As standardizable candles, they were critical to the detection of cosmic acceleration [37, 38]. The cosmic acceleration may be caused by a dark energy component of the universe. Recent efforts constrain the equation-of-state parameter w of dark energy to $\sim$10%, [1,2,11,21,26,28,45] using SN Ia. SN Ia have also been used to establish the extragalactic distance scale and measure the Hubble constant (recently [40–42]). Frieman et al. [12], Kirshner [27] and Howell [22] provide recent reviews of the use of Type Ia supernovae to constrain cosmology and dark energy.

The utility of SN Ia as distance indicators is based upon the standard candle principle: if SN Ia all had a single peak luminosity, then their relative distances can be determined from their observed apparent brightnesses. However, SN Ia are not exactly standard candles, and there are variations between different events. Statistical models for SN Ia as distance indicators exploit empirical correlations between peak optical luminosities of SN Ia and distance-independent measures such as light curve shape and color observed in the sample of nearby low-z SN Ia [5, 13, 19, 20, 23, 39]. One of the largest systematic uncertainties limiting the precision of distance estimates from rest-frame optical light curves is dust extinction in the host galaxy and the confounding of dust reddening with the intrinsic color variations of SN Ia [4]. Current approaches differ conceptually and practically on how apparent colors, intrinsic colors, and dust effects are modeled. While most methods make use of the optical luminosity-light curve width correlation, some methods, such as MLCS [24,38,43], attempt to separately model the intrinsic colors of the SN Ia and host galaxy dust reddening and extinction, whereas others model both effects with a single factor (e.g. SALT2; [17, 18]).

A quite promising approach towards mitigating the dust extinction problem is through the near-infrared (NIR). Observations of nearby SN Ia in the NIR revealed that the peak NIR luminosities of SN Ia have a dispersion smaller than 0.20 mag [5, 6, 30, 31, 36, 44], and could be utilized to estimate distance with a precision competitive with those derived from optical light curve shapes. The effect of dust extinction is significantly diminished at NIR wavelengths, relative to the optical. The combination of optical and NIR observations of SN Ia light curves could lead to even better estimates of SN Ia distances [29].

To address some of the challenges in the statistical modeling of Type Ia supernova, I have introduced a fully Bayesian, hierarchical or multi-level, framework for modeling the population of SN Ia and individual events. This is a natural, intuitive, and principled statistical approach: our observed data arise from multiple random and uncertain effects, such as measurement error, dust extinction, distances and peculiar velocities, acting on individual supernovae, but we wish to ultimately learn about the statistical characteristics, especially the intrinsic variations and correlations, of the SN Ia and dust populations and how best to use them to make predictions. I describe the conceptual framework and computational strategy in Sect. 20.2, and describe the application to the data and results in Sect. 20.3. Further details about this work can be found in Mandel et al. [34, 35] and Mandel [33].

20.2 Statistical Inference with Type Ia Supernovae

An example of a set of broadband SN Ia light curve observations is shown in Fig. 20.1. The left panel plots the time series of apparent magnitudes in the optical (*BVRI*) and NIR (*JH*) broadband filters of a nearby SN Ia designated SN 2005eq.

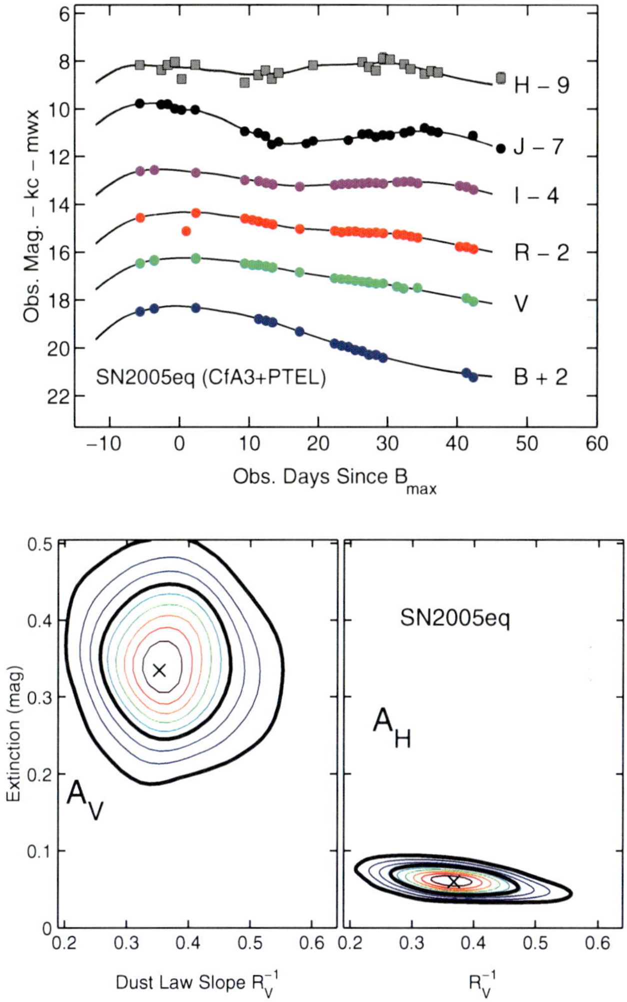

Fig. 20.1 (*Top*) Optical (*BVRI* from CfA3; [20]) and NIR (*JH* from PAIRITEL; [44]) light curve data of nearby ($z = 0.03$) Type Ia SN 2005eq, as observed by the CfA supernova group, are fitted with a non-parametric Gaussian process multi-band light curve model. (*Bottom*) Optical and NIR light curves of SN 2005eq are used to infer the host galaxy dust extinction properties. The hierarchical model enables coherent inference of host galaxy dust properties (A_V, R_V), while marginalizing over the posterior uncertainties in the dust and SN light curve populations. The *cross* indicates the marginal bivariate mode, and the two *black contours* contain 68% and 95% of the posterior probability. The inferred NIR extinction A_H is much smaller than the optical extinction A_V and has much smaller uncertainty

Observing the time series over an extended period of time is important for measuring and utilizing the population correlations between the shape of the light curves and the intrinsic properties of the SN Ia. Furthermore, the multi-wavelength observations provide constrains on the effects of the dust in the host galaxy of the SN Ia. The spectroscopic redshift ($z = 0.03$) is measured from the host galaxy spectrum.

20.2.1 Multiple Random Effects

The utility of SN Ia for cosmology inference rests upon empirical relations seen in the observed data of a sample of supernovae. The useful correlations between luminosity, color, and light curve shape are captured by statistical models that are learned from the data (as opposed to being set by theoretical models of supernova explosion physics). As a consequence, useful and realistic statistical models for SN Ia must deal with multiple sources of randomness and uncertainty.

The observed, apparent light curves of a SN Ia are the sum of its intrinsic, absolute light curves at multiple wavelengths, the effect of dust, the distance modulus and measurement error. The most obvious source of randomness is the photometric error in measuring the apparent brightness of the SN Ia in a series of images over time. The time series of measured brightnesses of each SN Ia in each broadband filter and its measurement errors are reported by the observers. The temporal or wavelength coverage of the data may not be uniform or complete for each SN Ia.

In addition to measurement error, SN Ia light curves have a component of intrinsic variation or randomness. The multi-wavelength absolute light curves of different SN Ia have different luminosities, different light curve shapes, and different colors. However, these properties of the supernovae are correlated in the population: e.g. SN Ia with fast-declining light curves tend to be intrinsically dimmer. SN Ia statistical models must capture and utilize these population-level correlations to infer the luminosities and distances to SN Ia.

The light originating from the SN Ia is attenuated by a random amount of interstellar dust in its host galaxy. The dust absorbs and scatters light to make the SN Ia appear dimmer. Since the effect of distance is also to make the SN Ia dimmer, these are partially confounding factors. They are not exactly degenerate effects, however. Dust tends to absorb and scatter the photons at short wavelengths more than at long wavelengths, and this causes a reddening effect in addition to the dimming. Importantly, the physical effect of dust along the line of sight to the SN is only to make it look dimmer and redder, not brighter and bluer, relative to the intrinsic SN colors.

To learn the statistical characteristics and absolute properties of a sample of nearby SN Ia, it is sufficient to constrain the distance by invoking the relation between the distance and recession velocity (or redshift) of a galaxy (Hubble's Law), for a given H_o, as it is relatively insensitive to the cosmological parameters in this regime. However, the relationship between the measured redshift of a supernova's host galaxy and its distance is noisy due to the random motions of galaxies with respect to the overall expansion. Hence, when "training" the model on a nearby sample, the variance around the distance-redshift relation due to these *peculiar* velocities must be accounted for in the likelihood function. When using the model to predict the distances of high-z SN Ia from the light curves alone, the random peculiar velocities must also be accounted for when fitting the distance-redshift relation as a function of the cosmological parameters.

20.2.2 A Hierarchical Bayesian Approach

A hierarchical Bayesian, or multi-level modeling, strategy is an ideal framework in which to express structured probability models describing multiple, physical random effects latent in the observed data. It allows one to coherently model and make inferences at both the level of an ensemble or population of objects as well as at the level of individuals from the ensemble. This statistical approach is well-known in the statistics literature (e.g. [14]) and has been discussed previously in the astro-statistics literature (e.g. [32]).

I have applied the hierarchical Bayesian paradigm to the statistical modeling of the multiple random and uncertain effects underlying the SN Ia data: measurement error, intrinsic variation and correlation of absolute light curve properties, host galaxy dust extinction and reddening, peculiar velocities, redshifts and distances. This approach enables the coherent estimation of the populations and individuals underlying the ensemble of SN Ia data, i.e. the parameters describing the intrinsic properties, dust effects, and distances of individual SN Ia, and the hyperparameters describing population of the intrinsic SN Ia and the dust distribution. Inference with the hierarchical model may be thought of as a probabilistic deconvolution of the observed SN data into the multiple latent random effects generating it.

The global posterior probability density, derived from the modeling assumptions and Bayes' Theorem, provides a unified measure of the joint uncertainties in the unknowns given the observed data and a clear objective function for the analysis. It quantifies the trade-offs and degeneracies in inference between competing effects, for example, the intrinsic color and extrinsic dust effects, and allows one to marginalize over these trade-offs between "nuisance" parameters when making inferences and predictions of other parameters of interest, for example, distances. The hierarchical approach naturally deals with the missing data problem (SN Ia data is sometimes sparse, and incomplete!), and its modularity simplifies model expansion.

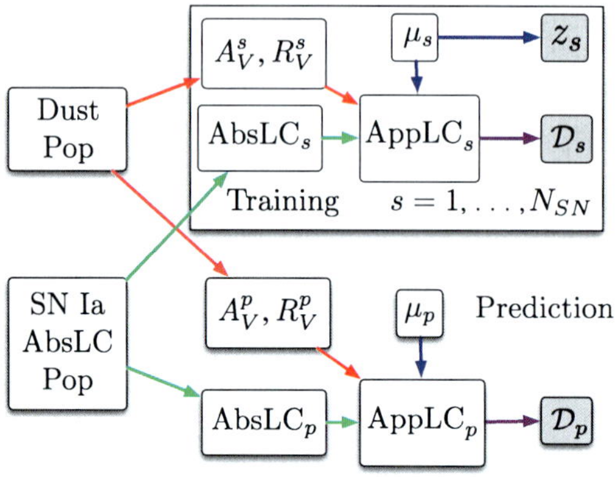

Fig. 20.2 Hierarchical framework for statistical inference with SN Ia light curves. The global posterior density of the hierarchical model parameters given the full SN data set is represented formally with a directed acyclic graph. Unknown parameters are represented by *open nodes*. Observed data (redshifts z and measured light curves $\mathscr{D}$) are represented by *shaded nodes*. Each *arrow* or *link* describes a relationship of conditional probability

20.2.3 A Generative Model

The hierarchical model I have developed can be visually expressed and intuitively understood in the form of a probabilistic graphical model called a directed acyclic graph. The overall structure of the hierarchical Bayesian model is depicted by a directed acyclic graph shown in Fig. 20.2. The graph can be understood as a generative model for the data. The hierarchical model coherently incorporates randomness and uncertainties due to measurement error (purple), intrinsic SN variations (green), dust extinction and reddening (red), peculiar velocities and distances (blue) into inferences about individual SN and the population. "SN Ia AbsLC Pop" represents hyperparameters describing the population of SN Ia light curves, including intrinsic variations and correlations in shape, color and luminosity across multiple wavelengths. From this population, each SN randomly draws a set of multi-wavelength light curves "AbsLC." The box "Dust Pop" represents hyperparameters governing the population distribution of host galaxy dust values. Each SN randomly draws dust parameters A_V (the amount of dust extinction), R_V (the wavelength dependence of the extinction) from this distribution. These dust parameters combine with the individual absolute light curves and distance modulus μ to generate an apparent light curve "AppLC," which is sampled with noise to produce the observed multi-wavelength light curve data $\mathscr{D}$. In the nearby universe, the distance modulus is related to the observed recession velocity or redshift through the Hubble law plus a noise term representing random peculiar velocities of host galaxies. This generative process is conceptually repeated for each SN in the data set. The difference between "training" and distance prediction is that the latter does not condition on the redshift-distance likelihood information of the SN (bottom).

20.2.4 Computing Inferences with the Hierarchical Bayes Model

To compute estimates for the unknown parameters of the hierarchical model, conditional on the observed data, I have constructed and implemented an MCMC algorithm, BAYESN, designed for training statistical models for SN Ia light curves and generating coherent probabilistic predictions. This code generates a Markov chain that explores the parameter space of the properties of individual events (the dust extinction A_V, the slope of the dust-reddening law, R_V, the distance modulus μ, and the apparent magnitudes and light curve shape parameters) and the hyperparameters describing the dust population and intrinsic, absolute SN Ia light curve distribution. In the long run, the chains produce samples from the global posterior probability density of the unknowns given the data described by the graphical model in Fig. 20.2.

The BAYESN algorithm exploits the conditional independence structure of the graphical model to produce efficient moves in the global parameters space within a Gibbs sampling framework. A sketch and details of the implementation are presented in [34]. Typically, four to eight chains, starting in different initial locations, are run in parallel and convergence for each parameter is monitored using the Gelman-Rubin statistic [15].

20.3 Application to SN Ia in the Optical and Near-Infrared

Mandel et al. [34, 35] applied this hierarchical Bayesian methodology to inference with nearby SN Ia in observed the optical and NIR wavelengths. The data set consisted of the apparent light curves and spectroscopic redshifts of 127 nearby SN Ia observed by the CfA Supernova Group [20, 44], the Carnegie Supernova Project [5], and others the literature.

Under "training," the redshifts are used to constrain the nearby supernova distances by invoking Hubble's Law. In this phase, the quantities of interest are the intrinsic and dust parameters of individual SN Ia as well as the hyperparameters of their population distributions. In the hierarchical model, the intrinsic covariance structure of SN Ia absolute light curves and colors over time from optical through NIR wavelengths was explicitly modeled coherently with the dust population. By examining the marginal posterior estimates of the population variances of the peak absolute magnitudes of SN Ia, it was found that the intrinsic scatter of SN Ia peak absolute magnitudes was small in the NIR (in the H-band, $\sigma(M_H) \approx 0.11$ mag, [35]), confirming previous findings that SN Ia are good standard candles in the NIR. The marginal posterior estimates of the population correlations between the absolute magnitudes at different wavelengths indicated a low correlation between the optical and NIR [34].

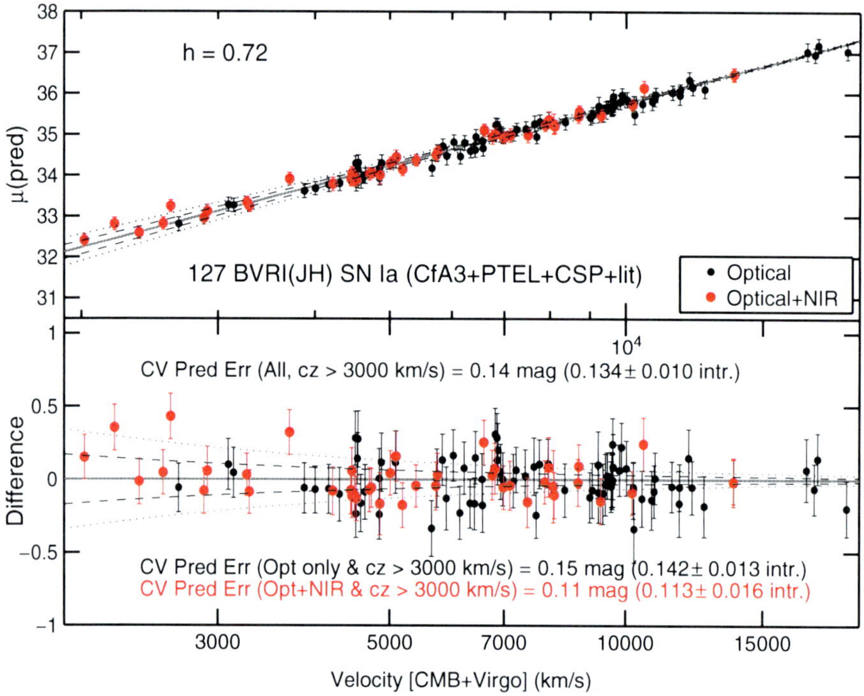

Fig. 20.3 Cross-validated Hubble diagram computed with BAYESN for the low-z nearby training set using CfA, CSP and literature SN. *Red points* indicate the SN with joint optical *BVRI* and NIR *JH* data. *Black points* are SN with only optical data. The *dashed (dotted) line* indicates the magnitude uncertainty in $\mu(z)$ for $\sigma_{\text{pec}} = 150\,(300)$ km s^{-1}. We perform bootstrap cross-validation to estimate the out-of-sample prediction error

The hierarchical model was used to test for a potential correlation between the amount of dust extinction to SN Ia (A_V) and the slope of the reddening law as a function of wavelength, R_V. It was found that SN Ia with high dust extinction had low values of the dust law slope $R_V \approx 1.7$, while those at low extinction favored $R_V \approx 2.8$. This suggests that the light from the majority of SN Ia at low extinction are affected by interstellar dust of a similar nature to Milky Way dust ($R_V = 3.1$) and the dust in nearby external galaxies ($R_V \approx 2.8$; [7,8]), while those at the highest extinctions may be obscured by a peculiar type of dust with a steeper extinction law. One possibility is that the low R_V values indicate circumstellar dust though which the SN photons undergo multiple scatterings, leading to a stepper effective extinction profile as a function of wavelength [16]. This differential trend in R_V vs. A_V contrasts with previous studies that assumed that one value of R_V applied equally to all SN Ia, and found peculiarly low values $R_V < 1.7$ [4,9,21].

Using cross-validation, I found that the distance moduli to SN Ia with joint optical and NIR data could be predicted with greater accuracy (rms = 0.11 mag) than those with optical data only (rms = 0.15 mag, Fig. 20.3). This improvement, approximately a doubling of the precision (inverse variance), can be traced to three features. First, the NIR luminosity (especially H-band) by itself has low population variance and thus provides an excellent standard candle that is relatively insensitive to dust extinction. Secondly, the intrinsic, absolute magnitudes in the optical and NIR are nearly uncorrelated in the population, indicating that the NIR provides additional, independent information about the SN Ia distance. Third, the combination of the optical and NIR extends the wavelength span over which the extinction law can be fit, improving the estimates of the extinction at both optical and NIR wavelengths. By combining and properly weighting all these statistical and physical effects in a single statistical model, the accuracy and precision of SN Ia distances are improved over those obtained from the optical light curves alone, the current standard. The improved distances to SN Ia from combining optical and NIR data have significant consequences for obtaining the best cosmological inferences about dark energy.

The constructed hierarchical model is conceptually and practically modular and flexible, allowing one to easily change the assumptions about the model components or incorporate additional information into the inference in a coherent way. For example, Mandel [33] expanded the hierarchical model to include measurements of SN Ia spectral features that have potential population correlations with the intrinsic SN Ia properties. Applying the expanded hierarchical Bayesian model to light curve and spectroscopic data for a sample of nearby SN Ia, I estimated significant population correlations between the peak optical intrinsic colors and ejecta velocities, while accounting for dust effects (Mandel et al., 2011, in preparation).

These applications demonstrate the power of the hierarchical Bayesian approach for principled, coherent estimation and prediction with SN Ia data sets. Further applications of these methods include the extension to spectral data (e.g. [3, 10]), host galaxy properties (e.g. [25]), and the analysis of cosmologically distant, high-z SN Ia.

Acknowledgements Supernova research at Harvard College Observatory is supported in part by NSF grant AST-0907903. KM thanks R.P. Kirshner and the CfA Supernova Group for continued collaborations.

References

1. Amanullah, R., et al. 2010, *Astrophys. J.*, 716, 712
2. Astier, P., et al. 2006, *Astron. Astrophys.*, 447, 31
3. Blondin, S., Mandel, K. S., & Kirshner, R. P. 2011, *Astron. Astrophys.*, 526, A81+
4. Conley, A., Carlberg, R. G., Guy, J., Howell, D. A., Jha, S., Riess, A. G., & Sullivan, M. 2007, *Astrophys. J.*, 664, L13
5. Contreras, C., et al. 2010, *Astron. J.*, 139, 519
6. Elias, J. H., Matthews, K., Neugebauer, G., & Persson, S. E. 1985, *Astrophys. J.*, 296, 379
7. Finkelman, I., et al. 2008, MNRAS, 390, 969
8. Finkelman, I., et al. 2010, MNRAS, 409, 727
9. Folatelli, G., et al. 2010, *Astron. J.*, 139, 120

10. Foley, R. J. & Kasen, D. 2011, *Astrophys. J.*, 729, 55
11. Freedman, W. L., et al. 2009, *Astrophys. J.*, 704, 1036
12. Frieman, J. A., Turner, M. S., & Huterer, D. 2008, *Ann. Rev. Astron. Astrophs.*, 46, 385
13. Ganeshalingam, M., et al. 2010, *Astrophys. J. Suppl,*, 190, 418
14. Gelman, A., Carlin, J. B., Stern, H. S., & Rubin, D. B. 2003, Bayesian Data Analysis, Second Edition (Boca Raton, Fla.: Chapman & Hall/CRC)
15. Gelman, A. & Rubin, D. B. 1992, Statistical Science, 7, 457
16. Goobar, A. 2008, *Astrophys. J.,*, 686, L103
17. Guy, J., et al. 2007, *Astron. Astrophys.*, 466, 11
18. Guy, J., Astier, P., Nobili, S., Regnault, N., & Pain, R. 2005, *Astron. Astrophys.*, 443, 781
19. Hamuy, M., et al. 1996, *Astron. J.*, 112, 2408
20. Hicken, M., et al. 2009, *Astrophys. J.*, 700, 331
21. Hicken, M., et al. 2009, *Astrophys. J.*, 700, 1097
22. Howell, D. A. 2010, ArXiv e-prints 1011.0441
23. Jha, S., et al. 2006, *Astron. J.*, 131, 527
24. Jha, S., Riess, A. G., & Kirshner, R. P. 2007, *Astrophys. J.*, 659, 122
25. Kelly, P. L., Hicken, M., Burke, D. L., Mandel, K. S., & Kirshner, R. P. 2010, *Astrophys. J.*, 715, 743
26. Kessler, R., et al. 2009, *Astrophys. J. Suppl,*, 185, 32
27. Kirshner, R. P. 2010, in Dark Energy: Observational and Theoretical Approaches, ed. P. Ruiz-Lapuente (Cambridge, UK: Cambridge University Press), 151
28. Kowalski, M., et al. 2008, *Astrophys. J.*, 686, 749
29. Krisciunas, K., et al. 2007, *Astron. J.*, 133, 58
30. Krisciunas, K., Phillips, M. M., & Suntzeff, N. B. 2004, *Astrophys. J.,*, 602, L81
31. Krisciunas, K., et al. 2004, *Astron. J.*, 128, 3034
32. Loredo, T. J. & Hendry, M. A. 2010, in Bayesian Methods in Cosmology, ed. M. Hobson et al. (Cambridge: Cambridge University Press), 245
33. Mandel, K. S. 2011, Ph.D. thesis, Harvard University
34. Mandel, K. S., Narayan, G., & Kirshner, R. P. 2011, *Astrophys. J.*, 731, 120
35. Mandel, K. S., Wood-Vasey, W. M., Friedman, A. S., & Kirshner, R. P. 2009, *Astrophys. J.*, 704, 629
36. Meikle, W. P. S. 2000, MNRAS, 314, 782
37. Perlmutter, S., et al. 1999, *Astrophys. J.*, 517, 565
38. Riess, A. G., et al. 1998, *Astron. J.*, 116, 1009
39. Riess, A. G., et al. 1999, *Astron. J.*, 117, 707
40. Riess, A. G., et al. 2011, *Astrophys. J.*, 730, 119
41. Riess, A. G., et al. 2009, *Astrophys. J.*, 699, 539
42. Riess, A. G., et al. 2009, *Astrophys. J. Suppl,*, 183, 109
43. Riess, A. G., Press, W. H., & Kirshner, R. P. 1996, *Astrophys. J.*, 473, 88
44. Wood-Vasey, W. M., et al. 2008, *Astrophys. J.*, 689, 377
45. Wood-Vasey, W. M., et al. 2007, *Astrophys. J.*, 666, 694

Chapter 21
Bayesian Flux Reconstruction in One and Two Bands

Eric R. Switzer, Thomas M. Crawford, and Christian L. Reichardt

Abstract Astrophysical surveys in radio through sub-mm wavelengths have given rise to a variety of statistical methods for photometry and counts analysis. Here, we describe a Bayesian method for reconstructing the flux of individual astrophysical point sources of emission subject to prior information about their abundance as a function of flux and spectral properties.

21.1 Setting

We will consider the limit where the angular extent of the astrophysical sources of emission is much smaller than the resolution of the survey, and refer to these sources as "point sources". A multi-band survey instrument produces maps of regions of the sky with many such sources, typically. These maps report the intensity of radiation averaged in bands around some set of wavelengths. Each source in the maps is described by its position $\mathbf{x}$ and vector of fluxes in those wavelengths, $\mathbf{S}$. The catalog $\Theta = \{N_s, \{\mathbf{x}_1, \mathbf{S}_1\}, \{\mathbf{x}_2, \mathbf{S}_2\} \cdots \{\mathbf{x}_{N_s}, \mathbf{S}_{N_s}\}\}$ then describes the flux from N_s sources in the survey. For the sake of argument, we will describe the single-band case first where the survey produces a single map $\mathbf{d}$, and sources are described by a single flux, S. Let $R(\Delta\mathbf{x})$ be the point spread function (PSF) of the instrument's response, in some direction $\Delta\mathbf{x}$ off of its central pointing, normalized such at $R(0) = 1$. Let the observed map pixels $\mathbf{d}$ be indexed by j along pointings $\mathbf{x}_j$, and represent the instrumental response due to a source at position $\mathbf{x}_s$ as the vector

E.R. Switzer (✉) • T.M. Crawford
The Kavli Institute for Cosmological Physics, The University of Chicago, 933 East 56th Street, Chicago, IL 60637, USA
e-mail: switzer@kicp.uchicago.edu

C.L. Reichardt
Department of Physics, University of California, Berkeley, CA 94720, USA
e-mail: cr@berkeley.edu

E.D. Feigelson and G.J. Babu (eds.), *Statistical Challenges in Modern Astronomy V*, Lecture Notes in Statistics 209, DOI 10.1007/978-1-4614-3520-4_21,

$[\mathbf{R}(\mathbf{x}_s)]_j = R(\mathbf{x}_s - \mathbf{x}_j)$. The observed map pixels are then the combination of signal and noise maps, as

$$\mathbf{d} = \mathbf{s} + \mathbf{n} = \sum_{i=1}^{N_s} S_i \mathbf{R}(\mathbf{x}_i) + \mathbf{n}. \tag{21.1}$$

The Bayesian reconstruction problem is to find $P(\Theta|\mathbf{d}) = L(\mathbf{d}|\Theta)\pi(\Theta)/E(\mathbf{d})$, identifying the likelihood L, prior π and evidence E which support the posterior P. Prior knowledge of the abundance weighs the interpretation that the observed flux is a bright source versus the interpretation that it is a dimmer source superimposed on a positive noise fluctuation. In typical astrophysical settings, dim sources are more common, favoring the second interpretation, thus "deboosting" the inferred source flux. In a multi-band setting, the prior also represents our knowledge of the correlations of flux between bands.

For an unknown number of sources (potentially in the thousands) this space is too large to explore using MCMC or analytic approaches. Therefore, consider the posterior of one source at a time, $(\Theta = \{\mathbf{x}, S\})$, rather than the joint posterior, assuming for now that the detected sources have negligible influence on one another's parameters. Simplify further by assuming that the noise map is normally distributed with covariance $\mathbf{N} = \langle \mathbf{n}\mathbf{n}^T \rangle$, giving the log-likelihood

$$\ln L = C - \frac{1}{2}(\mathbf{d} - \mathbf{s})^T \mathbf{N}^{-1} (\mathbf{d} - \mathbf{s}), \tag{21.2}$$

where C is a constant that does not depend on source parameter choices. With the signal model $\mathbf{s} = S\mathbf{R}(\mathbf{x})$ and at a fixed position $\mathbf{x}$, the maximum likelihood flux is determined following Carvalho et al. [1]

$$\frac{dL}{dS} = \mathbf{R}(\mathbf{x})^T \mathbf{N}^{-1} (\mathbf{d} - S\mathbf{R}(\mathbf{x})) \Rightarrow S_{\mathrm{ML}}(\mathbf{x}) = \frac{\mathbf{R}(\mathbf{x})^T \mathbf{N}^{-1} \mathbf{d}}{\mathbf{R}(\mathbf{x})^T \mathbf{N}^{-1} \mathbf{R}(\mathbf{x})} = \mathbf{F}^T(\mathbf{x})\mathbf{d}, \tag{21.3}$$

where we interpret the action on the map $\mathbf{d}$ as a linear filter $\mathbf{F}(\mathbf{x})$, which coincides with the unbiased, minimum variance estimator common in literature (see e.g., [4]). One then uses the effective solid angle of the filtered PSF for photometry. Interpret the denominator as the noise variance in the filtered map at some position $\mathbf{x}$, $\sigma_f^2(\mathbf{x}) \equiv [\mathbf{R}(\mathbf{x})^T \mathbf{N}^{-1} \mathbf{R}(\mathbf{x})]^{-1}$.

21.2 Single-Band Flux Reconstruction

We would now like to find the posterior distribution by incorporating prior information about the source population, namely, that one expects bright sources to be rare. There are two outlooks in literature: (1) use the maximum likelihood approach as a first pass to a fully Bayesian approach (see e.g., [1]) which incorporates prior information and explores Θ, (2) use the maximum likelihood approach to measure the source flux, but then "de-boost" those fluxes in a one-dimensional posterior setting.

The discussion here considers the latter and is based on the methods developed in Crawford et al. [3] for the analysis of sources detected in the mm-wavelength survey by the South Pole Telescope [7]. Begin by writing the likelihood in terms of the difference between the posterior S and the maximum likelihood value, $S = S_{\rm ML} + \Delta S$. In the context of matched filters, we will simply call $S_{\rm ML}$ the "measured" flux S_m, and the posterior S the "intrinsic" flux S_i that we would like to reconstruct. Then, at fixed $\mathbf{x}$

$$P(S_i|S_m) \propto L(S_m|S_i)\pi(S_i) \propto \exp\left\{-\frac{(S_i - S_m)^2}{2\sigma_f^2}\right\}\pi(S_i). \tag{21.4}$$

The interpretation of the likelihood is that the measured flux of a source is simply the intrinsic flux of that source plus normally-distributed noise $S_m = S_i + n$ (where $n \sim N[0, \sigma_f^2(\mathbf{x})]$), reflecting the assumption that sources are isolated on the sky and can be modeled individually (e.g., assuming either infinite resolution or rare sources). Yet, in practical radio to sub-mm astronomy, we need to revisit this assumption. Consider a bright source with flux S_i at position $\mathbf{x}$. A nearby source at position $\mathbf{x}_s$ with flux S_s will contribute $S_s\mathbf{F}(\mathbf{x})^T\mathbf{R}(\mathbf{x_s})$ to measurement at $\mathbf{x}$, which is S_s times the convolution of PSF with the filter, giving an effective PSF R_f. The flux at $\mathbf{x}$ due to all sources at separations $\Delta\mathbf{x}$ is

$$S_m = S_i + \sum_{j\neq i} S_j R_f(\Delta\mathbf{x}_j) + n = S_i + S_b + n, \tag{21.5}$$

where we have identified the sum as the background flux of all other sources, S_b. In the regime where sources can overlap, the reconstruction problem is not unique.

A common approach in the literature (see e.g., [2]) is to reconstruct the total flux that contributes to a given pointing, rather than the flux of an individual source at that pointing. That is, to let $\pi(S_i)$ be the distribution of flux in a pointing in a signal-only realization, and use the posterior of (21.4). Another popular method, the "$P(D)$" solves this problem by reconstructing the abundance of a *population* as a function of flux which reproduces the measured probability distribution of S_m across the map pixels (see e.g., [5]).

When the goal is to catalog and categorize individual sources, one can break the ambiguity by writing a posterior distribution for the brightest individual source in a given pointing, $S_{\max}$, or

$$P(S_{\max}|S_m) = L(S_m|S_{\max})\pi(S_{\max}). \tag{21.6}$$

The prior probability density that the brightest source in a pixel has flux $S_{\max}$ is the probability of having that source times the probability that there are no brighter sources, or

$$\pi(S_{\max}) \propto \left.\frac{dN}{dS}\right|_{S=S_{\max}} \exp\left(-\Omega \int_{S_{\max}}^{\infty} \frac{dN}{dS'} dS'\right), \tag{21.7}$$

where dN/dS is the differential counts of the source population per solid angle, and Ω is the filtered PSF solid angle. The likelihood of measured flux given (21.5) is then the convolution of the PDFs of the contributing pieces

$$L(S_m|S_{\max}) = \delta(S_{\max}) * FT^{-1}\{e^{[r(\omega)-r(0)]}\} * \frac{1}{\sqrt{2\pi\sigma_f^2}} e^{-S_m^2/2\sigma_f^2}, \tag{21.8}$$

where $r(\omega)$ [6] is found from the characteristic function of the response-weighted source counts truncated at $S_{\max}$ through the Fourier transform (FT),

$$r(\omega) = FT_q\left\{\int_{<S_{\max}} \frac{d\Omega_{\Delta\mathbf{x}}}{|R_f(\Delta\mathbf{x})|} n\left(\frac{q}{R_f(\Delta\mathbf{x})}\right)\right\}. \tag{21.9}$$

Indeed, when there are many background sources, Crawford et al. [3] finds an accurate normal approximation to the likelihood, $L(S_m|S_{\max}) \sim N(S_m - S_{\max} - \overline{S_b}, \sigma_{\text{tot}}^2)$, where σ_{tot} is the sum of the instrumental noise fluctuation and the background source fluctuations and $\overline{S_b}$ is the mean flux contributed by background point sources.

21.3 Two-Band Flux Reconstruction

Let the spectral index α describe the spectral behavior between the two bands as $S(\lambda_2) = S(\lambda_1)(\lambda_2/\lambda_1)^{-\alpha}$, where λ_1 and λ_2 are the wavelengths. The measured fluxes in the two bands are then (following (21.5))

$$S_{\text{m}}^{(1)} = S_{\max}^{(1)} + S_b^{(1)} + n^{(1)} \text{ and } S_{\text{m}}^{(2)} = S_{\max}^{(2)} + S_b^{(2)} + n^{(2)}. \tag{21.10}$$

We will assume that the noise terms $n^{(1)}$ and $n^{(2)}$ are uncorrelated. Bayes theorem reads

$$P(S_{\max}^{(1)}, S_{\max}^{(2)}|S_{\text{m}}^{(1)}, S_{\text{m}}^{(2)}) \propto L(S_{\text{m}}^{(1)}, S_{\text{m}}^{(2)}|S_{\max}^{(1)}, S_{\max}^{(2)})\pi(S_{\max}^{(1)}, S_{\max}^{(2)}). \tag{21.11}$$

The posterior distribution must represent two new aspects: (1) the correlation in the background source fluxes $S_b^{(1)}$ and $S_b^{(2)}$ by virtue of sharing common physical sources of emission, and (2) that knowledge of the spectral index and $S_{\max,j}^{(1)}$ informs $S_{\max,j}^{(2)}$. In the absence of these, the joint posterior splits into replicas of (21.6) for each band. For the purpose of categorizing sources, we will apply $d\alpha/dS_{\max}^{(2)}$ to transform the posterior density in $S_{\max}^{(1)}, S_{\max}^{(2)}$ to a density of the flux in one band $S_{\max}^{(1)}$ and the spectral index α between bands.

While it is possible to compute for the two-dimensional likelihood analogous to (21.8), it is much simpler to consider survey regimes where there are a sufficient number of background sources to contribute an approximately correlated Gaussian term to the flux along a given pointing. There,

$$\ln L(S_{\mathrm{m}}^{(1)}, S_{\mathrm{m}}^{(2)} | S_{\mathrm{max}}^{(1)}, \alpha) = C' - \frac{1}{2}\, \mathbf{r}^T \mathbf{C}^{-1} \mathbf{r}, \tag{21.12}$$

where $\mathbf{C}$ is the noise covariance between the bands (including contributions from instrumental noise, the atmosphere, and sources fainter than S_{max}), and $\mathbf{r}$ is the vector

$$\mathbf{r} = \left\{ S_{\mathrm{m}}^{(1)} - S_{\mathrm{max}}^{(1)} - \overline{S_b^{(1)}}, S_{\mathrm{m}}^{(2)} - S_{\mathrm{max}}^{(2)}(\alpha) - \overline{S_b^{(2)}} \right\}. \tag{21.13}$$

Under the Gaussian likelihood assumption, this method can be trivially extended to multiple bands.

One can explicitly test the flux reconstruction by simulating realizations of maps in the various bands and comparing the reconstructed fluxes with the known flux which entered the simulations. By applying the single-band method to each band individually, Crawford et al. [3] found that the inferred spectral indices could be dramatically incorrect. This is because the signal in the weaker band can be effectively deboosted to the background flux level. This can lead to an incorrect categorization based on the spectral index. In contrast, in the multiband case, information from a stronger detection in another band is carried over to the weak band.

21.4 Concluding Remarks

The fluxes inferred from the procedures described here usually enter into three sorts of subsequent products: population abundance, spectral energy distributions, and categorized source catalogs. Source categorization is central to many of these goals, and here one can apply a cut that the spectral index exceed a discrimination threshold α_d for one population versus another, $P(\alpha > \alpha_d) > P_d$. In developing the population abundance, one now has a PDF of fluxes for each source, and would like to determine the underlying abundance as a function of flux which is consistent with that sample. At high flux, most of the abundance information comes from the data, while a lower flux, progressively more information comes from the prior. More work needs to be done to quantify this tradeoff. Further rigorous methods should also be developed to combine counts data from a variety of experiments and to properly include prior information in this setting.

References

1. Carvalho, P., Rocha, G. & Hobson, M. P. (2009) A fast Bayesian approach to discrete object detection in astronomical data sets - PowellSnakes I, *Mon. Not. Royal Astro. Soc.*, 393, 681–702
2. Coppin, K., Halpern, M., Scott, D., Borys, C. & Chapman, S. (2005) An 850-μm SCUBA map of the Groth Strip and reliable source extraction, *Mon. Not. Royal Astro. Soc.*, 357, 1022–1028
3. Crawford, T. M., Switzer, E. R., Holzapfel, W. L., Reichardt, C. L., Marrone, D. P. & Vieira, J. D. (2010) A Method for Individual Source Brightness Estimation in Single- and Multi-band Data, *Astrophys. J.*, 718, 513–521
4. Haehnelt, M. G. & Tegmark. M. (1996) Using the Kinematic Sunyaev-Zeldovich effect to determine the peculiar velocities of clusters of galaxies, *Mon. Not. Royal Astro. Soc.*, 279, 545
5. Pantanchon, G., et al. (2009), Submillimeter Number Counts from Statistical Analysis of BLAST Maps, *Astrophys. J.*, 707, 1750–1765
6. Scheuer, P. A. G. (1957) A statistical method for analysing observations of faint radio stars, *Proc. Cambridge Phil. Soc.*, 53, 764–773
7. Vieira, J. D., Crawford, T. M., Switzer, E. R., et al. (2010) Extragalactic Millimeter-wave Sources in South Pole Telescope Survey Data: Source Counts, Catalog, and Statistics for an 87 Square-degree Field, *Astrophys. J.*, 719, 793–783

Chapter 22
Commentary: Bayesian Analysis Across Astronomy

Thomas J. Loredo

Abstract This contribution is a commentary on seven papers presented in the session "Bayesian Analysis Across Astronomy" at the *Statistical Challenges in Modern Astronomy V* conference held at Pennsylvania State University in June 2011. I provide a perspective on the current state and future direction of Bayesian astrostatistics with an emphasis on the development of multilevel models to link astronomical data to astrophysical theory.

I am tasked with providing a commentary on seven of the eight papers presented in the "Bayesian analysis across astronomy" session at the fifth *Statistical Challenges in Modern Astronomy* conference (SCMA V). Of course, it is impossible to comment in detail on so diverse a set of papers in the brief allotted space. At the editors' suggestion, I will instead use my commentary as a sort of bully pulpit to provide a perspective on the current state and future direction of Bayesian astrostatistics. The seven papers provide a contemporary vantage point; my participation in the previous SCMA conferences, spanning two decades, provides a somewhat more historical vantage point.

22.1 Looking Back

The first SCMA conference was held in August 1991. Bayesian methods were both new and controversial in astronomy at that time. Of the 22 papers published in the proceedings volume [6], only two were devoted to Bayesian methods ([15, 27];

T.J. Loredo (✉)
Center for Radiophysics and Space Research, Cornell University, Ithaca, NY 14853-6801, USA
e-mail: loredo@astro.cornell.edu

E.D. Feigelson and G.J. Babu (eds.), *Statistical Challenges in Modern Astronomy V*, Lecture Notes in Statistics 209, DOI 10.1007/978-1-4614-3520-4_22,

see also the unabridged version of the latter, [16]).[1] Both papers had a strong pedagogical component (and a bit of polemic). Of the 131 SCMA I participants (about 60% astronomers and 40% statisticians), only two were astronomers whose research prominently featured Bayesian methods (Steve Gull and me).

Twenty years later, the role of Bayesian methods in astrostatistics research is dramatically different. At SCMA V, two sessions were devoted entirely to Bayesian methods in astronomy: "Bayesian analysis across astronomy" (BAA), with eight papers and two commentaries, and "Bayesian cosmology," including three papers with individual commentaries. Overall, 14 of 32 invited presentations (not counting commentaries) featured Bayesian methods, and the focus was on calculations and results rather than on pedagogy and polemic.

On the face of it, the changes seem to indicate that Bayesian methods are not only no longer controversial, but are in fact now widely used, even favored for some applications (most notably for parametric inference in cosmology). But how representative are the SCMA presentations of broader astrostatistical practice?

At the meeting, Martin Hendry, Roberto Trotta and I engaged in an unintentional friendly competition in amateur bibliometrics, each of us presenting statistics based on ADS queries aiming to provide a handle on the use of Bayesian methods "across astronomy." Figure 22.1 shows my entry; Hendry's and Trotta's were similar. The publication counts indicate significant and rapidly growing use of Bayesian methods in astronomy and physics.

It is tempting to conclude from the SCMA and bibliometric indicators that Bayesian methods are now well-established and well-understood across astronomy. But the SCMA metrics reflect the role of Bayesian methods in the astrostatistics research community, not in bread-and-butter astronomical data analysis. And as impressive as the trends in the bibliometric indicators may be, the absolute numbers remain small in comparison to all astronomy and physics publications, even limiting consideration to data-based studies. Although their impact is growing, Bayesian methods are not yet in wide use by astronomers.

My interactions with colleagues indicate that significant misunderstandings persist about the differences between Bayesian and more conventional frequentist approaches to scientific inference. I believe these play no small role in hindering broader adoption of Bayesian methods in routine data analysis. This opinion seems to be shared by a number of astronomers and statisticians at SCMA V who use Bayesian methods; there was a lively discussion at the end of the BAAsession about two particularly prevalent misconceptions: the notion that prior probabilities are the main thing distinguishing Bayesian and frequentist methods, and the notion that Bayesian computation is harder than frequentist computation for implementing methods with comparable capability; both notions are incorrect. Clearing up these and other misconceptions within the broader community of astronomical data analysts is an important pedagogical task for the future, potentially paving the way

[1] A third paper [24] had some Bayesian content but focused on frequentist evaluation criteria, even for the one Bayesian procedure considered.

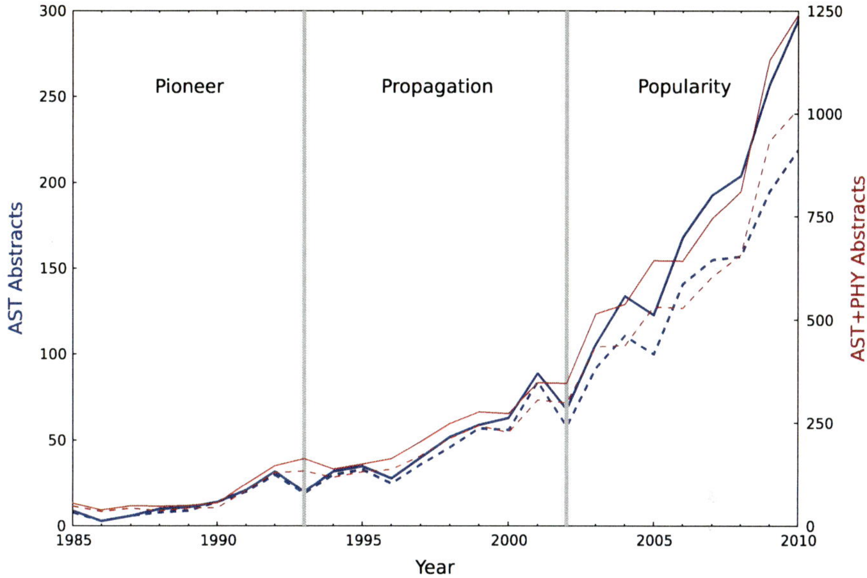

Fig. 22.1 Simple bibliometrics measuring the growing use of Bayesian methods in astronomy and physics, based on queries of the NASA ADS database in October 2011. Thick (*blue*) curves (against the *left axis*) are from queries of the astronomy database; thin (*red*) curves (against the *right axis*) are from joint queries of the astronomy and physics databases. For each case the *dashed lower curve* indicates the number of papers each year that include "Bayes" or "Bayesian" in the title or abstract. The *upper curve* is based on the same query, but also counting papers that use characteristically Bayesian terminology in the abstract (e.g., the phrase "posterior distribution" or the acronym "MCMC"); it is meant to capture Bayesian usage in areas where the methods are well-established, with the "Bayesian" appellation no longer deemed necessary or notable

to broader use of basic Bayesian methods by astronomers. And experience with basic methods will provide a bridge to understanding the more advanced methods astrostatistics researchers are developing.

22.2 Looking Forward

Now I will turn from the past to highlight an emerging theme in Bayesian astrostatistics research that is evident in the BAAApresentations. The theme harkens back to SCMA I, in particular to Mike West's commentary on my SCMA I paper [31]. In his closing remarks he pointed to an especially promising direction for future Bayesian work in astrostatistics:

> On possible future directions, it is clear that Bayesian developments during recent years have much to offer—I would identify prior modeling developments in *hierarchical* models as particularly noteworthy. Applications of such models have grown tremendously in biomedical and social sciences, but this has yet to be paralleled in the physical sciences.

> Investigations involving repeat experimentation on similar, related systems provide the archetype logical structure for hierarchical modeling... There are clear opportunities for exploitation of these (and other) developments by astronomical investigators....

However clear the opportunities may have appeared to West, there was very little work on hierarchical Bayesian modeling in astronomy for over a decade after SCMA I. A particularly promising application area is modeling of populations of astronomical sources, where hierarchical models can naturally account for measurement error, selection effects, and "scatter" of properties across a population. I discussed this at some length at SCMA IV in 2006 [18], but as of that time there was little work in astronomy using hierarchical Bayesian methods, and for the most part only the simplest such models were used.

SCMA V marks a changepoint in this respect. Several of the papers in the BAAAsession (and elsewhere) describe recent and ongoing research developing sophisticated hierarchical models for complex astronomical data. Other papers raise issues that may be addressed with hierarchical models. Together, these papers point to hierarchical Bayesian modeling as an important emerging research direction for astrostatistics.

To illustrate the notion of a hierarchical model—also known as a *multilevel model* (MLM)—we start with a simple parametric density estimation problem, and then promote it to a MLM by adding measurement error.

Suppose we would like to estimate parameters θ defining a probability density function $f(x;\theta)$ for an observable x. A concrete example might be estimation of a galaxy luminosity function, where x would be two-dimensional, $x = (L,z)$ for luminosity L and redshift z, and $f(x;\theta)$ would be the normalized luminosity function (i.e., a probability density rather than a galaxy number density). Consider first the case where we have a set of precise measurements of the observables, $\{x_i\}$ (and no selection effects). Panel (a) in Fig. 22.2 depicts this simple setting. The likelihood function for θ is $\mathscr{L}(\theta) \equiv p(\{x_i\}|\theta,M) = \prod_i f(x_i;\theta)$. Bayesian estimation of θ requires a prior density, $\pi(\theta)$, leading to a posterior density $p(\theta|\{x_i\},M) \propto \pi(\theta)\mathscr{L}(\theta)$.

An alternative way to write Bayes's theorem expresses the posterior in terms of the joint distribution for parameters and data:

$$p(\theta|\{x_i\},M) = \frac{p(\theta,\{x_i\}|M)}{p(\{x_i\}|M)}. \tag{22.1}$$

This "probability for everything" version of Bayes's theorem switches the goal of modeling from separate specification of a prior and likelihood, to specification of the joint distribution for everything; this proves helpful for building models with complex dependencies. Panel (c) depicts the dependencies in the joint distribution with a graph—a collection of nodes connected by edges—where each node represents a probability distribution for the indicated variable, and the directed edges indicate dependences between variables. Shaded nodes indicate variables whose values are known (here, the data); we may manipulate the joint to condition on these quantities. The graph structure visually displays how the joint distribution

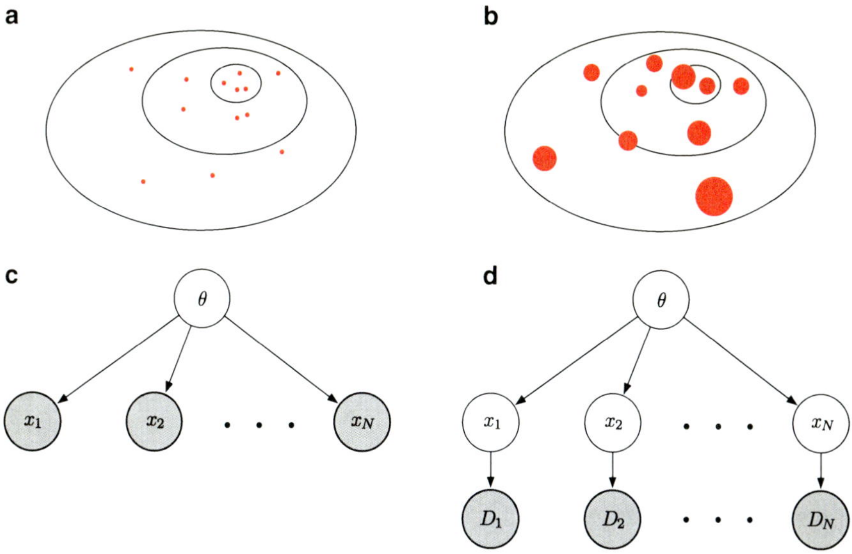

Fig. 22.2 Illustration of multilevel model approach to handling measurement error. (**a**) and (**b**) (*top row*): Measurements of a two-dimensional observable and its probability distribution (*contours*); in (**a**) the measurements are precise (*points*); in (**b**) they are noisy (filled *circles* depict uncertainties). (**c**) and (**d**): Graphical models corresponding to Bayesian estimation of the density in (**a**) and (**b**), respectively

may be factored as a sequence of independent and conditional distributions: the θ node represents the prior, and the x_i nodes represent $f(x_i;\theta)$ factors, dependent on θ but independent of other x_i values when θ is given (i.e., conditionally independent). The joint distribution is thus $p(\theta,\{x_i\}|M) = \pi(\theta)\prod_i f(x_i;\theta)$. In a sense, the most important edges in the graph are the *missing* edges; they indicate independence that makes factors simpler than they might otherwise be.

Now suppose that, instead of precise x_i measurements, for each observation we get noisy data, D_i, producing a measurement likelihood function $\ell_i(x_i) \equiv p(D_i|x_i,M)$ describing the uncertainties in x_i (we might summarize it with the mean and standard deviation of a Gaussian). Panel (b) depicts the situation; instead of points in x space, we now have likelihood functions (depicted as "1σ" error circles). Panel (d) shows a graph describing this measurement error problem, which adds a $\{D_i\}$ level to the previous graph; we now have a multilevel model.[2] The x_i nodes are now unshaded; they are no longer known, and have become *latent parameters*. From the graph we can read off the form of the joint distribution:

[2]The convention is to reserve the term for models with three or more levels of nodes, i.e., two or more levels of edges, or two or more levels of nodes for *uncertain variables* (i.e., unshaded nodes). The model depicted in panel (d) would be called a two-level model.

$$p(\theta, \{x_i\}, \{D_i\}|M) = \pi(\theta) \prod_i f(x_i; \theta) \ell_i(x_i). \tag{22.2}$$

From this joint distribution we can make inferences about any quantity of interest. To estimate θ, we use the joint to calculate $p(\theta, \{x_i\}|\{D_i\}, M)$ (i.e., we condition on the known data using Bayes's theorem), and then we marginalize over all x_i variables. We can estimate all the x_i values jointly by instead marginalizing over θ. Note that this produces a joint marginal distribution for $\{x_i\}$ that is not a product of independent factors; although the x_i values are conditionally independent given θ, they are *marginally* dependent. If we do not know θ, each x_i tells us something about all the others through what it tells us about θ. Statisticians use the phrase "borrowing strength" to describe this effect, from John Tukey's evocative description of "mustering and borrowing strength" from related data in multiple stages of data analysis (see [19] for a tutorial discussion of this effect and the related concept of shrinkage estimators).

The few Bayesian MLMs used by astronomers through the 1990s and early 2000s did not go much beyond this simplest hierarchical structure. For example, unbeknownst to West, at the time of his writing my thesis work had already developed a MLM for analyzing the arrival times and energies of neutrinos detected from SN 1987A; the multilevel structure was needed to handle measurement error in the energies (an expanded version of this work appears in [20]). Panel (a) of Fig. 22.3 shows a graph describing the model. The rectangles are "plates" indicating substructures that are repeated; the integer variable in the corner indicates the number of repeats. There are two plates because neutrino detectors have a limited (and energy-dependent) detection efficiency. The plate with a known repeat count, N, corresponds to the N detected neutrinos with times t and energies ε; the plate with an unknown repeat count, $\overline{N}$, corresponds to undetected neutrinos, which must be considered in order to constrain the total signal rate; $\overline{D}$ denotes the nondetection data, i.e., reports of zero events in time intervals between detections.

Other problems tackled by astronomers with two-level MLMs include: modeling of number-size distributions ("$\log N$–$\log S$" or "number counts") of gamma-ray bursts and trans-Neptunian objects (e.g., [21, 26]); performing linear regression with measurement error along both axes, e.g., for correlating quasar hardness and luminosity ([13]; see his paper in these proceedings for an introduction to MLMs for measurement error); accounting for Eddington and Malmquist biases in cosmology [19]; statistical assessment of directional coincidences with gamma-ray bursts [10, 22] and in large catalog cross-matching ([3]; see Budavári's paper and my commentary on it in these proceedings for discussion of the underlying MLM); and handling multivariate measurement error when estimating stellar velocity distributions from proper motion survey data [2]. Dobigeon, [4] developed a similar three-level MLM to tackle joint segmentation of astronomical arrival time series (a multivariate extension of Scargle's well-known Bayesian Blocks algorithm). van Dyk et al. [29] describe a many-level MLM for fitting *Chandra* X-ray spectral data; a host of latent parameters enable accurate accounting for uncertain backgrounds and instrumental effects such as pulse pile-up.

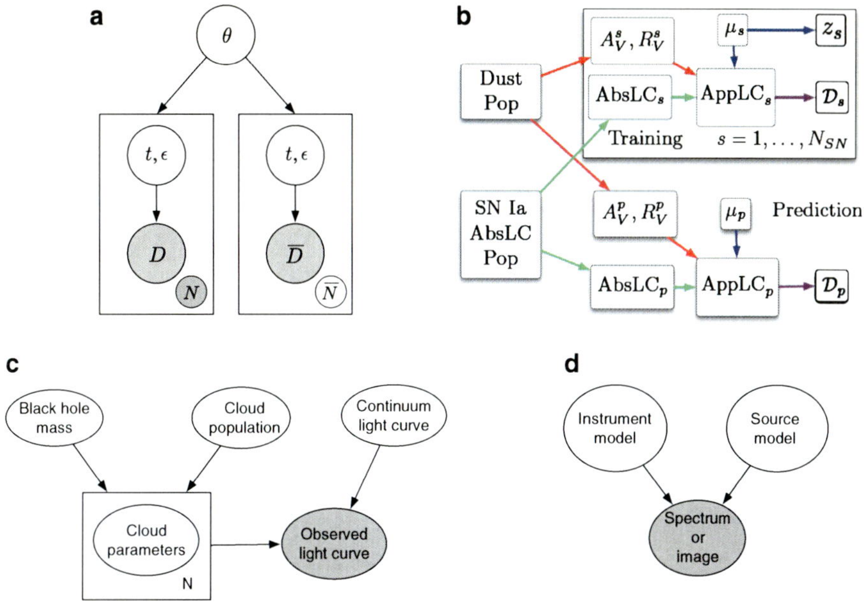

Fig. 22.3 Graphs describing multilevel models used in astronomy, as described in the text

Cosmology is a natural arena for multilevel modeling, because of the indirect link between theory and observables. For example, in modeling both the cosmic microwave background (CMB) and the large scale structure (LSS) of the galaxy distribution, theory does not predict a specific temperature map or set of galaxy locations (these depend on unknowable initial conditions), but instead predicts statistical quantities, such as angular or spatial power spectra. Modeling observables given theoretical parameters typically requires introducing these quantities as latent parameters. In Loredo [17] I described a highly simplified hierarchical treatment of CMB data, with noisy CMB temperature difference time series data at the lowest level, $l = 2$ spherical harmonic coefficients in the middle, and a single physical parameter of interest, the cosmological quadrupole moment Q, at the top. While a useful illustration of the MLM approach, I noted there were enormous computational challenges facing a more realistic implementation. It took a decade for such an implementation to be developed, in the pioneering work of [30]. And only recently have explicit hierarchical models been implemented for LSS modeling (e.g., [14]).

This brings us to the present, and the SCMA V contributions. Panel (b) of Fig. 22.3 shows the graph for one of the MLMs described in Mandel's presentation in the BAAAsession (the reader will have to consult his paper in these proceedings, or Mandel et al. [23], for a description of the variables and the model). With three levels, complex connections between latent variables (some of them random *functions*—light curves—rather than scalars), and three different types of data, this model leapfrogs nearly all previous astronomical MLMs in complexity.

One has to be careful with highly structured MLMs. Information gain from the data tends to weaken as one considers parameters at increasingly high levels [9]. On the one hand, if one is interested in quantities at lower levels, this weakens dependence on assumptions made at high levels. On the other hand, if one is interested in high-level quantities, sensitivity to the prior becomes an issue. The weakened impact of data on high levels has the effect that improper (i.e., non-normalized) priors that are safe to use in simple models (because the likelihood makes the posterior proper) can be dangerous in MLMs [7, 11]. The impact of the graph structure on the model's predictive ability becomes less intuitively accessible as complexity grows, making predictive tests of MLMs important, but also nontrivial [1, 8, 28]. An exemplary feature of Mandel's work is the use of careful predictive checks, implemented via a frequentist cross-validation procedure, to quantitatively assess the adequacy of the model.

Other BAAApresentations invoked interesting MLMs to tackle forefront astrostatistics problems. Brewer's treatment of reverbation mapping, aiming to estimate supermassive black hole masses from observation of light echoes from the broad line regions of active galactic nuclei, used a complex MLM with the structure schematically shown in the graph in panel (c). Kashyap showed how to use Bayesian methods to account for systematic error in instrument properties, such as the energy-dependent effective area or point spread function, when analyzing *Chandra* X-ray spectral and imaging data. Panel (d) is a deceptively simple graphical summary of his team's approach, showing two levels of an MLM. The novel feature here is the split upper level, with the left node, representing instrument properties, *not* gray.[3] Conventionally, astronomers fix the instrument description (corresponding to making this node gray in the graph). Kashyap's team instead assigns a probability distribution to the instrument properties, using a Monte Carlo code modeling interactions between instrument subcomponents (this node itself has multiple levels). They show how marginalizing over uncertain instrument properties can propagate systematic error throughout the whole analysis. In fact, in some of this team's work the bottom node in this graph is not simply data, but is itself a nontrivial MLM connecting the source and instrument inputs to the observables [29].

Other sessions also featured work using Bayesian MLMs. Wandelt's presentation in the Bayesian cosmology session described a hierarchical Bayes approach (albeit without explicit MLM language) for reconstructing the galaxy density field from noisy photometric redshift data. Contributed presentations also featured MLMs, including a treatment of number-size distributions by Baines et al., and an analysis of directional coincidences between ultra-high energy cosmic rays and local active galactic nuclei by Soiaporn et al., who tackled the problem with a three-level MLM combining marked point processes for modeling the population of cosmic ray sources and directional statistics for describing measurement errors.

[3]Drell et al. [5] used a similar structure, in a much simpler setting, to account for systematic error in supernova cosmology; in place of the instrument property node was a node describing possible evolution of supernova properties.

The BAAAsession also included presentations raising issues that could profitably be addressed with a multilevel approach. For example, Switzer described an approach for accounting for Eddington bias in estimating fluxes of point sources measured by multiband surveys. The underlying MLM resembles panel (d) of Fig. 22.2, but with the top node gray, corresponding to fixed specification of a population flux distribution. Flux estimates shift from their naive best-fit values because the flux distribution plays the role of a nonuniform prior. A hierarchical Bayesian treatment would parameterize the flux distribution, allowing for more adaptive adjustment of the fluxes. Several papers cited above on MLMs for number-size distributions adopt this approach, for single-band fluxes. The challenge Switzer's team faces is extending it to multiband data, where the distribution of spectral shapes across the population becomes important.

I will close this commentary with a provocative recommendation I have offered at meetings (including SCMA V) but not yet in print, born of my experience using multilevel models for astronomical populations. It is that astronomers cease producing catalogs of estimated fluxes and other source properties from surveys. This warrants explanation and elaboration.

As noted above, a consequence of the hierarchical structure of MLMs is that the values of latent parameters at low levels cannot be estimated independently of each other. In a survey context, this means that the flux (and potentially other properties) of a source cannot be accurately or optimally estimated considering only the data for that source. This may initially seem surprising, but at some level astronomers already know this to be true. We know—from Eddington, Malmquist, and Lutz and Kelker—that simple estimates of source properties will be misleading if we do not take into account information besides the measurements, i.e., specification of the population distribution of the property. The standard Malmquist and Lutz-Kelker corrections adopt a fixed (e.g., spatially homogeneous) population distribution, and produce an independent corrected estimate for each object. What the fully Bayesian MLM approach adds to the picture is the ability to handle uncertainty in the population distribution. After all, a prime reason for performing surveys is to learn about populations. When the population distribution is not well-known a priori, every source property measurement bears on estimation of the population distribution, and thus indirectly, each measurement bears on the estimation of the properties of every other source, via a kind of adaptive "bias correction."[4] This is Tukey's "mustering and borrowing of strength" at work.

To enable this mustering and borrowing, we have to stop thinking of a catalog entry as providing all the information needed to infer a particular source's properties (even in the absence of auxiliary information from outside a particular survey). Such a complete summary of information is instead provided by the marginal posterior

[4]It is worth pointing out that this is not a uniquely Bayesian insight. Eddington, Malmquist, and Lutz and Kelker used frequentist arguments to justify their corrections; Eddington even offered adaptive corrections. The large and influential statistics literature on shrinkage estimators leads to similar conclusions; see [18] for further discussion and references.

distribution for that source, which depends on the data from *all* sources—and on population-level modeling assumptions. However, in the MLM structure (e.g., panel (d) of Fig. 22.2), the *likelihood function* for the properties of a particular source may be independent of information about other sources. The simplest output of a survey that would enable accurate and optimal subsequent analysis is thus a *catalog of likelihood functions* (or possibly marginal likelihood functions when there are uncertain backgrounds or other "nuisance" effects the surveyor must account for).

For a well-measured source, the likelihood function may be well-approximated by a Gaussian that can be easily summarized with a mean and standard deviation. But these should not be presented as point estimates and uncertainties.[5] For sources near a nominal "detection limit," more complicated summaries may be justified. Counterpart surveys should cease reporting upper limits when a known source is not securely detected; instead they should report a more informative non-Gaussian likelihood summary. Discovery surveys (aiming to detect new sources rather than counterparts) could potentially devise likelihood summaries that communicate information about sources with fluxes *below* a nominal detection limit, and about uncertain source multiplicty in crowded fields. Recent work on maximum-likelihood fitting of "pixel histograms" (also known as "probability of deflection" or $P(D)$ distributions), which contain information about undetected sources, hints at the science such summaries might enable in a MLM setting (e.g., [25]).

In this approach to survey reporting, the notion of a detection limit as a decision boundary identifying sources disappears. In its place there will be decision boundaries, driven by both computational and scientific considerations, that determine what type of likelihood summary is associated with a particular candidate source location.

Coming at this issue from another direction, Hogg and Lang [12] have recently made similar suggestions, including some specific ideas for how likelihoods may be summarized. Multilevel models provide a principled framework, both for motivating such a thoroughgoing revision of current practice, and for guiding its detailed development. Perhaps at SCMA VI in 2016 we will be able to report on analyses of the first survey catalogs providing such more optimal, MLM-ready summaries.

But even in the absence of so revolutionary a development, I think one can place high odds in favor of a bet that Bayesian multilevel modeling will be flourishing in astrostatistics research by the time of SCMA VI. Whether Bayesian methods (multilevel and otherwise) will start flourishing *outside* the astrostatistics research community is another matter, dependent on how effectively astrostatisticians can rise to the challenge of making Bayesian methods more broadly used and understood. The abundance of young astronomers with enthusiasm for Bayesian astrostatistics, present both at SCMA V and in the Center for Astrostatistics summer schools, makes me optimistic.

---

[5] I am tempted to recommend that, even in this regime, the likelihood summary be chosen so as to deter misuse as an estimate, say by tabulating the $+1\sigma$ and -2σ points rather than means and standard deviations. I am only partly facetious about this!

Acknowledgements I gratefully acknowledge NSF and NASA for support of current research underlying this commentary, via grants AST-0908439, NNX09AK60G and NNX09AD03G. I thank Martin Weinberg for helpful discussions on information propagation within multilevel models. Finally, I congratulate the SCMA organizers, Eric Feigelson and Jogesh Babu, for reaching the 20-year milestone marked by SCMA V. The multidisciplinary SCMA conferences, and other outreach efforts of Babu's and Feigelson's Center for Astrostatistics, have nurtured the burgeoning astrostatistics community and sowed seeds for important collaborations between astronomers and statisticians. CASt activities have played no small role in encouraging the adoption and maturation of Bayesian methods in astronomy. I am grateful in particular for how Feigelson's and Babu's efforts have supported my own Bayesian astrostatistics research and educational efforts.

References

1. Bayarri, M.J., Castellanos, M.E.: Bayesian checking of the second levels of hierarchical models. Statist. Sci. **22**(3), 322–343 (2007). DOI 10.1214/07-STS235.
2. Bovy, J., Hogg, D.W., Roweis, S.T.: Extreme deconvolution: Inferring complete distribution functions from noisy, heterogeneous and incomplete observations. Ann. Appl. Stat. **5**(2B), 1657–1677 (2011)
3. Budavári, T., Szalay, A.S.: Probabilistic Cross-Identification of Astronomical Sources. *Astrophys. J.*, **679**, 301–309 (2008). DOI 10.1086/587156
4. Dobigeon, N., Tourneret, J.Y., Scargle, J.D.: Joint segmentation of multivariate astronomical time series: Bayesian sampling with a hierarchical model. IEEE Trans. Signal Process. **55**(2), 414–423 (2007). DOI 10.1109/TSP.2006.885768.
5. Drell, P.S., Loredo, T.J., Wasserman, I.: Type IA Supernovae, Evolution, and the Cosmological Constant. *Astrophys. J.*, **530**, 593–617 (2000). DOI 10.1086/308393
6. Feigelson, E.D., Babu, G.J. (eds.): Statistical Challenges in Modern Astronomy (1992)
7. Gelman, A.: Prior distributions for variance parameters in hierarchical models (comment on article by Browne and Draper). Bayesian Anal. **1**(3), 515–533 (electronic) (2006)
8. Gelman, A., Carlin, J.B., Stern, H.S., Rubin, D.B.: Bayesian data analysis, second edn. Texts in Statistical Science Series. Chapman & Hall/CRC, Boca Raton, FL (2004)
9. Goel, P.K., DeGroot, M.H.: Information about hyperparameters in hierarchical models. J. Amer. Statist. Assoc. **76**(373), 140–147 (1981).
10. Graziani, C., Lamb, D.Q.: Likelihood methods and classical burster repetition. In: R. E. Rothschild & R. E. Lingenfelter (ed.) High Velocity Neutron Stars, *American Institute of Physics Conference Series*, vol. 366, pp. 196–200 (1996). DOI 10.1063/1.50246
11. Hadjicostas, P., Berry, S.M.: Improper and proper posteriors with improper priors in a Poisson-gamma hierarchical model. Test **8**(1), 147–166 (1999). DOI 10.1007/BF02595867.
12. Hogg, D.W., Lang, D.: Telescopes don't make catalogues! In: EAS Publications Series, *EAS Publications Series*, vol. 45, pp. 351–358 (2011). DOI 10.1051/eas/1045059
13. Kelly, B.C.: Some Aspects of Measurement Error in Linear Regression of Astronomical Data. *Astrophys. J.*, **665**, 1489–1506 (2007). DOI 10.1086/519947
14. Kitaura, F.S., Enßlin, T.A.: Bayesian reconstruction of the cosmological large-scale structure: methodology, inverse algorithms and numerical optimization. MNRAS**389**, 497–544 (2008). DOI 10.1111/j.1365-2966.2008.13341.x
15. Loredo, T.J.: Promise of Bayesian inference for astrophysics. In: E. D. Feigelson & G. J. Babu (ed.) Statistical Challenges in Modern Astronomy, pp. 275–306 (1992)
16. Loredo, T.J.: The promise of bayesian inference for astrophysics (unabridged). Tech. rep., Department of Astronomy, Cornell University (1992). http://citeseerx.ist.psu.edu/viewdoc/summary?doi=10.1.1.56.1842CiteSeer DOI 10.1.1.56.1842

17. Loredo, T.J.: The return of the prodigal: Bayesian inference For astrophysics. In: J. M. Bernardo, J. O. Berger, A. P. Dawid and A. F. M. Smith (eds.) *Bayesian Statistics 5 Preliminary Proceedings*, volume distributed to participants of the 5th Valencia Meeting on Bayesian Statistics (1995). http://citeseerx.ist.psu.edu/viewdoc/summary?doi=10.1.1.55.3616 CiteSeer DOI 10.1.1.55.3616
18. Loredo, T.J.: Analyzing Data from Astronomical Surveys: Issues and Directions. In: G. J. Babu & E. D. Feigelson (ed.) Statistical Challenges in Modern Astronomy IV, *Astronomical Society of the Pacific Conference Series*, vol. 371, p. 121 (2007)
19. Loredo, T.J., Hendry, M.A.: Bayesian multilevel modelling of cosmological populations. In: Hobson, M. P., Jaffe, A. H., Liddle, A. R., Mukeherjee, P., & Parkinson, D. (ed.) Bayesian Methods in Cosmology, p. 245. Cambridge University Press (2010)
20. Loredo, T.J., Lamb, D.Q.: Bayesian analysis of neutrinos observed from supernova SN 1987A. *Physical Review D*, **65**(6), 063002 (2002). DOI 10.1103/PhysRevD.65.063002
21. Loredo, T.J., Wasserman, I.M.: Inferring the Spatial and Energy Distribution of Gamma-Ray Burst Sources. II. Isotropic Models. *Astrophys. J.*, **502**, 75 (1998). DOI 10.1086/305870
22. Luo, S., Loredo, T., Wasserman, I.: Likelihood analysis of GRB repetition. In: C. Kouveliotou, M. F. Briggs, & G. J. Fishman (ed.) American Institute of Physics Conference Series, *American Institute of Physics Conference Series*, vol. 384, pp. 477–481 (1996). DOI 10.1063/1.51706
23. Mandel, K.S., Narayan, G., Kirshner, R.P.: Type Ia Supernova Light Curve Inference: Hierarchical Models in the Optical and Near-infrared. *Astrophys. J.*, **731**, 120 (2011). DOI 10.1088/0004-637X/731/2/120
24. Nousek, J.A.: Source existence and parameter fitting when few counts are available. In: E. D. Feigelson & G. J. Babu (ed.) Statistical Challenges in Modern Astronomy, pp. 307–327 (1992)
25. Patanchon, G., et al.: Submillimeter Number Counts from Statistical Analysis of BLAST Maps. *Astrophys. J.*, **707**, 1750–1765 (2009). DOI 10.1088/0004-637X/707/2/1750
26. Petit, J.M., Kavelaars, J.J., Gladman, B., Loredo, T.: Size Distribution of Multikilometer Transneptunian Objects. In: Barucci, M. A., Boehnhardt, H., Cruikshank, D. P., Morbidelli, A., & Dotson, R. (ed.) The Solar System Beyond Neptune, pp. 71–87. University of Arizona Press (2008)
27. Ripley, B.D.: Bayesian methods of deconvolution and shape classification. In: E. D. Feigelson & G. J. Babu (ed.) Statistical Challenges in Modern Astronomy, pp. 329–346 (1992)
28. Sinharay, S., Stern, H.S.: Posterior predictive model checking in hierarchical models. J. Statist. Plann. Inference **111**(1-2), 209–221 (2003). DOI 10.1016/S0378-3758(02)00303-8.
29. van Dyk, D.A., Connors, A., Kashyap, V.L., Siemiginowska, A.: Analysis of Energy Spectra with Low Photon Counts via Bayesian Posterior Simulation. *Astrophys. J.*, **548**, 224–243 (2001). DOI 10.1086/318656
30. Wandelt, B.D., Larson, D.L., Lakshminarayanan, A.: Global, exact cosmic microwave background data analysis using Gibbs sampling. *Physical Review D*, **70**(8), 083511 (2004). DOI 10.1103/PhysRevD.70.083511
31. West, M.: Commentary. In: E. D. Feigelson & G. J. Babu (ed.) Statistical Challenges in Modern Astronomy, p. 328ff (1992)

Part III
Data Mining and Astroinformatics

Chapter 23
Sparse Astronomical Data Analysis

Jean-Luc Starck

Abstract Wavelets have been used extensively for several years now in astronomy for many purposes, ranging from data filtering and deconvolution, to star and galaxy detection or cosmic ray removal. More recent sparse representations such as ridgelets or curvelets have also been proposed for the detection of anisotropic features such as cosmic strings in the l microwave background. We review in this paper a range of methods based on sparsity that have been proposed for astronomical data analysis.

23.1 Introduction

The wavelet transform (WT) has been extensively used in astronomical data analysis during the last 10 years. A quick search with ADS (NASA Astrophysics Data System, adswww.harvard.edu) shows that around 1,000 papers contain the keyword "wavelet" in their abstract, and this holds for all astrophysical domains, from study of the sun through to CMB (Cosmic Microwave Background) analysis [29]. This broad success of the wavelet transform is due to the fact that astronomical data generally gives rise to complex hierarchical structures, often described as fractals. Using multiscale approaches such as the wavelet transform, an image can be decomposed into components at different scales, and the wavelet transform is therefore well-adapted to the study of astronomical data. Furthermore, since noise in the physical sciences is often not Gaussian, modeling in wavelet space of

J.-L. Starck (✉)
Laboratoire AIM, UMR CEA-CNRS-Paris 7, Irfu, SEDI-SAP, Service d'Astrophysique, CEA Saclay, F-91191 Gif-sur-Yvette CEDEX, France
e-mail: jstarck@cea.fr

E.D. Feigelson and G.J. Babu (eds.), *Statistical Challenges in Modern Astronomy V*, Lecture Notes in Statistics 209, DOI 10.1007/978-1-4614-3520-4_23,

many kinds of noise—Poisson noise, combination of Gaussian and Poisson noise components, non-stationary noise, and so on— has been a key motivation for the use of wavelets in astrophysics.

If wavelets represent well isotropic features, they are far from optimal for analyzing anisotropic objects. This has motived other constructions such as the curvelet transform [4]. More generally, the best data decomposition is the one which leads to the sparsest representation, i.e. few coefficients have a large magnitude, while most of them are close to zero. Hence, for specific astronomical data sets containing edges (planetary images, cosmic strings, etc.), curvelets should be preferred to wavelets.

In this paper, we review a range of astronomical data analysis methods based on sparse representations. We first introduce the concept of sparsity, and we present several sparse representations that have been used for astronomical data analysis. Then we present how sparse representations can be used in different applications.

23.2 Introduction to Sparsity

23.2.1 What Is Sparsity?

A signal x, $x = [x_1, \cdots, x_N]$, is sparse if most of its entries are equal to zero. For instance, a k-sparse signal is a signal where only k samples have a non-zero value. A less strict definition is to consider a signal as weakly sparse or compressible when only a few of its entries have a large magnitude, while most of them are close to zero.

If a signal is not sparse, it may be *sparsified* using a given data representation. For instance, if x is a sine, it is clearly not sparse but its Fourier transform is extremely sparse (i.e. 1-sparse). Hence we say that a signal x is sparse in the Fourier domain if its Fourier coefficients $\hat{x}[u]$, $\hat{x}[u] = \frac{1}{N}\sum_{k=-\infty}^{+\infty} x[k]e^{2i\pi\frac{uk}{N}}$, is sparse. More generally, we can model a vector signal $x \in \mathbb{R}^N$ as the linear combination of T elementary waveforms, also called *signal atoms*: $x = \Phi\alpha = \sum_{i=1}^{T}\alpha[i]\phi_i$, where $\alpha[i] = \langle x, \phi_i \rangle$ are called the decomposition coefficients of x in the dictionary $\Phi = [\phi_1, \cdots, \phi_T]^T$ (the $N \times T$ matrix whose columns are the atoms normalized to a unit ℓ_2-norm, i.e. $\forall i \in [1,T], \|\phi_i\|_{\ell^2} = 1$).

Therefore to get a sparse representation of our data we need first to define the dictionary Φ and then to compute the coefficients α. x is sparse in Φ if the sorted coefficients in decreasing magnitude have a fast decay; i.e. most of coefficients α vanish but a few.

23.2.2 What Is the Best Dictionary?

Obviously, the best dictionary is the one which leads to the sparsest representation. Hence we could imagine having a huge overcomplete dictionary (i.e. $T \gg N$), but we would be faced with prohibitive computation time cost for calculating the α coefficients. Therefore there is a trade-off between the complexity of our analysis step (i.e. the size of the dictionary) and the computation time. Some specific dictionaries have the advantage of having fast operators and are very good candidates for analyzing the data. The Fourier dictionary is certainly the most famous, but many others have been proposed in the literature such as wavelets [16], ridgelets [4], curvelets [6, 22], bandlets [15], to name only a few. Different approaches have also been recently proposed in order to build a dictionary directly from the data. This is the case in learned dictionaries [18], for instance using e.g. the KSVD algorithm [1], the grouplet decomposition [17] or the GMCA method for multichannel/hyperspectral data [3].

In astronomy, the most well known sparse representation, if we omit the Fourier transform, is certainly the isotropic undecimated wavelet transform (see next section).

23.3 Useful Dictionaries for Astronomical Data

23.3.1 *The Isotropic Undecimated Wavelet Transform*

The Isotropic undecimated wavelet transform (IUWT) [29] decomposes an $n \times n$ image c_0 into a coefficient set $W = \{w_1, \ldots, w_J, c_J\}$, as a superposition of the form

$$c_0[k,l] = c_J[k,l] + \sum_{j=1}^{J} w_j[k,l],$$

where c_J is a coarse or smooth version of the original image c_0 and w_j represents the details of c_0 at scale 2^{-j} (see [22,23] for more information). Thus, the algorithm outputs $J+1$ sub-band arrays of size $n \times n$. (The present indexing is such that $j = 1$ corresponds to the finest scale or high frequencies).

Hence, we have a *multi-scale pixel representation*, i.e. each pixel of the input image is associated with a set of pixels of the multi-scale transform. This wavelet transform is very well adapted to the detection of isotropic features, and this explains its success for astronomical image processing, where the data contain mostly isotropic or quasi-isotropic objects, such as stars, galaxies or galaxy clusters.

The decomposition is achieved using the filter bank $(h_{2D}, g_{2D} = \delta - h_{2D}, \tilde{h}_{2D} = \delta, \tilde{g}_{2D} = \delta)$ where h_{2D} is the tensor product of two 1D filters h_{1D} and δ is the dirac function. The passage from one resolution to the next one is obtained using the "à trous" algorithm [23]

$$\begin{aligned} c_{j+1}[k,l] &= \sum_m \sum_n h_{1D}[m] h_{1D}[n] c_j[k + 2^j m, l + 2^j n], \\ w_{j+1}[k,l] &= c_j[k,l] - c_{j+1}[k,l] \, , \end{aligned} \tag{23.1}$$

where h_{1D} is typically a symmetric low-pass filter such as the B_3 Spline filter: $h_{1D} = \left\{\frac{1}{16}, \frac{1}{4}, \frac{3}{8}, \frac{1}{4}, \frac{1}{16}\right\}$.

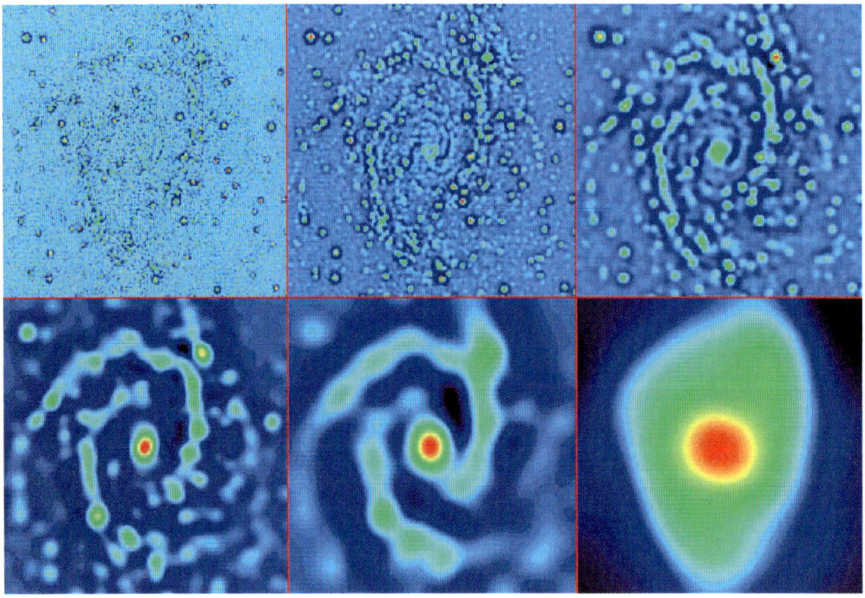

Fig. 23.1 Wavelet transform of NGC 2997 by the IUWT. The co-addition of these six images reproduces exactly the original image

Fig. 23.2 Galaxy NGC 2997

Figure 23.1 shows IUWT of the galaxy NGC 2997 displayed in Fig. 23.2. Five wavelet scales are shown and the final smoothed plane (lower right). The original image is given exactly by the sum of these six images.

23.3.2 Signal Detection in the Wavelet Space

Observed data Y in the physical sciences are generally corrupted by noise, which is often additive and which follows in many cases a Gaussian distribution, a Poisson distribution, or a combination of both. It is important to detect the wavelet coefficients which are "significant", i.e. the wavelet coefficients which have an absolute value too large to be due to noise. We defined the multiresolution support M of an image Y by:

$$M_j[k,l] = \begin{cases} 1 & \text{if } w_j[k,l] \text{ is significant} \\ 0 & \text{if } w_j[k,l] \text{ is not significant} \end{cases} \tag{23.2}$$

where $w_j[k,l]$ is the wavelet coefficient of Y at scale j and at position (k,l). We need now to determine when a wavelet coefficient is significant. For Gaussian noise, it is easy to derive an estimation of the noise standard deviation σ_j at scale j from the noise standard deviation, which can be evaluated with good accuracy in an automated way [24]. To detect the significant wavelet coefficients, it suffices to compare the wavelet coefficients $w_j[k,l]$ to a threshold level t_j. t_j is generally taken equal to $K\sigma_j$, and K is chosen between 3 and 5. The value of 3 corresponds to a probability of false detection of 0.27%. If $w_j[k,l]$ is small, then it is not significant and could be due to noise. If $w_j[k,l]$ is large, it is significant:

$$\begin{array}{ll} \text{if } \mid w_j[k,l] \mid \geq t_j & \text{then } w_j[k,l] \text{ is significant} \\ \text{if } \mid w_j[k,l] \mid < t_j & \text{then } w_j[k,l] \text{ is not significant} \end{array} \tag{23.3}$$

When the noise is not Gaussian, other strategies may be used:

- **Poisson noise:** if the noise in the data Y is Poisson, the transformation [2] $\mathscr{A}(Y) = 2\sqrt{I + \frac{3}{8}}$ acts as if the data arose from a Gaussian white noise model, with $\sigma = 1$, under the assumption that the mean value of I is sufficiently large. However, this transform has some limits and it has been shown that it cannot be applied for data with less than 20 photons per pixel. So for X-ray or gamma ray data, other solutions have to be chosen, which manage the case of a reduced number of events or photons under assumptions of Poisson statistics.
- **Gaussian + Poisson noise:** the generalization of variance stabilization [20] is:

$$\mathscr{G}((Y[k,l]) = \frac{2}{\alpha}\sqrt{\alpha Y[k,l] + \frac{3}{8}\alpha^2 + \sigma^2 - \alpha g}$$

where α is the gain of the detector, and g and σ are the mean and the standard deviation of the read-out noise.
- **Poisson noise with few events using the MS-VST:** For images with very few photons, one solution consists in using the Multi-scale Variance Stabilization Transform [35]. The MSVST combines both the Anscombe transform and

the IUWT in order to produce *stabilized* wavelet coefficients, i.e. coefficients corrupted by a Gaussian noise with a standard deviation equal to 1. In this framework, wavelet coefficients are now calculated by:

$$\begin{array}{c}\text{IUWT}\\ +\\ \text{MS-VST}\end{array} \left\{ \begin{array}{ll} c_j & = \sum_m \sum_n h_{1D}[m] h_{1D}[n] \\ & \quad c_{j-1}[k + 2^{j-1}m, l + 2^{j-1}n] \\ w_j & = \mathscr{A}_{j-1}(c_{j-1}) - \mathscr{A}_j(c_j) \end{array} \right. \tag{23.4}$$

where $\mathscr{A}_j$ is the VST operator at scale j defined by:

$$\mathscr{A}_j(c_j) = b^{(j)} \sqrt{|c_j + e^{(j)}|} \tag{23.5}$$

where the variance stabilization constants $b^{(j)}$ and $e^{(j)}$ only depends on the filter h_{1D} and the scale level j. They can all be pre-computed once for any given h_{1D} [35]. The multiresolution support is computed from the MSVST coefficients, considering a Gaussian noise with a standard deviation equal to 1. This stabilization procedure is also invertible as we have:

$$c_0 = \mathscr{A}_0^{-1} \left[\mathscr{A}_J(a_J) + \sum_{j=1}^{J} w_j \right] \tag{23.6}$$

For other kind of noise (correlated noise, non-stationary noise, etc.), other solutions have been proposed to derive the multiresolution support [29]. In the next section, we show how the multiresolution support can be used for denoising and deconvolution.

23.3.3 Curvelet

The 2D curvelet transform [4] was developed in an attempt to overcome some limitations inherent in former multiscale methods e.g. the 2D wavelet, when handling smooth images with edges i.e. singularities along smooth curves. Basically, the curvelet dictionary is a multiscale pyramid of localized directional functions with anisotropic support obeying a specific parabolic scaling such that at scale 2^{-j}, its length is $2^{-j/2}$ and its width is 2^{-j}. This is motivated by the parabolic scaling property of smooth curves. Other properties of the curvelet transform as well as decisive optimality results in approximation theory are reported in [5]. Notably, curvelets provide optimally sparse representations of manifolds which are smooth away from edge singularities along smooth curves. Several digital curvelet transforms [6, 22] have been proposed which attempt to preserve the essential properties of the continuous curvelet transform and several papers report on their successful application in astrophysical experiments [25, 26, 28].

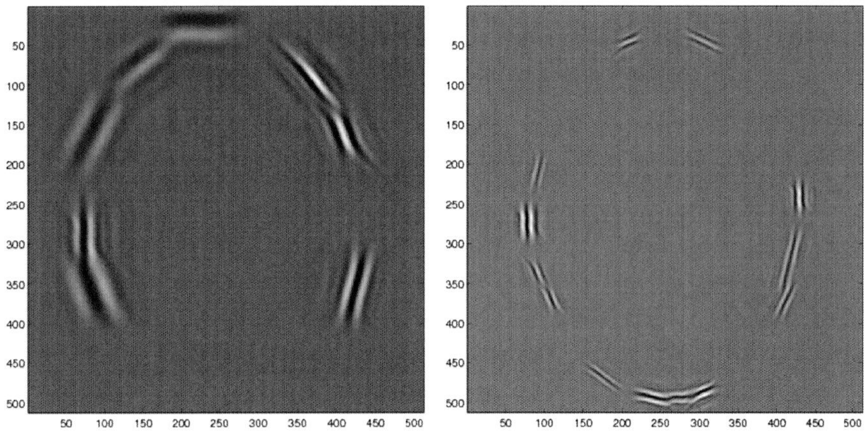

Fig. 23.3 A few first generation curvelets. Backprojections of a few curvelet coefficients at different positions and scales

Figure 23.3 shows a few curvelets at different scales, orientations and locations. It has been shown that the curvelet transform could be very useful for detecting weak anisotropic features such as cosmic strings [26].

Curvelets have been recently extended to the third dimension [32–34].

23.3.4 Sparsity on the Sphere

23.3.4.1 Introduction

Cosmic Microwave Background (CMB) observed data from WMAP and PLANCK satellite are both spherical (i.e. the whole sky is observed), and polarized. Full-sky CMB polarization data consists of measurements of the Stokes parameters so that in addition to the temperature T map, Q and U maps are given as well. The fourth Stokes parameter commonly denoted V is a measure of circular polarization. In the case of CMB which is not expected to have circularly polarized anisotropies, V vanishes. The former three quantities, T, Q and U then fully describe the linear polarization state of the CMB radiation incident along some radial line of sight. T is the total incoming intensity, Q is the difference between the intensities transmitted by two perfect orthogonal polarizers the directions of which define a reference frame in the tangent plane, and U is the same as Q but with polarizers rotated 45° in that tangent plane. To analyze CMB data (non Gaussinaty detection, etc), sparse decompositions have been recently developed. This section reviews how wavelet and curvelet transforms can be extended to spherical data and polarized spherical fields.

23.3.4.2 Wavelet and Curvelet on the Sphere

The undecimated isotropic transform on the sphere described in [28] is similar in many respects to the usual isotropic undecimated wavelet transform described previously. It is obtained using a zonal scaling function $\phi_{l_c}(\vartheta,\varphi)$ which depends only on colatitude ϑ and is invariant with respect to a change in longitude φ. It follows that the spherical harmonic coefficients $\hat{\phi}_{l_c}(l,m)$ of ϕ_{l_c} vanish when $m \neq 0$ which makes it simple to compute spherical harmonic coefficients $\hat{c}_0(l,m)$ of $c_0 = \phi_{l_c} * f$ where $*$ stands for convolution:

$$\hat{c}_0(l,m) = \widehat{\phi_{l_c} * f}(l,m) = \sqrt{\frac{4\pi}{2l+1}}\hat{\phi}_{l_c}(l,0)\hat{f}(l,m) \tag{23.7}$$

A possible scaling function [31], defined in the spherical harmonics representation, is $\phi_{l_c}(l,m) = \frac{2}{3}B_3(\frac{2l}{l_c})$ where B_3 is the cubic B-spline compactly supported over $[-2,2]$. Denoting $\phi_{2^{-j}l_c}$ a rescaled version of ϕ_{l_c} with cut-off frequency $2^{-j}l_c$, a multi-resolution decomposition of f on a dyadic scale is obtained recursively:

$$\begin{aligned} c_0 &= \phi_{l_c} * f \\ c_j &= \phi_{2^{-j}l_c} * f = c_{j-1} * h_{j-1} \end{aligned} \tag{23.8}$$

where the zonal low pass filters h_j are defined by

$$\hat{H}_j(l,m) = \sqrt{\frac{4\pi}{2l+1}}\hat{h}_j(l,m) = \begin{cases} \dfrac{\hat{\phi}_{\frac{l_c}{2^{j+1}}}(l,m)}{\hat{\phi}_{\frac{l_c}{2^j}}(l,m)} & \text{if } l < \frac{l_c}{2^{j+1}} \quad \text{and} \quad m=0 \\ 0 & \text{otherwise} \end{cases} \tag{23.9}$$

The cut-off frequency is reduced by a factor of 2 at each step so that in applications where this is useful such as compression, the number of samples could be reduced adequately. Using a pixelization scheme such as Healpix [14], this can easily be done by dividing by 2 the Healpix *nside* parameter when computing the inverse spherical harmonics transform. As in the undecimated isotropic algorithm, the wavelet coefficients can be defined as the difference between two consecutive resolutions, $w_{j+1}(\vartheta,\varphi) = c_j(\vartheta,\varphi) - c_{j+1}(\vartheta,\varphi)$. This defines a zonal wavelet function ψ_{l_c} as

$$\hat{\psi}_{\frac{l_c}{2^j}}(l,m) = \hat{\phi}_{\frac{l_c}{2^{j-1}}}(l,m) - \hat{\phi}_{\frac{l_c}{2^j}}(l,m) \tag{23.10}$$

With this particular choice of wavelet function, the decomposition is readily inverted by summing the coefficient maps on all wavelet scales

$$f(\vartheta,\varphi) = c_J(\vartheta,\varphi) + \sum_{j=1}^{J} w_j(\vartheta,\varphi) \tag{23.11}$$

where we have made the simplifying assumption that f is equal to c_0. Obviously, other wavelet functions ψ could be used just as well, such as the needlet [19]. Based on this undecimated wavelet transform on the sphere, it was shown in [28] that curvelet on sphere can be derived by applying ridgelets on the different wavelet scales.

23.3.5 *Polarized Data on the Sphere*

The spin-2 spherical harmonics basis denoted ${}_{\pm 2}Y_{\ell m}$:

$$Q \pm iU = \sum_{\ell,m} {}_{\pm 2}a_{\ell m \pm 2}Y_{\ell m} \tag{23.12}$$

and the E and B mode are defined on the sphere by

$$\begin{aligned} E &= \textstyle\sum_{\ell,m} a^E_{\ell m} Y_{\ell m} = \sum_{\ell,m} -\frac{{}_2a_{\ell m} + {}_{-2}a_{\ell m}}{2} Y_{\ell m} \\ B &= \textstyle\sum_{\ell,m} a^B_{\ell m} Y_{\ell m} = \sum_{\ell,m} i \frac{{}_2a_{\ell m} - {}_{-2}a_{\ell m}}{2} Y_{\ell m} \end{aligned} \tag{23.13}$$

where $Y_{\ell m}$ stands for the usual spin 0 spherical harmonics basis functions. The quantities E and B are derived by applying the spin lowering operator twice to $Q + iU$ and the spin raising operator twice to $Q - iU$ so that E and B are real scalar fields on the sphere, invariant through rotations of the local reference frame. From Cosmic Mircowave Background observations such those provided by WMAP and PLANCK, cosmological information can be directly obtained from the power spectra and cross-spectra of the fields T,E,B

Combining the wavelet transform on the sphere and the spin-2 spherical harmonics decomposition, it was shown in [30] that the we can derive coefficient maps w^T_j, w^E_j, w^B_j and the low resolution approximation maps c^T_J, c^E_J, c^B_J:

$$T = c^T_J + \sum_{j=1}^{J} w^T_j \qquad E = c^E_J + \sum_{j=1}^{J} w^E_j \qquad B = c^B_J + \sum_{j=1}^{J} w^B_j \tag{23.14}$$

where c^X_J stands for the low resolution approximation to component X and w^X_j is the map of wavelet coefficients of that component on scale j. Finally, we can easily construct an EB-curvelet transform by first computing the E-B wavelet transform, following by a ridgelet transform on the different scales of the decomposition. More details can be found in [30]. Figure 23.4 shows, on the left, backprojections of E-wavelet coefficients, and, on the right, backprojections of B-wavelet coefficients on the right hand side at different scales. Figure 23.5 shows the backprojection of a B-curvelet coefficient.

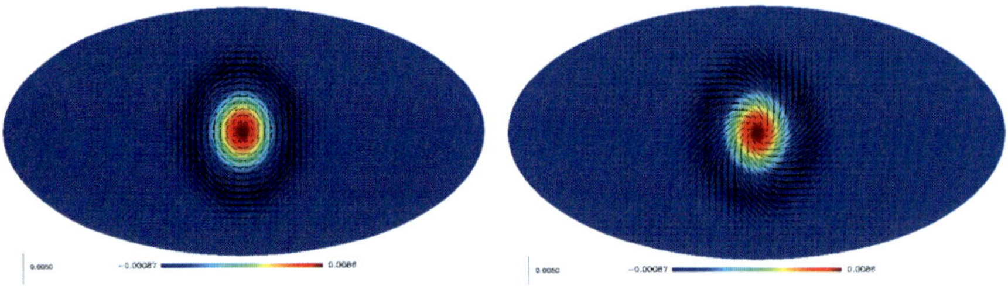

Fig. 23.4 E-wavelet (*left*) and B-wavelet atoms (*right*)

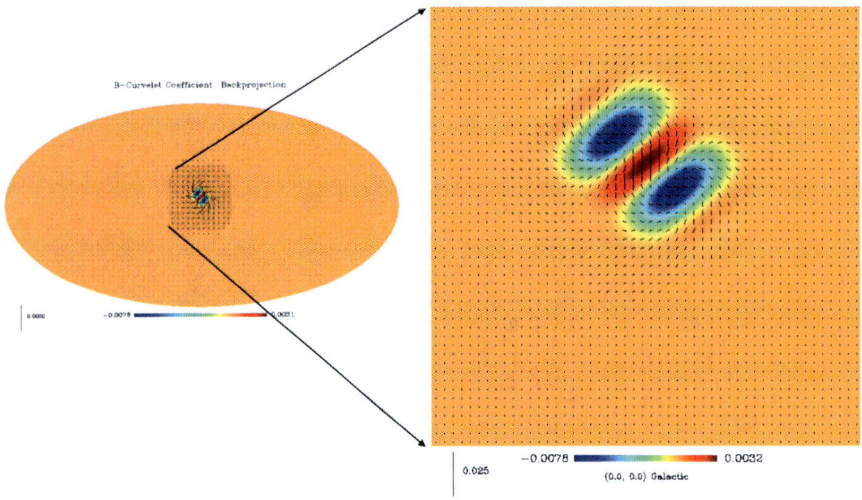

Fig. 23.5 B-curvelet atom

23.4 Inverse Problems and Sparsity

23.4.1 The Sparsity Prior

Many image problems in astronomy can be formalized as a linear inverse problem,

$$Y = AX + \varepsilon, \tag{23.15}$$

where Y are a set of noisy measurements, ε is an additive noise, X is the solution of our problem, and A is a linear operator. Finding X knowing the data Y and A is an inverse problem. When it has not a unique and stable solution, it is an *ill-posed problem*, and a regularization is necessary to reduce the space of candidate solutions. Once the dictionary Φ is chosen, inverse problems can be regularized using a sparsity penalty. Between all possible solutions, we want the one which has the sparsest representation in the dictionary Φ. Noting α the representation coefficients in Φ, the solution X can be reconstructed as $X = \Phi\alpha$, the sparsity can be measured through the $\|\alpha\|_{\ell^0}$ norm, which indicates the limit of ℓ^p when $p \to 0$. This counts in fact the number of non-zero elements in the sequence. This approach leads to the following minimization problem:

$$\min_{\alpha} \|\alpha\|_{\ell^0} \text{ s.t. } \|Y - A\Phi\alpha\|_{\ell^2} \leq \sigma. \tag{23.16}$$

It was proposed to convexify the constraint by substituting the convex ℓ_1 norm for the ℓ_0 norm leading to [7]:

$$\min_{\alpha} \|\alpha\|_{\ell^1} \text{ s.t. } \|Y - A\Phi\alpha\|_{\ell^2} \leq \sigma. \tag{23.17}$$

This equation can also be recast in its Lagrangian form:

$$\min_{\alpha} \lambda \|\alpha\|_{\ell^1} + \frac{1}{2}\|Y - A\Phi\alpha\|_{\ell^2}^2. \tag{23.18}$$

Depending on the A operator, there are several ways to obtain the solution of this equation.

23.4.2 Deconvolution

In a deconvolution problem, when the sensor is linear, A is the block Toeplitz matrix. A first iterative thresholding deconvolution method was proposed in [27] which consists in the following iterative scheme:

$$X^{(n+1)} = X^{(n)} + A^T \left(\mathbf{WDen}_{\Omega^{(n)}} \left(Y - AX^{(n)} \right) \right) \tag{23.19}$$

where **WDen** is an operator which performs a wavelet thresholding, i.e. applies the wavelet transform of the residual $R^{(n)}$ (i.e. $R^{(n)} = Y - AX^{(n)}$), threshold some wavelet coefficients, and applies the inverse wavelet transform. Only coefficients that belong to the so called *multiresolution support* $\Omega^{(n)}$ [27] are kept, while the others are set to zero. At each iteration, the multiresolution support $\Omega^{(n)}$ is updated by selecting new coefficients in wavelet transform of the residual which have an absolute value larger than a given threshold. The threshold is automatically derived assuming a given noise distribution such as Gaussian or Poisson noise.

More recently, it was shown ([8, 13], Daubechies et al. 2007) that a solution of (23.18) can be obtained through a thresholded Landweber iteration

$$X^{(n+1)} = \mathbf{WDen}_{\lambda} \left(X^{(n)} + A^T \left(Y - AX^{(n)} \right) \right), \tag{23.20}$$

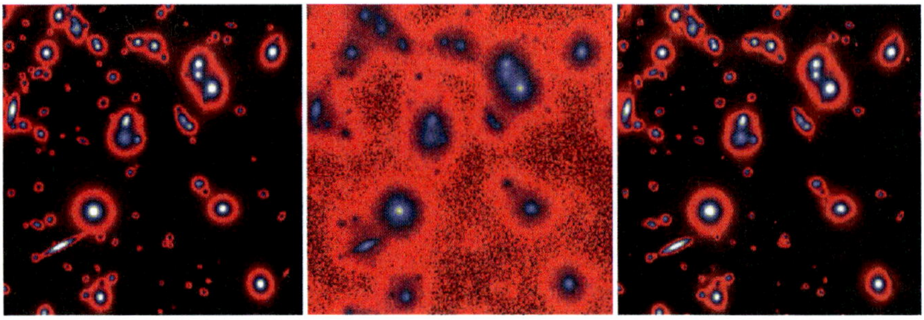

Fig. 23.6 Simulated Hubble Space Telescope image of a distant cluster of galaxies. *Left*: original, unaberrated and noise-free. *middle*: input, aberrated, noise added. *Right*, wavelet restoration wavelet

with $||A|| = 1$. In the framework of monotone operator splitting theory, it was shown that for frame dictionaries, a slight modification of this algorithm converges to the solution [8]. Extension to constrained non-linear deconvolution is proposed in [10].

A simulated Hubble Space Telescope image of a distant cluster of galaxies is shown in Fig. 23.6, middle. The simulated data are shown in Fig. 23.6, left. Wavelet deconvolution solution is shown Fig. 23.6, right. The method is stable for any kind of point spread function, and any kind of noise modeling can be considered.

23.4.3 Inpainting

The classical image inpainting problem can be defined as follows. Let X be the ideal complete image, Y the observed incomplete image and M the binary mask (i.e. $M_i = 1$ if we have information at pixel i, $M_i = 0$ otherwise). In short, we have: $Y = MX$. Inpainting consists in recovering X knowing Y and M. We thus want to minimize:

$$\min_X ||\Phi^T X||_0 \quad \text{subject to} \quad Y = MX. \tag{23.21}$$

Note that we now switch to an analysis-type prior in (23.21). It was shown in [11] that this optimization problem can be efficiently solved through an iterative thresholding algorithm called MCA:

$$X^{(n+1)} = \Delta_{\Phi,\lambda_n}(X^{(n)} + Y - MX^{(n)}). \tag{23.22}$$

where the nonlinear operator $\Delta_{\Phi,\lambda}(Z)$ consists in (1) decomposing the signal Z in the dictionary Φ to derive the coefficients $\alpha = \Phi^T Z$, (2) thresholding the coefficients: $\tilde{\alpha} = \rho(\alpha, \lambda)$, where the thresholding operator ρ can either be a hard thresholding or a soft thresholding, and (3) reconstructing $\tilde{Z}$ from the thresholded coefficients $\tilde{\alpha}$.

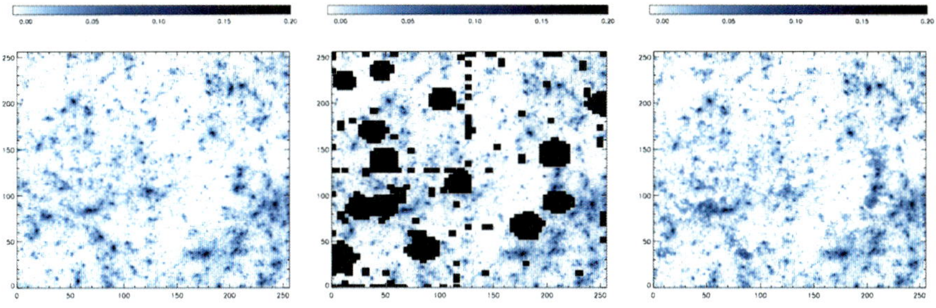

Fig. 23.7 *Left panel*, simulated weak lensing mass map, *middle panel*, simulated mass map with a standard mask pattern, *right panels*, inpainted mass map. The region shown is $1^\circ \times 1^\circ$

The threshold parameter λ_n decreases with the iteration number and it plays a role similar to the cooling parameter of the simulated annealing techniques, i.e. it allows the solution to escape from local minima. More details on optimization in inpainting with sparsity can be found in [12].The case where the dictionary is a union of subdictionaries $\Phi = \{\Phi_1, \ldots, \Phi_K\}$ where each Φ_i has a fast operator has also been investigated in [11, 12].

The experiment was conducted on a simulated weak lensing mass map masked by a typical mask pattern (see Fig. 23.7). The left panel shows the simulated mass map and the middle panel shows the masked map. The result of the inpainting method is shown in the right panel. We note that the gaps are undistinguishable by eye. More interesting, it has been shown that, using the inpainted map, we can reach an accuracy of about 1% for the power spectrum and 3% for the bispectrum [21].

Acknowledgements This work has been supported by the European Research Council grant SparseAstro (ERC-228261) and the French National Agency for Research (ANR -08-EMER-009-01).

References

1. Aharon, M., Elad, M. & Bruckstein, A. (2006) K-SVD: An Algorithm for Designing Overcomplete Dictionaries for Sparse Representation, *IEEE Trans. Signal Processing*, 54, 4311–4322
2. Anscombe, F. J. (1948), The transformation of Poisson, binomial and negative-binomial data, *Biometrika*, 15, 246–254
3. Bobin, J., Starck, J.-L., Moudden, Y. & Fadili, M. J. (2008) Blind source separation: The sparsity revolution, *Advances in Imaging & Electron Physics*, 152, 221–306
4. Candès, E. J. & Donoho, D. L. (1999), Ridgelets: the key to high dimensional intermittency?, *Phil. Trans. Royal Soc. London A*, 357, 2495–2509
5. Candès, E. J. & Donoho, D. L. (1999b) Curvelets—A surprisingly effective nonadaptive representation for objects with edges, in *Curve and Surface Fitting: Saint Malo 1999*, (A. Cohen et al., eds.), Vanderbilt Univ. Press

6. Candès, E. J., Demanet, L., Donoho, D. L. & Ying, L. (2006), Fast discrete curvelet transforms, *SIAM Multiscale Modeling & Simulation*, 5, 861–899
7. Chen, S. S., Donoho, D. L. & Saunders, M. A. (1999), Atomic decomposition by basis pursuit, *SIAM J. Scientific Computing*, 20, 33–61
8. Combettes, P. L. & Wajs, V. R. (2005) Signal recovery by proximal forward–backward splitting, *SIAM Multiscale Modeling and Simulation*, 4, 1168–1200
9. Daubechies, I., Tecschke, G. & Vese, L. (2007), Iteratively solving linear inverse problems under general convex constraints, *Inverse Problems and Imaging*, 1, 29–46
10. Dupé, F.-X., Fadili, M. J.& Starck, J.-L. (2009) A proximal iteration for deconvolving Poisson noisy images using sparse representations, *IEEE Trans. Image Processing*, 18, 310–321
11. Elad, M. (2005), Why simple shrinkage is still relevant for redundant representations, Tech. Rept., Dept. Computer Science, Technion
12. Fadili, M. J., Starck, J.-L. & Murtagh, F. (2009) Inpainting and zooming under sparse representations, *The Computer Journal*, 52, 64–79
13. Figueiredo, M. A. & Nowak, R. (2003), An EM Algorithm for wavelet-based image restoration, *IEEE Trans. Image Processing*, 12, 906–916
14. Górski, K.-M., et al. (2005) HEALPix: A framework for high-resolution discretization and fast analysis of data distributed on the sphere, *Astron. J.*, 122, 759–771
15. Le-Pennec, E. & Mallet, S. (2005) Sparse geometric image representation with bandelets, *IEEE Trans. Image Processing*, 14, 423–438
16. Mallat, S. C. (1998) *A Wavelet Tour of Signal Processing*, 2nd ed., Academic Press
17. Mallat, S. (2009) Geometrical grouplets, *Applied & Computational Harmonic Analysis*, 26, 161–180
18. Olshausen, B. A. & Field, D. J. (2008) Emergence of simple-cell receptive-field properties by learning a sparse code for natural images, *Nature*, 381, 607–609
19. Marinucci, D., et al. (2008) Spherical needlets for cosmic microwave background data analysis, *Mon. Notices Roy. Astro. Soc.*, 383, 539–545
20. Murtagh, F., Starck, J.-L. & Bijaoui, A. (1995) Image restoration with noise suppression using a multiresolution support, *Astron. Astrophys. Suppl.*, 112, 179–189
21. Pires, S., Starck, J.-L., et al. (2009) FAst STatistics for weak Lensing (FASTLens): fast method for weak lensing statistics and map making, *Mon. Not. Roy. Astro. Soc.*, 395, 1265–1279
22. Starck, J.-L., Candès, E. J. & Donoho, D. L. (2002) The curvelet transform for image denoising, *IEEE Trans. Image Processing*, 11, 131–141
23. Starck, J.-L., Murtagh, F. & Bijaoui, A. (1998) *Image Processing and Data Analysis: The Multiscale Approach*, Cambridge Univ Press
24. Starck, J.-L. & Murtagh, F. (1998) Automatic noise estimation from the multiresolution support, *Pub. Astron. Soc. Pacific*, 110, 193–199
25. Starck, J.-L., Nguyen, M. K. & Murtagh, F. (2003) Wavelets and curvelets for image deconvolution: A combined approach, *Signal Processing*, 83, 2279–2283
26. Starck, J.-L., Aghanim, N. & Forni, O. (2004) Detecting cosmological non-Gaussian signatures by multiscale methods, *Astron. Astrophys.*, 416, 9–17
27. Starck, J.-L., Bijaoui, A. & Murtagh, F. (1995) Multiresolution support applied to image filtering and deconvolution, *CVGIP: Graphical Models and Image Processing*, 57, 420–431
28. Starck, J.-L., Moudden, Y., Abrial, P. & Nguyen, M. (2006) Wavelets, ridgelets and curvelets on the sphere, *Astron. Astrophys.*, 446, 1191–1204
29. Starck, J.-L. & Murtagh, F. (2006) *Astronomical Image and Data Analysis*, 2nd ed., Springer
30. Starck, J.-L., Moudden, Y. & Bobin, J. (2009) *Astron. Astrophys.*, 497, 931–943
31. Starck, J.-L., Murtagh, F. & Bertero, M. (2010), The starlet transform in astronomical data processing: Application to source detection and image deconvolution, in *Handbook of Mathematical Models in Imaging*, Springer
32. Woiselle, A., Starck, J.-L. & Fadili, M. J. (2010) 3D curvelet transforms and astronomical data restoration, *Applied and Computational Harmonic Analysis*, 28, 171–188
33. Woiselle, A., Starck, J.-L. & Fadili, M. J. (2011) 3D data denoising and inpainting with the fast curvelet transform, *J. Mathematical Imaging & Vision*, in press

34. Ying, L., Demanet, L. & Candès, E. (2005) 3D discrete curvelet transform, in *Wavelets XI Conference*, SPIE
35. Zhang, B., Fadili, M. J. & Starck, J.-L. (2008) Wavelets, ridgelets and curvelets for Poisson noise retrieval, *IEEE Trans. Image Processing*, 17, 1093–1108

Chapter 24
Exploiting Non-linear Structure in Astronomical Data for Improved Statistical Inference

Ann B. Lee and Peter E. Freeman

Abstract Many estimation problems in astrophysics are highly complex, with high-dimensional, non-standard data objects (e.g., images, spectra, entire distributions, etc.) that are not amenable to formal statistical analysis. To utilize such data and make accurate inferences, it is crucial to transform the data into a simpler, reduced form. Spectral kernel methods are non-linear data transformation methods that efficiently reveal the underlying geometry of observable data. Here we focus on one particular technique: diffusion maps or more generally spectral connectivity analysis (SCA). We give examples of applications in astronomy; e.g., photometric redshift estimation, prototype selection for estimation of star formation history, and supernova light curve classification. We outline some computational and statistical challenges that remain, and we discuss some promising future directions for astronomy and data mining.

24.1 Introduction

The recent years have seen a rapid growth in the depth, richness, and scope of astronomical data. This trend is sure to accelerate with the next-generation all-sky surveys (e.g., Dark Energy Survey (DES),[1] Large Synoptic Survey Telescope (LSST),[2] Panoramic Survey Telescope and Rapid Response System (PanSTARRS),[3] Visible

[1] www.darkenergysurvey.org

[2] www.lsst.org [19].

[3] www.pan-starrs.ifa.hawaii.edu/public

A.B. Lee (✉) • P.E. Freeman
Department of Statistics, Carnegie Mellon University, Pittsburgh, PA, USA
e-mail: annlee@cmu.edu; pfreeman@cmu.edu

E.D. Feigelson and G.J. Babu (eds.), *Statistical Challenges in Modern Astronomy V*, Lecture Notes in Statistics 209, DOI 10.1007/978-1-4614-3520-4_24,

and Infrared Survey Telescope for Astronomy (VISTA),[4]) hence creating an ever increasing demand on sophisticated statistical methods that can draw fast and accurate inferences from large databases of high-dimensional data. From a data mining perspective, there are two general challenges one has to face. The first is the obvious *computational* challenge of rapidly processing and drawing inferences from massive data sets. The second is the *statistical* challenge of drawing accurate inferences from data that are high-dimensional and/or noisy.

Many of the estimation problems in astronomical databases are extremely complex, with observed data that take a form not amenable to analysis via standard methods of statistical inference. To utilize such data, it is crucial to encode them in a simpler, reduced form. The most obvious strategy is to hand-pick a subset of attributes based on prior domain knowledge. For example, ratios of known emission lines in galaxy spectra may aid in the classification of low-redshift galaxies into starburst, active galactic nuclei, and passive galaxies. In astrophysical data analysis, a widely used technique for statistical learning is *template fitting*, where observed data are compared with sets of simulated or empirical data from systems with known properties; see e.g., [1, 13, 18, 31] for some recent template-based work in a variety of astrophysical contexts. Another common data mining approach is

principal component analysis (PCA), a globally linear projection method that finds directions of maximum variance; see, e.g., ([26] and references therein; [5]).

Despite their wide popularity in astrophysical data analysis, the above strategies to statistical learning all have obvious draw-backs: When handpicking a few attributes, one may discard potentially useful information in the data. For template fitting, the final estimates depend strongly on the particular selection of templates as well as the quality of each of the templates. Finally, PCA works best when the data lie in a linear subspace of the high-dimensional observable space, and can perform poorly when this is not the case.

In this paper, we describe a more flexible approach to statistical learning that exploits the intrinsic (possibly non-linear) geometry of observable data with a minimum of assumptions. The idea is that naturally occurring data often have sparse structure due to constraints in the underlying physical process. In other words, the dimension d of the data space may be large but most of this space is empty. *Spectral kernel methods*, such as spectral clustering [25, 34], Laplacian maps [3], Hessian maps [14], and locally linear embeddings [30], analyze the data geometry by using certain differential operators and their corresponding eigenfunctions. These eigenfunctions provide a new coordinate system. For example, consider the emission spectra of astronomical objects. The original data with measurements at thousands of different wavelengths are not in a form amenable to traditional statistical analysis and nonparametric regression. Figure 24.1, however, shows a low-dimensional embedding of a sample of 2,793 SDSS galaxy spectra. The gray scale codes for redshift. The results indicate that by analyzing only a few dominant eigenfunctions of this highly complex data set, one can capture the main variability

[4]www.vista.ac.uk

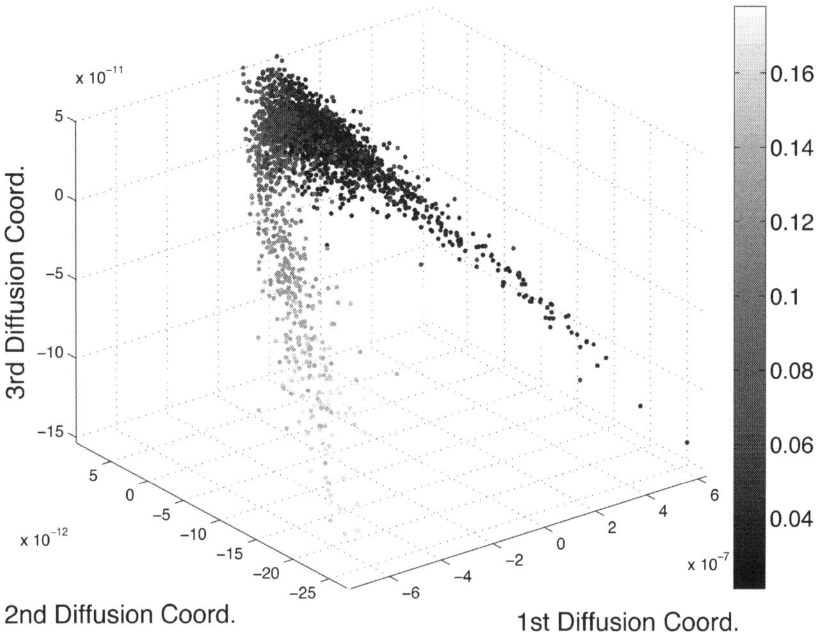

Fig. 24.1 Embedding of a sample of 2,793 SDSS galaxy spectra using the first three diffusion map coordinates. The *gray* scale codes for redshift (Reproduced from Richards et al. [26])

in redshift, although this quantity was not taken into account in the construction of the embedding. Moreover, the computed eigenfunctions are not only useful coordinates for the data. They form an orthogonal Hilbert basis for smooth functions of the data—a property that we utilize in [26] for redshift estimation.

More generally, the central goal of spectral kernel methods can be described as follows:

> Find a transformation $Z = \Psi(X)$ such that the structure of the distribution P_Z is simpler than the structure of the distribution P_X while preserving key geometric properties of P_X.

"Simpler" can mean lower dimensional but can also be interpreted much more broadly. For example, for redshift prediction using photometric data [17], we transform the original 16 variables (for magnitude differences between five broad wavelength bandpasses, as measured using four different magnitude systems) to a 150-dimensional space. For the transformed data, we then fit an additive model of the form

$$Y = \sum_{i=1}^{p} \beta_i \psi_i(\mathbf{x}) + \varepsilon$$

where Y denotes observed redshift, x is the original data object (galaxy), $\psi_i(x)$ is the i:th coordinate after the transformation, and ε is some random noise.

In this work, we focus on one particular non-linear data transformation called *diffusion maps* [11, 22], which is an approach to spectral connectivity analysis (SCA; [23]). SCA analyzes the higher-order connectivity of the data by defining a Markov process on a graph, where each graph node is an observable object, such as a spectrum, galaxy image, or set of light curves for a supernova, etc. The data are then transformed to a metric space where distances reflect the connectivity of the data. In Sect. 24.2, we describe the method. In Sect. 24.3, we give examples of some applications in astronomy. Finally, in Sect. 24.4, we discuss computational and statistical challenges for estimation for large astronomical databases, and outline some promising future directions.

24.2 Spectral Connectivity Analysis

There are several data transformation methods that aim to find a low-dimensional embedding $Z = \Psi(X)$ of the data while preserving key geometric properties of the data distribution P_X in local neighborhoods. Examples of locality-preserving methods are local linear embedding, Laplacian eigenmaps, Hessian eigenmaps, local tangent space alignment (LTSA; [35]), and diffusion maps. While the exact details vary, the optimal r-dimensional embedding (where $r < d$) is provided as the solution to an eigenvalue problem, where the first r eigenvectors $(\psi_1, \ldots, \psi_r)$ provide the new data coordinates.

Here we elaborate on diffusion maps—a specific approach to spectral connectivity analysis; Euclidean Commute Time maps is a closely related SCA technique discussed in e.g., [16]. Assume we observe data $\mathscr{X}_{\text{obs}} = \{\mathbf{x}_1, \ldots, \mathbf{x}_n\}$, where $\mathbf{x} \in \mathbb{R}^d$. The basic idea is to create a distance $D(\mathbf{x}_i, \mathbf{x}_j)$ that measures "connectivity" or how easily information "flows" from point $\mathbf{x}_i$ to $\mathbf{x}_j$ in a Markov chain on the observed data. (The data "points" $\mathbf{x}_i$ and $\mathbf{x}_j$ represent *entire* observable objects; for example, the full emission spectra of two astronomical objects, images of two galaxies, or light curves of two supernovae; D is a measure of distance between the objects.) High flow occurs in high-density regions, and points that are connected by many high-flow paths are close with respect to the diffusion metric. The transition matrix A of the Markov chain is based on a user-defined pairwise distance $\Delta(\cdot, \cdot)$ that is a good measure of dissimilarity in local neighborhoods; a common choice is the Euclidean distance in $\mathbb{R}^d$ but other dissimilarity measures that incorporate prior knowledge and measurement errors can also be used. We define the transition probability from $\mathbf{x}_i$ to $\mathbf{x}_j$ in one step by $A(\mathbf{x}_i, \mathbf{x}_j) = \frac{\exp(-\Delta(\mathbf{x}_i, \mathbf{x}_j)/\varepsilon)}{\sum_k \exp(-\Delta(\mathbf{x}_i, \mathbf{x}_k)/\varepsilon)}$, where $\varepsilon > 0$ is a tuning parameter that determines the local neighborhood size. Let $A_t(\mathbf{x}_i, \mathbf{x}_j)$ denote the t-step transition probability; the parameter t determines the amount of smoothing along high-density regions and the "scale" of the analysis. The diffusion distance between points $\mathbf{x}_i$ and $\mathbf{x}_j$ is defined as

$$D(\mathbf{x}_i, \mathbf{x}_j) = \sum_{\mathbf{z} \in \mathscr{X}_{\text{obs}}} \frac{(A_t(\mathbf{x}_i, \mathbf{z}) - A_t(\mathbf{x}_j, \mathbf{z}))^2}{\phi_0(\mathbf{z})}, \tag{24.1}$$

where the sum is over all points $\mathbf{z}$ in the data set $\mathscr{X}_{\text{obs}}$, and $\phi_0(\mathbf{z})$ is the stationary distribution of the Markov chain as $t \to \infty$. In practice, we never explicitly implement the Markov chain but instead solve an eigenproblem for an n-by-n matrix. Let $\lambda_1 \geq \lambda_2 \geq \ldots$ and $\{\psi_i\}$ be the eigenvalues and corresponding right eigenvectors of the 1-step transition matrix A. The diffusion map $\Psi_t : \mathscr{X}_{\text{obs}} \to \mathbb{R}^r$ (where $r < n$) is given by

$$\Psi_t(\mathbf{x}) = (\lambda_1^t \psi_1(\mathbf{x}), \ldots, \lambda_r^t \psi_r(\mathbf{x})). \tag{24.2}$$

As shown in [10,22], it holds that $D_t(\mathbf{x}_i, \mathbf{x}_j) \approx \|\Psi_t(\mathbf{x}_i) - \Psi_t(\mathbf{x}_j)\|$, i.e., the Euclidean distance in the new coordinate system approximates the diffusion distance in the original coordinate system. Because all connections between data points are simultaneously considered, diffusion maps are robust to noise and outliers and they return embeddings where metrics are explicitly defined.

Incorporating data geometry via Ψ and SCA can lead to radically improved inference algorithms. For details on the statistical properties of SCA refer to [23]. In Sect. 24.3, we give examples of some specific applications in astronomy.

24.2.1 *Out-of-Sample Extensions of Empirical Data Sets*

Let $\mathscr{X}$ denote the space of all data. One can show that the random walk and the eigenvectors $\{\psi_j\}$ derived from the finite set $\mathscr{X}_{\text{obs}}$ have meaningful limits as the sample size $n \to \infty$. Hence, we can think of the eigenvectors of the discrete random walk as estimates of eigenfunctions $\{\psi_j(\mathbf{x})\}_{j \in \mathbb{N}}$, defined on $\mathscr{X}$, at the *observed* values $\mathbf{x}_1, \ldots, \mathbf{x}_n$. We estimate the function $\psi_j(\mathbf{x})$ at values of $\mathbf{x}$ not corresponding to one of the $\mathbf{x}_i$'s in the data set by the kernel-smoothed estimate

$$\hat{\psi}_j(\mathbf{x}) = \frac{1}{\lambda_j} \sum_{i=1}^{n} A(\mathbf{x}, \mathbf{x}_i) \psi_j(\mathbf{x}_i), \tag{24.3}$$

where $A(\mathbf{x}, \mathbf{x}_j) = \frac{\exp(-\Delta(\mathbf{x}, \mathbf{x}_j)/\varepsilon)}{\sum_k \exp(-\Delta(\mathbf{x}, \mathbf{x}_k)/\varepsilon)}$. This expression is known in the applied mathematics literature as the Nyström approximation. These out-of-sample extensions allow us to make predictions for new data points that are not in the sample using diffusion maps and, for example, adaptive regression (as in Sec. 24.3.1).

24.3 Applications in Astronomy

In this section, we give some examples of applications of SCA to astrophysical problems. Among other things, diffusion maps can be used to estimate parameters in a regression framework, build classification models, and select prototypes for parameter estimation in complex models. The details are described in separate papers.

24.3.1 Adaptive Regression and Redshift Estimation

In [26], we show how one can take advantage of the underlying data structure in non-parametric regression such as redshift prediction. The main idea is to describe the intrinsic data geometry in terms of fundamental eigenmodes. These eigenmodes can be viewed both as (1) *coordinates* of the data, as in Fig. 24.1, and as (2) *orthogonal basis functions* for curve estimation. The latter insight can be used to develop a general regression framework for sparse, complex data.

Let $\mathscr{X} \subset \mathbb{R}^d$ denote the space of all observed data. In regression, the goal is to predict a real-valued function $f(\mathbf{x})$ for data $\mathbf{x} \in \mathscr{X}$, when given a sample of known pairs $(\mathbf{x}, Y)$ where the response $Y = f(\mathbf{x}) + \varepsilon$. If $f \in L^2(\mathscr{X})$ and $\{\psi_1, \psi_2, \ldots,\}$ is an orthonormal basis, then we can write

$$f(\mathbf{x}) = \sum_{j=1}^{\infty} \beta_j \psi_j(\mathbf{x}),$$

where the expansion coefficients $\beta_j = \int f(\mathbf{x})\psi_j(\mathbf{x})d\mathbf{x}$. The standard approach in non-parametric curve estimation [15] is to choose a fixed known basis (e.g., Fourier or wavelet bases) for, for example, $L^2([0,1])$, and then extend the basis to two or three dimensions by a tensor product. Such an approach quickly becomes intractable in higher dimensions. In astrophysical problems, the response Y may be the redshift, age or metallicity of galaxies, and $\mathbf{x}$ is often a high-dimensional, non-standard data object, such as the emission spectrum measured at $p > 1{,}000$ wavelength bins, or photometry data in a color space with $p > 10$ dimensions.

In [28], we suggest a new, adaptive approach to non-parametric curve estimation, which utilize the data-driven (orthogonal) eigenfunctions $\{\psi_1, \psi_2, \ldots,\}$ computed by PCA or spectral kernel methods. The regression function estimate $\hat{r}(\mathbf{x})$ is then given by

$$\hat{f}(\mathbf{x}) = \sum_{j=1}^{p} \hat{\beta}_j \psi_j(\mathbf{x}),$$

where the coefficients $\hat{\beta}_j$ are estimated from the data, and p is a smoothing parameter determined by cross-validation and a mean-squared error prediction risk. The method is computationally efficient, making it appropriate for large databases such as the SDSS. One can use the predictions to speed up more computationally expensive estimation techniques by narrowing down the relevant parameter space; e.g., the redshift range or the set of templates in cross-correlation techniques. Adaptive regression also provides a useful tool for quickly identifying outliers; e.g., misclassified spectra, spectra with anomalous features, etc. In Richards et al., we consider a sample of 3,835 galaxy spectra from the SDSS database. For this data, the estimates based on eigenmodes from SCA (diffusion maps) lead to markedly better predictions than estimates from PCA, indicating the importance of non-linear geometries.

The development of fast and accurate methods of photometric redshift estimation is a crucial step towards being able to fully utilize the data of next-generation surveys for precision cosmology. In [17], we apply adaptive regression and SCA to the problem of *photometric redshift estimation* for three different data sets: 350,738 SDSS main sample galaxies, 29,816 SDSS luminous red galaxies, and 5,223 galaxies from DEEP2 with CFHTLS *ugriz* photometry. For computational speed, we first derive diffusion coordinates for training sets limited to about 10^4 galaxies, and then extend these coordinates to the full data sets by the Nyström method. The final redshift predictions achieve an accuracy on par with that of existing ML-based techniques, e.g., artificial neural networks [12] and k-nearest neighbors [2].

24.3.2 Prototype Selection for Estimation of Star Formation History

Parameter estimation in astronomy and cosmology often requires the use of complex physical models. In a typical application the mapping from the parameter space to the observed data space is built on sophisticated physical theory or simulation models or both. In [27, 28], we study one such scenario: the problem of estimating star formation history (SFH) in galaxies given SDSS high-resolution spectra. A common technique in the astronomy literature, called *empirical population synthesis* (see e.g., Cid Fernandes et al. [9] and references within), is to model each galaxy as a mixture of stars from different simple stellar populations (SSPs), where an SSP is defined as a group of stars with the same age and metallicity. The principle behind this method is that each galaxy consists of multiple subpopulations of stars of different ages and compositions so that the integrated observed light from each galaxy is a mixture of the light contributed by each SSP. By estimating the mixture coefficient of each SSP, one can then reconstruct the star formation rate and composition as a function of time throughout the life of that galaxy.

In our work, we use theoretical SSP models by Bruzual and Charlot [6]. For the galaxy spectra, we adopt the empirical population synthesis model in [8]:

$$\mathbf{Y}_\lambda(\gamma, M_{\lambda_0}, A_V, v_*, \sigma_*) = M_{\lambda_0}\left(\sum_{j=1}^{N} \gamma_j \mathbf{X}_{j,\lambda} r_\lambda(A_V)\right) \otimes G(v_*, \sigma_*) \tag{24.4}$$

where $\mathbf{Y}_\lambda$ is the light flux at wavelength λ; $\mathbf{X}_j$ is the normalized jth SSP spectrum; and $\gamma_j \in [0,1]$ is the proportion of luminosity contributed by the jth SSP. (The remaining model parameters describe the flux normalization and observational noise, such as the amount of reddening due to foreground dust, spectral distortions due the movement of stars within the observed galaxy, etc.) We fit the signal model in (24.4) to observed noisy galaxy data with maximum likelihood estimation and MCMC. We then derive various physical parameters of interest from the SSP parameters (which are known) and the component weights in the signal model (which are estimated). For example: the average log age of the stars in a galaxy, $\langle \log t \rangle = \sum_{j=1}^{N} \gamma_j \log t(\mathbf{X}_j)$, where $t(\mathbf{X}_j)$ is the age of the j:th SSP; similarly, the average log metallicity $\langle \log Z \rangle = \sum_{j=1}^{N} \gamma_j \log Z(\mathbf{X}_j)$, where $Z(\mathbf{X}_j)$ is the metallicity of the j:th SSP.

An important question is: How should one choose the set of SSPs? Though the parameters that define each SSP are continuous, optimizing the signal model over a large set of SSPs on a fine parameter grid is computationally infeasible and inefficient. As we shall see, it also leads to poor statistical estimates. In [28], we introduce a principled approach of choosing a small basis of SSP *prototypes* for optimal SFH parameter estimation. The basic idea is to explore the underlying geometry of the SSP observable data, and quantize the vector space and effective support of these model components. In addition to greater computational efficiency, we achieve better estimates of the SFH target parameters. In simulations, our proposed quantization method obtains a substantial improvement in estimating the target parameters over the common method of employing a parameter grid. The main reason for the improvement is that under the presence of noise, components with similar functional forms will be indistinguishable. Hence, it is more advantageous to choose prototypes that are approximately evenly spaced in the space of model data rather than evenly spaced in the parameter space. By replacing the theoretical models in each neighborhood by their local average in diffusion space ("Diffusion K-means"; Fig. 24.2), the model quantization approach is optimal for treating degeneracies because it allows a slight increase in bias to achieve a large decrease in variance of the target parameter estimates. See Fig. 24.3 for a plot of two SSP spectral bases with K prototypes chosen by a regular parameter grid and by our proposed quantization method, respectively.

24.3.3 Supernova Classification

In many astronomical problems, classification is of paramount importance. For instance, one may be interested in determining which of a collection of light curves is associated with RR Lyrae stars, or Cepheids, etc. Depending on the problem, classification may be done in an unsupervised manner, to uncover hidden structure in the data, or, if at least some of the data labels are known, a classifier can be trained and then used to predict the classes of unlabeled data.

The next generation of survey telescopes will observe hundreds of thousands of noisy and irregular photometric SN light curves, from which astronomers will want to construct highly pure and efficient Type Ia SN samples for use in testing cosmological theories. In [29], we apply a semi-supervised approach to supernova classification. In the unsupervised step, we fit regression splines to each of a set of light curves, then via diffusion map place them in a lower-dimensional embedding space that capture the geometry of the underlying data distribution. In that space,

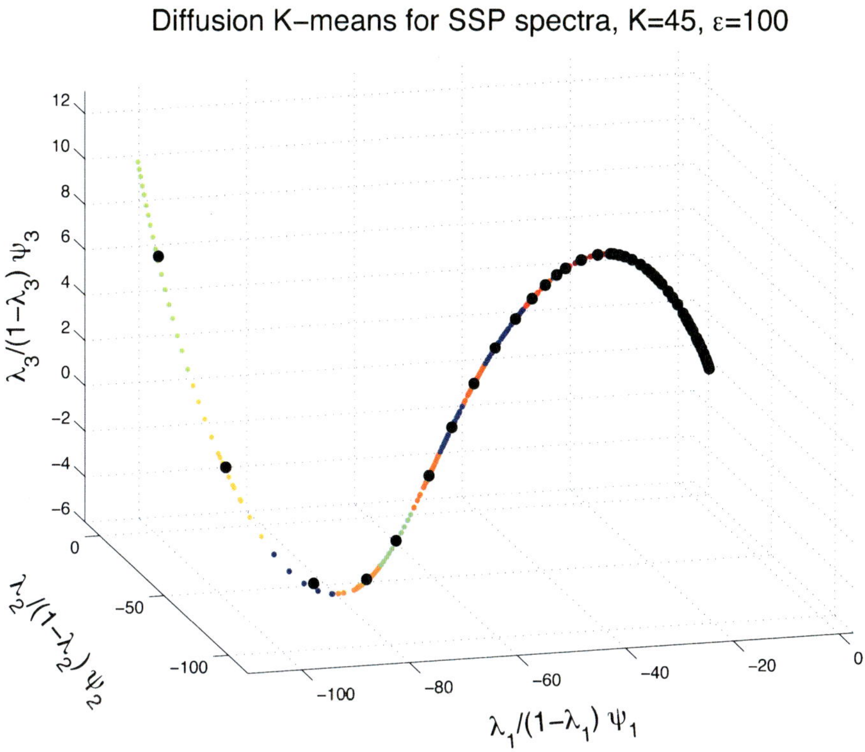

Fig. 24.2 Prototyping of SSP spectra by Diffusion K-means. Representation of 1,278 SSP spectra in three-dimensional diffusion space. *Large black dots* denote the $K = 45$ centroids. Individual SSPs are colored by cluster membership. The theoretical SSPs reside on a simple, low-dimensional manifold which is captured by the K prototypes (Reproduced from Richards et al. [27])

we then take the supervised step of training a random forest classifier with only the labeled data, with the results used to classify the unlabeled data. Applied to the data of the Supernova Photometric Classification Challenge (Kessler et al. 2010), we achieve 96% purity and 86% efficiency when labeling the training set; for the test set, the figures are 56% and 48% respectively. As the sample sizes (of unlabeled and/or labeled data) increase, our semi-supervised approach will yield progressively more accurate classifications, in contrast to template-based approaches which do not benefit from larger data sets. We also explore how different spectroscopic followup strategies affect these figures, finding that deeper surveys yielding fewer labeled SNe can produce better results than shallower surveys. Determination of an optimal labeling strategy is an important component of *active learning*, a topic we will return to in the discussion below.

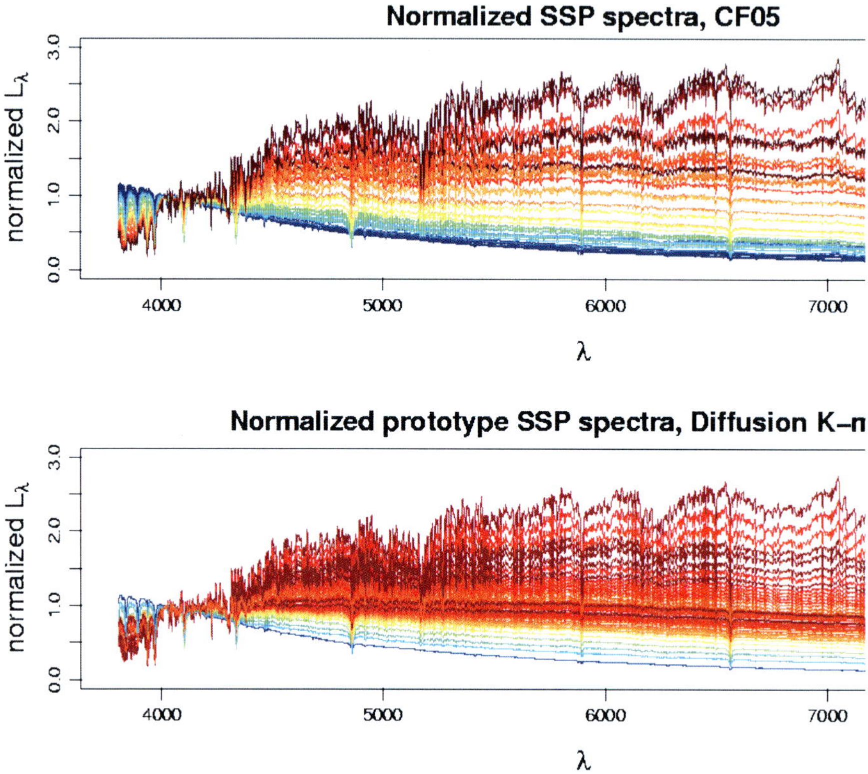

Fig. 24.3 Basis spectra for CF05 and Diffusion K-means, colored by $\log t$. All spectra are normalized to 1 at $\lambda_0 = 4{,}020$ Å. The diffusion K-means basis covers the spectral range in relatively uniform increments, while the CF05 basis *oversamples* spectra from young stellar populations and *undersamples* the spectral range of older populations (Reproduced from Richards et al. [27])

24.4 Discussion and Future Directions

In this review, we have described SCA—a statistical technique for transforming complex, data objects into a coordinate system that reveals the structure of the underlying data distribution. Such a transformation may improve the performance of classification, regression, clustering and parameter estimation. We have seen applications of SCA in redshift prediction, estimation of star formation history and photometric supernova classification. Currently, we are working with Chad Schafer to develop SCA as a tool for combining theoretical modeling and observational evidence into optimal constraints on the parameters of physical models. The idea is to map observed data (e.g., light curves of Type Ia supernova) as well as distributions for the observable data, constrained by physical theory (e.g., cosmological models) into a simpler encoding space. The shared representation of data and distributions is then exploited to achieve optimal constraints on physical theories, in the form of set estimators on the distribution space; see Schafer's SCMA 2011 talk and paper for details.

Another promising direction of SCA is *semi-supervised learning* (SSL), in particular, in combination with *active learning*. Suppose that we have a regression or classification problem. The typical scenario in SSL is that we have access to a large database of unlabeled examples (e.g., photometric data with unknown redshift), but relatively few labeled examples (e.g., data with spectroscopically confirmed redshift). Classical regression and classification techniques only take advantage of labeled data, but the central idea behind SSL is that one can make use of the unlabeled data to improve predictions; see e.g., [4, 21, 33] for theoretical results on SSL. In our supernova classification application, we showed that learning a low-dimensional coordinate system using unlabeled data improves subsequent classification by trees. We also found evidence that the exact choice of training examples has a large effect on the results. In future work, we plan to explore whether we can achieve greater accuracy in classification and regression problems with fewer training labels if a so-called active learner is allowed to repeatedly pose *queries*, in the form of unlabeled data instances to be labeled by an oracle. In the machine learning literature, there are many variants of active learning; see, e.g., [32] for a literature survey. All these models involve a search through the hypothesis space. Such searches and subsequent queries could potentially be better adapted to the underlying data distribution via an unsupervised technique such as SCA that exploit clusters and groupings in data.

Finally, there are the computational challenges of efficiently constructing weighted graphs and performing eigencomputations for very large databases. We are currently exploring several solutions—most notably, fast approximate nearest neighborhood searches via trees, eigencomputations via streaming PCA [7], very large-scale algebraic computations via matrix randomization [24], and subsampling combined with Nyström extensions to reduce the size of the distance matrix that is effectively eigendecomposed.

Acknowledgements Part of this work is joint with Joseph W. Richards, Chad M. Schafer, Jeffrey A. Newman, and Darren W. Homrighausen. We would also like to acknowledge ONR grant #00424143, NSF grant #0707059, and NASA AISR grant NNX09AK59G.

References

1. Bailer-Jones, C. A. L. (2010, March). The ILIUM forward modelling algorithm for multivariate parameter estimation and its application to derive stellar parameters from Gaia spectrophotometry. *Monthly Notices of the Royal Astronomical Society 403*, 96–116.
2. Ball, N. M., R. J. Brunner, A. D. Myers, N. E. Strand, S. L. Alberts, and D. Tcheng (2008, August). Robust Machine Learning Applied to Astronomical Data Sets. III. Probabilistic Photometric Redshifts for Galaxies and Quasars in the SDSS and GALEX. *Astrophysical Journal 683*, 12–21.

3. Belkin, M. and P. Niyogi (2003). Laplacian eigenmaps for dimensionality reduction and data representation. *Neural Computation 6*(15), 1373–1396.
4. Belkin, M. and P. Niyogi (2005). Semi-supervised learning on Riemannian manifolds. *Machine Learning 56*, 209–239.
5. Boroson, T. A. and T. R. Lauer (2010, August). Exploring the Spectral Space of Low Redshift QSOs. *The Astronomical Journal 140*, 390–402.
6. Bruzual, G. and S. Charlot (2003). Stellar population synthesis at the resolution of 2003. *Monthly Notices of the Royal Astronomical Society 344*, 1000–1028.
7. Budavári, T., V. Wild, A. S. Szalay, L. Dobos, and C.-W. Yip (2009, April). Reliable eigenspectra for new generation surveys. *Monthly Notices of the Royal Astronomical Society 394*, 1496–1502.
8. Cid Fernandes, R., Q. Gu, J. Melnick, E. Terlevich, R. Terlevich, D. Kunth, R. Rodrigues Lacerda, and B. Joguet (2004). The star formation history of Seyfert 2 nuclei. *Monthly Notices of the Royal Astronomical Society 355*, 273–296.
9. Cid Fernandes, R., L. Sodré, H. R. Schmitt, and J. R. S. Leão (2001, July). A probabilistic formulation for empirical population synthesis: sampling methods and tests. *Monthly Notices of the Royal Astronomical Society 325*, 60–76.
10. Coifman, R. and S. Lafon (2006). Diffusion maps. *Applied and Computational Harmonic Analysis 21*, 5–30.
11. Coifman, R., S. Lafon, A. Lee, M. Maggioni, B. Nadler, F. Warner, and S. Zucker (2005). Geometric diffusions as a tool for harmonics analysis and structure definition of data: Diffusion maps. *Proc. of the National Academy of Sciences 102*(21), 7426–7431.
12. Collister, A. A. and O. Lahav (2004, April). ANNz: Estimating Photometric Redshifts Using Artificial Neural Networks. *Publ. of the Astronomical Society of the Pacific 116*, 345–351.
13. Dahlen, T., B. Mobasher, M. Dickinson, H. C. Ferguson, M. Giavalisco, N. A. Grogin, Y. Guo, A. Koekemoer, K.-S. Lee, S.-K. Lee, M. Nonino, A. G. Riess, and S. Salimbeni (2010, November). A Detailed Study of Photometric Redshifts for GOODS-South Galaxies. *Astrophysical Journa 724*, 425–447.
14. Donoho, D. and C. Grimes (2003, May). Hessian eigenmaps: new locally linear embedding techniques for high-dimensional data. *Proc. of the National Academy of Sciences 100*(10), 5591–5596.
15. Efromovich, S. (1999). *Nonparametric curve estimation: methods, theory and applications*. Springer series in statistics. Springer.
16. Fouss, F., A. Pirotte, and M. Saerens (2005). A novel way of computing similarities between nodes of a graph, with application to collaborative recommendation. In *Proc. of the 2005 IEEE/WIC/ACM International Joint Conference on Web Intelligence*, pp. 550–556.
17. Freeman, P. E., J. Newman, A. B. Lee, J. W. Richards, and C. M. Schafer (2009). Photometric redshift estimation using SCA. *Monthly Notices of the Royal Astronomical Society 398*, 2012–2021.
18. Hayden, B. T., P. M. Garnavich, et al. (2010, March). The Rise and Fall of Type Ia Supernova Light Curves in the SDSS-II Supernova Survey. *Astrophysical Journa 712*, 350–366.
19. Ivezic, Z., J. A. Tyson, and for the LSST Collaboration (2008, May). LSST: from Science Drivers to Reference Design and Anticipated Data Products. *ArXiv e-prints*.
20. Kessler, R., Bassett, B., et al. (2010) Results from the Supernova Photometric Classification Challenge, *Publ. Astro. Soc. Pacific*, *122*, 1415–1431.
21. Lafferty, J. and L. Wasserman (2007). Statistical analysis of semi-supervised regression. In *Adv. in Neural Inf. Processing Systems*.
22. Lafon, S. and A. Lee (2006). Diffusion maps and coarse-graining: A unified framework for dimensionality reduction, graph partitioning, and data set parameterization. *IEEE Trans. Pattern Anal. and Mach. Intel. 28*, 1393–1403.
23. Lee, A. B. and L. Wasserman (2010). Spectral connectivity analysis. *Journal of the American Statistical Association 105*(491), 1241–1255.
24. N. Halko, P. M. and J. Tropp (2011). Finding structure with randomness: Probabilistic algorithms for constructing approximate matrix decompositions. *SIAM Review 53*(2).

25. Ng, A. Y., M. I. Jordan, and Y. Weiss (2001). On spectral clustering: Analysis and an algorithm. In *Adv. in Neural Inf. Processing Systems*.
26. Richards, J. W., P. E. Freeman, A. B. Lee, and C. M. Schafer (2009a). Accurate parameter estimation for star formation history in galaxies using SDSS spectra. *Monthly Notices of the Royal Astronomical Society 399*, 1044–1057.
27. Richards, J. W., P. E. Freeman, A. B. Lee, and C. M. Schafer (2009b). Exploiting low-dimensional structure in astronomical spectra. *Astrophysical Journal 691*, 32–42.
28. Richards, J. W., P. E. Freeman, A. B. Lee, and C. M. Schafer (2011a). Prototype selection for parameter estimation in complex models. Submitted; arXiv:1105.6344.
29. Richards, J. W., D. Homrighausen, P. E. Freeman, C. M. Schafer, and D. Poznanski (2011b). Semi-supervised learning for photometric supernova classification. Submitted; arXiv:1103.6034.
30. Roweis, S. and L. Saul (2000). Nonlinear dimensionality reduction by annalsly linear embedding. *Science 290*, 2323–2326.
31. Sesar, B., Ž. Ivezić, et al. (2010, January). Light Curve Templates and Galactic Distribution of RR Lyrae Stars from Sloan Digital Sky Survey Stripe 82. *Astrophysical Journal 708*, 717–741.
32. Settles, B. (2010). Active learning literature survey. Technical Report 1648, Dept. of Computer Science, University of Wisconsin-Madison.
33. Singh, A., R. Nowak, and X. Zhu (2008). Unlabeled data: Now it helps, now it doesn't. In *Adv. in Neural Inf. Processing Systems*.
34. von Luxburg, U. (2007). A tutorial on spectral clustering. *Statistics and Computing 17*(4), 395–416.
35. Zhang, Z. and H. Zha (2002). Principal manifolds and nonlinear dimension reduction via local tangent space alignement. Technical Report CSE-02-019, Department of computer science and engineering, Pennsylvania State University.

Chapter 25
Commentary: Exploiting Non-linear Structure in Astronomical Data for Improved Statistical Inference

Didier Fraix-Burnet

Abstract As discussed in the paper by Ann Lee, both dimensionality reduction and classification seek a reduced simpler form of the data. The first one works with the parameter space, while classification works with the object space. Ideally, one wishes to find a parameter space in which the points are naturally gathered into distinct groups and, as a physicist more particularly, data points can fit our model curves. I want to point out that dimensionality reduction methods and classification approaches are highly complementary and should even be carried out together. Astrophysical objects are complex, so that numerical simulations are now a common tools to do physics. Model fitting has thus become a comparison between populations (the observed ones and the synthetic ones) rather than plotting a curve onto data points. This is exactly the role of statistics.

25.1 Structures in the Data Space and Classification

The paper by Ann Lee and Peter Freeman deals with the difficulty of infering anything meaningful from astrophysical data that are complex and of high-dimensionality (and non-standard). Dimensionality reduction aims at easing statistical inference and simplifying interpretation through a simpler form of the data. In astrophysics, where technological achievements provide us with a growing number of different kinds of observables, extracting the most influencial parameters also serves as a guide for future investigations and even telescope/detector design. A reduced parameter space is essential for modeling especially if analytical calculations are carried out. However, the numerical simulations become most

D. Fraix-Burnet (✉)
Université Joseph Fourier – Grenoble 1/CNRS, Institut de Planétologie et d'Astrophysique de Grenoble, BP 53, F-38041 Grenoble cedex 9, France
e-mail: fraix@obs.ujf-grenoble.fr

E.D. Feigelson and G.J. Babu (eds.), *Statistical Challenges in Modern Astronomy V*, Lecture Notes in Statistics 209, DOI 10.1007/978-1-4614-3520-4_25,

often unavoidable because of the complexity of the astrophysical objects. Then, the number of parameters here also must be synthetized to the most important ones.

The general purpose of classification is to ease memory and discover the relationships between classes. It is easier to recall properties for tens of classes rather than a million objects. It is also much easier (and less computer intensive) to fit models on a limited number of representatives of classes than to many not so different objects. But obtaining classes is not sufficient if we are not able to understand why they are composed as they are and why they are different. Finding relationships is thus essential.

Dimensionality reduction reduces the number of parameters

Classification reduces the number of objects

	Par1	Par2	Par3	Par4	...
Object1	.	.	.	.	...
Object2	.	.	.	.	...
Object3	.	.	.	.	...
Object4	.	.	.	.	...
...	...	...	...	...	...

In summary, both dimensionality reduction and classification share the same goal. In simple words, the common ideal objective is to find a parameter space in which the points are naturally gathered into distinct groups and data points can fit our models. Ann Lee has shown us how dimensionality must care about structures in the data space. I would like to show that classification is also very concerned with these same structures.

Traditional classification in astrophysics makes heavy use of scatter plots and hard limits, most often linear. Parameters are chosen according to the observational means (infrared or radio galaxies, X-ray objects...), their "obviousness" (elliptical, Lyman-α or compact galaxies...) or an a priori understanding of the underlying physics (star-forming or massive galaxies...). Such classifications are thus limited by the use of very few properties and cannot reflect the real complexity of astrophysical objects.

Multivariate classifications are just beginning to be used in astrophysics [1–3]. Clustering analyses are generally based on distance matrices, principally using euclidian distances, thus assuming a linear multivariate space. More sophisticated methods use a priori knowledge to implement a particular geometry of the data space and use an adapted distance definition. On the contrary parameter-based (or character-based) approaches, using the coordinates of the objects and not their pairwise distances, explore the geometry of the data space. As one can easily understand, distance-based methods are generally much more computationaly efficient.

It appears to me that the diffusion maps technique described in the paper by Ann Lee and Peter Freeman, and the spectral connectivity analysis more generally, is of the second kind, These methods explore the geometry of the data space even though they assume an euclidian metrics *locally* (any curved geometry can be locally approximated by euclidian spaces). This works well because the data space is expected to be sparse due to the physical relations that explain the diversity of objects.

Transformation processes that cause properties to evolve are all continuous in astrophysics. The distribution of data points in a multivariate space is thus mainly continuous. For sparsity to occur, that is for structures to be differentiated with voids in between, the variables must be constrained by some underlying phenomena.

Classifying objects in a continuous data space is not that easy because fuzziness is unavoidable: limits cannot be hard and overlaps are possible. Even if gaps are observed, it is generally impossible to guarantee that they will not be filled by newly discovered objects. So classification in a continuous data space must be understood as an ordered organisation. Distance-based or character-based methods establish relationships between the objects, most easily depicted on a hierarchical representations like trees or split-networks (a generalization of trees). The relationships so revealed allow for a flexible classification, the number of groups depending on the level where the tree is cut.

However, when does a parameter matrix or a distance matrix be represented on a tree-like scheme? It can be shown [4] that this is the case when the objects define a convex structures in the data space This is very similar to the salesman problem, a classical question in algorithmics that seeks to optimize the journey of a salesman through different cities. The solution is easy when the cities are arranged on a single convex hull, then the tree is linear. When several complex hulls are present, the tree becomes more complicated and can take the form of a split-network.

Hence, the geometry of the data space is crucial to organize the objects in an intelligible way. This data space cannot be any, it is defined by the parameters with which the convex hulls appear.

In conclusion, to reduce the number of objects, we need to be in the right data space. We thus need to select the right parameters, To do that objectively and extensively, the methods to reduce the dimensionality are extremely useful since they can identify the most discriminant axes of variability. But they must preserve the main geometrical properties of the data space. This is a quality of the spectral connectivity analysis method used by Ann Lee and Pete Freeman.

25.2 Finding the Right Data Space

There is thus a parallel and complementary search of the right data space both by using dimensionality reduction techniques, to probe the parameters, and by using multivariate classification, to probe the robustness and the interest of the groups that can be defined from these parameters. Starting from the initial parameter space, one constructs a sub-parameter space with the first kind of approach, and then check whether a classification can be obtained. From this second analysis, some information is gathered on the structuring properties of the parameters, then further iterations can lead to a final sub-parameter space from which a final classification is proposed. Then, and only then, the interpretation can begin.

25.3 Model Fitting and Populations

Would we envisage to put living organisms into equations and follow their evolution? Biologists rather use statistical laws to model the evolutions and relationships of *populations*.

Model fitting in astrophysics still often means plotting a curve onto data points. Unfortunately, the observations and their parameters are too many, so that most scatter plots are merely clouds of points in which many curves can fit equally well. In addition, without a proper classification, the chance is weak that the right population of galaxies has been picked up for the test.

But there is more. Ann Lee presents an application of the spectral connectivity analysis to obtain prototypes of synthetic galaxy spectra. The reason is that it would take too much time to find the best values for the many variables of the models by fitting each of the million observed spectra. It is simpler to only use a limited number of model prototypes selected from the synthetic population of models. We have here a good example where the search of the most influencing parameters (reduction of dimensionality in the model space) leads to a classification (the prototypes).

I however find it amusing to use individual observed objects against prototypes of models, and not using "prototypes" of observed objects. This reflects the radical evolution of contemporary astrophysics. On one side we have a huge amount of observations, with many objects described by many parameters. On the other side, computers allow us to investigate a detailed and complicated physics. Numerical simulations produce huge populations of synthetic objects. The question is how use them to compare with the observed populations?

Model fitting nowadays clearly appears as a comparison between populations, not any more fitting a curve for an individual galaxy. Classification becomes crucial, but not with the old fascioned way of segregating objects according to their most obvious properties. This is real statistics that astronomers must use. Physicists in general are not formed at all to this way of thinking, of doing science. This is cultural, and certainly explains why astrostatistics is still not widely popular in astrophysics. It will certainly take some time, but change is coming.

References

1. Ellis, S.C., Driver, S.P., Allen, P.D., Liske, J., Bland-Hawthorn, J., De Propris, R.: The millennium galaxy catalogue: on the natural sub-division of galaxies. Monthly Notices of the Royal Astronomical Society **363**, 1257–1271 (2005) (astro-ph/0508365).

2. Chattopadhyay, A.K., Chattopadhyay, T., Davoust,, E., Mondal, S., Sharina, M.: Study of NGC 5128 globular clusters under multivariate statistical paradigm. Astrophysical Journal **705**, 1533–1547 (2009) (arXiv:0909.4161)
3. Fraix-Burnet, D., Dugué, M., Chattopadhyay, T., Chattopadhyay, A. K., Davoust, E.: Structures in the fundamental plane of early-type galaxies. Month. Not. Royal. Astron. Soc. **407**, 2207–2222 (2010) (arXiv:1005.5645)
4. Thuillard M., Fraix-Burnet D., 2009, Evolutionary Bioinformatics, 5, 33 (arXiv:0905.2481)

Chapter 26
Surprise Detection in Multivariate Astronomical Data

Kirk D. Borne and Arun Vedachalam

Abstract Astronomers systematically study the sky with large sky surveys. A common feature of modern sky surveys is that they produce hundreds of terabytes (TB) up to 100 (or more) petabytes (PB) both in the image data archive and in the object catalogs. For example, the LSST will produce a 20–40 PB catalog database. Large sky surveys have enormous potential to enable countless astronomical discoveries. Such discoveries will span the full spectrum of statistics: from rare one-in-a-billion (or one-in-a-trillion) object types, to complete statistical and astrophysical specifications of many classes of objects (based upon millions of instances of each class). The growth in data volumes requires more effective knowledge discovery and extraction algorithms. Among these are algorithms for outlier (novelty/surprise/anomaly) detection. Outlier detection algorithms enable scientists to discover the most "interesting" scientific knowledge hidden within large and high-dimensional datasets: the "unknown unknowns". Effective outlier detection is essential for rapid discovery of potentially interesting and/or hazardous events. Emerging unexpected conditions in hardware, software, or network resources need to be detected, characterized, and analyzed as soon as possible for obvious system health and safety reasons, just as emerging behaviors and variations in scientific targets should be similarly detected and characterized promptly in order to enable rapid decision support in response to such events. We have developed a new algorithm for outlier detection (KNN-DD: K-Nearest Neighbor Data Distributions). We have derived results from preliminary experiments in terms of the algorithm's precision and recall for known outliers, and in terms of its ability to distinguish between characteristically different data distributions among different classes of objects.

K.D. Borne (✉) • A. Vedachalam
Astronomy, & Computational Science, School of Physics, George Mason University, 4400 University Drive, Fairfax, VA 22030, USA
e-mail: kborne@gmu.edu; avedacha@masonlive.gmu.edu

E.D. Feigelson and G.J. Babu (eds.), *Statistical Challenges in Modern Astronomy V*, Lecture Notes in Statistics 209, DOI 10.1007/978-1-4614-3520-4_26,

26.1 Introduction

Novelty and surprise are two of the more exciting aspects of science—finding something totally new and unexpected. This can lead to a quick research paper, or it can make your career, or it can earn the discoverer a Nobel Prize. As scientists, we all yearn to make a significant discovery. Petascale databases potentially offer a multitude of such opportunities. But how do we find that surprising novel thing? These come under various names: interestingness, outliers, novelty, surprise, anomalies, or defects (depending on the application). We are investigating various measures of interestingness in large databases and in high-rate data streams (e.g., the Sloan Digital Sky Survey [SDSS], 2-μm All-Sky Survey [2MASS], and GALEX sky survey), in anticipation of the petascale databases of the future (e.g., the Large Synoptic Survey Telescope [LSST]), in order to validate algorithms for rapid detection and characterization of events (i.e., changes, outliers, anomalies, novelties).

In order to frame our investigation of these algorithms, we focus on a specific extragalactic research problem. We explore the environmental dependences of hierarchical mass assembly and of fundamental galaxy parameters using a combination of large multi-survey (multi-wavelength) object catalogs, including SDSS (optical) and 2MASS (NIR: near-infrared). We generated and are now studying a sample of over 100,000 galaxies that have been identified and catalogued in both SDSS and 2MASS. The combination of multi-wavelength data in this cross-matched set of 100,000 galaxies from these optical and NIR surveys will enable more sophisticated characterization and more in-depth exploration of relationships among galaxy morphological and dynamical parameters. The early results are quite tantalizing. We have sliced and diced the data set into various physically partitioned large subsamples (typically 30 bins of more than 3,000 galaxies each). We initially studied the fundamental plane of elliptical galaxies, which is a tight correlation among three observational parameters: radius, surface brightness, and velocity dispersion [11,12]. This well known relation now reveals systematic and statistically significant variations as a function of local galaxy density [7]. We are now extending this work into the realm of outlier/surprise/novelty detection and discovery.

26.2 Motivation

The growth in massive scientific databases has offered the potential for major new discoveries. Of course, simply having the potential for scientific discovery is insufficient, unsatisfactory, and frustrating. Scientists actually do want to make real discoveries. Consequently, effective and efficient algorithms that explore these massive datasets are essential. These algorithms will then enable scientists to mine and analyze ever-growing data streams from satellites, sensors, and simulations—to discover the most "interesting" scientific knowledge hidden within large and high-dimensional datasets, including new and interesting correlations, patterns, linkages, relationships, associations, principal components, redundant and surrogate attributes, condensed representations, object classes/subclasses and their classification rules, transient events, outliers, anomalies, novelties, and surprises. Searching for the "unknown unknowns" thus requires unsupervised and semisupervised learning algorithms. This is consistent with the observation that "*unsupervised exploratory analysis plays an important role in the study of large, high-dimensional datasets*" [28].

Among the sciences, astronomy provides a prototypical example of the growth of datasets. Astronomers now systematically study the sky with large sky surveys. These surveys make use of uniform calibrations and well engineered pipelines for the production of a comprehensive set of quality-assured data products. Surveys are used to collect and measure data from all objects that are visible within large regions of the sky, in a systematic, controlled, and repeatable fashion. These statistically robust procedures thereby generate very large unbiased samples of many classes of astronomical objects. A common feature of modern astronomical sky surveys is that they are producing massive catalogs. Surveys produce hundreds of terabytes (TB) up to 100 (or more) petabytes (PB) both in the image data archive and in the object catalogs. These include the existing SDSS and 2MASS, plus the future LSST in the next decade (with a 20–40 PB database). Large sky surveys have enormous potential to enable countless astronomical discoveries. Such discoveries will span the full spectrum of statistics: from rare one-in-a-billion (or one-in-a-trillion) type objects, to the complete statistical and astrophysical specification of a class of objects (based upon millions of instances of the class).

With the advent of large rich sky survey data sets, astronomers have been slicing and dicing the galaxy parameter catalogs to find additional, sometimes subtle, inter-relationships among a large variety of external and internal galaxy parameters. Occasionally, objects are found that do not fit anybody's model or relationship. The discovery of Hanny's Voorwerp by the Galaxy Zoo citizen science volunteers is one example [20, 21]. Some rare objects that are expected to exist are found only after deep exploration of multi-wavelength data sets (e.g., Type II QSOs [25, 33]; and Brown Dwarfs [3, 27]). These two methods of discovery (i.e., large-sample correlations and detection of rare outliers) demonstrate the two modes of scientific discovery potential from large data sets: (1) the best-ever statistical evaluation and parametric characterization of major patterns in the data, thereby explicating scaling relations in terms of fundamental astrophysical processes; and (2) the detection of rare one-in-a-million novel, unexpected, anomalous outliers, which are outside the expectations and predictions of our models, thereby revealing new astrophysical phenomena and processes (the "unknown unknowns"). Soon, with much larger sky surveys, we may discover even rarer one-in-a-billion objects and object classes.

LSST (www.lsst.org) is the most impressive astronomical sky survey being planned for the next decade. Compared to other sky surveys, the LSST survey will deliver time domain coverage for orders of magnitude more objects. The project is expected to produce ~15–30 TB of data per night of observation for 10 years.

The final image archive will be ~100 PB, and the final LSST astronomical object catalog (object-attribute database) is expected to be ~20–40 PB, comprising over 200 attributes for 50 billion objects and ~20 trillion source observations.

Many astronomy data mining use cases are anticipated with the LSST database [4], including:

- Provide rapid probabilistic classifications for all 10,000–100,000 LSST events each night;
- Find new correlations and associations of all kinds from the 200+ science attributes;
- Discover voids in multi-dimensional parameter spaces (e.g., period gaps);
- Discover new and exotic classes of objects, or new properties of known classes;
- Discover new and improved rules for classifying known classes of objects;
- Identify novel, unexpected behavior in the time domain from time series data;
- Hypothesis testing verify existing (or generate new) astronomical hypotheses with strong statistical confidence, using millions of training samples;
- Serendipitous discovery of very rare type of objects through outlier detection.

We are testing and validating exploratory data analysis algorithms that specifically support many of these science user scenarios for large database exploration.

26.3 Related Work

Various information theoretic measures of interestingness and surprise in databases have been studied in the past. Among these are Shannon entropy, information gain [17], Weaver's Surprise Index [32], and the J-Measure [30]. In general, such metrics estimate the relative information content between two sets of discrete-valued attributes. These measures can be used to identify interesting events in massive databases and data streams (through efficient interestingness metrics).

We have used PCA to identify outliers [14, 15]. In particular, we have been studying cases where the first two PC vectors capture and explain most (>90%) of the sample variance in the fundamental plane of elliptical galaxies. Consequently, in such a case, the component of a data record's attribute feature vector that projects onto the third PC eigenvector will provide a measure of the distance z_3 of that data record from the fundamental plane that defines the structure of the overwhelming majority of the data points. Simply sorting the records by z_3, and then identifying those with the largest values, will result in an ordered set of outliers [13] from most interesting to least interesting. We have tested this technique on a small cross-matched test sample of ellipticals from SDSS and 2MASS [14]. We will research the scalability of this algorithm to larger dataset sizes, to higher dimensions (i.e., number of science parameters), and to a greater number of principal components.

In many cases, the first test for outliers can be a simple multivariate statistical test of the "normalcy" of the data: is the location and scatter of the data consistent with a normal distribution in multiple dimensions? There are many tests for univariate data, but for multivariate data, we will investigate the Shapiro–Wilk test for normalcy and the Stahel-Donoho multivariate estimator for outlyingness [22, 29]. The Stahel-Donoho outlyingness parameter is straightforward to calculate and assign for each object: it is simply the absolute deviation of a data point from the centroid of the data set, normalized to the scale of the data. These tests should not be construed as proofs of non-normalcy or outlyingness, but as evidence. For petascale data, even simple tests are non-trivial in terms of computational cost, but it is essential to apply a range of algorithms in order to make progress in mining the data.

Several other algorithms and methods have been developed, and we will investigate these for their applicability and scalability to the large-data environment anticipated for LSST: *"Measures of Surprise in Bayesian Analysis"* [1], *"Quantifying Surprise in Data and Model Verification"* [2], and *"Anomaly Detection Model Based on Bio-Inspired Algorithm and Independent Component Analysis"* [31]. Such estimators can also be used in visual data mining—to highlight the most interesting regions of the dataset—this provides yet another tool for visual exploration and navigation of large databases for outliers and other interesting features [16, 23]; cf. [9, 19].

26.4 New Algorithm for Outlier Detection: KNN-DD

We have implemented a new algorithm for outlier detection that has proven to be effective at detecting a variety of novel, interesting, and anomalous data behaviors [5]. The *"K-Nearest Neighbor Data Distributions"*(KNN-DD) outlier detection algorithm evaluates the local data distribution around a test data point and compares that distribution with the data distribution within the sample defined by its K nearest neighbors. An outlier is defined as a data point whose distribution of distances between itself and its K-nearest neighbors is measurably different from the distribution of distances among the K-nearest neighbors alone (i.e., the two sets of distances are not drawn from the same population). In other words, an outlier is defined to be a point whose behavior (i.e., the point's location in parameter space) deviates in an unexpected way from the rest of the data distribution.

Our algorithm has these advantages: it makes no assumption about the shape of the data distribution or about "normal" behavior, it is univariate (as a function only of the distance between data points), it is computed only on a small-N local subsample of the full dataset, and as such it is easily parallelized when testing multiple data points for outlyingness. The algorithm is specified by the following rules, slightly updated from our previous results [5], as a consequence of our new experimental results (Sect. 8.1):

Here, $f(d,x)$ is the distribution of distances d between point x and a sample of data points, $f_K(d,O)$ is the distribution of distances between a potential outlier O and its K-nearest neighbors, and $f_K(d,K)$ is the distribution of distances among the K-nearest neighbors. The algorithm compares the two distance distribution functions $f_K(d,O)$ and $f_K(d,K)$ by testing if the two sets of distances are drawn from the same population.

Algorithm Outlier detection using K-nearest neighbor data distributions (KNN-DD)

1. Find the set S(K) of K nearest neighbors to the test data point O.
2. Calculate the K distances between O and the members of S(K). These distances define $f_K(d,O)$.
3. Calculate the K(K-1)/2 distances among the points within S(K). These distances define $f_K(d,K)$.
4. Compute the cumulative distribution functions $C_K(d,O)$ and $C_K(d,K)$, respectively, for $f_K(d,O)$ and $f_K(d,K)$.
5. Perform the K-S Test on $C_K(d,O)$ and $C_K(d,K)$. Estimate the p-value of the test.
6. Calculate the Outlier Index = $1 - \text{p}$.
7. If Outlier Index > 0.98, then mark O as an "Outlier". The Null Hypothesis is rejected.
8. If $0.90 <$ Outlier Index < 0.98, then mark O as a "Potential Outlier".
9. If $\text{p} > 0.10$, then the Null Hypothesis is accepted: the two distance distributions are drawn from the same population. Data point O is not marked as an outlier.

According to the definition of the KNN-DD algorithm, an outlier is defined as a data point whose distribution of distances between itself and its K-nearest neighbors is measurably different from the distribution of distances among the K-nearest neighbors alone (i.e., the two sets of distances are *not* drawn from the same population). We tested the effectiveness of this algorithm on a variety of synthetic idealized data streams (Sect. 26.5).

Our algorithm takes advantage of the two-sample K-S (Kolmogorov-Smirnov) statistical test, which is a classical non-parametric test used to estimate the likelihood that two sample distributions are drawn from the same population (= the Null Hypothesis). There is no assumption of normalcy or any other functional form for the distance distribution functions—this is an important and essential criterion in order to avoid introducing any bias in the estimation of outlier probability. We initially attempted to apply the Mann–Whitney (Wilcoxon) U Test to compare the two distance distribution functions, but this test failed to detect true outliers effectively—the primary reason is that the U Test essentially measures the difference in the median of the two distributions, which demonstrates that a single parameter is often a completely insufficient indicator of true outlyingness in multivariate data. The p-value derived from the K-S statistic (= the maximum difference between the two samples' cumulative density functions) measures the likelihood that the two samples satisfy the Null Hypothesis. We define the *Outlier Index* as $(1 - \text{p})$ = the probability that the Null Hypothesis is invalid (i.e., that the data distributions are not drawn from the same population). Consequently, the *Outlier Index* measures the likelihood that the test data point deviates from the behavior of the remainder of the data stream. Our algorithm has the advantage that it makes no assumption about the shape of the data distribution or about "normal" behavior.

The KNN-DD algorithm is different from the Distribution of Distances algorithm for outlier detection [26], in which the comparison is between the local data distribution around a test data point and a uniform data distribution. Our algorithm is also different from the k-Nearest Neighbor Graph algorithm for outlier detection [18], in which data points define a directed graph and outliers are those connected graph components that have just one vertex. Furthermore, our algorithm appears similar but is actually different in important ways from the incremental LOF (Local Outlier Factor) algorithms [6,24], in which the outlier estimate is density-based (determined as a function of the data point's local reachability density), whereas our outlier estimate is based on the full local data distribution. Finally, we believe that the KORM (K-median OutlieR Miner) approach to outlier detection in dynamic data streams [10] is most similar to our algorithm, except that their approach is cluster-based (based on K-medians) whereas our approach is statistics-based (based on the distribution function of distances).

To test the KNN-DD algorithm and to evaluate its effectiveness, we conducted two levels of experimentation. First, we tested the algorithm with a variety of simple synthetic data sets with and without outliers (Sect. 26.5). Second, we tested the algorithm on a set of actual scientific data, extracted from two astronomy databases (the data sets are described in Sect. 26.6 and the results of our outlier detection experiments are summarized in Sect. 26.7).

26.5 Synthetic Data Experiments

For our initial tests of the KNN-DD algorithm for outlier detection and its effectiveness, we performed a sequence of experiments on idealized synthetic data series. We synthesized three types of data streams:

- Linear data streams
- V-shaped data reversals (i.e., the "normal" data trend suddenly changes direction)
- Circular-shaped data distributions

The next step in the experiments was to insert test data points at varying distances from the "normal" data stream: ranging from "true normal" (TN) to "soft outlier" (SO) to "hard outlier" (HO), for which the test point was placed progressively farther and farther from the "normal" data. We finally applied our algorithm and measured the *Outlier Index* for the test data points, which estimates the likelihood that the test points are outliers. In each experiment, there were 25 points in the data stream, from which we identified the K = 9 nearest neighbors. Therefore, the 36 distances between these 9 points were calculated and used as an estimate for $f_K(d,K)$. Similarly, the nine distances between the test data point and K nearest neighbors were calculated and used as an estimate for $f_K(d,O)$. In each of the scatter plots shown below (Figs. 26.1–26.3), the outlier is identified as a filled brown square, the K nearest neighbors are identified as filled green circles, and the remaining (non-nearest neighbor) points in the data stream are identified as filled blue diamonds.

Table 26.1 presents our experimental results: the KS Test p-value, the Outlier Index, and the Outlier Flag for the nine experiments described above. It is clear from this table that the K-Nearest Neighbor Data Distribution algorithm for outlier detection is very effective at identifying outliers and at quantitatively estimating

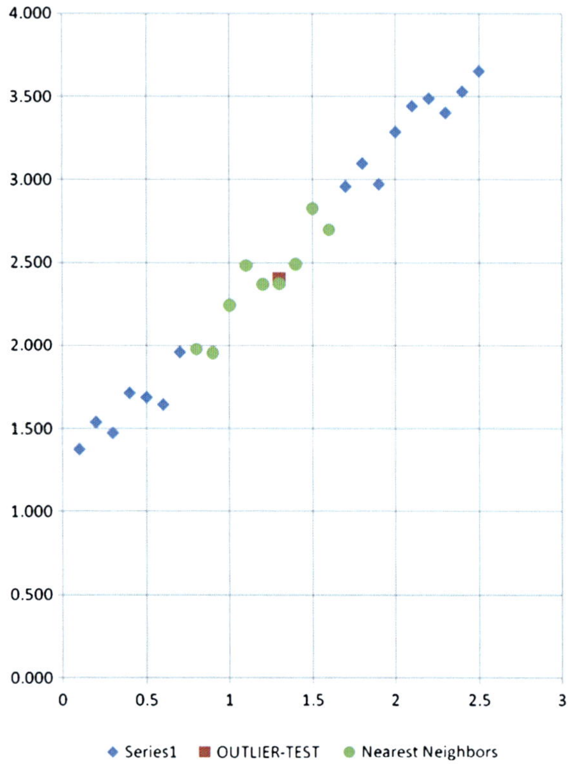

Fig. 26.1 Experiment L-TN (see Table 26.1)

their likelihood of "outlyingness". These results provide confidence that our new algorithm can be used to detect a variety of anomalous deviations in topologically diverse data streams.

26.6 Experimental Scientific Data Set

For the preliminary experiments reported here, we used a very small set of elliptical galaxies and stars from the combined SDSS+2MASS science data catalogs. We used 1,000 galaxies as the training set (i.e., as the set that represents "normal" behavior). We then used 114 other galaxies and 90 stars as test points (i.e., to measure and test for outlyingness). The galaxies represent "normal" behavior, and the stars are intended to represent outlier behavior. We chose seven color attributes as our feature vector for each object. The seven colors are essentially the seven unique (distance-independent, hence intrinsic) flux ratios (i.e., "colors") that can be generated from the eight (distance-dependent, hence extrinsic) flux measures from SDSS and 2MASS: the ugriz+JHK flux bands (which astronomers call "magnitudes"). Hence, we are exploring outlier detection in a 7-dimensional

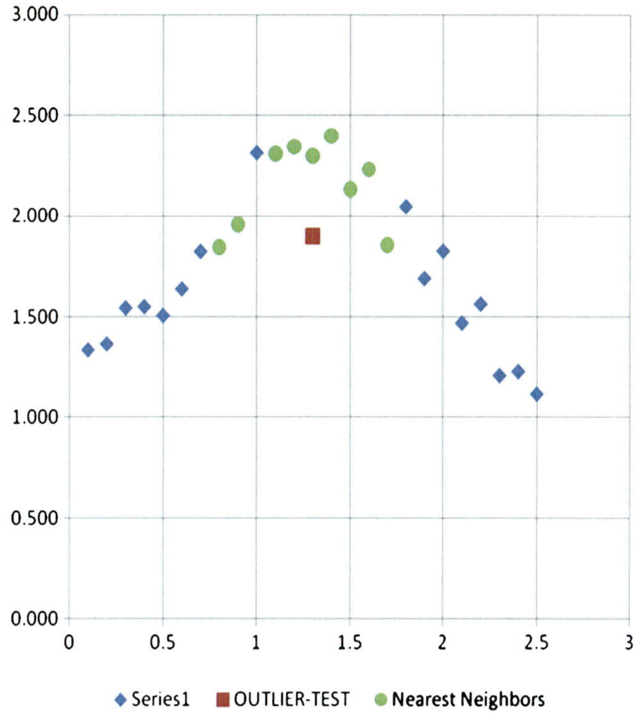

Fig. 26.2 Experiment V-SO (see Table 26.1)

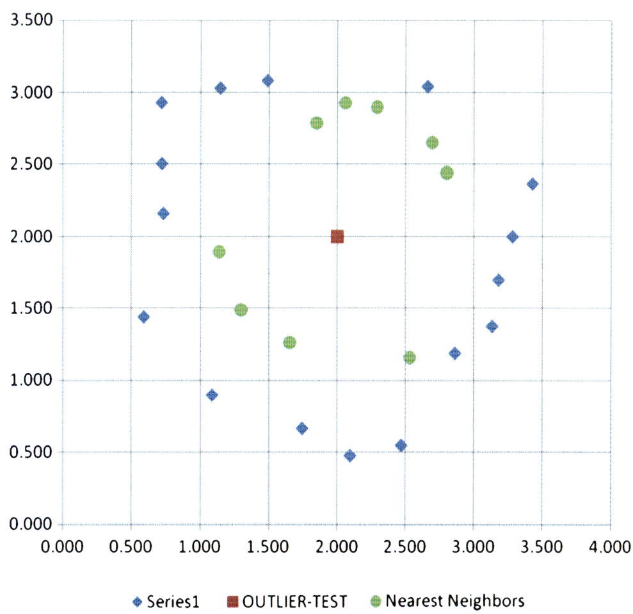

Fig. 26.3 Experiment C-HO (see Table 26.1)

Table 26.1 Results of experiments on the effectiveness of the K-Nearest Neighbor Data Distributions algorithm for outlier detection

Experiment ID	Short description of experiment	KS test p-value	Outlier index = 1 − p = Outlyingness likelihood (%)	Outlier flag ($p < 0.05$?)
L-TN (Fig. 26.1)	Linear data stream, true normal test	0.590	41.0	False
L-SO	Linear data stream, soft outlier test	0.096	90.4	Potential outlier
L-HO	Linear data stream, hard outlier test	0.025	97.5	TRUE
V-TN	V-shaped stream, true normal test	0.366	63.4	False
V-SO (Fig. 26.2)	V-shaped stream, soft outlier test	0.063	93.7	Potential outlier
V-HO	V-shaped stream, hard outlier test	0.041	95.9	TRUE
C-TN	Circular stream, true normal test	0.728	27.2	False
C-SO	Circular stream, soft outlier test	0.009	99.1	TRUE
C-HO (Fig. 26.3)	Circular stream, hard outlier test	0.005	99.5	TRUE

parameter space. In reality, there is some overlap in the colors of galaxies and stars, since galaxies are made up of stars, which thereby causes the stars to have much less outlyingness measure than we would like. On the other hand, this type of star-galaxy lassification/segregation is a standard and very important astronomy use case for any sky survey, and therefore it is a useful outlier test case for astronomy. The data distribution overlap among the stars and galaxies in our 7-dimensional parameter is somewhat ameliorated by the following fact. The flux of a galaxy GAL(flux) in one waveband is approximately the linear combination of its ten billion constituent stars' fluxes SUM*(flux) in that same waveband (modulo other effects, such as dust absorption and reddening, which are minimal in elliptical galaxies). Hence the colors of a galaxy are formed from the ratios of these linearly combined SUM*(flux) values. Consequently, the 7-dimensional feature vector of a galaxy need not align with any particular combination of stars' feature vectors. To illustrate this point, we consider a "toy" galaxy comprised of just two stars, with red band fluxes R*1 and R*2 and ultraviolet band fluxes U*1 and U*2. The U-R color (i.e., flux ratio) of the galaxy (modulo a logarithm and scale factor that astronomers like to use) is essentially (U*1+U*2)/(R*1+R*2), which cannot be represented by any simple linear combination of the stars' U-R colors: U*1/R*1 and U*2/R*2. Therefore, the actual distributions of stars and galaxies in our parameter space are sufficiently non-overlapping to allow us to perform reasonable outlier tests using stars as the outlier test points with regard to the "normal" galaxy points.

For our distance metric, we used a simple Euclidean (L2-norm) distance calculated from the seven feature vector attributes. Since they are all flux ratios, the seven attributes are already physically similar in terms of their mean, variance, and scale factor. No further data normalization or transformation is required.

Though the total numbers of galaxies and stars in our experiments are quite small, especially compared to the massive databases from which they were extracted, they actually do represent a somewhat typical data stream use case, in which a small number of incoming observations are tested against a small comparison set of "local" measurements, to search for and to detect outlier behaviors among the incoming measurements. In the future, we will expand our experiments to much greater numbers of test and training points.

26.7 Results from Scientific Data Set Experiments

We found the following results for the KNN-DD algorithm [5]. We measured standard Recall and Precision metrics along with the ROC curve as a function of continuously varying p-values (1 − p is the Outlier Index, as defined in the Algorithm definition in Sect. 26.4). In these experiments, Recall is calculated from the ratio of (number of stars correctly classified as outliers)/(total number of stars), and Precision is calculated from the ratio of (number of stars correctly classified as outliers)/[(number of galaxies misclassified as outliers)+(number of stars correctly classified as outliers)]. The variation in Precision as a function of p-value is illustrated in Fig. 26.4. The maximum precision (99%) for our test dataset is reached when the p-value reaches the limiting value 0.02. We establish this p-value (0.02) as the critical value used in the KNN-DD algorithm definition (Sect. 26.4).

We found that the "knee" in the ROC curve (i.e., the discrimination point that maximizes the combined Precision and Recall) occurs at p-value $\approx$ 0.05, which is the limiting value for outlier detection that we used in the KNN-DD algorithm [5]. We found that the Recall is almost exactly constant (approximately 100%) over most of the range of p values greater than 0.05. This clearly corroborates the point that we made in the first part of Sect. 26.6, that the data distribution of stars in our 7-dimensional feature space is mostly distinct from the data distribution of galaxies in that same parameter space. We tested this hypothesis by applying the DBI (Davies-Bouldin Index, [8]) as an evaluation metric for measuring the distinctness (i.e., separation) of the star and galaxy data distributions. In most cases, the DBI index verified that the star and galaxy data distributions were in fact well separated, though there were some interesting counter-examples that we will study in future work.

For p-value = 0.02, we find the following results: (1) for the 114 test galaxies, 89 are correctly classified (78%), and 25 are incorrectly classified as outliers (22%); and (2) of the 90 stars, 89 are correctly classified as outliers (99%), and one is misclassified as "normal". Hence, in this case, Recall = 99% and Precision = 78% $(= 89/(89+25))$.

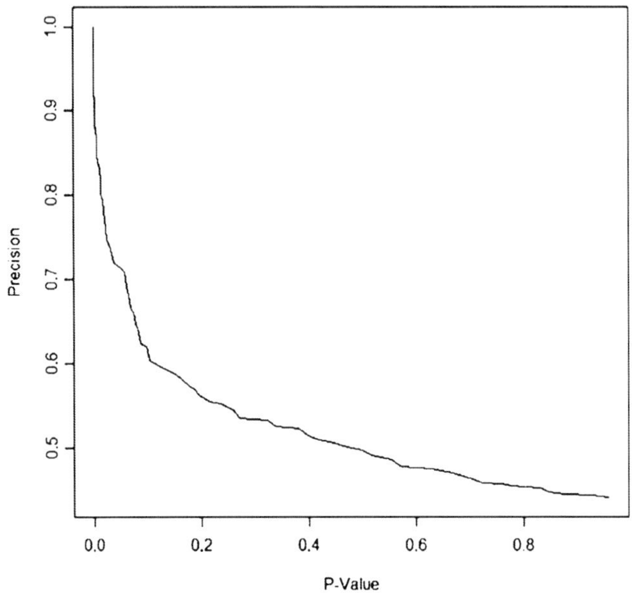

Fig. 26.4 Variation in the Precision of the outlier experiments using the KNN-DD algorithm, as a function of the p-value (where Outlier Index = 1 – p; see Sect. 26.4)

26.8 Concluding Remarks and Future Work

We find that our new KNN-DD algorithm is an effective and efficient algorithm for outlier detection. It has reasonable Precision and Recall accuracies, and it operates efficiently on small-N local data points, compared to other algorithms (e.g., PC-Out, [16]) that operate intensively on the full (large-N) set of global data. We therefore see the value of further experimentation with the KNN-DD algorithm on larger, more complex data streams. We also found some interesting behavior in high-dimension feature spaces regarding the region occupied by the outlier stars, compared with the region occupied by the outlier galaxies, compared with the region occupied by normal (non-outlier) galaxies. Further investigation of these surprising results is also warranted, which may already be yielding some scientific discoveries from these simple experimental test cases. We will also extend our KNN-DD comparison tests to include additional published outlier detection algorithms.

Our algorithm's success is based on the assumption that the distribution of distances between a true outlier and its nearest neighbors will be different from the distribution of distances among those neighbors by themselves. This assumption relies on the definition of an outlier as a point whose behavior (i.e., the point's location in parameter space) deviates in an unexpected way from the rest of the data distribution.

The main advantages of our KNN-DD algorithm are:

- It is based on the non-parametric K-S test.
- It makes no assumption about the shape of the data distribution or about "normal" behavior (of non-outliers).
- It compares the cumulative distributions of the test data (i.e., the set of inter-point distances), without regard to the nature of those distributions.
- It operates on multivariate data, thus solving the curse of dimensionality.
- It is algorithmically univariate, by estimating a function that is based entirely on the scalar distance between data points (which themselves occupy high-dimensional parameter space).
- It is simply extensible to higher dimensions.
- The KNN-DD distance distributions are computed only on small-K local sub-samples of the full dataset of N data points (K<<N).
- The algorithm is easily (embarrassingly) parallelizable when testing multiple data points for outlyingness.

The major deficiencies of the KNN-DD algorithm that need attention, as the algorithm is currently defined, and areas for future work include:

- The choice of K (see Sect. 26.4) is not determined or justified. We need to validate our choice of K, or else find a justifiable selection criterion for particular values.
- The choice of p (Sect. 26.4) is only weakly determined.
- We need to measure the learning times of the KNN-DD algorithm.
- We need to determine (and validate) the complexity of the KNN-DD algorithm.
- We need to compare the KNN-DD algorithm against a larger set of other outlier detection algorithms.
- We need to evaluate KNN-DD algorithm's effectiveness and efficiency on much larger datasets.
- We aim to demonstrate the usability of the KNN-DD algorithm on streaming data, not just with static data (as used in this paper's experiments).

Acknowledgements This research is supported in part by NASA AISR grant number NNX07AV70G and in part by NASA through the American Astronomical Society's Small Research Grant Program.

References

1. M. J. Bayarri and J. O. Berger. Measures of Surprise in Bayesian Analysis. Downloaded from http://citeseerx.ist.psu.edu/viewdoc/summary?doi=10.1.1.21.6365, 1997.
2. M. J. Bayarri and J. O. Berger. Quantifying Surprise in the Data and Model Verification. Downloaded from http://citeseer.ist.psu.edu/old/401333.html, 1998.
3. B. Berriman, D. Kirkpatrick, R. Hanisch, A. Szalay, and R. Williams. Discovery of Brown Dwarfs with Virtual Observatories. IAU Joint Discussion 8: Large Telescopes and Virtual Observatory: Visions for the Future. http://adsabs.harvard.edu/abs/2003IAUJD...8E..60B

4. K. Borne. Scientific Data Mining in Astronomy. Next Generation Data Mining. CRC Press: Taylor & Francis, Boca Raton, FL, pp. 91–114, 2009.
5. K. Borne. Effective Outlier Detection using K-Nearest Neighbor Data Distributions: Unsupervised Exploratory Mining of Non-Stationarity in Data Streams. Submitted to the Machine Learning Journal, March 2010.
6. M. Breunig, H.-P. Kriegel, R. Ng, and S. Sander. LOF: Identifying Density-Based Local Outliers. ACM SIGMOD Record, vol. 29, pp. 93–104, 2000.
7. K. Das, K. Bhaduri, S. Arora, W. Griffin, K. Borne, C. Giannella, and H. Kargupta. Scalable Distributed Change Detection from Astronomy Data Streams using Eigen-Monitoring Algorithms. 2009 SIAM International Conference on Data Mining (SDM09), 2009.
8. D. L. Davies and D. W. Bouldin. A Cluster Separation Measure. IEEE Transactions on Pattern Analysis and Machine Intelligence, 1(2): 224–227, 1979.
9. M. Debruyne. An Outlier Map for Support Vector Machine Classification. Annals of Applied Statistics, 3(4): 1566–1580, 2009.
10. P. Dhaliwal, M. Bhatia, and P. Bansal, P. A Cluster-based Approach for Outlier Detection in Dynamic Data Streams (KORM: K-median OutlieR Miner). Journal of Computing, vol. 2, pp. 74–80, 2010.
11. S. G. Djorgovski and M. Davis. Fundamental Properties of Elliptical Galaxies. Astrophysical Journal, vol. 313, pp. 59–68, 1987.
12. A. Dressler, D. Lynden-Bell, D. Burstein, R. L. Davies, S. M. Faber, R. Terlevich, and G. Wegner. Spectroscopy and Photometry of Elliptical Galaxies. I - A New Distance Estimator. Astrophysical Journal, vol. 313, pp. 42–58, 1987.
13. H. Dutta. Empowering Scientific Discovery by Distributed Data Mining on the Grid Infrastructure. Ph.D. dissertation, UMBC, 2007.
14. H. Dutta, C. Giannella, K. Borne, and H. Kargupta. Distributed Top-K Outlier Detection from Astronomy Catalogs using the DEMAC System. 2007 SIAM International Conference on Data Mining, 2007.
15. H. Dutta, C. Giannella, K. Borne, H. Kargupta, and R. Wolff. Distributed Data Mining for Astronomy Catalogs. IEEE Transactions in Knowledge and Data Engineering, 2009.
16. P. Filzmoser, R. Maronna, and M. Werner. Outlier Identification in High Dimensions. Computational Statistics and Data Analysis, 52, pp. 1694–1711, 2008.
17. A. Freitas On Objective Measures of Rule Surprisingness. LNCC, 1510, pp. 1–9, 1998.
18. V. Hautamaki, I. Karkkainen, and P. Franti. Outlier Detection Using k-Nearest Neighbour Graph. Proceedings of the 17th International Conference on Pattern Recognition (ICPR'04), 2004.
19. C. R. Johnson, M. Glatter, W. Kendall, J. Huang, and F. Hoffman. Querying for Feature Extraction and Visualization in Climate Modeling. ICCS 2009, Part II, LNCS 5545, pp. 416–425, 2009.
20. G. I. G. Jozsa, M. A. Garrett, T. A. Oosterloo, H. Rampadarath, Z. Paragi, H. van Arkel, C. Lintott, W. C.Keel, K. Schawinski, and E. Edmondson. Revealing Hanny's Voorwerp: Radio Observations of IC 2497. Astronomy and Astrophysics, vol. 500, pp. L33–L36, 2009.
21. C. J. Lintott, et al. Galaxy Zoo: Morphologies Derived from Visual Inspection of Galaxies from the Sloan Digital Sky Survey. Monthly Notices of the Royal Astronomical Society, vol. 389, pp. 1179–1189, 2008.
22. R. A. Maronna and V. J. Yohai. The Behavior of the Stahel-Donoho Robust Multivariate Estimator. Journal of the American Statistical Association, vol. 90, pp. 330–341, 1995.
23. D. Pena and F. J. Prieto. Multivariate Outlier Detection and Robust Covariance Matrix Estimation. Technometrics, vol. 43, pp. 286–301, 2001.
24. D. Pokrajac, A. Lazarevic, and L. Latecki, L. Incremental Local Outlier Detection for Data Streams. IEEE Symposium on Computational Intelligence and Data Mining (CIDM), 2007.
25. G. T. Richards, et al. Eight-Dimensional Mid-Infrared/Optical Bayesian Quasar Selection. Astronomical Journal, vol. 137, pp. 3884–3899, 2009.
26. V. Saltenis. Outlier Detection Based on the Distribution of Distances between Data Points. Informatica, 15(3): 399–410, 2004.

27. R.-D. Scholz, M. J. McCaughrean, N. Lodieu, and B. Kuhlbrodt. Epsilon Indi B: A New Benchmark T Dwarf. Astronomy and Astrophysics, vol. 398, pp. L29–L33, 2003.
28. A. A. Shabalin, V. J. Weigman, C. M. Perou, and A. B. Nobel. Finding Large Average Submatrices in High Dimensional Data. Annals of Applied Statistics, 3(3): 985–1012, 2009.
29. S. S. Shapiro and M. B. Wilk. An analysis of variance test for normality (complete samples). Biometrika, vol. 52, pp. 591–611, 1965.
30. P. Smyth and R. M. Goodman. Rule Induction Using Information Theory. Knowledge Discovery in Databases, pp 159–176, AAAI/MIT Press, 1991.
31. S. Srinoy and W. Kurutach. Anomaly Detection Model Based on Bio-Inspired Algorithm and Independent Component Analysis. TENCON 2006, IEEE Region 10 Conference proceedings, pp. 1–4, 2006.
32. Weaver's Surprise Index. Encyclopedia of Statistical Sciences (Wiley), vol. 9, pp. 104–109, 1988.
33. N. Zakamska, et al. Candidate Type II Quasars from the Sloan Digital Sky Survey. I. Selection and Optical Properties. Astronomical Journal, vol. 126, pp. 2125–2143, 2003.

Chapter 27
On Statistical Cross-Identification in Astronomy

Tamás Budavári

Abstract The association of independent detections of the same objects is one of the most important fundamental challenges of observational astronomy today. Multicolor datasets have proven to provide great insight into large-scale structure, galaxy evolution, and multi-epoch observations are becoming mainstream with the upcoming next-generation sky surveys. The cross-identification is, however, a difficult problem scientifically, computationally and statistically. We will discuss a probabilistic approach that applies Bayesian hypothesis testing to decide whether a given set of detections truly belong to the same source. Studying the ensemble statistics of the datasets we can assign probabilities to the matches. The algorithms are shown to perform well in simulations and real observations. We extend the method to stars with unknown proper motion and discuss further applications to transients events. Also we visit some of the issues that arise in the online aggregation of catalogs.

27.1 Introduction

The problem of source identification in separate observations is as old as astronomy itself. When the early astronomers re-observed a celestial object, they performed crossmatching to past detections by pointing the telescope to the previously measured direction and verified (by eye) that identify of the source. Today we automatically collect the detections in catalogs and identify hundreds of million sources at once. When the observations have similar selection functions and clearly isolate the sources, any reasonable criterion will yield correct associations. One can

---

T. Budavári (✉)
Department of Physics and Astronomy, Johns Hopkins University, 3400 N. Charles St., Baltimore, MD 21218, USA
e-mail: budavari@jhu.edu

E.D. Feigelson and G.J. Babu (eds.), *Statistical Challenges in Modern Astronomy V*, Lecture Notes in Statistics 209, DOI 10.1007/978-1-4614-3520-4_27,

choose to measure pairwise distances and threshold at "some" angular separations or derive the maximum likelihood of a best-fit common direction for the detections. In the general case, most of the ad hoc methods break down. Also the constraining power of the measured directions might not be fully sufficient to distinguish between the good and bad candidates, and we want to fold in other type of measurements into the process, for example, the brightness and/or color of the sources. In this talk we discuss a Bayesian approach to tackle the core problem and look at its applicability beyond the simplest positional matching of static sources.

27.2 The Real Question: Same or Not?

Instead of asking misleading meta-questions, e.g., about the angular separation of the sources, we can directly address the real question: are the given detections belong to the same astronomical object or not. In Bayesian hypothesis testing we can compare these two possibilities using the Bayes factor, which is simply the ratio of the likelihoods of the two complement hypotheses,

$$B = \frac{L_{\text{same}}}{L_{\text{not}}} \tag{27.1}$$

The interpretation seems straightforward at first. If $B = 1$, we cannot decide. When $B > 1$, the data suggest a match but $B < 1$ argues otherwise. In reality nothing is black or white and the decision is more complicated. This is the topic of Sect. 27.3. Before that we should familiarize ourselves with the Bayes factor, its calculation and the issues with its direct interpretation.

To evaluate the likelihood of the hypotheses we have to consider the entire domain of their parameter space and sum up all the possibilities. Here we first use an analogous problem to illustrate the simplicity of the calculation and to gain further insight into the challenges of the interpretation. We use playing dice. A die is an object analogous to an astronomical source, whose detection corresponds to rolling the die. The measured position is the side that the die shows. Of course, we have lots of object/dice. If the dice are fair and have no preference for any particular side, we learn nothing from the outcome of the rolling. If the dice, however, are loaded and do prefer certain sides, the outcomes of the rolling can help us make decisions about their identity. The astronomical cross-identification problem is exactly same as the following thought experiment. First let us consider the two-way case. From a bag of loaded dice, we draw twice with replacement. First we roll a ⚀ followed by a ⚃. Is it the same die? We just need a model for how the dice are loaded (Fig. 27.1). For example, a die with loadedness of $l = 1$ will prefer the side ⚀ as described by some known probabilities, e.g.,

$$P_1(\text{⚀}) = \frac{3}{12}, \quad P_1(\text{⚁}) = \frac{2}{12}, \quad \ldots, \quad P_1(\text{⚅}) = \frac{1}{12}$$

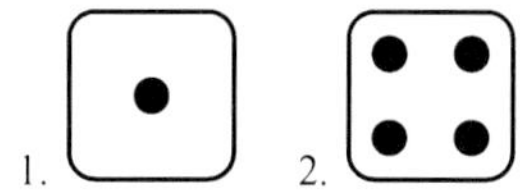

Fig. 27.1 From a bag of loaded dice that prefer different sides, we draw twice with replacement. First we roll a ⚀ followed by a ⚃. Is it the same die? The measurements can help us decide. If we roll the same side ⚀ with the die drawn for the second time, the match is more likely but clearly not guaranteed. This thought experiment is analogous to the cross-identification problem of astronomical detections

Similarly,

$$\begin{array}{ccc} P_2(\text{⚀}) = \frac{2}{12}, & P_2(\text{⚁}) = \frac{3}{12}, & \ldots, \quad P_2(\text{⚅}) = \frac{2}{12} \\ \vdots & \vdots & \vdots \\ P_6(\text{⚀}) = \frac{1}{12}, & P_6(\text{⚁}) = \frac{2}{12}, & \ldots, \quad P_6(\text{⚅}) = \frac{3}{12} \end{array}$$

This matrix of probabilities is the analog of the known astrometric accuracy model on the sky: the probability (density) of the possible outcomes for a given true direction.

If the dice drawn with replacement are indeed the same, their loadedness have to naturally be the identical. It is the same die after all. The likelihood of a given loadedness l is the product of the $P_l(\text{⚀})$ and $P_l(\text{⚃})$ probabilities. But we do not know what l is. We could use maximum likelihood estimation to figure out its best guess value(s) but now we are not interested in that. Instead we have to consider all the l values to account for all possibilities in our hypothesis. The uniform prior on l is 1/6, as it can take six possible values. The result is the likelihood of the dice being the same

$$L_{\text{same}} = \frac{1}{6} \sum_l P_l(\text{⚀}) P_l(\text{⚃}) \tag{27.2}$$

The sum is directly calculated from our data, in this case, the rolled faces. For the cross-identification of more than two dice, we can use the same calculation only the product in the likelihood will contain more terms; one for each observation.

The complement hypothesis claims that the two dice are different, hence their loadedness could differ. We need two variables l_1 and l_2 to parameterize the model. Now the sum conveniently falls apart as

$$L_{\text{not}} = \left[\frac{1}{6} \sum_{l_1} P_{l_1}(\text{⚀}) \right] \left[\frac{1}{6} \sum_{l_2} P_{l_2}(\text{⚃}) \right] \tag{27.3}$$

Similarly this formula also works with multiple observations and not just for two. In case of fair dice with all $P_l(\cdot)$ probabilities equal to 1/6, we can verify that $B = 1$, i.e., we did not learn anything from the observations. For loaded dice and real astronomical observations the ratio will be typically more conclusive.

Real directional measurements are continuous and so are the parameters of the models that we have to integrate out. Let us consider n detections, one from each survey i. Are they the same object or not? Our data consist of $D = \{x_i\}$ unit vectors of the observed directions. The astrometric model is usually described by the "normal distribution". On the surface of the sphere the Fisher distribution [3] is the simplest analog to the familiar Gaussian distribution;

$$F(x|m,w) = \frac{w\,\delta(|x|-1)}{4\pi\sinh(w)}\,\exp(w\,m{\cdot}x) \tag{27.4}$$

where m is the model direction (three dimensional unit vector) and $\delta(\cdot)$ is the Dirac δ. In the limit of high accuracies, the Fisher and Gaussian distributions become interchangeable, and the precision parameter w of the Fisher distribution is related to the σ of the Gaussian by the $w = 1/\sigma^2$ equality, where σ is in radians. With the Fisher distribution, the Bayes factor is analytically calculated. Assuming a uniform prior density on the entire sky,[1] the result takes the following simple form [1],

$$B = \frac{\sinh w}{w}\prod_{i=1}^{n}\frac{w_i}{\sinh w_i} \qquad \text{with} \qquad w = \left|\sum_{i=1}^{n} w_i x_i\right| \tag{27.5}$$

which is the same formula that also arose in an earlier study of GRB repeatability in the limit of substantial directional uncertainties [6]. The $w_i = 0$ values mean no spatial constraint (like fair dice) and the Bayes factor is $B = 1$. If all positional measurements are highly accurate ($w_i \gg 1$), we get back a more familiar exponential expression,

$$B = 2^{n-1}\frac{\prod w_i}{\sum w_i}\exp\left\{-\frac{\sum_{i<j} w_i w_j \varphi_{ij}^2}{2\sum w_i}\right\} \tag{27.6}$$

where φ_{ij} is the angle between x_i and x_j unit vectors. In the two-way case, the dimensionless Bayes factor simplifies to

$$B = \frac{2}{\sigma_1^2+\sigma_2^2}\exp\left\{-\frac{\varphi^2}{2(\sigma_1^2+\sigma_2^2)}\right\} \tag{27.7}$$

where all quantities are in radians, and $\sigma_i^2 = 1/w_i$ as before. The top panel of Fig. 27.2 illustrates (27.7) on a logarithmic scale for fixed 0.1" and 0.5" uncertainties that roughly correspond to the precision of the Sloan Digital Sky Survey (SDSS) and the Galaxy Evolution Explorer (GALEX). Note that for constant accuracies, a cut on the Bayes factor $B = B(\varphi;\sigma_1,\sigma_2)$ is equivalent to thresholding the angular separation as B only depends on φ.

[1] The formula is different for observations with limited field of view, however, the scaling eventually cancels out in the probability and only the density matters within the footprint [1].

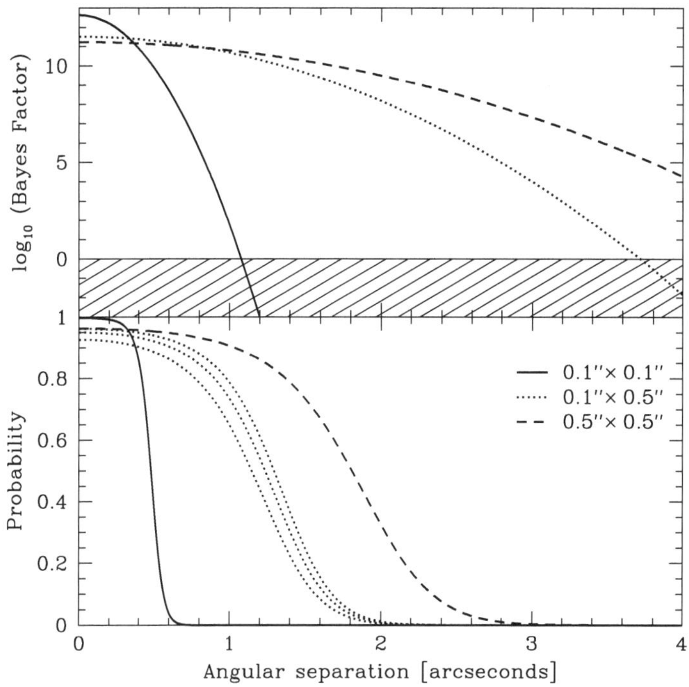

Fig. 27.2 The Bayes factor is shown in the *top panel* as a function of angular separation for three different matching scenarios. The *solid line* corresponds to detections with $\sigma = 0.1$" accuracy and the *dashed line* is for 0.5". These roughly corresponds to the precision of the SDSS and GALEX surveys, respectively. The *dotted line* is the analog of the SDSS-GALEX matching. The *bottom panel* illustrates the probability using the same line styles. Here we use 25,000 SDSS sources per square degree, and assume that the GALEX density is 50% of that. The three *dotted lines* are based on estimates of the selection functions that represent overlaps of 100%, 75% and 50% of GALEX. As the intersection decreases, so does the probability, but the large posteriors are fairly insensitive to (small) variations in the prior

When B is very large, the data suggest a good match, and when B is close to 0, the evidence points to separate objects. While in practice these extrema certainly occur very frequently, the interesting regime is in between at intermediate values. What these values really correspond to is difficult to see right away.

27.3 Probability

The Bayes factor is the fundamental quantity we rely on. Its interpretation, however, may not be obvious at first. The Bayes factor, also, does not capture the full complexity of the problem. Our goal is to assign probabilities that we can relate to.

Let us consider the bag of loaded dice in our previous thought experiment. Having drawn two dice with replacement and seen the rolled sides of ⚀ and ⚃, how can we determine or even estimate the probability? The answer is that we cannot unless we study the content of the bag and count the dice. If we only have a single die in the bag, the probability is 1 regardless of the value of B. Also if the bag contains a very large number of dice, we expect a low probability. Based on the number of dice in the bag, we can calculate the prior probability of drawing (with replacement) the same die twice before looking at the results of the rolls.

The Bayes factor is the missing link that connects the prior and the posterior probabilities. For two complement hypotheses (same or not), B tells us how the prior probability P_0 is updated based on the data to yield the posterior,

$$P = \left[1 + \frac{1 - P_0}{B\,P_0}\right]^{-1} \tag{27.8}$$

If we have N dice in the bag, the probability of drawing the same die for the second time is $P_0 = 1/N$. When we draw k times, the prior is $P_0 = N^{k-1}$. Astronomy is just a little bit different from this. The added complication comes from the fact that separate observations may have different selection functions, e.g., would detect different sources at different wavelengths. In Fig. 27.3, several SDSS, GALEX and 2MASS sources illustrate the different selections. In the general case, the prior is

$$P_0 = \frac{N_\star}{\prod N_i} \tag{27.9}$$

where N_i are the number of sources in the ith dataset,[2] and $N_\star$ is the number of sources that are seen in all. The latter is unknown but an educated guess usually works reasonably well. The bottom panel of Fig. 27.2 shows the SDSS-GALEX matching scenarios with the same styles as the Bayes factors in the top panel. For this illustration we assume N values that correspond to 25,000 and 12,500 detections per square degree for SDSS and GALEX, respectively. We see that the exquisite SDSS astrometry provides great constraints. When we match observations with SDSS-like accuracies (solid line), probabilities peak at around 1. The larger uncertainty of GALEX means lower maximum posterior for the GALEX-GALEX matching at 0 separation, and a slower drop as function of the angular distance. The SDSS-GALEX crossmatch shown in dotted lines can only be calculated with an estimate of the overlap. Here we plot the results $N_\star$ values that correspond to 100%, 75% and 50% of the GALEX density. We can clearly see the aforementioned robustness of the posterior against small changes in the prior.

[2] Here we again assume all sky coverage; see the previous footnote and the reference therein.

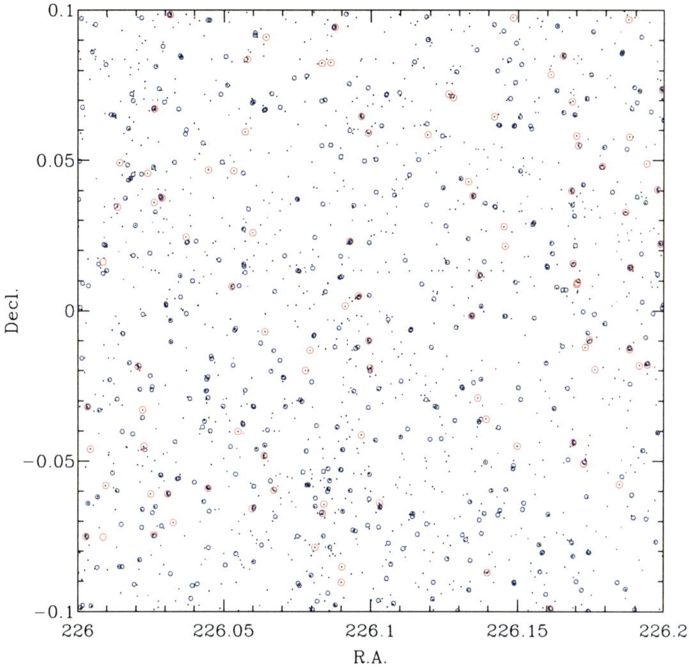

Fig. 27.3 Datasets contain different sets of objects with varying accuracy. Their densities on the sky are different, as well, and only a fraction of the sources appear in all

The prior can be accurately determined from the ensemble statistics of the input datasets [1]. Iteratively solving a set of two simple equations takes the guesswork completely out of the problem and provides a self-consistent result. Using a constant prior P_0, we can rewrite (27.9) as a sum over all possible $\prod N_i$ combinations in the catalogs

$$\sum P_0 = N_\star \tag{27.10}$$

Similarly this equality holds for the sum of the posteriors

$$\sum P = N_\star \tag{27.11}$$

Considering that P values are determined from corresponding Bayes factors and the prior, this is an equation for $N_\star$ that we can efficiently solve numerically in just a few steps of iterations. It is initialized with an estimate of $N_\star$, e.g., $\min\{N_i\}$, then calculates the prior, and sums up the corresponding posteriors to obtain a new estimate of $N_\star$.

27.4 Simulations for SDSS and GALEX

Using a simulated object in a mock universe with realistic mixture of sources and galaxy clustering, we can evaluate the performance of the new method [4]. We assign a uniform [0,1] random number to each mock object representing its properties, and express the selection function of the simulated surveys as intervals. The density of sources seen in an actual survey is matched by tuning the length of the interval. The intersection of the two surveys' selection function is adjusted by the overlap of the two intervals. We create catalogs that match the SDSS and GALEX observations and simulate their observed positions by drawing from realistic error distribution around the true positions of the mock sources.

Figure 27.4 shows the value of the prior as a function of the iteration. We see convergence in just a few steps, which is a result of the large posterior's robustness against small variations in the prior. In comparison to the true input overlap represented by the constant *solid line*, the measured *points* appear to go to a somewhat higher value. This is due to the proximity of objects by chance. This confusion is significantly lower for the high signal-to-noise GALEX detections that are shown with a *dashed line* and *crosses*. The value of the prior for these S/N>3 sources is higher because their density on the sky is lower and more of those sources are seen in both surveys, larger $N_{\star}$. Also the points approximate the constant better.

That said, these small deviations in the prior estimate are practically negligible in most cases, except for extremely confused observations. By overestimating the prior, we assign slightly high posterior probabilities. In case when this is unacceptable, a simulation similar to the presented SDSS-GALEX case can be performed to calibrate the discrepancy from the confusion.

Matches selected based on the Bayes factor are different than those from simple cuts on angular separation. The reason is that GALEX has varying astrometric

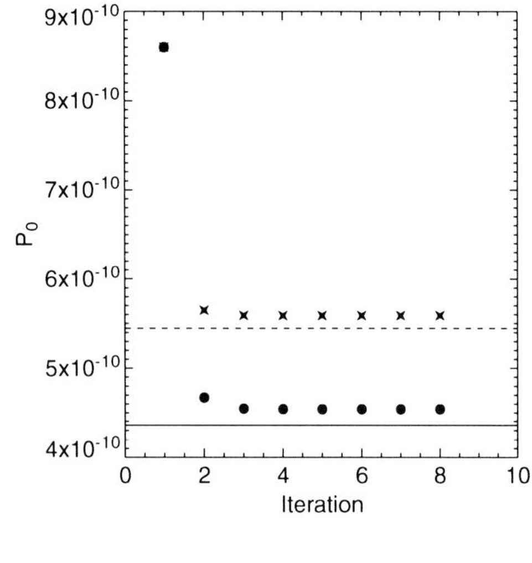

Fig. 27.4 The iterative procedure accurately determines the prior in just a few steps. The empirical solution converges to a slightly higher value due to the proximity of random objects. This discrepancy is even lower for the high signal-to-noise subset

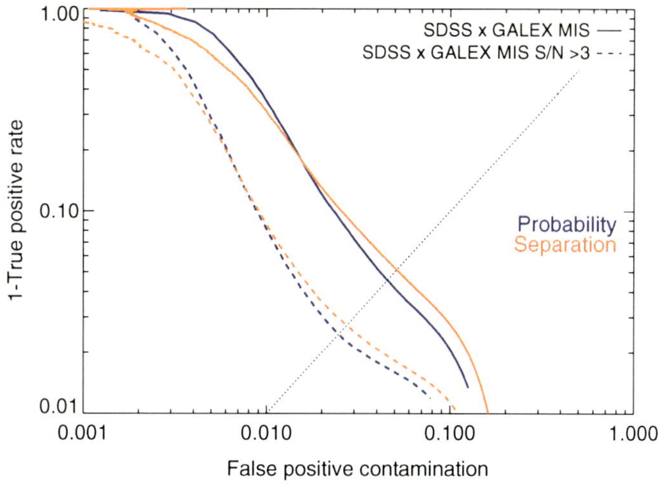

Fig. 27.5 Thresholding on the Bayes factor is superior to simply using angular separations even for two-way matching when the astrometric errors vary from detection to detection as in the GALEX catalog

errors, which is accurately emulated in the above example. In the simulation we know the truth and can compare the performance of the two methods. In Fig. 27.5 we show the false positive contamination versus the true positive rate. The thick solid lines illustrates the performance of the probabilistic approach in comparison to the think dashed line for the angular separation. The cross-over at large separations is an artifact of the astrometric model (truncated estimate). The thin lines show the same for the high signal-to-noise, S/N>3 subsets.

Another interesting outcome of the simulation is the explanation of multiple matches. In the real datasets, the bulk of the associations consists of 1-to-1 matches, but we also observe that a fraction of the GALEX detections are assigned to two or more SDSS detections, and vice versa. Early speculations were inclined to blame this on the image processing pipelines but our simple model of point sources with realistic astrometric errors can indeed accurately reproduce the correct fractions. We can hence conclude that these multiple matches are not artifacts of the data processing but are purely statistical in nature.

27.5 Proper Motion

We briefly mention the possibility of extending the new method to more complicated situations. For example, we can associate stars that move on the sky. The naive approach to this problem would be to artificially make the uncertainties larger,

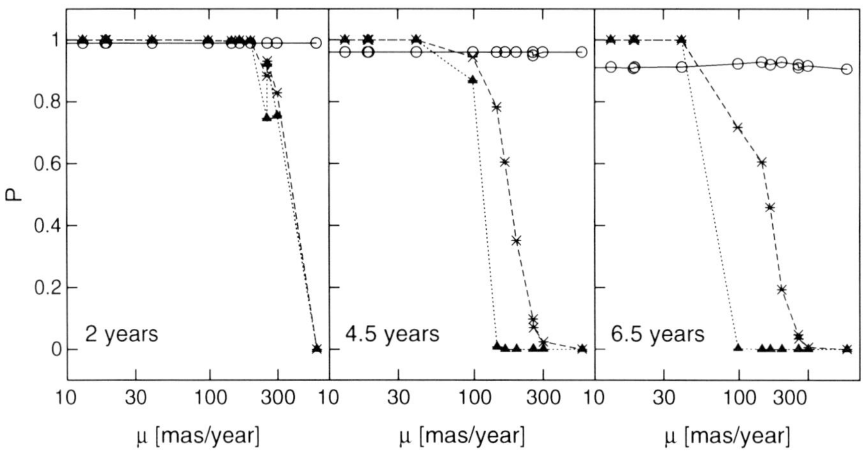

Fig. 27.6 Probability of two-way associations as a function of proper motion. As we look at longer time differences between the epochs, our models yield increasing different answers

so we consider objects farther away on the sky. This, however, would yield loads of false matches and would not account for the type of motion we are looking for.

The cross-identification of stars with unknown proper motion is done properly by computing the likelihood of the same two hypotheses. Same or not? The difference is that the parameters now not only consist of the position on the sky but also the parameters of the model for the motion. In a recent paper [5], we explore different approaches on select stars in the repeated SDSS observations of Stripe 82. Our models differ only in the prior on the proper motion: we compare the static model to a constant out to $\mu_{\max}$=600 mas/year and a more elaborate empirical prior based the SDSS statistics and matching Besançon simulations. In Fig. 27.6 we plot the two-way matching with varying separations in time. As we increase the time baseline (from left to right) the static model (triangles) starts to reject the faster stars but the proper motion models assign finite probabilities, even if not too large. The constant prior yields a constant posterior (open circles) that becomes lower with time. Perhaps even more interesting is Fig. 27.7 where we use the same time baseline but introduce intermediate epochs in our datasets. The previous two-epoch results quickly jump to high probabilities for three- and four-epoch observations. While their superior quality might be somewhat surprising at first, it is actually easy to understand: Proper motion is well-approximated by a movement along a great circle. The big difference between two and more epochs is the fact that it is always possible to precisely fit a great circle to two points but not to three or more, hence such configurations are very much rewarded.

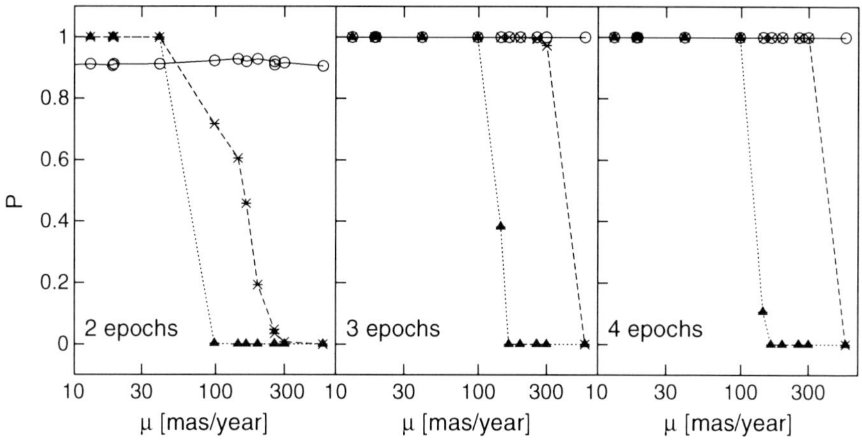

Fig. 27.7 Probability of two-, three-, and four-way star associations as a function of proper motion. Introducing a third intermediate epoch improves significantly the cross-identification. The effect of the fourth measurement is not so dramatic

27.6 Discussion

The power and elegance of the presented Bayesian approach to cross-identification shine most in the non-trivial cases. Thresholding two-way matches based on angular separations is equivalent to cutting on Bayes factors (or posterior probabilities) when the astrometric uncertainties are assumed to be constant (and the prior does not change as a function of position on the sky.) Even in that simple case, the statistical method provides guidance on where to draw the line. For more than two datasets and/or varying astrometric errors any other approach becomes substantially more complicated. Our probabilistic treatment completely takes the guesswork out of the association and delivers reliable results. The Bayes factor is directly calculated from the data and the prior is estimated in a self-consistent manner. While the fully Bayesian solution would include a hierarchical model and a hyper-prior for the prior or $N_\star$, the computational expenses cannot be really justified for the typical observations today. The reason is the insensitivity of large posteriors to small variations in the prior.

The beauty of the method is that it is not limited to positional information. Any kind of data can be naturally folded into the associations, which is often needed to break the degeneracies in uncertain data. One obvious example is the spectral energy distributions (SEDs) of the sources, that we usually can model reasonably well. One can calculate separate Bayes factors for all available observations [1] and simply multiply them.

$$B = B_{\text{direction}} \cdot B_{\text{photometry}} \cdots B_{\text{others}} \tag{27.12}$$

Generalizations of the presented approach are fairly straightforward, yet, exceptionally powerful. In addition to the aforementioned proper motion example, we can also extend the solution to associate cosmic events [2] or extended sources. As long as we can model the data, we can evaluate the likelihoods to decide whether the detections are the "same" or "not".

The matches can be obtained efficiently in an incremental algorithm [1], where we add new datasets or catalogs to partially matched associations. Aggregations of a large number of datasets, however, can still pose certain computational challenges. Conceptually the problem with such data is that we are not allowed to throw away partial matches in early steps that might turn out to be valid matches in case all the subsequent detections fall into the "ideal" directions. The theoretically correct, *pessimistic* approach will have too many tuples early on that are only pruned later in the process. A heuristic, *optimistic* approach can compensate for this by assuming certain errors in the subsequent data and hence can reduce the partial matches at the beginning at the risk of loosing exceptionally unlucky (and unlikely) alignments. Alternatively one can only aggregate detections that are "guaranteed" to belong to each other and apply a multi-pass solution to arrive at the final results. These strategies are currently being explored in the context of the Hubble Legacy Archive and the upcoming time-domain surveys.

Cross-identification has always been a hard problem scientifically, computationally and statistically. Part of the reason is that science and statistics are inherently interwoven in such settings. This explicit Bayesian approach shows the way out of this Catch-22 by offering a computable and consistent solution.

References

1. Budavári, T., & Szalay, A. S., 2008, *Astrophys. J.*,, 679, 301
2. Budavári, T. 2011, *Astrophys. J.*,, 736, 155B
3. Fisher, R., 1953, Proceedings of the Royal Society of London, Series A, Mathematical and Physical Sciences, Vol. 217, No. 1130., pp.295–305
4. Heinis, S., Budavári, T., & Szalay, A. S. 2009, *Astrophys. J.*,, 705, 739
5. Kerekes, G., Budavári, T., Csabai, I., Connolly, A. J., & Szalay, A. S. 2010, *Astrophys. J.*,, 719, 59
6. Luo, S., Loredo, T., & Wasserman, I. 1996, American Institute of Physics Conference Series, 384, 477

Chapter 28
Commentary: On Statistical Cross-Identification in Astronomy

Thomas J. Loredo

Abstract This contribution is a commentary on Tamás Budavári's paper, "On statistical cross-identification in astronomy," presented at the *Statistical Challenges in Modern Astronomy V* conference held at Pennsylvania State University in June 2011. I describe multilevel Bayesian treatments of this problem developed for identifying gamma ray burst counterparts.

Budavári's paper reviews the key concepts of a recent body of research by him and his colleagues on Bayesian cross-matching of astronomical object catalogs. When object directions have quantified uncertainties (e.g., error circles with confidence levels), this approach offers significant advantages over more conventional approaches that attempt to assess directional coincidences using ad hoc statistics (e.g., nearest neighbor angles, counts in cones, χ^2-based statistics, likelihood ratios) and p-values. In the mid-1990s gamma-ray burst (GRB) astronomers developed essentially the same approach for assessing evidence for repetition of GRBs [4, 6, 8], and for association of GRBs with special supernovae [5]. This work pre-dates the discovery of GRB X-ray afterglows; the available GRB data provided direction estimates with large uncertainties (5–25° error circles for directions from BATSE data; many-arc-minute error boxes for interplanetary network direction estimates). Budavári seeks to cross-match optical and UV catalogs that are much larger in size than GRB catalogs, and with much more accurate directions. This is a complementary regime, raising unique challenges for Bayesian cross-matching—especially computational challenges—that Budavári's team is addressing with innovative techniques just briefly touched on in his paper (e.g., see [1–3, 7]).

For this commentary I am taking a cue from Budavári's Discussion section, where he states that a "fully Bayesian solution would include a hierarchical model"

T.J. Loredo (✉)
Center for Radiophysics & Space Research, Cornell University, Ithaca, NY 14853-6801, USA
e-mail: loredo@astro.cornell.edu

E.D. Feigelson and G.J. Babu (eds.), *Statistical Challenges in Modern Astronomy V*,
Lecture Notes in Statistics 209, DOI 10.1007/978-1-4614-3520-4_28,

but that "the computational requirements do not justify the extra work." Luo et al. [8] developed a hierarchical (multilevel) Bayesian framework for assessing directional and temporal coincidences in a GRB catalog; an exact calculation was indeed impossible, and LLW96 had to rely on an unsatisfying approximate treatment. However, for an important issue that Budavári discusses, a simple multilevel model is both computationally accessible, and illuminating.

Specifically, Budavári highlights the important role of the prior probability for association, P_0, in Bayesian cross-matching; it is needed to convert marginal likelihoods (or Bayes factors) for association to posterior probabilities (or odds). But Budavári's P_0 is determined using the data; it cannot really be a prior probability. A multilevel model not only enables estimation of P_0, but also can account for its uncertainty, which should play a role in assessing the method's performance in simulation studies (e.g., determining whether the discrepancy between estimated and true association fractions in Budavári's Fig. 4 is acceptable). A multilevel treatment also illuminates other issues important for probabilistic cross-matching. In the limited space available here I describe a simple example calculation illustrating the main idea; a more complete and general treatment will be published elsewhere.

Suppose we have a "target" catalog of N_t newly detected objects, and we would like to determine if some or all of them are associated with any of N_c previously detected objects in a candidate host or counterpart catalog spanning the same region of the sky. From the target observations, analysis of the data associated with object number i produces a likelihood function, $\ell_i(\omega)$, for its direction, ω; the target catalog provides summaries of these likeihood functions (e.g., best-fit directions and error circles when uncertainties can be accurately described by Fisher distributions). Similarly, $m_k(\omega)$ is the likelihood function for the direction to object k in the candidate host catalog. Suppose the cataloged hosts are a sample from a large population with a known (or well-estimated) directional distribution, $\rho_c(\omega)$, e.g., an isotropic distribution with $\rho_c = 1/4\pi$. Then the posterior distribution for the direction to candidate host object k is $\rho_c(\omega)m_k(\omega)/Z_k$, where the normalization constant $Z_k = \int d\omega\, \rho_c(\omega)m_k(\omega)$. The marginal likelihood that target i is associated with host k (thus sharing a common direction) is

$$h_{ik} = \int d\omega \frac{\rho_c(\omega)m_k(\omega)}{Z_k}\ell_i(\omega). \tag{28.1}$$

The marginal likelihood that target i is instead from a background population of hosts with direction distribution $\rho_0(\omega)$ is

$$g_i = \int d\omega\, \rho_0(\omega)\ell_i(\omega). \tag{28.2}$$

The Bayes factor in favor of association of target i with host k versus a background source is $b_{ik} = h_{ik}/g_i$. When the direction likelihoods are proportional to Fisher distributions and the host and background densities are isotropic, this corresponds to Budavári's B (also derived earlier by LLW96 and [4]).

Of course, we do not know a priori which candidate host to assign to each target. The marginal likelihood that target i is associated with *one* of the candidate hosts must account for this uncertainty by introducing a prior probability for the host choice, say $1/N_c$, and marginalizing over k; the resulting marginal likelihood is

$$h_i = \frac{1}{N_c}\sum_k h_{ik} = \frac{1}{N_c}\sum_k \int d\omega \frac{\rho_c(\omega)m_k(\omega)}{Z_k}\ell_i(\omega). \tag{28.3}$$

Budavári introduced a prior probability for association, P_0, in order to convert marginal likelihoods (or Bayes factors) to posterior probabilities (or odds). Using intuitively appealing arguments, he develops equations to determine a value for P_0, but they use the data, and thus P_0 is not really a prior probability, and his posterior probabilities are not formally valid. To better motivate and extend Budavári's appealing results, we make the association model a multilevel model, introducing a population parameter that we will estimate from the data.

Define the target population association parameter, α, as the probability that a randomly selected target comes from the population of cataloged candidate hosts (so $1-\alpha$ is the probability that a target comes from the background). Were α known, the posterior probability that target i is associated with one of the hosts would be

$$p_i(\alpha) = \frac{\alpha h_i}{(1-\alpha)g_i + \alpha h_i}. \tag{28.4}$$

But typically α will *not* be known a priori; in fact, estimating α may be a significant scientific goal. The likelihood function for α is the probability for the target data, given α and the host catalog information; using the above results, it is

$$\mathscr{L}(\alpha) = \prod_{i=1}^{N_t}\left[(1-\alpha)g_i + \alpha h_i\right]. \tag{28.5}$$

A straightforward calculation shows that the maximum-likelihood value of α, $\hat{\alpha}$, satisfies the following equation:

$$\sum_i p_i(\hat{\alpha}) = \hat{\alpha}N_t. \tag{28.6}$$

This is an intuitively appealing result: for the maximum-likelihood value of α, the sum of the association probabilities is equal to the expected number of targets with associations.

To see the connection with Budavári's rule for assigning P_0 (his equation (10)), suppose the data provide direction estimates with uncertainties that are small compared with the angles between hosts. Then the sum in the marginal likelihood for association for a target object, (28.3), will typically be dominated by just one term, so

$$h_i \approx \frac{1}{N_c}\int d\omega \frac{\rho_c(\omega)m_{k(i)}(\omega)}{Z_{k(i)}}\ell_i(\omega), \tag{28.7}$$

where $k(i)$ specifies the index of the host that is the nearest neighbor to target i (in the sense of having the largest marginal likelihood term). If we use this approximation for h_i in (28.4) for $p_i(\alpha)$, then (28.6) becomes equivalent to Budavári's equation (10) (identifying $\hat{\alpha}$ with his P_0, p_i with his P, and $\hat{\alpha}N_t$ with his $N_\star$), for the case of two catalogs.

This calculation does more than simply justify Budavári's intuitive arguments for setting P_0. One concrete benefit is that it enables accounting for uncertainty in α. Combined with a prior for α, the likelihood function in (28.5) produces a posterior for α. If the prior is not highly informative, the posterior will be asymptotically normal, with a mean close to $\hat{\alpha}$ and a variance, σ_α^2, that can be found by calculating the second derivative of $\ln[\mathscr{L}(\alpha)]$ at $\hat{\alpha}$; the result is

$$\frac{1}{\sigma_\alpha^2} = \frac{1}{\hat{\alpha}(1-\hat{\alpha})}\sum_i(\hat{p}_i-\hat{\alpha})^2 = \frac{N_t}{\hat{\alpha}(1-\hat{\alpha})}\left[\frac{1}{N_t}\sum_i\hat{p}_i^2-\left(\frac{\sum_i\hat{p}_i}{N_t}\right)^2\right], \tag{28.8}$$

where $\hat{p}_i \equiv p_i(\hat{\alpha})$. Two limiting cases are illuminating. Suppose first that the target positions have very large uncertainties. In the limit where $\ell_i(\omega) \to C$, a constant, we have $g_i = h_i = C$. The Bayes factor for association of each object is unity (indicating the data provide no information to alter prior probabilities), and the likelihood function for α is flat, so there is no unique $\hat{\alpha}$ value. The right hand side of (28.8) vanishes, implying divergence of the variance (actually, the asymptotic approximation is not valid with a flat likelihood function). The data provide no information about the association fraction in this case, as one would expect. Now consider the opposite limit where the direction uncertainties are small, leading to unambiguous associations (very large Bayes factors), so that for values of α away from zero or unity, $p_i \approx 0$ or 1. In this case, (28.6) tells us that $\hat{\alpha} = N_+/N_t$, where N_+ is the number of targets with $p_i \approx 1$. Equation 28.8 indicates that in this limit, $\sigma_\alpha \to 1/\sqrt{N_t}$, again an intuitively reasonable result. For intermediate cases, where there is evidence for associations but with some ambiguity, the uncertainty in α will be larger than "root-N," by an amount depending on the variance between the $\hat{p}_i$ values and $\hat{\alpha}$. Calculating σ_α for the SDSS–GALEX example in Budavári's Sect. 4 may be helpful in assessing the discrepancy between the estimated and input values of P_0.

In the SDSS–GALEX example, P_0 was over-estimated; Budavári attributes this to confusion due to chance proximity of objects in each catalog. But one of the aims of probabilistic modeling of directional coincidences is to account for this sort of confusion. An accurate Bayesian calculation will account for it, resulting in no significant bias in estimation of the association fraction, but possibly increased α uncertainty when the directional uncertainties lead to significant counterpart confusion. When a particular target has multiple plausible associations, the probability for association will be split across them. One way to see how the Bayesian calculation handles counterpart ambiguity is to rewrite the likelihood function to more explicitly display how it accounts for each possible association. First introduce unifying notation for the components in the likelihood factor for a particular target:

define weights w_k, with $w_0 = 1 - \alpha$ and $w_k = \alpha / N_c$ for $k = 1$ to N_c, and let $h_{ik} = g_i$ when $k = 0$. Also introduce target labels λ_i that take values from 0 to N_c. Then (28.5) can be written as

$$\mathscr{L}(\alpha) = \prod_{i=1}^{N_t} \sum_{\lambda_i=0}^{N_c} w_{\lambda_i} h_{i\lambda_i} = \sum_{\lambda_1 \ldots \lambda_{N_t}} \left(\prod_k w_k^{m_k(\lambda)} \right) \prod_i h_{i\lambda_i}, \tag{28.9}$$

where the last sum is over all label assignments, and $m_k(\lambda)$ is the multiplicity for host k, counting the number of targets with $\lambda_i = k$ in a particular term of the sum. This sum-of-products decomposition displays the likelihood as a weighted sum of terms considering every possible assignment of targets to candidate hosts. If we adopt the best-candidate approximation of (28.7), only a small fraction of the terms is considered; when confusion is important, additional terms in h_i should be kept so that the calculation accounts for all plausible associations.

Equation 28.9 also reveals an unsatisfactory aspect of the model I have described here: it allows for all possible host multiplicities, in particular, it allows for assigning two targets to the same host. In some settings this is desirable, e.g., for constraining GRB repetition, or for determining whether ultra-high energy cosmic rays come from nearby active galaxies. But in many settings—including the SDSS–GALEX case—it is only meaningful to assign targets to distinct hosts. This argues that the sum-of-products version of the likelihood function is the more fundamental representation to use for building coincidence assessment models; for the SDSS–GALEX case, the sum over labels would be constrained to ensure distinct associations. This is why LLW96 adopted this representation for developing a general framework for spatio-temporal coincidence assessment.

As a final remark on the value of an explicit multilevel model for associations, recall that we needed to assign a prior probability for the host choice, taken as $1/N_c$ in (28.3); in the sum-of-products version of the likelihood function, this assignment appears in the w_k factors. More generally, the candidate host prior may not be constant; it could depend, for example, on host distances and luminosities, and this affects estimation of α. It is straightforward to account for this in a multilevel model, though it can complicate the calculations. The paper in these proceedings by Soiaporn et al. briefly describes work by my team based on just such a model, developed to assess evidence for association of ultra-high energy cosmic rays with local active galaxies.

Budavári developed his Bayesian approach from scratch, unaware of earlier work on the problem in the GRB literature. In fact, that work was well-hidden, tersely presented in short papers in conference proceedings. More extensive treatments did not follow because it proved extremely difficult to get funding to further develop the approach; reviewers expressed strong skepticism of Bayesian methods. To cite one ironically relevant example, the report from a 2005 NVO proposal review panel asserted that the Bayesian approach offered "nothing new" for the problem, and that its implementation "would not be much more than a 'few-liner' addition to `Xmatch`," the χ^2-based NVO cross-match algorithm now made obsolete by

Budavári's Bayesian algorithm. With this frustrating history, it has been a delight to see Budavári's team not only rediscover the approach, but also make significant and highly nontrivial statistical and computational innovations mating it to the needs of VAO users.

Acknowledgements Ira Wasserman helped me develop a framework for Bayesian coincidence assessment in the mid 1990s; that work was partially supported by NASA grant NAG 5-2762. Shan Luo helped us with early calculations. Currently, David Chernoff and statisticians David Ruppert and Kunlaya Soiaporn are helping us take the approach much further; our work together is funded by an interdisciplinary NSF grant, AST-0908439. I am grateful to all of these collaborators and funding agencies for their contributions to the research informing this commentary.

References

1. Budavári, T.: Probabilistic Cross-identification of Cosmic Events. *Astrophys. J.* **736**, 155–+ (2011). 10.1088/0004-637X/736/2/155
2. Budavári, T., Heinis, S., Szalay, A.S., Nieto-Santisteban, M., Gupchup, J., Shiao, B., Smith, M., Chang, R., Kauffmann, G., Morrissey, P., Schiminovich, D., Milliard, B., Wyder, T.K., Martin, D.C., Barlow, T.A., Seibert, M., Forster, K., Bianchi, L., Donas, J., Friedman, P.G., Heckman, T.M., Lee, Y.W., Madore, B.F., Neff, S.G., Rich, R.M., Welsh, B.Y.: GALEX-SDSS Catalogs for Statistical Studies. *Astrophys. J.* **694**, 1281–1292 (2009). 10.1088/0004-637X/694/2/1281
3. Budavári, T., Szalay, A.S.: Probabilistic Cross-Identification of Astronomical Sources. *Astrophys. J.* **679**, 301–309 (2008). 10.1086/587156
4. Graziani, C., Lamb, D.Q.: Likelihood methods and classical burster repetition. In: R. E. Rothschild & R. E. Lingenfelter (ed.) High Velocity Neutron Stars, *American Institute of Physics Conference Series*, vol. 366, pp. 196–200 (1996). 10.1063/1.50246
5. Graziani, C., Lamb, D.Q., Marion, G.H.: Evidence against an association between gamma-ray bursts and Type I supernovae. *Astron. Astrophys. Suppl.* **138**, 469–470 (1999). 10.1051/aas:1999313
6. Graziani, C., Lamb, D.Q., Quashnock, J.M.: Are the four gamma-ray bursts of 1996 October 27–29 due to repetition of a single source? In: C. A. Meegan, R. D. Preece, & T. M. Koshut (ed.) Gamma-Ray Bursts, 4th Hunstville Symposium, *American Institute of Physics Conference Series*, vol. 428, pp. 161–165 (1998). 10.1063/1.55314
7. Kerekes, G., Budavári, T., Csabai, I., Connolly, A.J., Szalay, A.S.: Cross Identification of Stars with Unknown Proper Motions. *Astrophys. J.* **719**, 59–66 (2010). 10.1088/0004-637X/719/1/59
8. Luo, S., Loredo, T., Wasserman, I.: Likelihood analysis of GRB repetition. In: C. Kouveliotou, M. F. Briggs, & G. J. Fishman (ed.) American Institute of Physics Conference Series, *American Institute of Physics Conference Series*, vol. 384, pp. 477–481 (1996). 10.1063/1.51706

Chapter 29
Data Compression Methods in Astrophysics

Raul Jimenez

Abstract Astrophysics is an observational science and as such it is gathering data from the whole sky. The availability of large CCD cameras on telescopes with large fields of view is permitting the collection of large amount of data. Eventually an observational astrophysicist would like to collect all information available in the sky. This however brings some problems as one usually has "too many" data and new techniques are required to analyse them. In this chapter I will provide a (biased) view of how the problem can be addressed using new statistical tools to achieve data compression of the data. Note that when I talk about data compression in astrophysics I will always refer to algorithms that are able to massively accelerate likelihood computations of comparing data with models and not about "throwing" data away. In particular I will illustrate how to deal with data from galaxy surveys, exoplanet light-transit searches and direct gravitational wave searches.

29.1 Introduction

There are many instances where objects consist of many data, whose values are determined by a small number of parameters. Often, it is only these parameters which are of interest.

Such a problem is very general, and has been attacked in the case of parameter estimation in large-scale structure and the microwave background (e.g. [1]). Previous work has concentrated largely on the estimation of a single parameter; the main advance of this paper is that it sets out a method for the estimation of multiple parameters. The method provides one projection per parameter, with the consequent possibility of a massive data compression factor. Furthermore, if the noise in the

R. Jimenez (✉)
ICREA & ICC, University of Barcelona (UB-IEEC), Marti i Franques 1, Barcelona 08028, Spain
e-mail: raul.jimenez@icc.ub.edu

E.D. Feigelson and G.J. Babu (eds.), *Statistical Challenges in Modern Astronomy V*, Lecture Notes in Statistics 209, DOI 10.1007/978-1-4614-3520-4_29,

data is independent of the parameters, then the method is entirely lossless. i.e. the compressed dataset contains as much information about the parameters as the full dataset, in the sense that the Fisher information matrix is the same for the compressed dataset as the entire original dataset. An equivalent statement is that the mean likelihood surface is at the peak locally identical when the full or compressed data are used.

29.2 MOPED

A data-compression method was developed by Heavens et al. [2]. We review it here.

Describe data by a vector $\mathbf{x}_i$, $i = 1,\ldots,N$ (e.g. a set of fluxes at different wavelengths). These measurements include a signal part, which we denote by μ, and noise, $\mathbf{n}$:

$$\mathbf{x} = \mu + \mathbf{n} \tag{29.1}$$

Assuming the noise has zero mean, $\langle \mathbf{x} \rangle = \mu$, the signal will depend on a set of parameters $\{\theta_\alpha\}$, which we wish to determine. For galaxy spectra, the parameters may be, for example, age, magnitude of source, metallicity and some parameters describing the star formation history. Thus, μ is a noise-free spectrum of a galaxy with certain age, metallicity etc.

The noise properties are described by the noise covariance matrix, $\mathbf{C}$, with components $C_{ij} = \langle n_i n_j \rangle$. If the noise is gaussian, the statistical properties of the data are determined entirely by μ and $\mathbf{C}$. In principle, the noise can also depend on the parameters. For example, in galaxy spectra, one component of the noise will come from photon counting statistics, and the contribution of this to the noise will depend on the mean number of photons expected from the source.

The aim is to derive the parameters from the data. If we assume uniform priors for the parameters, then the a posteriori probability for the parameters is the likelihood, which for gaussian noise is

$$\begin{aligned}\mathscr{L}(\theta_\alpha) = &\frac{1}{(2\pi)^{N/2}\sqrt{\det(\mathbf{C})}} \\ &\times \exp\left[-\frac{1}{2}\sum_{i,j}(x_i - \mu_i)\mathbf{C}_{ij}^{-1}(x_j - \mu_j)\right].\end{aligned} \tag{29.2}$$

One approach is simply to find the (highest) peak in the likelihood, by exploring all parameter space, and using all N pixels. The position of the peak gives estimates of the parameters which are asymptotically (low noise) the best unbiased estimators. This is therefore the best we can do. The maximum-likelihood procedure can, however, be time-consuming if N is large, and the parameter space is large. The aim of this paper is to see whether we can reduce the N numbers to a smaller number, without increasing the uncertainties on the derived parameters θ_α. To be specific, we try to find a number $N' < N$ of linear combinations of the spectral data $\mathbf{x}$ which encompass as much as possible of the information about the physical parameters. We find that this can be done losslessly in some circumstances; the spectra can be reduced to a handful of numbers without loss of information. The speed-up in parameter estimation is about a factor $\sim$100.

In general, reducing the dataset in this way will lead to larger error bars in the parameters. To assess how well the compression is doing, consider the behaviour of the (logarithm of the) likelihood function near the peak. Performing a Taylor expansion and truncating at the second-order terms,

$$\ln \mathscr{L} = \ln \mathscr{L}_{\text{peak}} + \frac{1}{2} \frac{\partial^2 \ln \mathscr{L}}{\partial \theta_\alpha \partial \theta_\beta} \Delta \theta_\alpha \Delta \theta_\beta . \tag{29.3}$$

Truncating here assumes that the likelihood surface itself is adequately approximated by a gaussian everywhere, not just at the maximum-likelihood point. The actual likelihood surface will vary when different data are used; on average, though, the width is set by the (inverse of the) Fisher information matrix:

$$\mathbf{F}_{\alpha\beta} \equiv - \left\langle \frac{\partial^2 \ln \mathscr{L}}{\partial \theta_\alpha \partial \theta_\beta} \right\rangle \tag{29.4}$$

where the average is over an ensemble with the same parameters but different noise.

For a single parameter, the Fisher matrix $\mathbf{F}$ is a scalar F, and the error on the parameter can be no smaller than $F^{-1/2}$. If the data depend on more than one parameter, and all the parameters have to be estimated from the data, then the error is larger. The error on one parameter α (marginalised over the others) is at least $\left[(\mathbf{F}^{-1})_{\alpha\alpha}\right]^{1/2}$. There is a little more discussion of the Fisher matrix in [1], hereafter TTH. The Fisher matrix depends on the signal and noise terms in the following way (TTH, equation 15)

$$\mathbf{F}_{\alpha\beta} = \frac{1}{2} \mathrm{Tr} \left[\mathbf{C}^{-1} \mathbf{C}_{,\alpha} \mathbf{C}^{-1} \mathbf{C}_{,\beta} + \mathbf{C}^{-1} (\mu_{,\alpha} \mu^t_{,\beta} + \mu_{,\beta} \mu^t_{,\alpha}) \right] . \tag{29.5}$$

where the comma indicates derivative with respect to the parameter. If we use the full dataset $\mathbf{x}$, then this Fisher matrix represents the best that can possibly be done via likelihood methods with the data.

In practice, some of the data may tell us very little about the parameters, either through being very noisy, or through having no sensitivity to the parameters. So in principle we may be able to throw some data away without losing very much information about the parameters. Rather than throwing individual data away, we can do better by forming linear combinations of the data, and then throwing away the combinations which tell us least. To proceed, we first consider a single linear combination of the data:

$$y \equiv \mathbf{b}^t \mathbf{f} x \tag{29.6}$$

for some weighting vector $\mathbf{b}$ (t indicates transpose). We will try to find a weighting which captures as much information about a particular parameter, θ_α. If we assume we know all the other parameters, this amounts to maximising $\mathbf{F}_{\alpha\alpha}$. The dataset (now consisting of a single number) has a Fisher matrix, which is given in TTH (equation 25) by:

$$\mathbf{F}_{\alpha\beta} = \frac{1}{2}\left(\frac{\mathbf{b}^t\mathbf{C}_{,\alpha}\mathbf{b}}{\mathbf{b}^t\mathbf{C}\mathbf{b}}\right)\left(\frac{\mathbf{b}^t\mathbf{C}_{,\beta}\mathbf{b}}{\mathbf{b}^t\mathbf{C}\mathbf{b}}\right) + \frac{(\mathbf{b}^t\mu_{,\alpha})(\mathbf{b}^t\mu_{,\beta})}{(\mathbf{b}^t\mathbf{C}\mathbf{b})}. \tag{29.7}$$

Note that the denominators are simply numbers. It is clear from this expression that if we multiply $\mathbf{b}$ by a constant, we get the same $\mathbf{F}$. This makes sense: multiplying the data by a constant factor does not change the information content. We can therefore fix the normalisation of $\mathbf{b}$ at our convenience. To simplify the denominators, we therefore maximise $\mathbf{F}_{\alpha\alpha}$ subject to the constraint

$$\mathbf{b}^t\mathbf{C}\mathbf{b} = 1. \tag{29.8}$$

The most general problem has both the mean μ and the covariance matrix $\mathbf{C}$ depending on the parameters of the spectrum, and the resulting maximisation leads to an eigenvalue problem which is nonlinear in $\mathbf{b}$. We are unable to solve this, so we consider a case for which an analytic solution can be found. TTH showed how to solve for the case of estimation of a single parameter in two special cases: (1) when μ is known, and (2) when $\mathbf{C}$ is known (i.e. doesn't depend on the parameters). We will concentrate on the latter case, but generalise to the problem of estimating many parameters at once. For a single parameter, TTH showed that the entire dataset could be reduced to a single number, with no loss of information about the parameter. We show below that, if we have M parameters to estimate, then we can reduce the dataset to M numbers. These M numbers contain just as much information as the original dataset; i.e. the data compression is lossless.

We consider the parameters in turn. With $\mathbf{C}$ independent of the parameters, $\mathbf{F}$ simplifies, and, maximising $\mathbf{F}_{11}$ subject to the constraint requires

$$\frac{\partial}{\partial b_i}\left(b_j\mu_{,1\,j}b_k\mu_{,1\,k} - \lambda b_j C_{jk} b_k\right) = 0 \tag{29.9}$$

where λ is a Lagrange multiplier, and we assume the summation convention ($j,k \in [1,N]$). This leads to

$$\mu_{,1}(\mathbf{b}^t\mu_{,1}) = \lambda\mathbf{C}\mathbf{b} \tag{29.10}$$

with solution, properly normalised

$$\mathbf{b}_1 = \frac{\mathbf{C}^{-1}\mu_{,1}}{\sqrt{\mu^t_{,1}\mathbf{C}^{-1}\mu_{,1}}} \tag{29.11}$$

and our compressed datum is the single number $y_1 = \mathbf{b}_1^t\mathbf{x}$. This solution makes sense—ignoring the unimportant denominator, the method weights high those data which are parameter-sensitive, and low those data which are noisy.

To see whether the compression is lossless, we compare the Fisher matrix element before and after the compression. Substitution of $\mathbf{b}_1$ into (29.7) gives

$$\mathbf{F}_{11} = \mu_{,1}^t \mathbf{C}^{-1} \mu_{,1} \tag{29.12}$$

which is identical to the Fisher matrix element using the full data (29.5) if $\mathbf{C}$ is independent of θ_1. Hence, as claimed by TTH, the compression from the *entire* dataset to the single number y_1 loses no information about θ_1. For example, if $\mu \propto \theta$, then $y_1 = \sum_i \mathbf{x}_i / \sum_i \mu_i$ and is simply an estimate of the parameter itself.

It is important to note that y_1 contains as much information about θ_1 only if all other parameters are known, and also provided that the covariance matrix and the derivative of the mean in (29.11) are those at the maximum likelihood point. We turn to the first of these restrictions in the next section, and discuss the second one here.

In practice, one does not know beforehand what the true solution is, so one has to make an initial guess for the parameters. This guess we refer to as the fiducial model. We compute the covariance matrix $\mathbf{C}$ and the gradient of the mean ($\mu_{,\alpha}$) for this fiducial model, to construct $\mathbf{b}_1$. The Fisher matrix for the compressed datum is (29.12), but with the fiducial values inserted. In general this is not the same as Fisher matrix at the true solution. In practice one can iterate: choose a fiducial model; use it to estimate the parameters, and then repeat, using the estimate as the estimated parameters as the fiducial model.

29.2.1 Estimation of Many Parameters

The problem of estimating a single parameter from a set of data is unusual in practice. Normally one has several parameters to estimate simultaneously, and this introduces substantial complications into the analysis. How can we generalise the single-parameter estimate above to the case of many parameters? We proceed by finding a second number $y_2 \equiv \mathbf{b}_2^t\mathbf{x}$ by the following requirements:

- y_2 is uncorrelated with y_1. This demands that $\mathbf{b}_2^t\mathbf{C}\mathbf{b}_1 = 0$.
- y_2 captures as much information as possible about the second parameter θ_2.

This requires two Lagrange multipliers (we normalise $\mathbf{b}_2$ by demanding that $\mathbf{b}_2^t\mathbf{C}\mathbf{b}_2 = 1$ as before). Maximising and applying the constraints gives the solution

$$\mathbf{b}_2 = \frac{\mathbf{C}^{-1}\mu_{,2} - (\mu_{,2}^t\mathbf{b}_1)\mathbf{b}_1}{\sqrt{\mu_{,2}\mathbf{C}^{-1}\mu_{,2} - (\mu_{,2}^t\mathbf{b}_1)^2}}. \tag{29.13}$$

This is readily generalised to any number M of parameters. There are then M orthogonal vectors $\mathbf{b}_m$, $m = 1,\ldots M$, each y_m capturing as much information about parameter α_m which is not already contained in y_q; $q < m$. The constrained maximisation gives

$$\mathbf{b}_m = \frac{\mathbf{C}^{-1}\mu_{,m} - \sum_{q=1}^{m-1}(\mu_{,m}^t\mathbf{b}_q)\mathbf{b}_q}{\sqrt{\mu_{,m}\mathbf{C}^{-1}\mu_{,m} - \sum_{q=1}^{m-1}(\mu_{,m}^t\mathbf{b}_q)^2}}. \tag{29.14}$$

This procedure is analogous to Gram-Schmidt orthogonalisation with a curved metric, with $\mathbf{C}$ playing the role of the metric tensor. Note that the procedure gives precisely M eigenvectors and hence M numbers, so the dataset has been compressed from the original N data down to the number of parameters M.

Since, by construction, the numbers y_m are uncorrelated, the likelihood of the parameters is obtained by multiplication of the likelihoods obtained from each statistic y_m. The y_m have mean $\langle y_m\rangle = \mathbf{b}_m^t\mu$ and unit variance, so the likelihood from the compressed data is simply

$$\ln\mathscr{L}(\theta_\alpha) = \text{constant} - \sum_{m=1}^{M}\frac{(y_m - \langle y_m\rangle)^2}{2} \tag{29.15}$$

and the Fisher matrix of the combined numbers is just the sum of the individual Fisher matrices. Note once again the role of the fiducial model in setting the weightings $\mathbf{b}_m$: the orthonormality of the new numbers only holds if the fiducial model is correct. Multiplication of the likelihoods is thus only approximately correct, but iteration could be used if desired.

Under the assumption that the covariance matrix is independent of the parameters, reduction of the original data to the M numbers y_m results in no loss of information about the M parameters at all. In fact the set $\{y_m\}$ produces, on average, a likelihood surface which is locally identical to that from the entire dataset—no information about the parameters is lost in the compression process. With the restriction that the information is defined locally by the Fisher matrix, the set $\{y_m\}$ is a set of sufficient statistics for the parameters $\{\theta_\alpha\}$. A proof of this for an arbitrary number of parameters is given in the appendix.

29.3 The General Case

In general, the covariance matrix does depend on the parameters, and this is the case for galaxy spectra, where at least one component of the noise is parameter-dependent. This is the photon counting noise, for which $\mathbf{C}_{ii} = \mu_i$. TTH argued that it is better to treat this case by using the n eigenvectors which arise from assuming the mean is known, rather than the single number (for one parameter) which arises if we assume that the covariance matrix is known, as above. We find that, on the contrary, the small number of eigenvectors $\mathbf{b}_m$ allow a much greater degree of compression than the known-mean eigenvectors (which in this case are simply individual pixels, ordered by $|\mu_{,\alpha}/\mu|$). For data signal-to-noise of around 2, the latter allow a data compression by about a factor of 2 before the errors on the parameters increase substantially, whereas the method here allows drastic compression from thousands of numbers to a handful. To show what can be achieved, we use a set of simulated galaxy spectra to constrain a few parameters characterising the galaxy star formation history.

In the case when the covariance matrix is independent of the parameters, it does not matter which parameter we choose to form y_1, y_2, etc, as the likelihood surface from the compressed numbers is, on average, locally identical to that from the full dataset. However, in the general case, the procedure does lose information, and the amount of information lost could depend on the order of assignment of parameters to m. If the parameter estimates are correlated, the error in both parameters is dominated by the length of the likelihood contours along the 'ridge'. It makes sense then to diagonalise the matrix of second derivatives of $\ln \mathscr{L}$ at the fiducial model, and use these as the parameters (temporarily). The parameter eigenvalues would order the importance of the parameter combinations to the likelihood. The procedure would be to take the smallest eigenvalue (with eigenvector lying along the ridge), and make the likelihood surface as narrow as possible in that direction. One then repeats along the parameter eigenvectors in increasing order of eigenvalue.

Specifically, diagonalise $\mathbf{F}_{\alpha\beta}$ in (29.5), to form a diagonal covariance matrix $\Lambda = \mathbf{S}^t\mathbf{F}\mathbf{S}$. The orthogonal parameter combinations are $\psi = \mathbf{S}^t\theta$, where $\mathbf{S}$ has the normalised eigenvectors of $\mathbf{F}$ as its columns. The weighting vectors $\mathbf{b}_m$ are then computed from (29.14) by replacing $\mu_{,\alpha p}$ by $\mathbf{S}_{pr}\mu_{,\alpha r}$.

29.4 Extension to MOPED Using an Ensemble of Fiducial Models

Unlike the case of galaxy spectra [2], in cases when the signal is very sparsly populated among the full data (e.g. light transit of an exoplanet), the fiducial model will weigh some data high, very erroneously if the fiducial model is way off from the true model. This is because the derivatives of the fiducial model with respect to the parameters are large near the walls of the box-like shape of the model.

In this section we present an alternative approach to find the best fitting transit model to a light-curve. Although the method is illustrated for the case of exo-planet searches [3], it is fully general and can be applied to other cases like gravitational wave detection [4]. The method is based on using an ensemble of randomly chosen fiducial models. For an arbitrary fiducial model the likelihood function will have several maxima one of which is guaranteed to be the correct solution. This is the case where the values of the free parameters ($\mathbf{q}$) are close to the true one; thus $\mu(\mathbf{q})$ is similar to $\mathbf{x}$. For a different arbitrary fiducial model there are also several maxima, but only one will be guaranteed to be a maximum, the true one. Therefore by using several fiducial models one can eliminate the spurious maxima and keep the one that is common to all the fiducial models which is the true one. We combine the MOPED likelihoods for different fiducial models by simply averaging them[1]

The new measure Y is defined:

$$Y(\mathbf{q}) \equiv \frac{1}{N_f} \sum_{\{\mathbf{q}_f\}} \mathscr{L}(\mathbf{q};\mathbf{q}_f) \ , \tag{29.16}$$

where $\mathbf{q}$ and $\mathbf{q}_f$ are the parameter vectors $\{T, \eta, \theta, \tau\}$ and their fiducial values $\{T_f, \eta_f, \theta_f, \tau_f\}$ and N_f is the number of fiducial models. The summation is over an ensemble of fiducial models $\{\mathbf{q}_f\}$. $\mathscr{L}(\mathbf{q};\mathbf{q}_f)$ is the MOPED likelihood, i.e.

$$\mathscr{L}(\mathbf{q};\mathbf{q}_f) = \sum_m \left[\boldsymbol{b}_m(\mathbf{q}_f) \cdot \boldsymbol{x} - \boldsymbol{b}_m(\mathbf{q}_f) \cdot \boldsymbol{\mu}(\mathbf{q})\right]^2 \tag{29.17}$$

Figure 29.1 shows the Y as a function of period T for a different size sets of fiducial models for a synthetic light-curve with $S/N = 3$ and 2,000 observations. The top panel shows the value of Y using an ensemble of three fiducial models. As it can be seen from the figure there are more than few minima. Using an ensemble of ten fiducial models (shown in the next panel) reduces the number of minima. In the last panel we used an ensemble of 20 fiducial models and there is only one obvious minimum, the true one.

Figure 29.2 shows the value of Y as a function of each free parameter for a synthetic light-curve. We set the values of 3 of the parameters to the "correct" values (used to construct the light-curve) and we let the fourth free for each panel. Note that the shape of the Y as a function of η, θ and τ is smooth, however the dependency on T is erratic suggesting that efficient minimization techniques are not applicable.

29.4.1 Confidence and Error Analysis

To confidently determine that the minimum found is not spurious the likelihood of the candidate solution must be compared to the value and distribution of Y derived from a set of light-curves with no transit signal. One can simulate a set of null light-curves and build a distribution by calculating the value of Y for each point in the parameter space for each simulated "null" light-curve; a real expensive computational task. Alternatively this null distribution can be analytically derived.

[1] This is chosen ad hoc. We have tried other approaches all of which work similarly well. Averaging turned out to be the functional form in which, error and confidence level of the measurement, could be easily and analytically calculated.

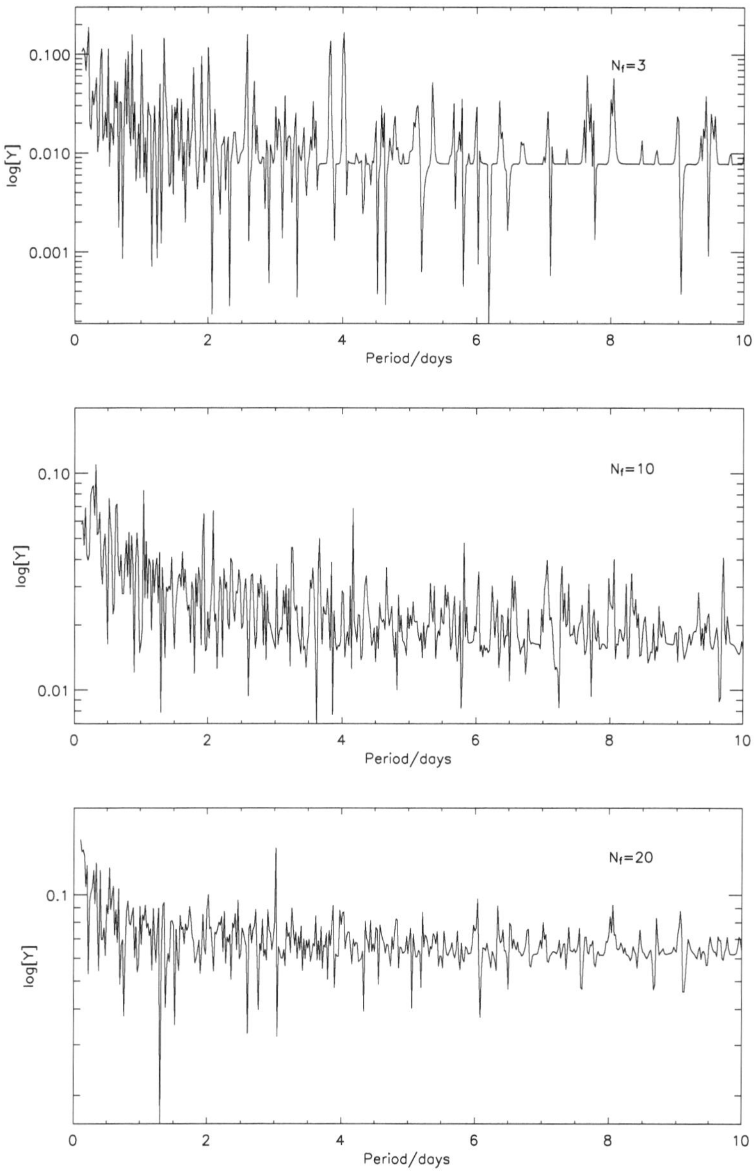

Fig. 29.1 Y as a function of period T for a set of fiducial models for a synthetic light-curve with $S/N = 3$ and 2,000 observations and $T = 1.3$ days. The *top panel* shows the value of Y using three randomly selected fiducial models, the *middle panel* 10 and the *bottom* using 20. As the number of fiducial models used increases the number of minima decreases. At $N_f = 20$ there is only one obvious minima at $T = 1.3$ days

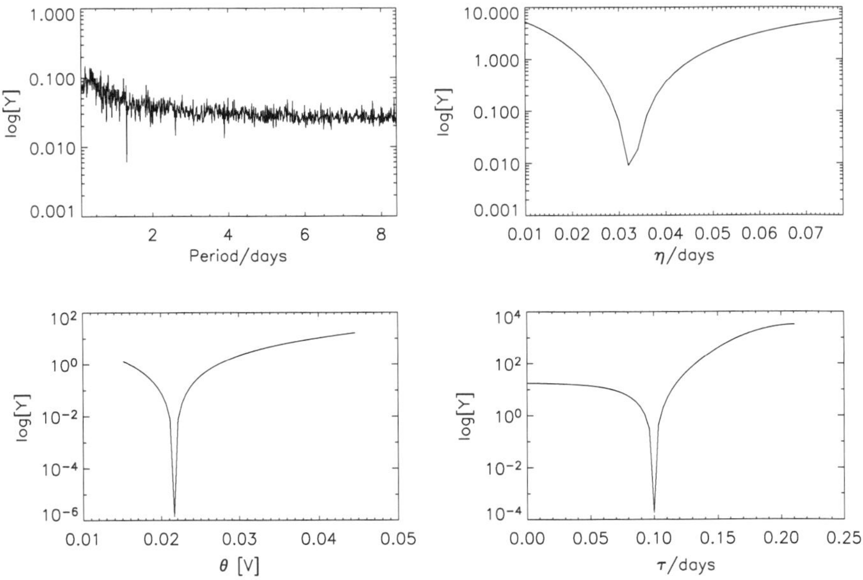

Fig. 29.2 *Top panel left*: Likelihood as a function of period T. *Top panel right*: Likelihood as a function of transit duration η. *Bottom panel left* : Likelihood as a function of θ and *bottom panel right* : Likelihood as a function of τ. In all parameters the correct value is found. Note that for T the topology of the Likelihood surface is fairly complicated with many local minima, thus making efficient minimization techniques not applicable

Since $x \sim N(\langle x \rangle, \sigma_x)$ and all other variables are deterministic, then it can be shown that $Y(\mathbf{q})$ follows a non-central $\mathscr{X}^2$ distribution $Y(\mathbf{q}) \sim \mathscr{X}^2(r, \lambda)$ where r is the number of degrees of freedom and λ is the non centrality of the distribution. The non-central $\mathscr{X}^2$ distribution has mean and variance according to:

$$\mu = r + \lambda \ , \tag{29.18}$$

$$\sigma^2 = 2(r + 2\lambda) \ , \tag{29.19}$$

where $r = 4$ and λ is given by

$$\lambda = \frac{\mathrm{E}^2\left[\mathscr{X}\right]}{\mathrm{var}\left[\mathscr{X}\right]} \ . \tag{29.20}$$

The square of the expectation value is,

$$\mathrm{E}^2\left[\mathscr{X}\right] = \sum_{\mathrm{m}} \left[\langle \mathrm{x} \rangle \, \mathrm{B_m}(\mathbf{q}_\mathrm{f}) - \mathrm{C_m}(\mathbf{q}; \mathbf{q}_\mathrm{f})\right]^2 \tag{29.21}$$

where we define

$$B_m(\mathbf{q}_f) \equiv \sum_t b^t_m(\mathbf{q}_f), \quad \text{and} \quad D_m(\mathbf{q}; \mathbf{q}_f) \equiv \boldsymbol{b}_m(\mathbf{q}_f) \cdot \boldsymbol{\mu}(\mathbf{q}) \tag{29.22}$$

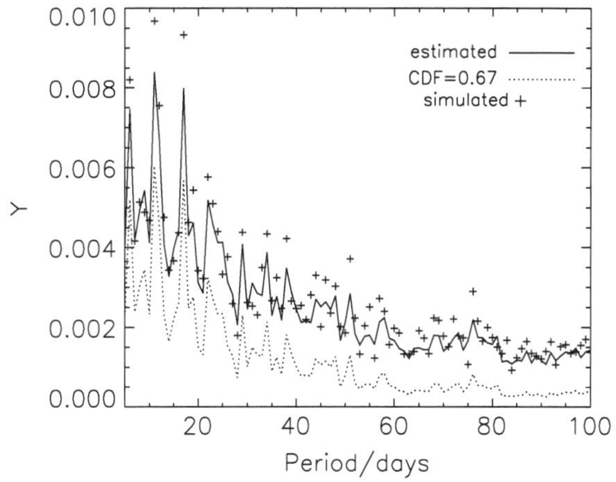

Fig. 29.3 Values of $Y(T)$ for the null case (i.e. a light-curve without a transit) both simulated (*crosses*) and analytically calculated (see Sect. 29.4.1) (*solid line* is the expected value and *dotted line* is the 67% confidence level). It is clear that the simulated values agree well with the theoretical ones

and the variance is given by

$$\begin{aligned}\mathrm{var}\left[\mathscr{X}\right] &= \mathrm{var}\left[\sum_m \boldsymbol{b}_m(\mathbf{q}_f)\cdot\boldsymbol{x} - \sum_m \boldsymbol{b}_m(\mathbf{q}_f)\cdot\boldsymbol{\mu}(\mathbf{q})\right] \\ &= \sum_m \left|\boldsymbol{b}_m(\mathbf{q}_f)\right|^2 \mathrm{var}\left[x^t\right] = \sigma_x^2 \beta_m(\mathbf{q}_f)\end{aligned} \tag{29.23}$$

where we define $\beta_m(\mathbf{q}_f)$ to be:

$$\beta_m(\mathbf{q}_f) \equiv \boldsymbol{b}_m(\mathbf{q}_f)\cdot\boldsymbol{b}_m(\mathbf{q}_f)\ . \tag{29.24}$$

Combining the above equations we get

$$\lambda = \frac{\sum_m \left[\langle x\rangle\, B_m(\mathbf{q}_f) - D_m(\mathbf{q};\mathbf{q}_f)\right]^2}{\sigma_x^2\, \beta_m(\mathbf{q}_f)} \tag{29.25}$$

To compute confidence levels for a particular Y we integrate a non-central $\mathscr{X}^2$ distribution with non centrality given by (29.25) from $Y(\mathbf{q})$ to infinity. This is done numerically, still this is a very quick operation. Furthermore, this will only be performed few times per light curve.

Figure 29.3 shows the values of $Y(T)$ for the null case (i.e. a light-curve without a transit) both simulated (crosses) and theoretically calculated using the equations above (solid line is the expected value and dotted line is the 80% confidence level) (Fig. 29.4). It is clear that the simulated values agree well with the theoretical ones. Note that because the confidence can be calculated analytically we do not have to simulate null light-curves and recalculate the Y for each light-curve thus gaining computational speed.

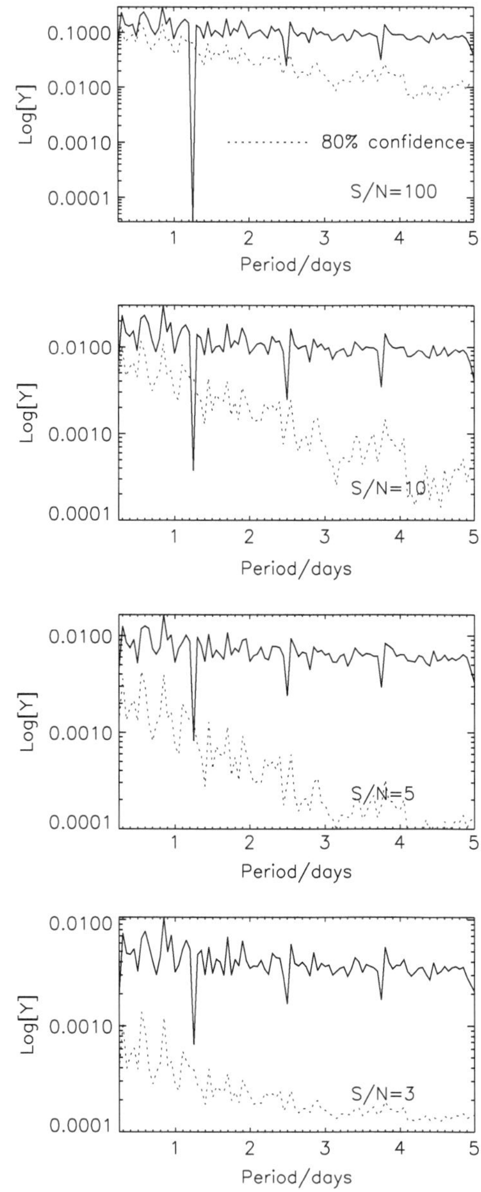

Fig. 29.4 The value of Y is shown as a function of period for a synthetic light-curve with a transit at 1.25 days. The *different panels* show different values of S/N. Note that there is a well defined minimum at the right period. The *dotted line* shows the 80% confidence level. Note that at this level there is only one single minimum at the right period even for S/N as low as 5

References

1. Tegmark M., Taylor A., Heavens A., 1997. ApJ, 480, 22.
2. Heavens A., Jimenez R., Lahav O., 2000, MNRAS, 317, 965
3. Protopapas, P., Jimenez, R., & Alcock, C. 2005, MNRAS, 362, 460
4. Graff, P., Hobson, M. P., & Lasenby, A. 2011, MNRAS, 413, L66

Chapter 30
Commentary: Data Compression Methods in Astrophysics

Ann B. Lee

Abstract A common problem in astrophysics is how to efficiently estimate key parameters for objects when given a very large data set. In his paper, Jimenez describes an approach to data compression that can massively accelerate likelihood computations without losing information about the parameters of interest. The method is known as MOPED (Heavens et al. Mon Not R Astron Soc 317(4):965–972, 2000) and has previously been used for the estimation of star formation history from galaxy spectra (Panter et al. Mon Not R Astron Soc 343(4):1145–1154, 2003; Panter et al. Mon Not R Astron Soc 378(4):1550–1564, 2007), and identification of planetary transits from light curves (Protopapas et al. Mon Not R Astron Soc 362(2):460–468, 2005). Here we discuss the set-up and some of the assumptions of the MOPED approach. We then describe a few alternative or complementary methods of compression and parameter estimation when these assumptions are violated, and discuss their pros and cons.

30.1 Introduction

The general setting is that we observe a set of N measurements $\mathbf{x} = (x_1, \ldots, x_N)$, such as the flux measurements at different wavelengths for a spectrum, or the measurements at different time points for a light curve. We have a parametric model for the distribution of observables, $f_\theta(\mathbf{x})$, where $\theta = (\theta_1, \ldots, \theta_k)$ are parameters of interest; e.g. the age and metallicity of a galaxy, or the period, depth, duration and epoch of planetary transits. The goal is to estimate such parameters given a data set $\mathbf{x}$.

A.B. Lee (✉)
Department of Statistics, Carnegie Mellon University, Pittsburgh, PA, USA
e-mail: annlee@cmu.edu

E.D. Feigelson and G.J. Babu (eds.), *Statistical Challenges in Modern Astronomy V*, Lecture Notes in Statistics 209, DOI 10.1007/978-1-4614-3520-4_30,

In his work, Jimenez describes the MOPED model [4] for this problem which has been used for astronomical problems such as modeling galaxy star formation histories [5, 6] and detecting planetary transits in photometric time series [7]. This model makes the additional assumption of a noiseless signal with additive Gaussian noise:

$$\mathbf{x} = \mu(\theta) + \mathbf{n},$$

where $\mathbf{n} \sim \mathrm{MVN}(0, C)$. The covariance matrix C is assumed to be "known", meaning that it does not depend on the model parameter θ.

In many applications, the number of measurements, N, could be large—which would make likelihood fits very slow. Hence, the question: Can we find a lower-dimensional linear transformation, $\mathbf{y} = B\mathbf{x}$, without losing information on the parameters θ? The definition of "lossless" here is that the Fisher matrix at the maximum likelihood (ML) point is the same whether we use the full data set or the compressed version. Jimenez' answer is yes. Under the previous assumptions, MOPED computes k projection indices $\{y_1, \ldots, y_k\}$ that form a set of sufficient statistics for the k parameters $\{\theta_1, \ldots, \theta_k\}$. If $k \ll N$, then the computational speed-up is considerable.

How about cases where MOPED may run into trouble? In the next section, we list and discuss some situations that could be challenging for MOPED and ML-based methods.

30.2 Discussion of Model Assumptions and Challenging Cases

1. *The transformation matrix B requires computations at the ML point which is unknown. In other words, it is important to have a good "fiducial model".*

   A common criticism of likelihood methods is that the Fisher Information matrix is evaluated at the ML point which is unknown, and that different fiducial models can lead to highly varying results. In [7], the authors offer a clever solution. They suggest a scheme where one combines the MOPED likelihoods for different models by simple averaging. Figure 1 in Jimenez' paper shows the averaged likelihood function for a synthetic light curve when averaging over $N = 3, 10, 20$ fiducial models. As the number of fiducial models increases the number of minima decreases, with one obvious minimum for $N = 20$. So far, there is no theoretical analysis of the method, and all data are simulated according to the assumed signal-to-noise model. Nonetheless, the empirical results are appealing.
2. *Estimates rely on the assumptions of the parametric model: noiseless signal plus Gaussian noise, and noise independent of the parameters.*

   This comes down to the age-old discussion of whether to use parametric or non-parametric methods. Well, how comfortable are you with the model assumptions for the application of interest? MOPED, for example, may not work well when the computed eigenvectors depend very sensitively on the model parameters, as in e.g. redshift determination where the observed wavelength $\lambda_{\text{obs}} = (1+z)\lambda_{\text{emit}}$ for emitted wavelengh λ_{emit} and redshift z. In such situations, one could instead use nonparametric data compression methods which derive new coordinates from an *ensemble* of data objects $\mathbf{x}_1, \ldots, \mathbf{x}_n \sim P$. The idea is to learn the structure of the underlying data distribution P for e.g. several galaxy spectra with different redshifts, and then find a compression scheme that retains the key properties of P. These methods are not "lossless" from an information point of view but they require no parametric model and no fiducial model for the data. In [8] and [2], we show how Principal Component Analysis (PCA) and nonlinear eigenmap methods such as Diffusion Maps can be used for nonparametric regression and redshift estimation of SDSS galaxy spectra.

3. *The parameter space can be very large and/or degenerate.*

   There are several reasons why this may be a problem: Likelihood methods do not perform well for large parameter spaces; see e.g. [3] for a discussion of maximum likelihood estimation in an infinite-dimensional parameter space. There is also no computational gain in using MOPED when the number of parameters K is of the same order or larger than the dimension N. In such situations, however, it may be possible to improve the performance of MLE and MOPED by first reducing the parameter space. Here is an example: In [1], the authors adopt an empirical population synthesis model to estimate star formation history in galaxies using SDSS spectra. Each galaxy is modeled as a mixture of stars from K different simple stellar populations (SSPs), where an SSP is defined as a group of stars with the same age and metallicity. By fitting the galaxy signal model to observed galaxy data with MLE, and by estimating the mixture coefficients of a set of K SSPs, one can reconstruct the star formation rate of a galaxy and its composition as a function of time. A key problem however is how to to choose the set of K SSPs. Though the parameters that define each SSP are continuous (i.e. K is infinite), optimizing the signal model over a large set of SSPs on a fine parameter grid is computationally infeasible and inefficient. SSP bases on regular age and metallicity grids also lead to poor estimates due to degeneracies (many prototypes with similar spectra). In [9], we introduce a principled approach of choosing a small basis of q SSP prototypes for optimal SFH parameter estimation. The basic idea is to explore the underlying geometry of the SSP observable data (the parameter space), and quantize the vector space and effective support of these model components. We showed that the quantization leads to improved ML estimates of parameters and greater computational efficiency. Now an interesting question is if one can use MOPED with the q chosen prototypes in parameter space for an even larger computational gain.

4. *The link between parameter space and observables may be so complex that there is no simple analytical form for the likelihood function.*

   Due to the complexity of the physical process or due to complex observational effects, one may not have an explicit expression for the distribution $f_\theta(\mathbf{x})$ of observables, but one may be able to simulate data under different values of θ. In his SCMA talk "Addressing the Challenges of Luminosity Function Estimation via Likelihood-Free Inference", Chad Schafer describes likelihood-free approaches for such situations—the main idea is to explore the parameter space by a Monte Carlo scheme involving sampling from the distribution space, followed by sampling from the data space and comparing observed data with the output of complex simulation models; see Schafer's SCMA paper in this proceeding for details.

30.3 Conclusions

MOPED is a lossless compression scheme for parameter estimation under certain model and signal assumptions. Use the method if you are comfortable with the model assumptions and if the number of parameters is relatively small compared to the number of observations. For very large or degenerate parameter spaces, one may benefit from first reducing the parameter space by e.g. quantization or a "method of sieves" [3] before applying likelihood methods or MOPED.

When model assumptions are violated, there are several alternative or complementary data reduction methods; e.g. dimension reduction methods such as principal component analysis and diffusion maps that explore the distribution of all data from an ensemble. These compression methods are not "optimal" for parameter estimation, but they assume no specific theoretical model. There also exist model-free methods for parameter estimation, as in e.g. redshift determination via nonparametric regression, and "likelihood-free" inference that utilize output from complex simulation models.

Acknowledgements The author is grateful to Peter Freeman and Chad Schafer for helpful discussions.

References

1. Cid Fernandes, R., Q. Gu, J. Melnick, E. Terlevich, R. Terlevich, D. Kunth, R. Rodrigues Lacerda, and B. Joguet (2004). The star formation history of Seyfert 2 nuclei. *Mon. Not. Royal Astro. Soc. 355*, 273–296.
2. Freeman, P. E., J. Newman, A. B. Lee, J. W. Richards, and C. M. Schafer (2009). Photometric redshift estimation using SCA. *Mon. Not. Royal Astro. Soc. 398*, 2012–2021.
3. Geman, S. and C.-R. Hwang (1982). Nonparametric maximum likelihood estimation by the method of sieves. *Annals of Statistics 10*(2), 401–414.
4. Heavens, A. F., R. Jimenez, and O. Lahav (2000). Massive lossless data compression and multiple parameter estimation from galaxy spectra. *Mon. Not. R. Astron. Soc. 317*(4), 965–972.
5. Panter, B., A. F. Heavens, and R. Jimenez (2003). Star formation and metallicity history of the SDSS galaxy survey: unlocking the fossil record. *Mon. Not. R. Astron. Soc. 343*(4), 1145–1154.
6. Panter, B., R. Jimenez, A. F. Heavens, and S. Charlot (2007). The star formation histories of galaxies in the Sloan Digital Sky Survey. *Mon. Not. R. Astron. Soc. 378*(4), 1550–1564.

7. Protopapas, P., R. Jimenez, and C. Alcock (2005). Fast identification of transits from lightcurves. *Mon. Not. R. Astron. Soc. 362*(2), 460–468.
8. Richards, J. W., P. E. Freeman, A. B. Lee, and C. M. Schafer (2009). Exploiting low-dimensional structure in astronomical spectra. *Astrophys. J. 691*, 32–42.
9. Richards, J. W., P. E. Freeman, A. B. Lee, and C. M. Schafer (2011). Prototype selection for parameter estimation in complex models. *Annals of Applied Statistics*. To appear; arXiv:1105.6344.

Part IV
Image and Time Series Analysis

Chapter 31
Morphological Image Analysis and Sunspot Classification

David Stenning, Vinay Kashyap, Thomas C.M. Lee, David A. van Dyk, and C. Alex Young

Abstract The morphology of sunspot groups is predictive both of their future evolution and of explosive associated events higher in the solar atmosphere, such as solar flares and coronal mass ejections. To aid in this prediction, sunspot groups are manually classified according to one of a number of schemes. This process is both laborious and prone to inconsistencies stemming from the subjective nature of the classification. In this paper we describe how mathematical morphology can be used to extract numerical summaries of sunspot images that are relevant to their classification and can be used as features in an automated classification scheme. We include a general overview of basic morphological operations and describe our ongoing work on detecting and classifying sunspot groups using these techniques.

D. Stenning (✉) • D.A. van Dyk
Department of Statistics, University of California, Donald Bren Hall, 2nd Floor, Irvine, CA 92697-1250, USA
e-mail: dstennin@ics.uci.edu; dvd@ics.uci.edu

V. Kashyap
Smithsonian Astrophysical Observatory, 60 Garden St., Cambridge, MA 02138, USA
e-mail: vkashyap@cfa.harvard.edu

T.C.M. Lee
Department of Statistics, University of California, 4118 Mathematical Sciences Building, One Shields Avenue, Davis, CA 95616, USA
e-mail: tcmlee@ucdavis.edu

C.A. Young
ADNET Systems, Inc., NASA/GSFC, Mail Code 671, Greenbelt, MD 20771, USA
e-mail: c.alex.young@nasa.gov

E.D. Feigelson and G.J. Babu (eds.), *Statistical Challenges in Modern Astronomy V*, Lecture Notes in Statistics 209, DOI 10.1007/978-1-4614-3520-4_31,

31.1 Scientific Background and Motivation

The Sun's *photosphere* is the region that emits the light that we see. The deeper regions are opaque and the higher and much less dense *corona* is only one-millionth as bright as the photosphere in visible light. Sunspots are dark areas on the photosphere that result from intense magnetic fields. The magnetic fields inhibit convection, cooling the corresponding surface regions. Areas on the photosphere where the surface temperature has been reduced then appear as dark spots when viewed in optical light. Sunspots can also be seen in *magnetograms* which are images that represent variations in the strength of magnetic fields in the Sun's photosphere [3]. In magnetograms, sunspots correspond to high flux regions that appear as areas of opposite magnetic polarity.

The classification and tracking of sunspots is an active undertaking of solar-physicists hoping to untangle connections between sunspot activity and various solar phenomena. Recent studies, for example, suggest that solar flares are related to the magnetically active regions around sunspot groups [5]. As a result, various sunspot classification schemes aim to characterize magnetic flux content in the active-regions on the solar disk [4]. One particular scheme—the Mount Wilson classification—puts solar active-regions into four classes based on the complexity of magnetic flux distribution. When combined with space weather data, this scheme can be used to predict activity in the solar corona such as highly energetic solar flares and massive bursts of solar wind known as coronal mass ejections [4]. While precise predictions remain elusive, the complexity of the magnetic flux distribution of sunspot groups can be used to infer trends and tendencies in the patterns of solar flares and coronal mass ejections.

Recently launched NASA missions such as the Solar Dynamics Observatory—with its continuous science data downlink rate of 130 Megabits per second—are producing an unprecedented volume of solar data. Nonetheless the majority of sunspot classification is still performed through visual inspection by experts [2]. This is a laborious process and, as with all manual procedures, is susceptible to bias from the human observer [4]. Since the morphology of sunspot groups form a continuous spectrum rather than a set of discrete and obvious classes, there is a level of subjectivity in manual classification. One of the attractions of the Mount Wilson scheme is its reliance on a relatively simply set of classification rules. While this may aid manual classification it introduces artificial dichotomies that may hinder scientific understanding. Even with the relatively straightforward Mount Wilson scheme, trained experts do not always agree on classifications. As a result, there is a need for an automated, objective and reliable procedure for detecting and classifying sunspot groups.

The Mount Wilson classification scheme divides sunspot groups into four classes. The simplest morphologically is the α class which consists of groups that are dominated by a single *unipolar* sunspot, i.e., a sunspot with a magnetic field that is dominated either by a positive or a negative polarity. The second class, the β class, is made up of groups with both polarities, but with a simple and distinct spatial

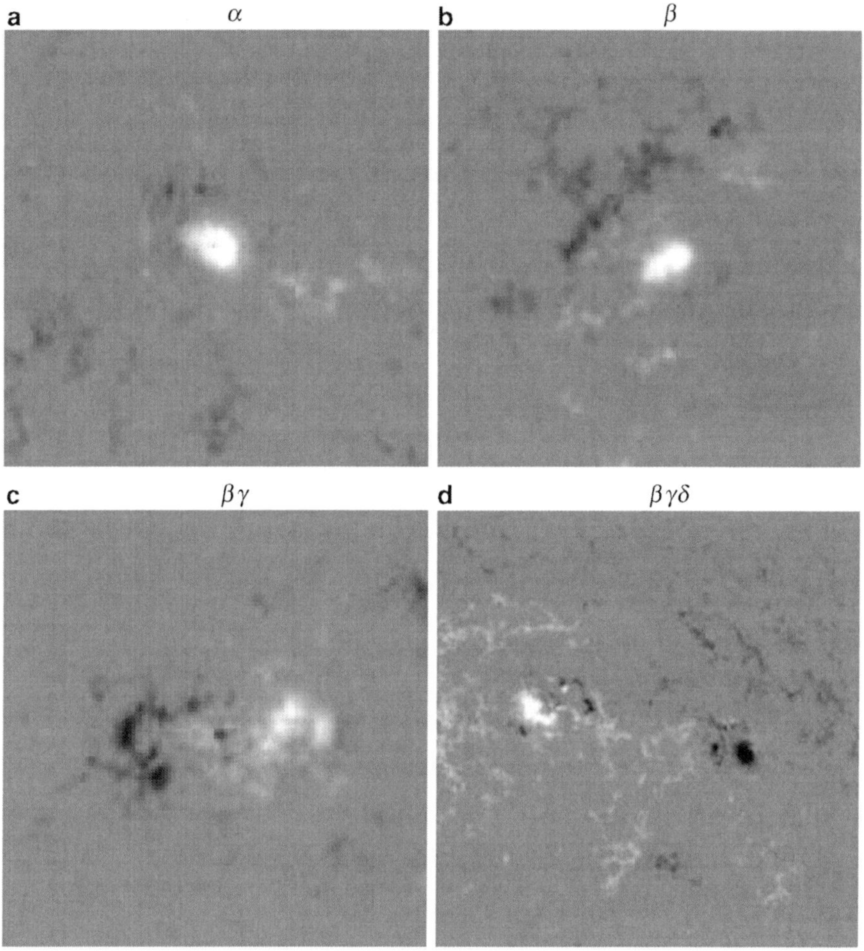

Fig. 31.1 Examples of the four classes of sunspot groups used in the Mount Wilson scheme. The α class (**a**) is dominated by a single pole that appears *black* or *white* in the magnetogram, depending on the polarity (positive or negative). The β class (**b**) has regions of both positive and negative polarity that can be separated by a *straight line*. The $\beta\gamma$ class (**c**) also exhibits both polarities but they cannot be easily separated into two regions. In the $\beta\gamma\delta$ class (**d**) the two polarities are scattered throughout the region

division between the polarities. In particular a straight line can be drawn through the group that nearly divides the negative from the positive polarities. Groups in the third class, $\beta\gamma$, are also *bipolar*, but are sufficiently complex that a straight line cannot divide the positive and the negative polarities. Finally, in the fourth class, $\beta\gamma\delta$, the positive and negative polarities are scattered throughout the region and cannot be easily separated. Example of the sunspot groups from the four classes appear in Fig. 31.1.

Because this classification scheme is defined in terms of the morphology of the sunspots, we propose to use methods from mathematical morphology to extract features from the magnetograms that can be used in an automated classification technique, such as a classification tree, support vector machine or some other common method, to reconstruct the Mount Wilson classification. We use a data set that consists of magnetogram images collected by the Solar and Heliospheric Observatory/Michelson Doppler Imager (SOHO/MDI). Each magnetogram includes the date and time the image was taken, the location on the solar disk, and the identification number of the sunspot group given jointly by the U.S. Air Force and the National Oceanic and Atmospheric Administration (USAF/NOAA). The manual classification of the sunspot group by USAF/NOAA according to the Mount Wilson scheme is also provided.

The primary goal of this article is to make progress toward an automatic sunspot classification method that relies on features extracted using techniques from mathematical morphology. We begin in Sect. 31.2 with an overview of the mathematical morphology methods that we employ. In Sect. 31.3 we describe how we compute relevant numerical summaries of the magnetogram images using mathematical morphology and methods for using these summaries for classification. Finally in Sect. 31.4 we discuss the road forward toward automated sunspot classification.

31.2 Mathematical Morphology

Mathematical morphology is a powerful tool for image analysis, which was developed about 40 years ago. Unlike other tools (e.g., Fourier methods), morphological operators relate directly to shape. When used appropriately, morphological operations can simplify images by preserving their essential shapes and eliminating noise. For detailed descriptions of the subject, see [6, 7].

31.2.1 Binary and Greyscale Images

Objects in digitized images are only approximations to their counterparts in the real world. One reason is simply because their domains are defined in different spaces: images are pixelated and thus "discrete" while the object itself is "continuous" in nature. We will use $\mathbb{Z}^2$ to denote the space of objects in binary images. That is $\mathbb{Z}^2$ can be thought of as a two dimensional grid of pixels that is infinitely tall and infinitely wide. We can treat $\mathbb{Z}^2$ as the discrete version of the Euclidean plane $\mathbf{R}^2$, and represent it as a two dimensional Cartesian square grid.[1]

[1]Originally mathematical morphology was defined in the d-dimensional Euclidean space $\mathbf{R}^d$, but there is no great difficulty in translating the this theory from $\mathbf{R}^d$ to its discrete version $\mathbb{Z}^d$.

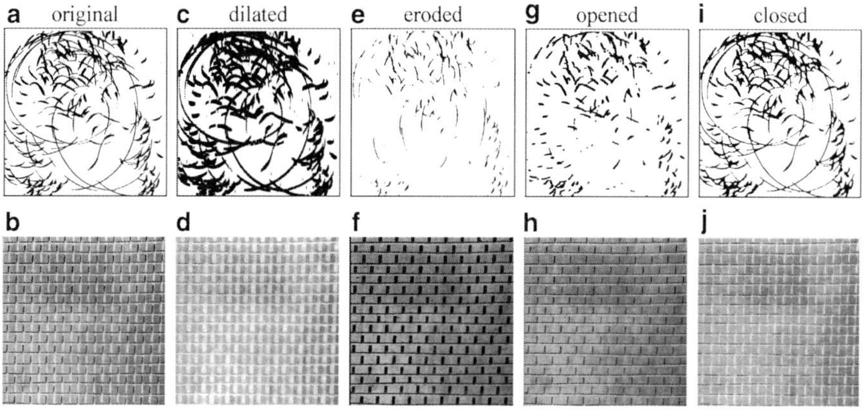

Fig. 31.2 *Top row*: (**a**) a *binary image* that has been (**c**) dilated, (**e**) eroded, (**g**) opened, and (**i**) closed. *Bottom row*: (**b**) a *greyscale image* that has been (**d**) dilated, (**f**) eroded, (**h**) opened, and (**j**) closed. For the *binary image* a *vertical line* was used as the SE in the morphological operations. For the *greyscale image*, a *rectangle* was used

A *binary image*, f, is a image where each pixel is either black or white. For example we can assign the value 1 (i.e., black) to a pixel if it belongs to an object, otherwise the value 0 (i.e., white). Notice that we can always consider objects (i.e., the "black" parts) in a binary image as sets and the image itself as the union of all such sets. See Fig. 31.2a for a binary image. Mathematically, we can write a binary image as a mapping, which maps each pixel of a subset $\mathscr{D}_f$ of $\mathbb{Z}^2$ into the couple $\{0,1\}$:

$$f : \mathscr{D}_f \subset \mathbb{Z}^2 \longrightarrow \{0,1\},$$

where $\mathscr{D}_f$ is some subset of $\mathbb{Z}^2$ and is called the definition domain of f.

More generally, a *greyscale image*, f, is a mapping which maps each element in a subset $\mathscr{D}_f$ of $\mathbb{Z}^2$ into the set of non-negative integers $\mathbf{N}_0$:

$$f : \mathscr{D}_f \subset \mathbb{Z}^2 \longrightarrow \mathbf{N}_0.$$

Very often the set of non-negative integers under consideration is $\{0,\ldots,255\}$, where the larger the value, the brighter the pixel is. In mathematical morphology, it is useful to treat the pixel values of a greyscale image as the heights of a surface above the image plane. See Fig. 31.2b for a greyscale image.

In our discussion about mathematical morphology, we use $\mathbb{Z}^2$, but understand that the development works equally well for either $\mathbb{Z}^d$ or $\mathbf{R}^d$.

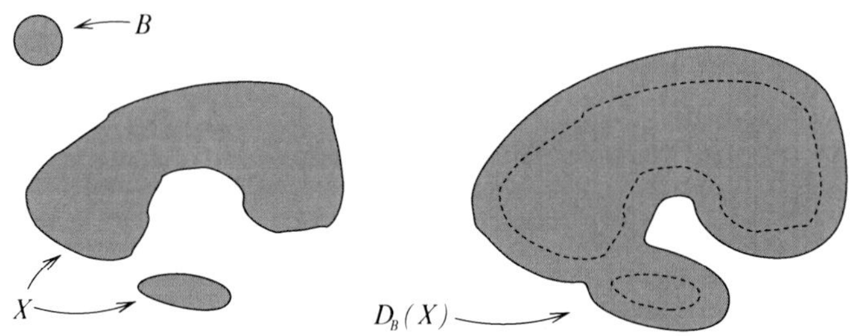

Fig. 31.3 Dilation of a set X by a disk-shaped structuring element B

31.2.2 Dilation and Erosion

In mathematical morphology there are two basic operations: *dilation* and *erosion*. These are the basic building blocks and many other morphological operations can be expressed in terms of dilation and erosion. We first define dilation.

Suppose we have a set $X \subset \mathbb{Z}^2$ and a curser B that scrolls across $\mathbb{Z}^2$. If we record the location of B whenever it intersects or "runs into" X the result is called the dilation of X by B, denoted by $D_B(X)$. This is illustrated in Figure 31.3. Notices that the dilation of X is a bloated version of X, where the degree and character of the bloating is determined by the shape and size of B. The dilation of X by B is the answer to the question: "What is the location of B when B *hits* X?" (We define A *hits* B as $A \cap B \neq \phi$.) In other words, $D_B(X)$ is the set of all points $\mathbf{x}$ such that B *hits* X when the location or *origin* of B is at $\mathbf{x}$.

We call B a *structuring element* (SE). Generally speaking a SE is a subset of $\mathbb{Z}^2$ with a known shape and origin. SE elements are used to examine or transform the image f under study. As with dilation, all morphological operators treat the image as a set (i.e., a binary image) and use one or more SEs to examine it. We could also say these operators use the shape(s) of the SE(s) to transform f. Notice that the SE B is arbitrary, hence one can always choose a suitable SE to perform the desired task. This gives the user a great flexibility in applying morphological methods. Usually SEs are regular and small in size when compared to the image. For example, in the case of a binary image in Fig. 31.3, B is a disk with a small radius and with its center as the origin.

The formal definition of dilation is:

$$D_B(X) \equiv \{\mathbf{x} \in \mathbb{Z}^d \mid B_{\mathbf{x}} \cap X \neq \phi\},$$

where $B_{\mathbf{x}}$ is the SE B placed with its origin at $\mathbf{x}$. Figure 31.2c, d show the dilation of the images displayed in Fig. 31.2a, b, respectively.

The erosion of X by B, denoted by $E_B(X)$, is the answer to the question: "Where is the origin of B when B *fits* wholly inside X?" That is, $E_B(X)$ is the set of points $\mathbf{x}$

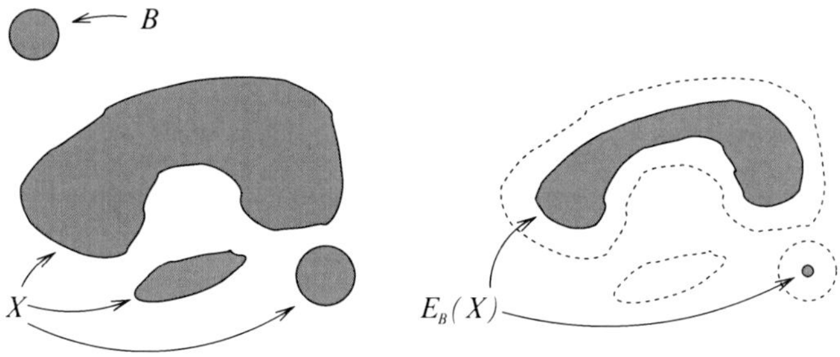

Fig. 31.4 Erosion of a set X by a disk-shaped structuring element B

such that B *fits* wholly inside X when the origin of B is at $\mathbf{x}$. The formal definition of erosion is:

$$E_B(X) \equiv \{\mathbf{x} \in \mathbb{Z}^d \mid B_{\mathbf{x}} \subset X\}.$$

See Fig. 31.4 for an example, and Fig. 31.2e, f for examples of eroded images.[2]

31.2.3 Opening and Closing

Dilation and erosion remove information and in general the lost information cannot be retrieved. The search for an operation that attempts to revert the effects of dilation and erosion leads to the definition of, respectively, morphological *closing* and *opening*. We first give the definition of opening, and for that, we define the reflection $\check{A}$ of a set A: $\check{A} \equiv \{-\mathbf{a} \mid \mathbf{a} \in A\}$. That is, $\check{A}$ is the mirror image of A about the origin.

The opening of X by B, denoted by $O_B(X)$, is defined as the erosion of X by B followed by the dilation by $\check{B}$. That is:

$$O_B(X) \equiv D_{\check{B}}\{E_B(X)\}.$$

Figure 31.5 is an example of opening. Notice that X has been rounded by B from the inside, and that those disks which are smaller in size than the SE B *vanish* after opening.

Also notice the filtering effect of opening: those image structures that cannot contain the SE B are removed from the image. Therefore the size and shape of B

[2]It is easy to verify that dilation and erosion form a pair of dual transformations:

$$D_B(X) \equiv \{E_B(X^c)\}^c.$$

This duality property means that, when using the same SE, the dilation of a set X is equivalent to the complement of the erosion of the complement (i.e., the "background") of the set X.

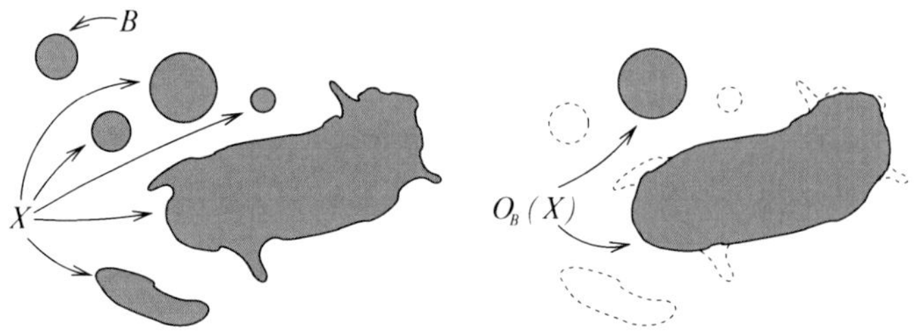

Fig. 31.5 Opening of a set X by a disk-shaped structuring element B

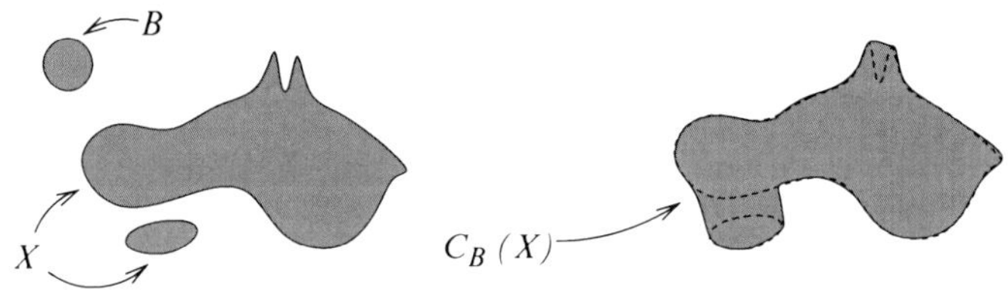

Fig. 31.6 Closing of a set X by a disk-shaped structuring element B

should be carefully chosen for the information to be extracted from the image. For example, if one wants to remove linear features but not disk shaped structures, B should be chosen as a disk of a suitable size. Examples of opened images can be found in Fig. 31.2g, h.

The closing of X by B, denoted by $C_B(X)$, is defined as the dilation of X by B followed by the erosion by $\check{B}$. That is:

$$C_B(X) \equiv E_{\check{B}}\{D_B(X)\}.$$

See Figure 31.6 for an example of closing. As opposite to opening, closing rounded the objects "from outside". See also Fig. 31.2i, j for examples of closed images.[3]

In practice the choice between opening or closing depends on the types of objects or noise to be extracted/removed. For example, the removal of "salt noise"—white dots in the image—requires opening, while "pepper noise"—black dots in the image—requires closing.

31.2.4 Other Morphological Operations

There are other useful morphological operators, but due to space limitation, we omit their detailed descriptions here. One such operation is *skeletonization*: the skeleton of an binary object is defined as the union of the centers of all the maximal balls

[3]Opening and closing also share a dual property: $O_B(X) = \{C_B(X^c)\}^c$.

inside the object. It is useful for extracting summary features to represent the object. Another useful operator for detecting object boundaries is *morphological gradient*, typically defined as the arithmetic difference $D_B(X) - E_B(X)$.

31.3 Detection and Classification of Sunspot Groups

We aim to develop an automatic procedure for detecting and classifying sunspot groups according to the Mount Wilson scheme. Given the complexity of the magnetogram images, we adopt an imaging-oriented modular approach. That is, the ultimate problem of detection and classification is broken into a sequence of sub-problems, and simple and effective imaging techniques are applied to sequentially solve these sub-problems.

Since the Mount Wilson scheme relies on characterizing the shape and distribution of magnetic flux in sunspot groups, mathematical morphology is utilized to extract scientifically meaningful features from the available magnetograms. That is, the morphological operations described in Sect. 31.2 are used to examine the distribution of positive and negative magnetic polarities visible in the magnetogram. In particular, we characterize the complexity of the sunspot group based on the scatter of magnetic flux and the separation of the two polarities. In this way, our procedure tailors a classifier to utilize expert knowledge in constructing an interpretable and effective classifier. Another approach to classification, at the other extreme, is to generate a large set of numerical summaries to use as features in a "blackbox" classifier. While this approach can also yield an effective classifier, the results tend to be much more difficult to interpret in terms of the underlying science.

31.3.1 Science-Driven Feature Extraction

In this section we describe the procedure that we employ to extract numerical summaries of the magnetogram images that will serve as features in the ultimate classification. Our strategy is to derive features that are tailored to distinguish between the four classes in the Mount Wilson scheme. Since all four classes are defined in terms of the distribution of the positively and negatively oriented magnetic fields, we begin by using morphological operators to identify the regions of positive and negative polarity in a magnetogram.

To do this we first clean the image using a morphological opening operation with a spherical structuring element of radius 2. This smooths the white sunspots—the regions of positively oriented magnetic field that appear white in the magnetograms—so that smooth boundaries can be obtained after thresholding. After cleaning we extract the white sunspot by selecting pixels with magnetogram intensity greater than a given threshold, namely greater than $\bar{x} + 2.5s$, where $\bar{x}$ and s are, respectively, the mean and sample standard deviation of all the pixel values in the image. Next we aim to extract the black sunspots—the regions of negatively oriented magnetic polarities that appear black in the magnetograms. To do this, we invert the original image by multiplying by negative one so that it looks like a film negative, and then clean and threshold the inverted image in exactly the same we did with the original image when extracting the white sunspots.

Figure 31.7 illustrates our feature extraction routine for α, β, $\beta\gamma$, and $\beta\gamma\delta$ sunspot groups. In this figure, the first row is the original magnetogram that appears in Fig. 31.1, the second row is the cleaned magnetogram, the third row is the extracted white sunspot, and the fourth row is the extracted black sunspot. The columns represent α, β, $\beta\gamma$, and $\beta\gamma\delta$ types, respectively. We will describe the final two rows below.

Given the extracted white and black sunspots, we are in a position to define a feature that aims to identify sunspot groups in the α class. Since this class is defined by "A unipolar sunspot group", an extreme ratio of the number of extracted pixels the white and black sunspots (N_W and N_B, respectively) should be indicative of an α group. This ratio is denoted $|N_W/N_B|$ and is given, for each representative magnetogram, beneath its respective column in Fig. 31.7.

The difference between the β, $\beta\gamma$, and $\beta\gamma\delta$ classes is the degree of separation between the white and black sunspots. In the β class they can be largely separated by a straight line, in the $\beta\gamma$ class they can be largely separated, but not by a straight line, and in the $\beta\gamma\delta$ class they are mixed. Thus to distinguish between these groups we aim to identify the best boundary between the white and black sunspots and to access the quality of this boundary. We do this by combining the extracted white and black sunspots into the same image and using a standard region growing operation to produce the separating boundary. In Fig. 31.7, the fifth row shows the combined image, with the white and black sunspots plotted in blue and yellow, and the sixth row illustrates the resulting separating boundary. Notice that the boundary becomes more complex for the $\beta\gamma$ group than β group and even more so for the $\beta\gamma\delta$ group.

A natural way to distinguish β groups and $\beta\gamma$ groups is to measure the "roughness" of the separating line. A good example of roughness measure is the averaged second derivative, which we compute using second differencing. In some cases the region growing operation results in more than one separating line, indicating poor separation between the white and black sunspots. In this case the group should be classified as a $\beta\gamma\delta$ group.

To help identify sunspot groups in the $\beta\gamma\delta$ class we must quantify the degree of scatter or mixture of the region's positive and negative polarities. In order to do this we introduce a spatial complexity measure. In particular, let $\mathscr{W}$ be the set of pixels in an extracted white sunspot. We then compute the center of mass, c, of $\mathscr{W}$. For each pixel $w \in \mathscr{W}$, the number of pixels that a line segment from w to c passes through is denoted $L(w)$ and of these, the number of blue pixels is denoted $l(w)$. (Recall that blue pixels correspond to the white sunspots.) The spatial complexity measure, $A(\mathscr{W})$, is computed as

$$A(\mathscr{W}) = \frac{1}{|\mathscr{W}|} \sum_{w \in \mathscr{W}} \frac{l(w)}{L(w)},$$

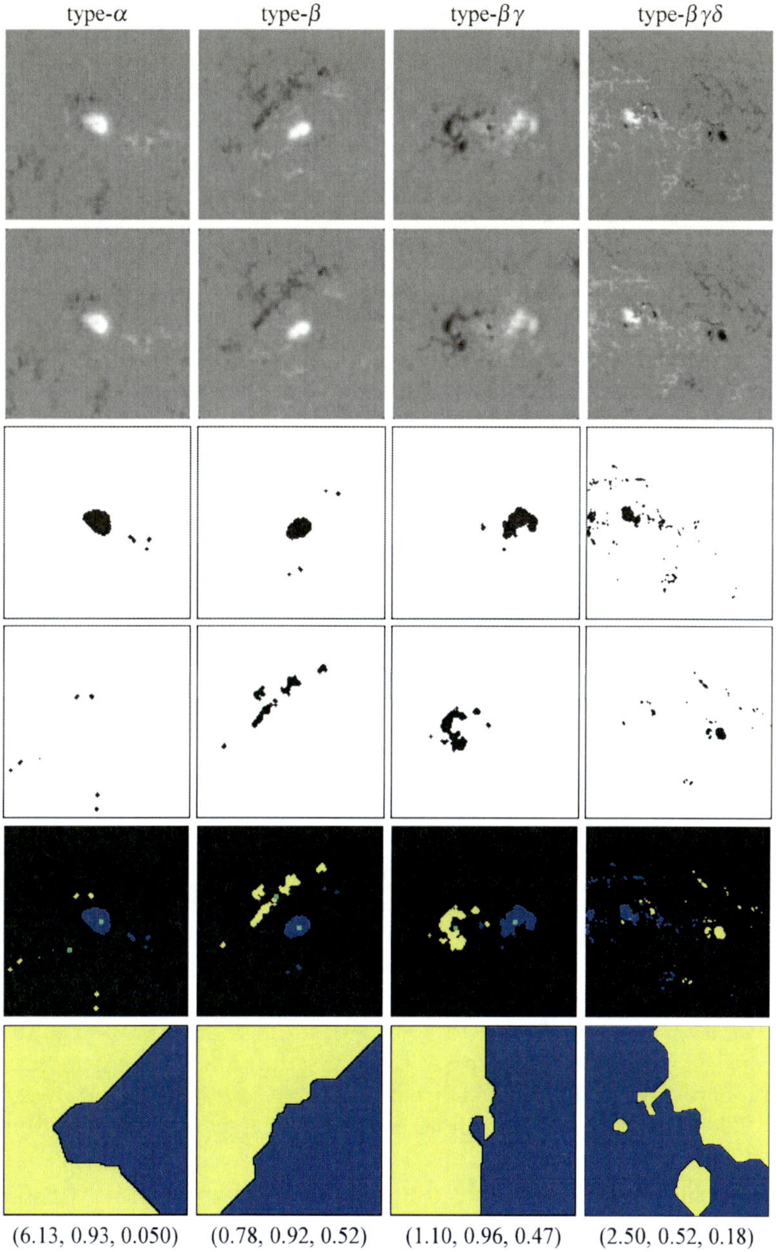

Fig. 31.7 *Top row*: original magnetograms for four types of sunspots. *Second row*: morphologically cleaned magnetograms. *Third row*: extracted *white* sunspot. *Fourth row*: extracted *black* sunspot. *Fifth row*: detected *white* (in *blue*) and *black* (in *yellow*) sunspots. The *green dots* are their centers of mass. *Bottom row*: separating line(s) between the *white* and *black* sunspots. The parenthetical summaries at the *bottom* are the area ratio of *white* to *black* sunspots and the spatial complexity measure $A(\cdot)$ values for the *white* and for the *black* sunspots. We expect the area ratio to be more extreme for α groups and the complexity measurements to be smaller for the $\beta\gamma\delta$ groups than for β or $\beta\gamma$ groups

where $|\mathscr{W}|$ is the number of pixels in $\mathscr{W}$. Notice that $L(w) \geq l(w)$ and $0 \leq A(\mathscr{W}) \leq 1$. To see why $A(\mathscr{W})$ can be used as a spatial complexity measure, observe that if the white sunspot pixels are scattered (and disconnected) around in the image, then for most $w \in \mathscr{W}$, $l(w)$ is small relative to $L(w)$, and thus a small value of $A(\mathscr{W})$ indicates high spatial complexity of $\mathscr{W}$.

A similar quantity $A(\mathscr{B})$ can be computed for the set of pixels in an extracted black sunspot $\mathscr{B}$. The $A(\mathscr{W})$ and $A(\mathscr{B})$ values for each of the representative magnetograms are given beneath the columns in Fig. 31.7. The green dots in the fifth row of Fig. 31.7 are the centers of mass of $\mathscr{W}$ and $\mathscr{B}$.

The full procedure for computing the features is as follows:

1. Clean the original magnetogram image using morphological operations.
2. Extract the "white sunspots" by thresholding the cleaned image.
3. Apply the above steps to the negative of the image to extract the "black sunspots".
4. Compare the relative areas of the white and black sunspots (for discriminating α from the other three types).
5. Compute the separating line for the white and black sunspots (for discriminating β and $\beta\gamma$).
6. Compute the complexity measures $A(\mathscr{W})$ and $A(\mathscr{B})$ (for discriminating $\beta\gamma\delta$ from the rest).

31.3.2 Classification

Given the set of four features described in Sect. 31.3.1 along with their quadratic and interaction terms, we can use a standard classification (supervised learning) technique to derive a classification rule. There are numerous possible method, but we focus mainly on the technique known as *random forests* [1] because it is relatively immune to over-fitting, meaning we have to worry less about the classifier being over-sensitive to spurious relationships in the data, even when including a large number of features. (Four features grows to 14 features if we include quadratic and interaction terms.)

A random forest is a state-of-the-art nonparametric classifier that is an ensemble of a set of *decision trees*. The individual trees are grown by finding the best split of the training cases into the classes based on a set of features. The classification in each of the resulting subgroups is improved using new separate classification rules. In a case with N training cases and p features, the number of features used to make a decision at each node of a tree is set at r, where r is much less than p (one common technique is to set $r = \sqrt{p}$). The ensemble of trees is created by randomly selecting N cases with replacement from the original N training cases. Each tree is grown by randomly choosing r features at each node and making a split based on the selected features. Each tree is grown to completion without pruning , and the random forest combines the individual decision trees based on the majority vote of the trees.

As an illustration we randomly divided a data set consisting of 128 magnetograms into a training set of 90 (70%) magnetograms and test set of 38 (30%) magnetograms. We fit a random forest of 250 trees using the `randomForest` routine in `R` to the training set and used the resulting classification rule to separately classify both the training and test sets. While the training set had a 100% correct classification rate, 58% of the test set was correctly classified, based on the USAF/NOAA classification. All of the misclassified sunspot groups were classified into a class neighboring the USAF/NOAA classification (i.e., all α sunspot groups were classified as either α or β, all β groups as β or $\beta\gamma$, all $\beta\gamma$ as β or $\beta\gamma$, and all $\beta\gamma\delta$ as $\beta\gamma$ or $\beta\gamma\delta$.)

A difficulty that arrises when we try to evaluate the quality of our proposed features for sunspot classification is that the USAF/NOAA classification is not particularly reliable. An examination of the magnetograms that appear to be missclassified by our method more often than not reveals that the USAF/NOAA classification is incorrect or that the sunspot groups is marginal and does not clearly belong to any one of the four classes. This is of course problematic not only for evaluating the classifier but also for training the classifier because the USAF/NOAA classifications in the training set are no more reliable than those in the test set. The problem stems from the lack of true discrete classes. There is a continuum between the α class that is "dominated by a single unipolar spot" and the bipolar β class, as the second polarity grows from negligible to equal in importance. Likewise there is a continuum from the β to the $\beta\gamma$ and to the $\beta\gamma\delta$ class as the bipolar group ranges from simple distinct regions of positive and negative polarity to a group with positive and negative polarities scattered throughout. The lack of a distinct underlying classification leads to subjective assessments as to the proper classification of a group and an inherent inconsistency in the human classification. It is both difficult and ultimately fruitless to automatically reproduce such a human classification.

31.4 Discussion

Our ultimate goal is to provide numerical descriptions and summaries of sunspot images that capture physical characteristics in sunspot development and evolution and can be use to predict turbulent events such as solar flares and coronal mass ejections. Research suggests that the morphology of the sunspot groups is relevant to the evolution of the group and predictive of such events. Thus our work has focused on developing morphological summaries that in the first place capture scientific theories about formation and evolution and secondly may be able to be used to reproduce existing classification schemes. An immediate goal is to develop new classification schemes and/or continuous numerical summaries that better represent the observed variability in sunspot images and are more correlated with solar activity. Current classification schemes are based on static sunspot groups. A more interesting classification would characterize not just the static morphology, but also the development, evolution, and track of the group. The goal is to automatically track the formation and evolution of sunspot groups using the massive solar data sets that are now coming online—and for this tracking to be in terms of sunspot features that are most pertinent to the ultimate scientific objectives.

Acknowledgements D. Stenning and D. van Dyk acknowledge support from NSF grant DMS-09-07522, V. Kashyap from NASA contract NAS-39073 to the Chandra X-Ray Center, T.C.M. Lee from NSF Grant 1007520,and C. A. Young from the NASA SESDA-2 contract. Finally we thank the editors of this volume and the organizers of SCMA V for the opportunity to participate in this conference.

References

1. L. Breiman. Random forests. *Mach. Learn.*, 45:5–32, October 2001.
2. T. Colak and R. S. R. Qahwaji. Automated mcintosh-based classification of sunspot groups using mdi images. *Solar Physics*, 248(2):277–296, 2009.
3. H. J. Hagenaar. Ephemeral regions on a sequence of full?disk michelson doppler imager magnetograms. *The Astrophysical Journal*, 555(1):448–461, 2001.
4. J. Ireland, C. A. Young, R. T. J. McAteer, C. Whelan, R. J. Hewett, and P. T. Gallagher. Multiresolution analysis of active region magnetic structure and its correlation with the mt. wilson classification and flaring activity. *Solar Physics*, 2008.
5. R. S. R. Qahwaji and T. Colak. Automatic short-term solar flare prediction using machine learning and sunspot associations. *Solar Physics*, 2009.
6. J. Serra. *Image Analysis and Mathematical Morphology*. Academic Press, 1982.
7. P. Soille. *Morphological Image Analysis: Principles and Applications*. Springer, Berlin, second edition, 2003.

Chapter 32
Commentary: Morphological Image Analysis and Sunspot Classification

Ricardo Vilalta

Abstract The paper by Stenning et al. discusses mathematical morphological classification of images of sunspots on the Sun's surface. Faced with complicated shapes and distributions of magnetic flux, they seek to reproduce an established classification scheme with four classes. The problem shares characteristics with the classification of elevation maps of the surface of Mars where six classes are present (plateau, crater, ridge, etc.).

32.1 Feature Representation, Bias, Variance, and Irreducible Error

Typical issues under consideration when selecting or designing a classification algorithm are the bias and variance components of error induced by the algorithm [1]. For example, one may choose a simple algorithm (e.g., linear combination of feature values, Naive Bayes, single logical rules, etc.) and draw a hypothesis from a small family of functions; the poor repertoire of functions may produce high bias (the best function may be far from the target function) but low variance (because of the sensitivity on local data irregularities). The alternative is to increase the degree of complexity by drawing a hypothesis from a large class of functions (e.g., neural networks with a large number of hidden units); here the hypothesis exhibits flexible decision boundaries (low bias) but becomes sensitive to small variations in the data (high variance).

A less explored—but perhaps more critical issue—is that of the feature representation, which can be the cause of a third component of error known as

---

R. Vilalta (✉)
Department of Computer Science, University of Houston, 501 Philip G. Hoffman, Houston, TX 77204-3010, USA
e-mail: vilalta@cs.uh.edu

E.D. Feigelson and G.J. Babu (eds.), *Statistical Challenges in Modern Astronomy V*, Lecture Notes in Statistics 209, DOI 10.1007/978-1-4614-3520-4_32,

Bayes (irreducible) error. This occurs when the feature representation leads to class overlap. While bias and variance can be traded off by varying the classification strategy, Bayes error remains immutable as soon as the feature representation is fixed. The importance of high quality features is crucial to attain accurate predictions and cannot be over-emphasized [2]. High quality features convey much information about the problem; in this case, even a simple hypothesis suffices to produce good results. In contrast, low quality features complicate the classification process. Features can bear poor correlation with the class, or interact in many ways, which calls for additional steps to discover important feature combinations.

32.2 A Commentary on Sunspot Classification

The paper by Stenning, et. al. (this volume) entitled "Morphological Image Analysis and its Application to Sunspot Classification" describes an interesting approach to sunspot classification using techniques from mathematical morphology and image processing. The authors describe a well-thought set of techniques to differentiate among four classes, following the Mount-Wilson classification scheme. Such classes vary according to the shape and distribution of magnetic flux in sunspots groups. It is clear from the paper that the task of extracting relevant features to differentiate among such classes is extremely difficult. The distribution of positive and negative magnetic polarities extracted from the magnetogram can exhibit multiple configurations, which makes it very difficult to point to the right class precisely. The paper gives a hint at the strong challenge of acquiring additional features to improve on accuracy performance (currently reported at around 58% on a testing set using random forests).

The problem of finding relevant features in images with spatial content appears in many other scientific domains. We have found that one key element to discover relevant features in these problems is to look for contextual information according to the precise nature of the classes under analysis. One particular domain is that of automatic classification of landforms on Mars, described next.

32.3 An Analogous Problem in Landform Classification on Mars

We follow our discussion with a brief description of a pattern recognition tool for mapping landforms on Mars [3, 4] that receives as input a DEM (Digital Elevation Map). It uses the values of elevations stored in the DEM to calculate additional geomorphometric features; we use the following cell-based features: slope, curvature, and flooding adjustment. At the end of our feature generation process we have a three-dimensional feature vector assigned to each cell in the raster. The raster is

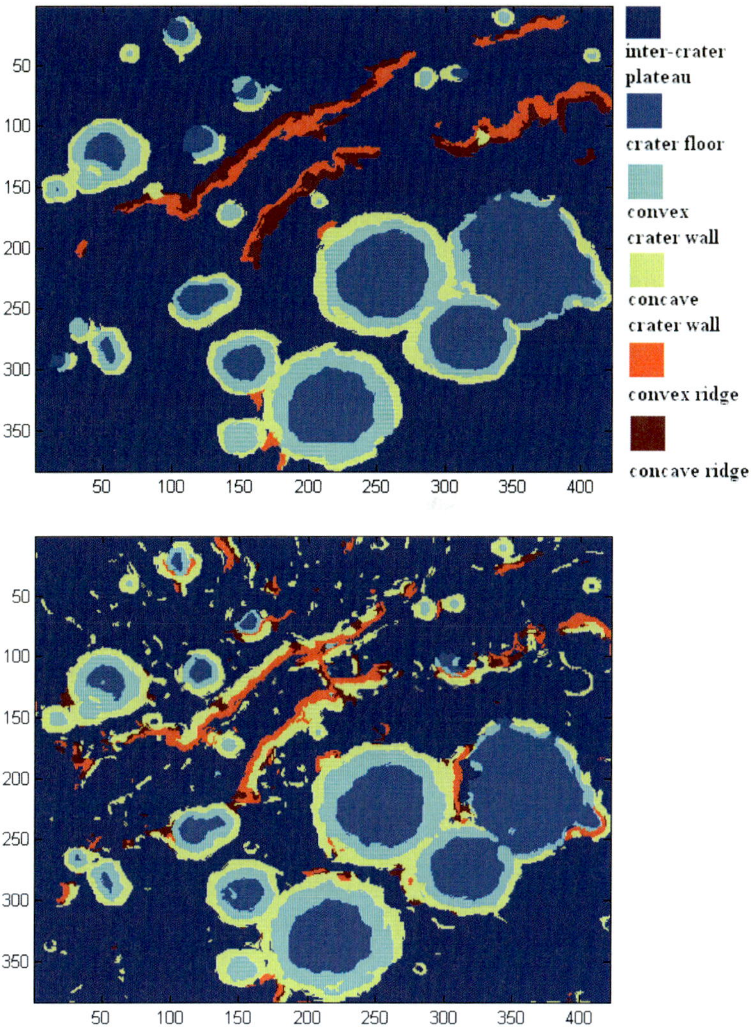

Fig. 32.1 *Top*: A color-labeling of a site on Mars (Tisia site) with perfect landform classification. *Bottom*: The approximation made by a learning algorithm (support vector machines) using local features

then segmented into spatially single-connected, feature vectors. After segmentation, the raster consists of a number of spatial patches; these patches are the objects of final classification based on six possible classes: inter-crater plateau, crater floor, convex crater wall, concave crater wall, convex ridge, and concave ridge.

Figure 32.1 shows two images of Mars. Figure 32.1-top describes the "ground-truth", where all segments have been correctly classified by an expert, with class labels displayed on the right side. Figure 32.1-bottom shows the result of automatically classifying that region of Mars using a small amount of segments for training, and using the rest for testing. The problem with this task is that there are semantically different landforms that display similar or even identical landscape elements. This is difficult because it requires domain knowledge of Martian topology as regards to structural shape. An example of two distinct landforms consisting of very similar landscape elements is the case of concave crater walls and concave ridges. Both landscape elements are rim-like surfaces, the difference is that in the case of the crater, a collection of rim structures forms a circle-like landform, and, in the case of the ridge, it forms a linear-like structure. This is again a problem where the feature representation is crucial to attain good results. In the case of Mars, additional features capturing the shape and global distribution of segments on each landform are necessary to overcome the irreducible error that comes from class overlapping.

References

1. Hastie, T., Tibshirani, R., Friedman, J.: *The Elements of Statistical Learning: Data Mining, Inference, and Prediction*. Springer, New York (2009)
2. Liu, H., Motoda, H.: *Computational Methods of Feature Selection*. Chapman & Hall/CRC Data Mining and Knowledge Discovery Series (2007)
3. Stepinski, T., VilaltaR.: Machine Learning Tools for Geomorphic Mapping of Planetary Surfaces. In: Zhang Y. (eds.) *Machine Learning*, pp. 251–266. In-Tech (2010)
4. Ghosh S., Stepinski T., Vilalta R.: Automatic Annotation of Planetary Surfaces with Geomorphic Labels. *IEEE Transactions on Geoscience and Remote Sensing*. 48, 175–185 (2009)

Chapter 33
Learning About the Sky Through Simulations

Andrew Connolly, John Peterson, Garret Jernigan, D. Bard and the LSST Image Simulation Group

Abstract With data sets that will soon reach petabytes in size, astrophysics is rapidly moving from a regime limited by statistical noise to one where scientific progress will be determined by our ability to understand and control systematic uncertainties. Simulations can play a critical role in this process by providing a better understanding of the capabilities and limitations of any observational system; enabling the development of new statistical techniques, testing the performance of a new design or optimization, and evaluating how an existing scientific analysis might scale to data volumes a thousand times larger than todays. We describe here an approach, adopted by the Large Synoptic Survey Telescope (LSST), to develop high-fidelity simulations at the scale and complexity of the LSST survey itself. These simulations comprise cosmological models of the universe including: galaxy populations, stellar distributions from Galactic structure models, and populations of asteroids within our Solar System. When coupled to simulations of sequences of LSST observations, and to a photon-based simulator that generates detailed images with the properties of the LSST system (accounting for the effects of the atmosphere, telescope and camera), we have an end-to-end system capable of addressing a broad range of astrophysical and statistical questions. We describe

A. Connolly (✉)
Department of Astronomy, University of Washington, Seattle, WA, USA
e-mail: ajc@astro.washington.edu,

J. Peterson
Department of Physics, Purdue University, West Lafayette, IN 47907, USA
e-mail: peters11@purdue.edu

G. Jernigan
Space Sciences Laboratory, University California, Berkeley, CA 94720, USA
e-mail: jgj@universe.sonoma.edu

D. Bard
SLAC National Accelerator Laboratory, 2575 Sand Hill Rd., Menlo Park, CA, USA
e-mail: djbard@slac.stanford.edu

E.D. Feigelson and G.J. Babu (eds.), *Statistical Challenges in Modern Astronomy V*, Lecture Notes in Statistics 209, DOI 10.1007/978-1-4614-3520-4_33,

here, using a study of image subtraction techniques, how the current generation of simulations can be used to develop new statistical techniques for processing and analyzing astronomical data sets.

33.1 Introduction

A new generation of astronomical survey telescopes, the Dark Energy Survey (DES), the Panoramic Survey Telescope and Rapid Response System (Pan-STARRS), EUCLID, the Visible and Infrared Survey Telescope for Astronomy (VISTA), and the Large Synoptic Survey Telescope (LSST) are now, or soon will be, surveying the universe in unprecedented detail. Repeated observations of the same part of the sky, with hundreds to thousands of observations over a period of 10 years, will enable a detailed study of the temporal universe (ranging from transient sources such as supernovae and optical bursters, to periodic variables such as Cepheids and RR-Lyrae stars, to moving sources such as asteroids and high proper motion stars). Combined, these observations will provide some of the deepest, large-scale surveys of the universe, ever undertaken and provide the ability to measure the nature of dark energy with figures of merit 10–100 times better than current surveys (DETF, [1]).

The stringent requirements on the statistical power of these telescopes means that we will soon no longer be limited by shot noise (i.e. the number of sources within a sample) but by how well we can understand systematic uncertainties within our data streams. These systematic effects can arise from the design of the telescope (e.g. ghosting of images or scatter light), from the response of the atmosphere (e.g. the stability of the point-spread-function or the variability in the transmissivity of the sky), from the strategy used to survey the sky (e.g. inhomogeneous sampling of astronomical light curves), or from limitations in our analysis algorithms (e.g. due to the finite processing power available approximations may need to be made when characterizing the properties of detected sources). Understanding which of these issues will impact the science from a given telescope is critical if we hope to maximize their scientific returns.

Simulations of the data flow from survey telescopes can provide a critical role in understanding the capabilities of an astronomical system and in optimizing its scientific returns. By providing data with the expected characteristics of a survey well in advance of first light, algorithms and statistical techniques can be optimized and scaled to the expected data volumes or new statistical approaches can be developed to improve the data analysis. In the following sections we describe an approach undertaken by the LSST to simulate its data flow. We describe the simulation framework, its requirements and design, the data generated to date, how these data are being used by the LSST, and how we can employ this simulation framework to explore optimal statistical techniques for the detection of transient or variable sources in noisy images.

33.2 Simulating an Astronomical Data Stream

The design of a framework [2] to simulate the data expected from the LSST requires flexibility and scalability (to enable data generation runs that range from a single CCD image of a gravitational lens to images from thousands of full focal planes that trace the expected observing cadence of the survey). This is accomplished by dividing the simulation workload into three separate components: a base component that stores a model of the universe (including the distribution of galaxies from a cosmological simulation, the distribution of stars from a Galactic Structure model that incorporates contributions from a thin disk, thick disk and halo, and a model for the asteroid populations within our Solar System), a system for querying the underlying model of the universe using simulations of sequences of LSST observations, and a framework for the generation of images via the ray-tracing of individual photons. The system as implemented today can generate large-scale astronomical catalogs, sequences of individual CCD images taken over periods of days or weeks, and large scale imaging runs that generate terabytes of images and associated reference catalogs.

Figure 33.1 shows an example of the flow of information through the LSST simulation framework. Simulations of sequences of LSST observations enable catalogs of LSST sources to be generated. These catalogs can be analysed for different science programs or passed to a photon based image generator that create input images for the data management analysis pipelines. The design enables the generation of a wide range of data products: from all-sky catalogs used in modeling the LSST calibration pipeline, to time domain data used to characterize variability as a function of signal-to-noise and temporal sampling, to sequences of images of gravitational lenses from which to measure cosmological time delays. To date the framework has been used for three LSST data challenges over the last 3 years and generated over 10TB of images ($>10^6$ amplifier images) and simulated over 10^{13} photons.

33.2.1 Simulating the Distribution of Sources to Faint Magnitudes

The underlying source catalogs within the LSST simulator extend to a depth or $r_{AB} = 28$. This limit is set by the requirement that sources extend below the detection limit of the LSST images even after the coaddition of 10 years of data (as the distribution of sources below the detection limit influences the statistics of sky background through their color distribution and clustering). Galaxy distributions are derived from the Millennium simulations of de Lucia et al. [3] and assume a standard Λ-CDM cosmology. These models extend dark matter N-body simulations to include gas cooling, star formation, supernovae, and AGN and are designed to reproduce the observed colors, luminosities, and clustering of galaxies as a function

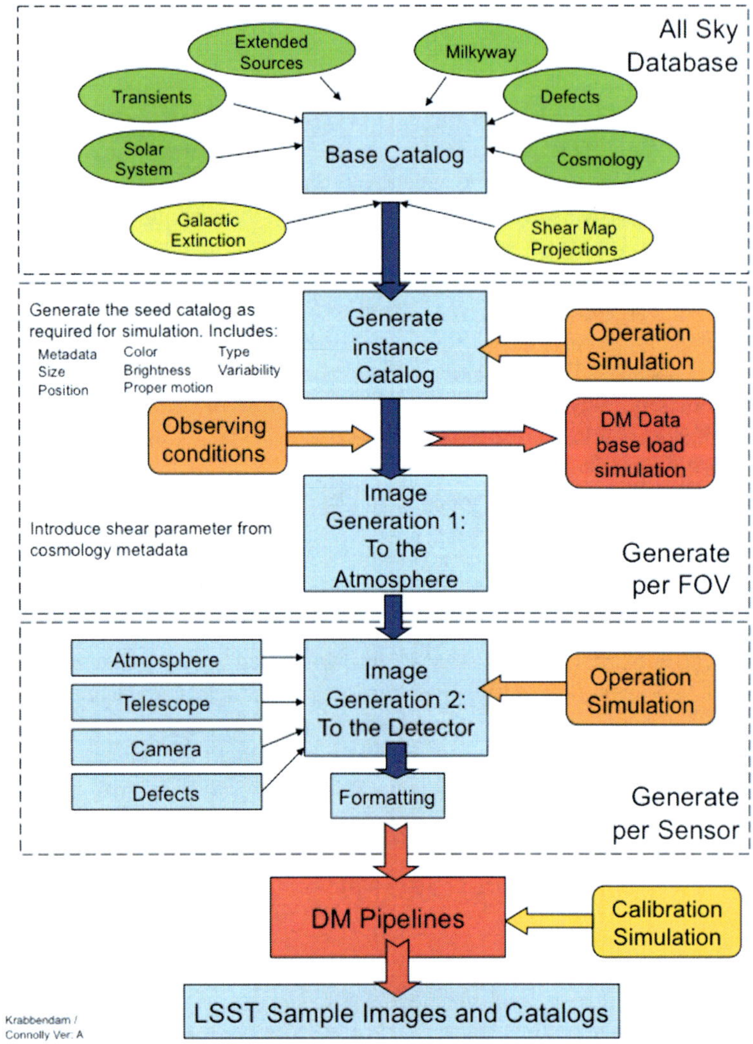

Fig. 33.1 The flow of information through the LSST simulation framework. Databases of astrophysical sources are populated with models of the cosmological distributions of galaxies, the distributions of stars within our Galaxy, and the populations of asteroids within our Solar System. Using historical records for the weather at Cerro Pachon and the observing cadences required by the science drivers for the LSST, sequences of simulated observations are generated by the Operations simulator. From these simulated pointings, catalogs and images of galaxies can be generated that match the expected properties of the LSST system. Comparing the catalogs derived by processing the LSST data with those used to generate the inputs we enable a full end-to-end test of the LSST system

of redshift. The LSST cosmological catalogs were generated by constructing a lightcone, covering redshifts $0 < z < 6$, from 58,500 h^{-1}Mpc simulation snapshots. This lightcone covers a 4.5×4.5 degree footprint on the sky and samples halo masses over the range 2.5×10^9 to 10^{12} $M_{\odot}$. Dynamically tiling this footprint across the sky enables the simulation of the full LSST footprint while keeping the underlying data volume small (but at the expense of introducing periodicity in the large scale structure). For all sources, a spectral energy distribution, is fit to the galaxy colors using Bruzual and Charlot spectral synthesis models. These fits include inclination dependent reddening and are undertaken independently for the bulge and disk components. Morphologies are modeled using two Sersic profiles and a single point source (for the AGN) with bulge-to-disk ratios and disk scale lengths from de Lucia et al. Half-light radii for bulges are estimated using the empirical absolute-magnitude vs half-light radius relation given by Gonzalez et al. [4]. Comparisons between the redshift and number-magnitude distributions of the simulated catalogs with those derived from deep imaging and spectroscopic surveys showed that the de Lucia models under-predict the density of sources at faint magnitudes and high redshifts. To correct for these effects, sources are "cloned" in magnitude and redshift space until their densities reflect the average observed properties. Figure 33.2 shows the resulting size distribution of galaxies within the simulations compared to observations from the Hubble Space Telescope.

The distribution of stars are based on the stellar structure models of Juric et al. [5]. These include thin-disk, thick-disk, and halo star components with colors that match those observed by the Sloan Digital Sky Survey (SDSS, [6]). Spectral energy distributions are fit to the predicted colors using the models of Kurucz [7] for main sequence stars and giants, Bergeron et al. [8] for white dwarfs, and a combination of spectral models and SDSS spectra for M, L, and T dwarfs (e.g., [9–13]). The dynamical models for the galaxy are taken from Bond et al. [14] and for each star a parallax and proper motion is derived.

Approximately 10% of the stellar sources are defined to be variable. Variability is modeled for sources within the base catalogs by defining a light curve, its amplitude, a period, and a phase. For queries that contain a time constraints the magnitude of the source is adjusted based on the properties of the light curve (the current implementation only allows for monochromatic variations in the fluxes). Variables modeled range from cataclysmic variables, flaring M-dwarfs, and micro-lensing events. For transient sources, the period of the light curve is set to >10 years such that the sources will not repeat within the period of the LSST observations.

For Galactic reddening, a value of E(B-V) is assigned to each star using the three-dimensional Galactic model of Amores and Lepine [15]. For consistency with extragalactic observations the dust model in the Milky Way is re-normalized to match the Schlegel et al. [16] dust maps at a fiducial distance of 100 kpc.

Solar System sources are derived from Grav et al. [17]. They include populations for Near Earth Objects (NEOs), Main Belt Asteroids, the Trojans of Mars, Jupiter, Saturn, Uranus, and Neptune, Trans Neptunian Objects, and Centaurs. For each

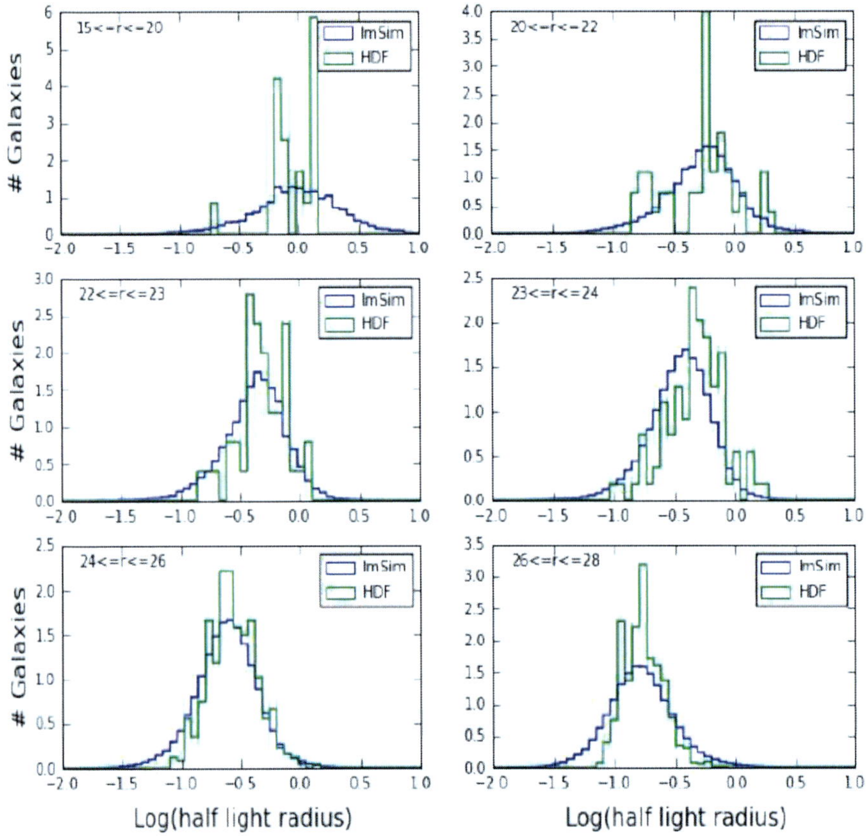

Fig. 33.2 A comparison of the sizes of galaxies in the simulated galaxy catalogs with observations of galaxy half light radii from the Hubble Deep Field. The *panels* show the distributions as a function of limiting magnitude and are in good agreement for magnitudes $r > 20$

source the full set of orbital parameters are defined. Integrating these orbital elements for every source for each observation is, however, computationally expensive if undertaken when querying the database. Chebychev polynomials are, therefore, used to interpolate between nightly positions [18]. This results in an astrometric precision of better than 5 mas for any predicted position and a query time of less that 2 s to identify the 8,000 asteroids found in each LSST pointing. Spectral energy distributions are assigned using the C and S type asteroids of DeMeo et al. [19]. In total there are 11 million asteroids within the simulation. For each asteroid the orbital positions are precomputed for each night of the 10 years of LSST operations.

33.2.2 Querying the LSST Simulations for Sources

The base catalog is stored as a SQL database (using a Microsoft SQLServer). Data are accessible through a Python interface that uses SQLalchemy to provide a database agnostic view of the sources. For any LSST pointing sources can be queried as a function of position and time with the returned data accounting for any change in brightness due to variability. For large scale runs, the base catalog is queried using sequences of observations derived from the Operations Simulator [20]. The Operation Simulator simulates LSST pointings that meet the cadence and depth requirements of the LSST science cases while accounting for historical weather patterns for Cerro Pachon and the visibility of the LSST footprint on the sky. Each simulated pointing provides a position and time of the observation together with the appropriate sky conditions (e.g. seeing, moon phase and angle, and sky brightness). Positions of sources are propagated to the time of observation (from the proper motion information for stars and orbits for Solar System sources). Magnitudes and source counts are derived using the atmospheric and filter response functions appropriate for the airmass of the observation and after applying corrections for source variability. The resulting catalogs (instance catalogs) can be formatted for use in a science application (e.g. measuring the proper motions of high velocity stars) or fed to the final component of the simulation framework, the image simulator. This component simulates images by ray-tracing individual photons through the atmosphere, telescope and camera systems. Photons are drawn from the spectral energy distributions that define the simulated data and are ray-traced through the optical system before being converted to electrons by simulating the camera physics. Images are read-out using a simulation of the camera electronics and amplifier layout and formatted for ingestion into the LSST data management system.

33.2.3 Simulating High-Fidelity Images Through Photon Sampling

Each LSST simulated image is generated by simulating the individual photons. The rational for this, as opposed to convolving an image with an analytic point-spread-function, is that the number of photons that must be simulated for an image is comparable to the number of pixels in the image. The number of operations for either approach is comparable whilst the photon based approach enables the simulation of wavelength dependent effects and the inclusion of perturbations in the optical surfaces in a natural way.

Photons are drawn from the spectral energy distribution of each source (scaled to the appropriate flux density based on the apparent magnitude of a source and accounting for the spatial distribution of light for extended sources). Each photon is ray-traced through the atmosphere, telescope and camera to generate a CCD image.

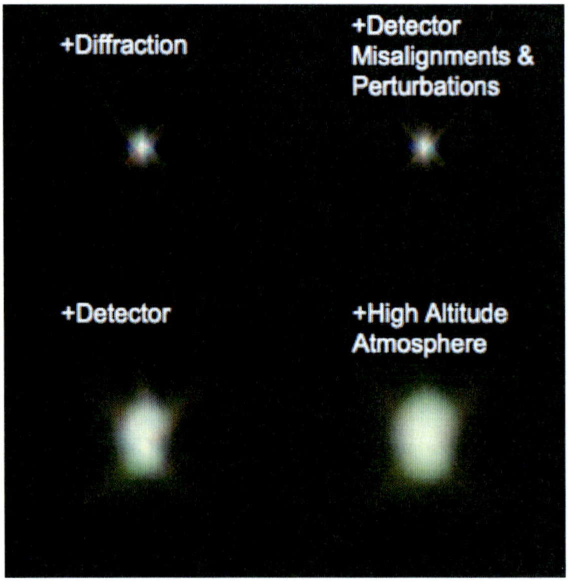

Fig. 33.3 The image simulation framework is flexible enough to switch off and on different optical components. From *left* to *right* and *top* to *bottom* we show the resulting PSF as we progressively add more components to the optical path (including perturbations in the optical surfaces and the addition of a single layer atmosphere)

The atmosphere is modeled using a Taylor frozen screen approximation (with the atmosphere described by six layers). The density fluctuations within these screens are described by a Kolmogorov spectrum with an outer scale (typically, 10–200 m) and an inner scale of 1 cm, set by the resolution of the atmospheric screens [21]. All screens move during an exposure with velocities derived from NOAA measurements of the wind velocities above the LSST site in Chile. Typical velocities are on the order of 20 ms^{-1} and are found to have a seasonable dependence that is modeled when generating the screens. Each photons trajectory is altered due to refraction as it passes through each screen. Figure 33.3 show the impact of the screens and the wind on the PSF and its homogenization.

Once through the atmosphere photons are reflected and refracted by the optical surfaces within the telescope and camera. The mirrors and lenses are simulated using geometric optics techniques in a fast ray-tracing algorithm and all optical surfaces include a spectrum of perturbations based on design tolerances. Each optic moves according to its six degrees of freedom within tolerances specified by the LSST system. Fast techniques for finding intercepts on the aspheric surface and altering the trajectory of a photon by reflection or wavelength-dependent refraction have been implemented to optimize the efficiency of the simulated images. Wavelength and angle-dependent transmission functions are incorporated within each of these techniques including simulation of the telescope spider and scattering off the optical spider.

Ray-tracing of the photons continues into the silicon of the detector. Conversion probability and refraction (a function of wavelength and temperature) and charge diffusion within the silicon are modeled for all photons. Photons are pixelated

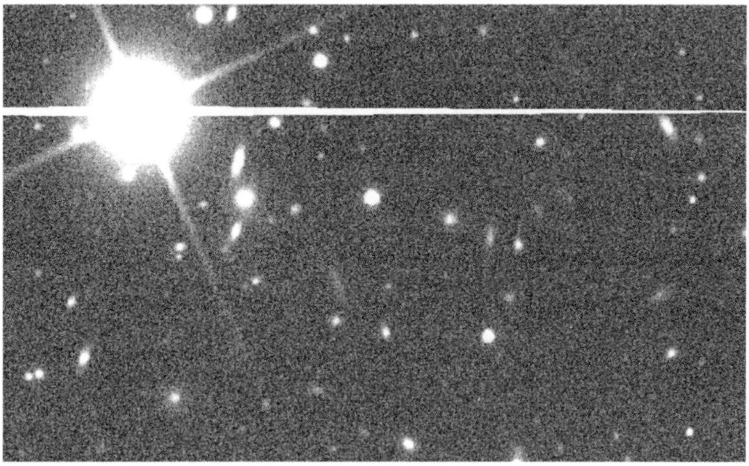

Fig. 33.4 An image from part of a single amplifier for the LSST focal plane (which contains 189 CCDs and 16 amplifier images per CCD). The distribution of galaxies, and the diffraction spikes and bleed trails of bright stars are visible within the image

and readout, simulating the effects of blooming, charge saturation, charge transfer inefficiency, gain and offsets, hot pixels and columns, and QE variations. Figure 33.3 shows the effect of these individual components within the image simulator.

The sky background is added as a post-processing step with the sky background generated, including Rayleigh scattering of the moons light, based on SEDs for the full moon and the dark sky. The background is vignetted according to the results of ray-trace simulations. The simulator generates ~300,000 photons per second on an average workstation and requires approximately 500 h to simulate a full focal plane. In Fig. 33.4 we zoom in on a single amplifier region and show the resulting distribution of stars and galaxies (including the simulated diffraction spikes and bleed trails).

33.3 Using Simulations to Define Science

The simulation framework has been used in a number of different studies to test both analysis algorithms and the efficiencies of different survey strategies. These applications range from studies of influence of the atmosphere on the distribution of source ellipticities [22], estimations of the ability of the LSST to measure light curves for strong gravitational lenses, searches for high proper motion stars within 10 years of observations, and the development and testing of techniques for tracking asteroids. In the following section we isolate one of these studies that focused on image subtraction and how we might characterize the kernel that maps one image to another.

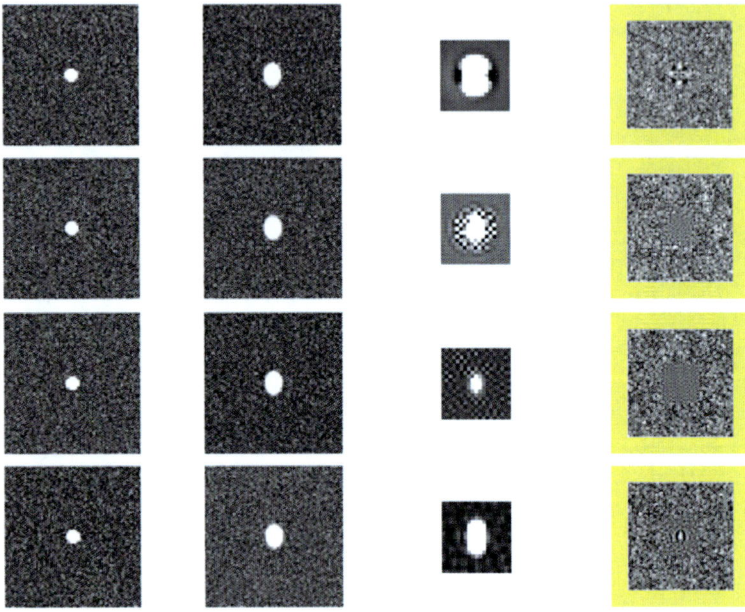

Fig. 33.5 From *left* to *right* the panels show a science observation of a single star, the template or reference image, the kernel that matches these two scenes and the resulting image subtraction. From *top* to *bottom* we simulate kernels generated using the Alard and Lupton method (for 30 terms), the Alard and Lupton method (320 terms), a delta function basis set, and a delta function basis with regularization

33.3.1 Characterizing Image Subtraction with Kernels

A common aspect of most modern imaging surveys is the desire to characterize the variability of astrophysical sources through multiple observations of the same part of the sky (variability can come in the form of transient events such as supernovae, periodic variability such as Cepheids and RR-Lyrae stars, or moving sources such as asteroids). Variability is typically measured through the application of image subtraction whereby a high signal-to-noise template image is subtracted from a recent science image and any residual flux is attributed to the variability of a source. In order for this to be a robust detection of variations in the source flux we must account for the non-astronomical difference between the two scenes (e.g. due to errors in the astrometric solutions, variation in the transmissivity of the atmosphere, and differences in the point-spread functions of the two images).

Following Alard and Lupton [23] we can consider two images taken at different times but in the same filter as representing the same scene but with different point-spread-functions (PSFs), i.e.

$$I(x,y) = (K \otimes T)(x,y) + \varepsilon(x,y). \tag{33.1}$$

where $I(x,y)$ is the observed image, $T(x,y)$ is a template image, $K(x,y)$ is the kernel that maps the template to the observed image, and $\varepsilon(x,y)$ is an additional noise component (usually assumed to be a Normal distribution).

If we model the kernel, K, as a linear sum of basis functions, $K(x,y) = \sum_i a_i K_i(x,y)$, we may solve for the coefficients of K_i through a standard least squares approach (i.e. minimizing the mean square errors between the observed and template images). These basis functions do not need to be complete nor orthogonal but should provide a compact representation of the mapping kernel. In Alard and Lupton (hereafter AL) the basis functions selected for the kernel, K, were assumed to be Gaussians modified by 2-dimensional polynomials,

$$K_i(u,v) = e^{-(u^2+v^2)/2\sigma_n^2} \; u^p v^q, \tag{33.2}$$

where i runs over all permutations of n,p,q. In this parameterization, n is the number of Gaussian components, of width σ_n and p and q are the orders of the Gauss–Hermite polynomials such that $0 \leq p+q \leq O_n$, with O_n the order of n. The total number of basis functions in the set is $\sum_n (O_n+1) \times (O_n+2)/2$ (for a typical application approximately 50 basis functions are used). It is worth nothing that the width, σ_n, number of Gaussians, and order of the polynomials are not fit in this process but defined through either a set of heuristics or via the application of cross-validation.

A question that remains open under this mapping is, do these Gauss-Hermite basis functions represent the optimal description of the mapping kernel or are there more appropriate bases. For example, in the presence of substantial astrometric errors between images, the centered Gauss-Hermite bases cannot efficiently map offsets between sources. In Becker et al. [24] we, therefore, investigated the advantages and disadvantages of building kernels using delta function basis functions ($K_{ij}(u,v) = \delta(u-i)\delta(v-j)$). The advantage of delta functions is their flexibility (e.g. their ability to compensate for astrometric misregistration) but this comes at the expense of a loss of compactness. Figure 33.5 illustrates this point by showing the image subtraction performance of Gauss-Hermite and delta function bases when subtracting pairs of stars simulated by the LSST simulation framework. From left to right the panels show the input science image, $I(x,y)$, the template image, $T(x,y)$, the derived mapping kernel, $K(x,y)$, and the residuals after subtracting the science image from the template (after convolving the template image with the mapping kernel). From top to bottom we consider the case of AL basis functions with 30 and then 320 terms, a delta function kernel, and a delta function kernel derived using a regularization term that ensures smoothness.

In all cases the residuals are comparable to or better than the variance derived from the image statistics. As noted by Becker et al., the delta function kernels outperform the AL bases in the case of significant astrometric shifts between the images (even for the case of 320 AL bases). The non-regularized kernels appear to perform as well as the regularize delta function kernels for the case of a single star. If, however, we use the mapping kernel derived for one star as a mapping kernel for its neighboring stars (even for the case of no variation in the PSF) the image subtraction residuals are much larger. This is because the number of parameters within this model (i.e. the number of pixels) is sufficiently large that the model is overfit (i.e. the kernel fits both the signal and the noise when mapping between image and template).

Regularization techniques can remove the problem of overfitting the data by adding a penalty term to the regression. In our case we consider regularization as a way to ensure smoothness in the resulting kernel (though this is by no means the only regularization term that could be applied) and penalize the second derivative by $\lambda \cdot \int\int |\nabla f(x,y)|^2 \mathrm{d}x\mathrm{d}y$. With λ as the only tuning parameter within this fit, a combination of cross-validation, risk estimation, and the ability to predict the appropriate kernel for neighboring sources can all be used to define an appropriate range of values for λ. For the example in Fig. 33.5 (4th row), a value of $\lambda \sim 0.5$ is sufficient to produce difference images with similar performance to those given by the sum–of–Gaussian AL bases while retaining predictive ability when applied to neighboring sources.

33.4 Conclusions

With a new generation of survey telescopes on the horizon (from PanSTARRS and DES to LSST) we now face the prospect of science being limited by our ability to model and correct for systematics within the data rather than simply counting statistics. Simulations can play a critical role in understanding both the nature of the data we must work with (e.g. the variation in the PSF as a function of color or position on the focal plane) and how we might scale our analyses to data volumes that are 100–1,000 times larger that todays. The LSST has, therefore, implemented a program to simulated the flow of data expected from this telescope (in the form of catalogs and images). We show, here, how these simulations can be used to develop new algorithms for processing imaging data. Within the near future simulated data sets will include more of the astronomically interesting sources for a deep, temporal survey (e.g. light curves from variable sources such as supernovae). Combining these tools with new statistical techniques will provide an opportunity to determine how well we can characterize the properties of the universe prior to a new generation of astronomical resources coming on line.

Acknowledgements AJC wishes to acknowledge his co-authors on the delta function kernel paper for kindly allowing some of the results of that paper to be discussed here. This work is supported by the Large Synoptic Survey Telescope and, in part, by the National Science Foundation under Grant Number AST-0709394.

References

1. Albrecht, A., et al., "Report of the Dark Energy Task Force", astro-ph/0609591 (2006)
2. Connolly, A.J, et al., "Simulating the LSST system", Proc. SPIE 7738, 77381O (2010)
3. De Lucia, G., Springel, V., White, S.D.M., Croton, D., & Kauffmann, G., The Formation History of Elliptical Galaxies, Monthly Notices of the Royal Astronomical Society, 366, 499–509 (2006)
4. Gonzlez, J.E., Lacey, C. G., Baugh, C. M., Frenk, C. S., & Benson, A. J., Testing model predictions of the cold dark matter cosmology for the sizes, colours, morphologies and luminosities of galaxies with the SDSS, Monthly Notices of the Royal Astronomical Society, 397, 1254–1274 (2009)
5. Juric, M., et al., The Milky Way Tomography with SDSS. I. Stellar Number Density Distribution, Astrophysical Journal, 673, 864–914 (2008)
6. York, D., et al., The Sloan Digital Sky Survey: Technical Summary, Astrophysical Journal, 120, 1579–1587, (2000)
7. Kurucz, R.L., CD-ROM No.13, Cambridge, Mass.,Smithsonian Astrophysical Observatory, (1993)
8. Bergeron, P., Wesemael, F., & Beauchamp, A., Wesemael, F., Photometric Calibration of Hydrogen- and Helium-Rich White Dwarf Models, Publications of the Astronomical Society of the Pacific, 107, 1047–1054 (1995)
9. Cushing, M.C., Rayner, J.T., & Vacca, W.D., An Infrared Spectroscopic Sequence of M, L, and T Dwarfs, Astrophysical Journal, 623, 1115–1140 (2005)
10. Bochanski, J.J., West, A.A., Hawley, S.L., & Covey, K.R., Low-Mass Dwarf Template Spectra from the Sloan Digital Sky Survey, Astronomical Journal, 133, 531–544 (2007)
11. Burrows, D., Sudarsky, D., & Hubeny, I., L and T Dwarf Models and the L to T Transition, Astrophysical Journal, 640, 1063–1077 (2006)
12. Pettersen, B.R., & Hawley, S.L., A spectroscopic survey of red dwarf flare stars, Astronomy & Astrophysics, 217, 187–200 (1989)
13. Kowalski, A., Hawley, S.L., Holtzman, J.A., Wisniewski, J.P., Hilton, E.J., A White Light Megaflare on the dM4.5e Star YZ CMi, Astrophysical Journal, 714, L98 (2010)
14. Bond, N., Ivezic, Z., Sesar, B., Juric, & Munn, J., The Milky Way Tomography with SDSS: III. Stellar Kinematics, arXiv:0909.0013 (2009)
15. Amres, E.B., & Lpine, J.R.D., Models for Interstellar Extinction in Galaxy, Astronomical Journal, 130, 659–673 (2005)
16. Schlegel, D.J., Finkbeiner, D.P., & Davis, M., Maps of Dust Infrared Emission for Use in Estimation of Reddening and Cosmic Microwave Background Radiation Foregrounds, Astrophysical Journal, 500, 525–553 (1998)
17. Grav, T., Jedicke, R., Denneau, L., Holman, M. J., & Spahr, T., The Pan-STARRS Synthetic Solar System Model and its Applications, BAAS, 211, 4721 (2007)
18. AlSayyad, Y., "Towards Efficient and Precise Queries Over Ten Million Asteroid Trajectory Models", in preparation (2012)
19. DeMeo, F.E., Binzel, R.P., Slivan, S.M., & Bus, S.J., An Extension of the Bus Asteroid Taxonomy into the Near-Infrared, Icarus, 202, 160–180 (2009)
20. Cook, Kem H, et al., "LSST: Cadence Design and Simulation", American Astronomical Society, AAS Meeting 213, 460.04; Bulletin of the American Astronomical Society, Vol. 41, p. 367 (2009)
21. Jernigan, J.G., in preparation, (2012)
22. Chang, C., et al., "Spurious Shear Systematics in Weak Lensing Investigations with LSST I: A Numerical Study with the LSST Image Simulator", in preparation, (2012)
23. Alard, C. & Lupton, R. H. 1998, Astrophysical Journal, 503, 325
24. Becker, A., et al., "Regularization Techniques for PSF–Matching Kernels. I. Choice of Kernel Basis", in preparation (2012)

Chapter 34
Commentary: Learning About the Sky Through Simulations

Michael J. Way

Abstract The Large Synoptic Survey Telescope (LSST) simulator being built by Andy Connolly and collaborators is an impressive undertaking and should make working with LSST in the beginning stages far more easy than it was initially with the Sloan Digital Sky Survey (SDSS). However, I would like to focus on an equally important problem that has not yet been discussed here, but in the coming years the community will need to address—can we deal with the flood of data from LSST and will we need to rethink the way we work?

34.1 Changing the Way We Work: From 2MASS to SDSS

Perhaps the best way to start things is to compare two large area sky surveys implementing the "standard way" of distributing data in their own time: The 1990s era Two Micron All Sky Survey [2MASS; 3] and the 2000s era Sloan Digital Sky Survey [SDSS; 4].

Initially if a researcher wanted to access the 2MASS[1] survey data one could obtain a five DVD set (double-sided) of the catalog data. The data were bar-delimited ascii text which could be easily read by everything from legacy scripting programs like awk to SQL databases like MySQL or Postgres. The ascii source catalog was about 43 GB in size if copied from the DVDs to local disk. The full-fidelity Atlas Images ($\sim$10 TB in size and not available via DVD) were later accessible via the 2MASS Image Services website.[2]

[1] http://www.ipac.caltech.edu/2mass

[2] http://irsa.ipac.caltech.edu/applications/2MASS/IM

M.J. Way (✉)
NASA Goddard Institute for Space Studies, 2880 Broadway, New York, NY, USA
e-mail: michael.j.way@nasa.gov

E.D. Feigelson and G.J. Babu (eds.), *Statistical Challenges in Modern Astronomy V*, Lecture Notes in Statistics 209, DOI 10.1007/978-1-4614-3520-4_34,

In essence, the average astronomer had to change almost nothing about the way that they or their Ph.D. advisor had worked over the previous 30 years. For example, instead of ordering nine-track (1/2 in. = 12.7 mm), exabyte (8 mm), or DDS/DAT (4 mm) tapes from the observatory (or bringing them along after an observing run) one simply ordered the 2MASS DVDs. This was possible due to increases in computer cpu speed and memory capacity combined with modest input/output (IO) improvements over the previous three to four decades.

All of this changed with the SDSS. It may have been possible to distribute DVD copies of the SDSS in a similar way to that of the 2MASS, but the scale had moved from gigabytes to terabytes. Having a few terabytes on a local computer in the early 2000s was not common, so the SDSS took a different route. Working with top-notch computer scientists such as Jim Gray of Microsoft they decided that much of the SDSS should be accessible via SQL query. There was certainly some anxiety amongst much of the community when they realized that their mode of obtaining data from the SDSS was going to be markedly different than in the past. Hence, it took the community some time to learn this new way of working, and certainly some early publications with SDSS data not published by the SDSS team were problematic because, for example, they did not realize that they could decide the quality of the photometry at a fine level, unlike that of the 2MASS which was relatively straightforward.

The SDSS is probably the most similar data set today compared with what the LSST will look like and how we will interact with it. Currently most users of SDSS use the casjobs[3] interface to obtain their data. The back end is a SQL database tied to a front end presented to the user as a web interface where SQL queries are entered. The database comes with a Schema Browser[4] that allows one to explore items of possible interest. There are also a host of on-line tutorials that one can go through to understand how make the queries, and many authors also publish their SDSS queries in the appendices of their publications. However, not enough authors do the latter in my opinion, and hence it is often impossible to replicate the data that people are using if the original author does not publish or cannot recall their original query.

In the 10 years since the creation of the SDSS disk data storage density has continued its inexorable rise (see Sect. 34.2) and today one could in fact host the entire SDSS database relatively easily and cheaply on a modestly sized desktop computer (e.g. an off-the-shelf workstation with 4×2 TB disks would do the trick). Again, one could (in theory) dump the entire casjobs catalog to a giant ascii file akin to that of the 2MASS and use awk or your favorite fortran program on it. One would need a system that can use 64 bit addressing, but that is fairly standard today (2011). Of course the IO will make things slow ($\sim$4 h to read a 1 TB disk sequentially), but nonetheless it is in theory quite possible. The question then arises, will one be able to work with LSST in the same manner as the SDSS given the inexorable rise of faster CPUs, Memory, and IO?

[3]http://casjobs.sdsss.org

[4]http://casjobs.sdss.org/dr7/en/help/browser/browser.asp

34.2 Changing the Way We Work: From SDSS to LSST

As we consider the move from SDSS sized data sets to LSST the questions that people like us might ask at this stage are fairly straightforward:

1. Will one be able to have a copy of the LSST data-set on a local desktop? This will allow researchers to continue their pre-SDSS era methods of data interrogation. This is what we like to call the *2MASS mode.*
2. Will one still utilize a casjobs type web-query interface to obtain LSST data of interest and then use legacy home-grown tools to work on the data? This is what we call the *SDSS mode.*
3. Must one completely change the way one works with LSST scale data sets? Will "data locality" be required—will one have to do all of the operations to obtain a project's scientific goals on the database directly? This may be called the *LSST mode*.

To attempt to answer these questions we have to look at several other factors discussed in the following subsections.

34.2.1 LSST Database Size and Possible Architecture

We heard from Andy Connolly and Kirk Borne at this conference that the LSST query database will be of order 10 petabytes (PB) in size, while there will be around 60 PB of images available after 10 years of operation. It turns out that query scales almost linearly with the size of the database. Given historical trends in CPU, memory, storage and IO this means one should be able to derive catalogs and do joins on tables in a future LSST database as we do today with SDSS casjobs. However, there are caveats related to IO that will be discussed later.

While query scales linearly with database size, the same cannot be said of the kinds of operations that scientists would prefer to do on the data. For example, classification, clustering, density estimation are all normally $O(N^2)$ or worse. However, earlier today Alex Gray showed us that his group has managed to make a host of algorithms O(N) that are normally considered to be $O(N^2)$.

Regardless, this points to some interesting issues. Assuming petabyte database sizes, the needs to do operations that are $O(N^2)$, and the need to look at a large fraction of the stored data (that will not fit in RAM) how are researchers going to do these things on the LSST database of tomorrow? Let's touch on the possible need for "data locality". Normally one should only consider moving the data from the source of the data if one needs more than 100,000 CPU cycles per byte of data [1]. The kinds of applications this brings to mind are Seti@HOME or cryptography. Thankfully most science applications are more information intensive with CPU to IO ratios less than 10,000 to 1. This means that we may have two reasons to consider the possibility that we will not actually download the data to our local machine/data-center: The size of the database is too large (petabyte scale) and we have CPU to IO

Table 34.1 Storage cost historical trends[a]

Year	Cost/GB	Cost/TB	Cost/PB
2000	\$19.00000	\$19,000	\$19 million
2010	\$00.06000	\$62	\$62,000
2020	\$00.00002	\$0.2	\$200

[a] Extrapolated from http://www.mkomo.com/cost-per-gigabyte

ration of less than 100,000 to 1. We will address these issues in some detail in the next section, but for now let's assume we will need to do some calculations at the site of the database itself.

The LSST has teamed up with several industry partners to develop a new database to host the LSST called SciDB.[5] The current plan is to host this database in several different geographic locations (obviously to avoid a single point of failure and to handle the anticipated load), but they also currently plan to have an R interface for "expert users". As I mentioned during my talk, I think this is an excellent idea, but I hope the designers will consider adding additional languages such as Python which currently has wrappers to support a host of useful tools commonly used by the current generation of younger astronomers.[6]

34.2.2 Moving the Data Around: Can I Have a Local Copy and Make Use of It?

Will one be able to download and store the LSST database to a desktop computer in 10 years time? If one wishes to download 1 PB over a *dedicated* 1 gigabit/s line (in common use today) it will take $\sim$100 days. In 9 years let's assume everyone will have 10 gigabit/s connections (the growth in desktop network speed has not grown at the same accelerated rate of storage or CPU) so that means it will only take about 10 days. That doesn't sound unreasonable. Now one has to ask, how much will it cost to own 1 PB of storage? One can look at historical trends documented in several places on the internet to get some idea.[7] In Table 34.1 you can see what disks costs were 10 years ago, today and by extrapolation in 10 years time.

Looking at Table 34.1 one comes to the conclusion that if one wants to keep a copy of the LSST data locally it should not be a problem given the drop in price over time. After all, who would have imagined 15 years ago that they would be able to purchase a 1 TB drive for their desktop computer for under \$100?

Unfortunately things are not this simple. To illustrate my point I want to recall some more of Amdahl's rules of thumb for a balanced system.

[5] http://www.scidb.org

[6] For example, numpy, scipy, Rpy (R interface), mlabwrap (MatLab), etc.

[7] e.g. http://www.mkomo.com/cost-per-gigabyte

Table 34.2 Conclusions from Amdahl's rules of thumb for a balanced system today

Operations per second	RAM	Disk I/O bytes/s		No. of disks for that BW at 100 MB/s/disk
Giga/10^9	GB	10^8	→	1
Tera/10^{12}	TB	10^{11}	→	1,000
Peta/10^{15}	PB	10^{14}	→	1,000,000

1. Bandwidth (BW): 1 bit of sequential IO/s per instruction/s
2. Memory: $\alpha = 1 = \mathrm{MB}/\mathrm{MIPS}^8$: 1 byte of memory per one instruction/s
3. One IO operation per 50,000 instructions

Looking at Table 34.2 in the context of Amdahl's ROT perhaps the biggest problem with high performance computer centers today and into the near future is that they are CPU rich, but IO poor. The cpus may spend a lot of time sitting idle while waiting for IO to send them more to work on because not everything can be stored in RAM. This problem is not going to go away, but there are (thankfully) people aware of the issue who believe that it is currently possible to keep power consumption low while increasing sequential read IO throughput by an order of magnitude using what are called Amdahl blades [5]. Note that power consumption is becoming an issue for mid-level data centers found at Universities and some government research labs. Most of these don't have Google's electricity budget for powering them and in fact many government data centers are even being shutdown to save money.[9]

Table 34.2 tells one a couple of other interesting things.[10] First, for a Peta-scale balanced system 100 TB/s of IO bandwidth (last row of column three = 10^{14}) would be required. It will take approximately 1,000,000 disks to deliver this bandwidth *today* assuming they are capable of 100 MB/s/disk. Note that the rate of disk IO growth has not been remarkable in the past 10 years [see 5].

34.3 Conclusions

In the beginning of Sect. 34.2 I posed three questions and I would like to pose some answers:

1. Will one be able to have a copy of the LSST data-set on a local desktop? Yes, I think the average researcher will be able to have a copy of the data on their local system assuming disk storage density continues its historical trend (note that there are a number of technical arguments against this, as there are for continuing

[8]Million Instructions Per Second

[9]http://www.nytimes.com/2011/07/20/technology/us-to-close-800-computer-data-centers.html

[10]This table is a modified version of one found in a talk by Alex Szalay that the author recently became aware of.

Moore's law into the future [2].[11] However, even if one has a copy of the LSST it is unlikely one will be able to make much use of it using traditional *2MASS mode* tools given the issues with sequential IO that were outlined above.
2. Will one still utilize a casjobs type web-query interface to obtain LSST data of interest and then use legacy home-grown tools to work on the data (The *SDSS mode.*)? Yes, the LSST team seems interested in continuing the use of a casjobs type interface with an SQL backend. Whether a researcher will then be able to use their traditional home-grown tools will depend on the data sizes they download as discussed above.
3. Must one completely change the way one works with LSST scale data sets? I believe that many of the scientific goals will only be achievable by running on the database locally as an "expert user". This points to the need to have a multitude of robust data/computational centers hosting the LSST data. Today the best place for these (in the United States) would probably be at the national level supercomputing centers such as PSC,[12] NSCA[13] or NAS[14] to name a few in the USA.

Acknowledgements Thanks to Andy Connolly for taking the time to discus his LSST simulator with me prior to the conference and for encouraging me in my belief that a commentary focused on computational challenges would be appropriate. I would also like to thank the Astrophysics Department at Uppsala University in Sweden for their gracious hospitality while part of this manuscript was being completed.

References

1. Bell, G., Gray, J., & Szalay, A. (2006) Petascale Computational Systems. In: Computer, vol 39, no 1, pp. 110–112, doi:10.1109/MC.2006.29
2. Esmaeilzadeh, H., Blem, E., St. Amant, R., Sankaralingam, K. & Burger, D. (2011) Dark Silicon and the end of Multicore Scaling. In: Proc. of the 38th International Symposium on Computer Architecture, doi:10.1145/2000064.2000108
3. Skrutskie, M.F. et al. (2006) The Two Micron All Sky Survey (2MASS). In: The Astronomical Journal, 131, 1163, doi:10.1086/498708
4. York, D.G., et al. (2000) The Sloan Digital Sky Survey: Technical Summary. In: The Astronomical Journal, 120, 1579, doi:10.1086/301513
5. Szalay, A., Bell, G.C, Huang, H.H., Terzis, A. & White, A. (2009) Low-Power Amdahl-Balanced Blades for Data Intensive Computing. In: ACM SIGOPS Operating Systems Review archive, vol 44, issue 1, ACM New York, NY, USA doi:10.1145/1740390.1740407

[11] http://www.nytimes.com/2011/08/01/science/01chips.html

[12] Pittsburgh Supercomputing Center

[13] National Center for Supercomputer Applications in Illinois

[14] NASA Advanced Supercomputing center at NASA/Ames in California

Chapter 35
Statistical Analyses of Data Cubes

Erik Rosolowsky

Abstract I review the statistical analyses that have been applied to position–position-velocity (PPV) data cubes derived from observations. I focus on the PPV data cubes derived for observations of the star forming interstellar medium. These techniques separate into the study of sparse data cubes, statistical analysis of turbulent flows and observationally motivated analyses with an emphasis on object recognition. I discuss some of the difficulties in object recognition algorithms and present two observationally motivated tools: the spectral correlation function (SCF) and a dendrogram analysis. I argue that the comparison of data sets must be made in the observational domain. Both the SCF and dendrograms show utility at making these differential measurements, highlighting room for improvement in modern simulations.

35.1 Introduction

Nearly all observational data in astronomy is the quantitative measurement of electromagnetic radiation from astronomical sources. These data are intrinsically four dimensional data as a function of four coordinate axes. The radiation field is characterized by four parameters (the Stokes parameters describing both the total intensity and polarization properties of the radiation). The Stokes parameters can be measured as a function of four coordinates: polar angles of the incident radiation (2 dimensions), frequency of the radiation (1 dimension) and the time of the observation (1 dimension). Observational limitations restrict the data products to appropriate averages over the different dimensions and polarization properties. For example, the most commonly encountered form of astronomical data is imaging

E. Rosolowsky (✉)
University of British Columbia, Okanagan Campus, 3333 University Way,
Kelowna, BC V1V 1V7, Canada
e-mail: erik.rosolowsky@ubc.ca

E.D. Feigelson and G.J. Babu (eds.), *Statistical Challenges in Modern Astronomy V*, Lecture Notes in Statistics 209, DOI 10.1007/978-1-4614-3520-4_35,

data, which presents the total intensity (one of the polarization properties) as a function of incident angles (e.g., right ascension and declination) averaged over a single frequency band. Instrumental limitations and attention to particular physical phenomena dictate the subset of polarization properties and coordinates present in the data.

This contribution deals with data sets with three coordinate axes: two angular coordinates and a frequency axis and one polarization property: total intensity. Such data are historically the product of mapping with radio telescopes; however, more recent advances in instrumentation have allowed mapping of spectral data across the electromagnetic spectrum. Radio astronomy remains a focus of data cube study since heterodyne receiver technologies routinely allow frequency resolution of $\Delta\nu/\nu \sim 10^{-9}$ (optical spectroscopy can approach this precision, but $\Delta\nu/\nu \sim 10^{-6}$ remains typical). For observations of spectral line features (with frequency ν_0), the frequency resolution determines the precision that can be obtained in velocity measurements due to the Doppler effect: $\Delta\nu/\nu_0 = -v_r/c$ where v_r is the radial component of the velocity and c is the speed of light. For observations that resolve the frequency structure of a spectral line, the frequency axis is typically converted to a radial velocity axis via the Doppler effect, and the angular coordinates can be scaled via trigonometry to physical coordinates provided the distance to the mapped region is known. Thus, three dimensional radio astronomical mapping data are often referred to as position-position-velocity (PPV) data cubes.

The characteristics of the intensity distribution within these PPV data sets determines the most useful statistical approaches. In what follows, I broadly divide the statistical analyses of data cubes into regimes which have attracted different statistical techniques. Finally, this contribution emphasizes statistical analyses that are unique to or most interesting when applied to three dimensional datasets and studies which reduce to lower-dimensions studies are ignored. This review of the statistics of three-dimensional data sets focuses on the fields of astronomy where the statistical study is richest and best developed, namely the physics of the interstellar medium in our own Galaxy.

Through this contribution, I use five PPV data sets as examples. Three of the PPV data cubes come from observations made by the COMPLETE survey of star forming regions [26]. The observed data cubes are from the FCRAO 14-m telescope observations of ^{13}CO ($1 \rightarrow 0$) emission from nearby star-forming molecular clouds. I extracted three subcubes from the full data set for the star forming regions of NGC 1333, IC 348, and Ophiuchus A. These regions are all forming small clusters of stars in different stages of the formation. The clouds are nearby ($d = 120$ pc for Ophiuchus and $d = 260$ pc for Perseus) such that the $50''$ telescope resolution projects to 0.3 pc and 0.6 pc respectively. The velocity resolution of the observations is 0.066 km/s, comparable to the thermal line width expected for ^{13}CO in molecular clouds (0.05 km/s). The other two PPV data sets are mock observations of ^{13}CO ($1 \rightarrow 0$) emission generated from numerical simulations. The first data set is a subset of the simulation from Padoan et al. [20]. The simulation box has a scale of 6 pc and a mean density of $n = 10^3$ cm^3. The simulation is conducted using the Enzo code to simulate a $1{,}024^3$ box using MHD with an initially uniform density and periodic boundary conditions. The simulation is isothermal and turbulence is driven in Fourier space at large scales. The mean Mach number in the simulation is $\mathscr{M} = 6$ and the simulation neglects self-gravity. The second simulated data set is ^{13}CO ($1 \to 0$) data generated from the simulations of Offner et al. [18]. The simulation uses the Orion code, which employs adaptive mesh refinement to attain a spatial resolution of 0.008 pc in a 2 pc box. The physics includes hydrodynamics, including self-gravity, representing collapsed regions with sink particles, and is also isothermal. The simulated data have had their resolution degraded and noise added to mimic telescope effects. I am grateful to both Paolo Padoan and Stellar Offner for making these simulated data available for analysis.

35.2 Sparse Data Sets

Sparse PPV data sets are those for which most of the data volume contains no astrophysical information. The remaining emission is well separated in the data set, and commonly well described by point-like objects with some prior information about their shapes and distribution. In this case, a statistical analysis usually reduces to cataloging the sparse information and generating low-order descriptors of the emission distributions found in the data set. A typical example is searching for the 21-cm emission line from atomic hydrogen in deep, extragalactic surveys (e.g., [5, 6]). Because of the Hubble flow expansion of the Universe, the emission from galaxies is well separated in frequency (velocity) space and the galaxies are well separated from each other. The primary statistical problem is the identification of weak emission features of significance comparable to the instrumental noise. This search is complicated by the characteristics of the instrumental noise, which is typically well described by a normal distribution: $\mathscr{N}(\mu = 0, \sigma^2)$. The noise variance is typically unknown, but can be estimated from instrumentation principles and the data itself, and can change both spatially and spectrally. Several algorithms for identifying and cataloging sources have been forwarded from thresholding [34] to more complex search techniques. The search and cataloging process reduces the three dimensional data set to summary data of the objects observed where the same analyses as typify cataloging of other data can be applied (distribution functions, searching for correlation among properties, etc.).

35.3 Turbulence

A more challenging statistical problem lies in the analysis of dense data sets where astrophysically significant emission fills the data volume. An example of such a data set is shown in Fig. 35.1. This figure depicts a data set which is the product of multiple physical processes, a complex radiative transfer problem, and contamination by the imperfect instrumental response to radiation. Despite the

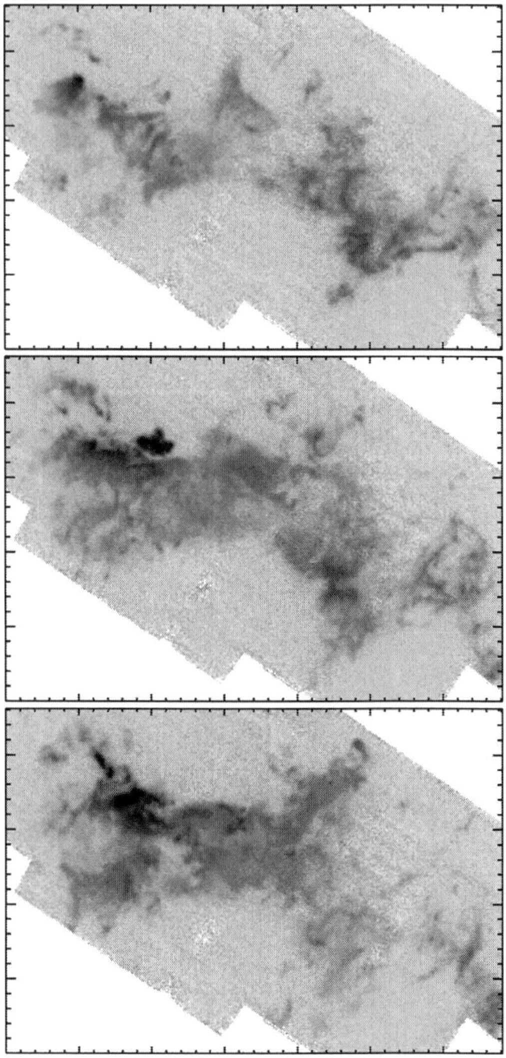

Fig. 35.1 Three velocity planes extracted from the PPV data set. The planes are extracted from the IC 348 data set of ^{13}CO $(1 \rightarrow 0)$ molecular line emission and are separated by 0.6 km/s in velocity. The figure illustrates the observational manifestation of a variety of physics processes. The emission is selected from around the IC 348 star forming region and the radiation from the region sculpts the matter and regulates the emissivity of the matter in the *upper left* region of each plane. In the *lower right*, farther from the star forming region, turbulent flows control the structure of the gas. The gas structure and physical conditions set the emission conditions and the emergent light that is measured here is a function of the emissivities integrated along the line of sight

complexities, it has been possible to investigate these data sets and discover the underlying astrophysical processes that generate such data. The clearest statistical link between the observational data sets and the underlying physics lies in the study of turbulent flows.

Statistical descriptions of fluid motion characterize a full six-dimensional phase space of the motions, but PPV observations only offer insight into half of these dimensions. Nonetheless, the velocity information that PPV data provide, coupled with assumptions regarding the degree and nature of anisotropies in the flow offer a path forward. Astrophysical fluid flows are typified by turbulent motions owing to extremely high Reynolds numbers ($Re > 10^8$). Owing to non-linear terms in the fluid equations, turbulent motions are not analytically tractable and have long been studied using their statistical properties beginning with Kolmogorov [12]. These approaches, first generated in the theoretical study of turbulence, have been applied to observational data in order to assess the underlying characteristics of the turbulent flow. A classic example is the velocity structure function as a function of vector separation **l**, as defined by Kolmogorov is

$$S_p(\mathbf{l}) \equiv \langle |v(\mathbf{r}) - v(\mathbf{r}+\mathbf{l})|^p \rangle \tag{35.1}$$

where the angle brackets indicate average over a data volume with position defined by **r**, v is a velocity component (usually the line-of-sight component). The power p is referred to as the order of the structure function. Under incompressible turbulent flows, Kolmogorov demonstrated $S_2(\mathbf{l}) \propto l^{2/3}$. Thus, observational estimates of the structure function can be made in order to determine whether fluid flows are consistent with Kolmogorov flows (e.g., [2,11,16,30], for molecular gas). In general, such studies have found that astrophysical turbulence is *not* well characterized by the comparatively simple assumptions that underpin the Kolmogorov model, showing structure functions that are significantly steeper. This discrepancy has driven the study of fluid turbulence into a wealth of other statistical methods in order to assess the properties of flows. Similarly, the theoretical study of turbulence has also turned from analytical work to numerical simulation.

Turbulence studies also admit a far simpler description than high order structure functions, namely probability distribution functions (PDFs) of the physical variables (e.g., velocity scales, density, magnetic field, etc.). The PDFs of these functions are also predicted by models, though often with substantial guidance from simulation work. For example, in isothermal turbulence, the density distribution is represented by a log-normal distribution [17], a result arrived at after substantial numerical study. Detailed examination of higher-order moments of emission PDFs (skewness, kurtosis, etc.) in both real and synthetic data show significant departures from the initially expected Gaussianity (e.g., [3]).

Despite the richness of possible statistical analyses, these studies are commonly frustrated because emission in a PPV data cube cannot be translated directly into a density in three dimensions of phase space. Relating emission to the underlying structure functions of turbulence with realistic radiative transfer requires significant theoretical effort to accomplish analytically [14, 15]. In this work, Lazarian and Pogosyan [14] argue that the Velocity Channel Analysis reliably recovers the structure function of the turbulent motions even in the face complications due to radiative transfer. This method constructs a power spectrum of intensity fluctuations for a set of two dimensional images, each of which is generated by integrating a PPV data cube over successively larger slices in velocity. The behavior of the power spectrum as a function of slice thickness can be related analytically to the underlying properties of turbulence. Parallel work by [8] uses principal component analyses to study the underlying structure functions of the turbulent flow using observational data.

Statistical analyses of PPV data cubes from turbulent gas has yielded substantial insight into the nature of turbulence in the ISM. The interaction has been bilateral: the study of turbulence suggests particular statistics to apply to data and the results of the statistical analyses of observations have driven our understanding of turbulence in the ISM.

35.4 Observationally Motivated Analyses

The tight relationship between the statistical study of turbulence and the analysis of PPV data sets has been fruitful for the study of turbulent motions. Unfortunately, many astrophysical observations cover regions where gravitation, bulk motions, radiation fields, and chemical effects complicate the tight relationship to theoretical study. Thus, the statistical study of these data sets cannot be directly related to a physical theory. Many of the tools used in the analysis of these complicated regions derive from the tools used to study turbulence, but these tools are often adapted to best represent the structure seen in real data. Lacking the direct connection to a physical theory, I classify these as "Observationally Motivated" diagnostics of PPV cube structure. Such descriptors of the data are primarily reductions of the PPV data to more representative numbers.

I note that many of the tools discussed are informed by and used with statistical descriptors developed for two dimensional (velocity-integrated) data. While beyond the scope of this contribution, these methods have been fruitful and illuminating since they share many theoretical problems with the general statistical description of images and the broader field of machine vision. Examples of such techniques include two dimensional structure-trees [9], fractal analysis [4], Δ-variance [32]. Compared to these two-dimensional studies, comparatively few methods analyze three-dimensional PPV data.

35.4.1 Correlation Functions in PPV

One approach to PPV data is to generate observational analogs to the theoretically motivated statistics. Designed specifically for PPV data, the Spectral Correlation Function (SCF, [23, 28]) was developed to quantify similarities between the shapes of spectra as a function of separation. The construction mimics a two-dimensional correlation function as used in turbulence studies. Given a data cube $I(\mathbf{r}, v)$, the SCF at a given scale and order p is

$$S_p(l) = \left\langle 1 - \sqrt{\frac{\sum_v |I(\mathbf{r},v) - I(\mathbf{r}+\mathbf{l},v)|^p}{\sum_v |I(\mathbf{r},v)|^p + \sum_v |I(\mathbf{r}+\mathbf{l},v)|^p}} \right\rangle . \quad (35.2)$$

The order p is typically taken as 2 and the average is taken over all positions in the data set. The SCF ranges from 0 for completely dissimilar spectra and 1, which is obtained for identical spectra. Empirically, the SCF was found to be noise robust with appropriate corrections and followed a power law relationship as a function of separation between trial spectra: $S_2(l) \propto l^{\alpha}$. In particular, α proved to be a useful discriminant between different regions and between simulations and observations. The scale at which the SCF departed from a power-law behavior in large-scale (100 pc) studies is conjectured to trace the scale at which turbulent flow transitions between regimes, such as from three-dimensions to two-dimensions [21]. An example application of the Spectral Correlation Function is presented in Sect. 35.5 below.

35.4.2 Object Recognition Algorithms and Catalog Statistics

Object recognition algorithms represent the dominant approach for exploring PPV data sets. In studies of star formation and the molecular ISM, these algorithms are commonly used since gravitation and excitation effects from young, high-mass stars create coherent structures that can be characterized as objects (see Fig. 35.1). The principal difficulty in object recognition studies of such data is that the nature of the objects under study is ill-defined. Unlike stars, or even galaxies, the molecular ISM shows similar structure on a large range of scales. At the largest scales ($>$10 pc), there appear to be *clouds* which are demarcated by the chemical transition from the atomic ISM to the molecular ISM and manifest as a change in the observational tracers of the gas. Even the definition of clouds is arbitrary as external conditions such as the radiation field, transient effects, and chemical enrichment can affect where the transition occurs. At the smallest scales ($\sim$0.1 pc), there well-defined gas structures in molecular clouds, called *cores*, which are thought to be the progenitors of stellar systems [1].

The search for meso-scale structure between these two regimes, often termed *clumps*, has driven the object recognition and cataloging as methods of studying molecular cloud structure. The Clumpfind [35] and Gaussclumps [33] algorithms serve as the primary means of identifying structures in a PPV data set (see Fig. 35.2). Gaussclumps iteratively fits and subtracts three-dimensional Gaussians from a data cube, using a modified χ^2 statistic to favor fits consistent with expectations of cloud structure. In particular, the fits favor objects that have less intensity than are in the data (so as not to over-subtract the emission) and to stay near the maximum of intensity in the data. Clumpfind is a watershed segmentation algorithm that identifies objects based off the structure of the data. The data are contoured with a set of user-defined levels. Beginning with the highest contour, objects are identified as connected regions of PPV space. These objects are extended to lower contour levels. When a volume element is contested between two or more objects, it is assigned to the 'closest' using a set of criteria defined by the algorithm. These methods produce catalogs of objects which can be studied as a population, and these populations can

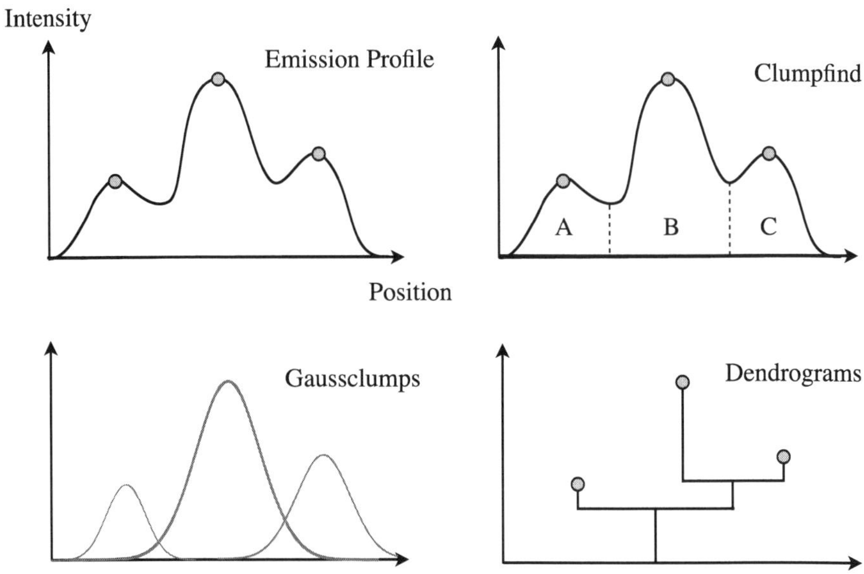

Fig. 35.2 Comparison of analysis algorithms for PPV data. The Clumpfind algorithm is a watershed algorithm that separates emission into regions based on the proximity to local maxima. The Gaussclumps algorithm uses a custom goodness-of-fit statistic to iteratively fit and subtract three-dimensional Gaussians to the data. Finally, the dendrogram algorithm is illustrated for the same emission profile. The dendrogram is constructed by drawing successively lower contour levels and identifying the thresholds at which distinct regions join into composite objects. The graph (*bottom right*) is a diagram of the representing on the intensity level of the contour (*y-axis*) and the associated branch of the image (*x-axis*)

be compared between different regions in an attempt to correlate their variations with observed variations in the star formation process seen in other wavebands.

These algorithms function well when applied to data sets with well-defined and separated objects, but this essentially reduces the problem to that of a sparse data set (Sect. 35.2). In the dense/blended data case (Fig. 35.1), the action of these well used algorithms becomes subject to their assumptions, and they may not reliably extract the physically relevant objects [24, 27]. This is due, in part, to astronomers not having a good understanding of what the physically relevant objects actually are.

There are substantial concerns for both of these methodologies, primarily concerning their lack of robustness. In this context, a *robust* structure identification algorithm identifies the same set of objects from a given true emission structure irrespective of observational considerations. If the same object is reobserved producing a different realization of noise, or even at a different resolution, the same objects should be extracted with the same physical properties. Designing such an algorithm requires prior knowledge of the population of objects being observed. A robust algorithm cannot be obtained since we lack that knowledge. For example, Gaussclumps relies on fitting a particular functional form to the data. Thus, deviation from the Gaussian profile makes the algorithm unstable. The algorithm fits are iterative: in each iteration, the brightest source is identified and a Gaussian clump is fit and subtracted. The residuals of bright, blended sources thus eventually become clumps in their own right. The optimization for the fit is over ten parameters (three centroid position coordinates, two angles for the principal axes of the Gaussian, three widths, an amplitude and a constant offset). The optimization is, not surprisingly, sensitive to initial conditions and the implementation of the minimization of the fit statistics, yielding different sets of objects with different minimization algorithms and prescriptions for generating initial parameter estimates. These concerns are mitigated by avoiding blended data sets and well-resolved objects where the telescope beam is small compared to the structure of the source.

The Clumpfind algorithm addresses many of the concerns about the Gaussclumps approach, since it assumes no specific functional form for the objects extracted. Instead, it relies on the shape of the data itself to govern how the results are partitioned (see Fig. 35.2). Clumpfind does not allow overlapping objects, so a data element cannot be assigned to multiple clouds. The algorithm's behavior is governed by two parameters: (1) the brightness steps at which new objects are identified or elements in the cube are assigned to identified objects and (2) the minimum brightness level to consider for assignment into objects. The Clumpfind algorithm suggests using 2σ where σ is the noise rms as the value for both parameters. The population of identified objects has its details regulated by the choice of the step size (parameter 1), which must be chosen to be small enough to track the shape of the data but large enough to reduce the effects of noise on the algorithm.

The Clumpfind algorithm also lacks robustness as defined above. In Fig. 35.3, I compare the results of the Clumpfind algorithm applied to the a set of ^{13}CO ($1 \rightarrow 0$) data imaging the COMPLETE survey. I processed these data using Clumpfind with the step size set to the recommended value (2σ), which labels every element in the data cube with an object label or as background. I repeated the analysis with the step size 25% larger and 1% larger, producing labeled data sets in each case. These labeled cubes were compared to the fiducial data set and the *correspondence* between individual objects is compared. The correspondences statistic for object i in the fiducial data set compared to a population of objects in a test data set (indexed with j)

$$C_i = \max_j \sqrt{\frac{(V_i \cap V_j)^2}{V_i V_j}}, \tag{35.3}$$

where the notation $V_i \cap V_j$ is the volume of the overlap between objects i and j measured relative to the volumes of the objects in their respective catalogs: V_i and V_j. The statistic thus takes values between 0 and 1 and quantifies the maximum correspondence between an object in the fiducial catalog across all the objects in different catalogs.[1] The results displayed in Fig. 35.3 show that changes in algorithm parameters do yield variations in the objects identified as would be expected.

[1] I am grateful for the suggestions of Chris Beaumont regarding the formulation of this statistic.

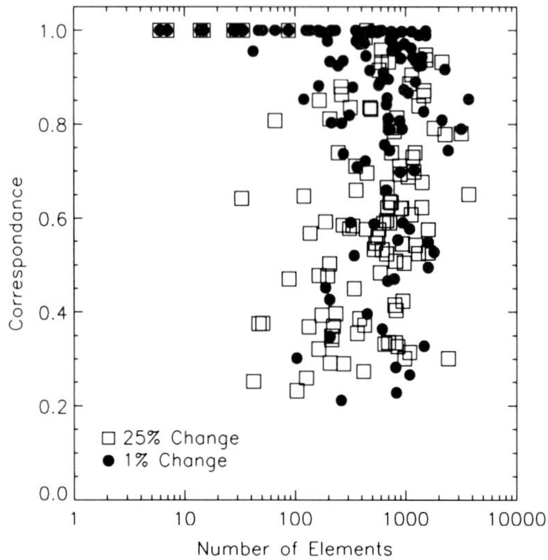

Fig. 35.3 Comparison of the Clumpfind algorithm objects identified for variations in the algorithm parameters. For each case (25% vs. 1%), the figure compares the correspondence measure (see text) between objects found for a fiducial population generated using the recommended algorithm parameters and the population generated for a 25% or 1% change in the algorithm parameters. The similarity is plotted vs. the number of data elements comprising the clump in the fiducial data set. Even for trivial changes in the algorithm parameters, very different sets of objects are extracted from the data

However, even trivial changes in the step size parameter (1%) produce substantial variations in many of the objects. The figure shows that the changes occur for large objects ($>10^2$ elements) which can have small correspondence to objects in other catalogs generated from the data. A 1% change produces correspondence statistics <0.75 for 30% of the elements used in the object catalog.

Despite the concerns about the reliability of the actual object lists produced by these approaches, there remains utility in viewing these algorithms as a reduction of the complexity of the data and making differential measurements between and within regions. The classic approach is to compare the intensity PDFs of the catalogs produced within regions [13, 33, 35]. The intensity distributions are usually characterized in terms of power-law (Pareto) or log-normal distributions. These distributions are examined because of their connection to other aspects of the star formation process. Power-law distributions are observed for the mass distributions of stars and a primary aim of the subfield is the connection of gas structures to the stellar population that is produced. Log-normal distributions of density and mass structures are produced by supersonic flows and this property has been linked to stellar mass distribution as well [22]. Even in this application, these algorithms must be used with care as the statistical results, such as the index of the power-law distribution of intensities for Clumpfind catalogs, is a function of algorithm parameters [24].

Given our limited understanding of the objects that actually should be cataloged and the interpretation of their physical properties, a less prescriptive approach appears merited. Viewed as a three-dimensional data volume, it becomes possible to characterize the data as a scalar valued function in $\mathbf{R}^3$ for which many paths of study from other disciplines present themselves. My collaborators and I have forwarded one method well suited for PPV analysis, which is currently under development. We have called the approach "dendrograms," reflecting the common visualization technique in statistics and biology for the output [7, 29]. The "dendrogram" technique is inspired by work on hierarchical structure in two-dimensional images where it was called "structure trees" [10]. In this case, *hierarchy* refers to the hierarchy of level-sets (contours) within an image. Because of continuity, the higher (brighter) level sets are necessarily contained within the lower (fainter) sets. The physically important information is how the numbers of distinct objects, as defined by the contour value change as a function of the contour. The original and most general description of the method is that of Reeb Graphs [25] within topology and Morse theory.

Figure 35.2 also illustrates the construction of the dendrogram. The dendrogram is generated by contouring the data set at successively lower intensity thresholds and tracking the objects defined by each level. The dendrogram tracks the contour level at which each pair of objects in the data set merges, graphically illustrating this by connecting the two branches in the dendrogram. The process is simplified since the structure of emission surfaces in three-dimensional PPV data is nearly always purely hierarchical: lower intensity emission is only rarely contained entirely within the contours of a higher intensity region and regions are defined to be simply connected. The representation of the emission trees as a dendrogram implies the number of objects never increasing with decreasing intensity threshold. Relaxing this criterion means only means that a proper *dendrogram* cannot represent the emission surface structures and the more general Reeb graph is needed. The dendrogram process is computationally simplified by pre-identifying the highest intensity *leaves* in the data set and suppressing those leaves which are likely to be generated by noise. This suppression is effected by requiring the leaves included in the analysis to be a minimum distance in position and velocity away from other leaves, usually set to two resolution elements. Furthermore, for a rms value of the noise of σ, a candidate leaf with brightness I_0 if removed from consideration if there exists another candidate with brightness $I_i > I_0$ and the highest value of the contour containing both leaves has a brightness levels $I_{node} > I_0 - n\sigma$. This latter condition ensures that the candidate leaves included in the dendrogram are significant with respect to the noise fluctuations in the data. The nature of the dendrogram algorithm makes it more robust than the Clumpfind or Gaussclumps algorithms since the tree structure for a given data set is entirely determined from the data alone. The noise suppression parameters *do* prune the tree shape, but all possible choices are just pruned versions of a single true tree. The full tree is usually not characterized due to computational time restrictions. The algorithm remains sensitive to various realizations of observational noise which can change the topology of the emission surfaces and affect tree structure. We present a dendrogram analysis of several data sets in Sect. 35.5 below.

35.5 Simulation vs. Observational Domain

The goal of understanding the physics of star formation from PPV observations is easily frustrated by the complexities of physics that shape the observed line emission. The studies of other sorts of PPV data, such as H I 21-cm observations or stellar spectral data cubes, usually observe emission that is (effectively) optically thin and the observed spectral line data can directly interpreted as due to the motions of the gas. Even so, the relationship between PPP and PPV space is complex and structures in PPV cannot be directly inverted to characterize structures in PPP [19, 31]. For star forming gas, the common molecular line tracers used are usually optically thick, the molecules themselves are subject to depletion and time varying chemical effects, and the emission properties depend sensitively on excitation conditions. Some statistical approaches even include opacity effects [15], but none account for the diverse chemical and thermal conditions that shape the molecular emission. The characterization of the emitting objects is further confused by telescope effects. For example, telescopes have finite resolution in all dimensions and can impose strong spatial filtering on a true emission distribution. The instrumental noise that is imposed upon observational data can dominate the signal and carry with it significant correlations or spatially varying statistical properties. Viewed in terms of a domain mapping, the actual physical objects in the Universe map into an "observational" domain through radiative transfer and telescope effects (Fig. 35.4). Inverting this mapping is nearly impossible except for the simplest of structures.

Since the physical processes in the star forming ISM are multifarious, the primary means we have of understanding their simultaneous action on a system is through numerical simulation. These simulations provide a means of including all of the relevant physics and mimicking structures in the physical domain. Since mapping observations into the physical domain is essentially impossible, the only means

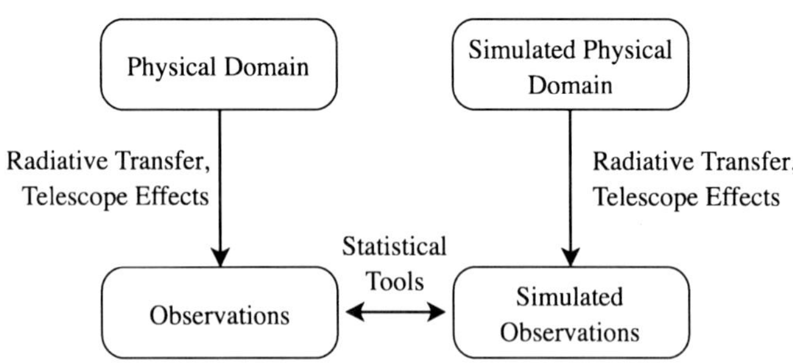

Fig. 35.4 Representing the comparison of observations and simulated observations via domain mapping. The translations from the physical objects into observational data involve radiative transfer and telescope effects. Inverting this mapping is complicated, and for dense data sets, not unique. Thus, a meaningful comparison of simulations and physical objects can only be made in the observational domains using statistical tools

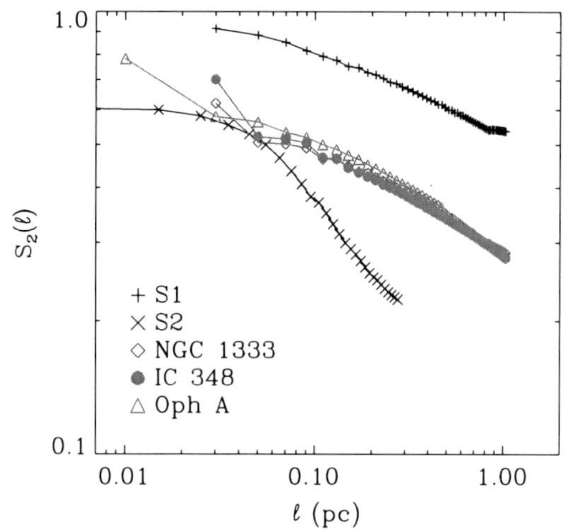

Fig. 35.5 The average Spectral Correlation Function as a function of separation l as measured for simulated observations (*black*) and observational data (*gray*). All curves present ^{13}CO (1 → 0) data and have been corrected for the presence of noise in the data. The simulated ^{13}CO data are from the simulations of Padoan (S1) and Offner (S2). The observational data are from the COMPLETE survey of star forming regions. All observational data show similar scaling with separation whereas the simulations have very different properties

of evaluating simulations is to apply radiative transfer and telescope effects to these data. Thus, the data can be mapped into a simulated observational domain, and the simulated observations can be compared to actual observations using the statistical methods described in Sect. 35.4. The similarity of simulations can be used to evaluate which methodologies and sets of physical parameters produce the data most similar to actual observations. These best cases are suggestive of the physical parameters that actually lead to the observed data.

By means of an example, Fig. 35.5 shows the results of the spectral correlation function applied to the five sample data cubes described above. The SCF is generated by averaging the correlation function in the data for all the positions in the data set following (35.2). The correlation function has been corrected for the decorrelating effects of noise following [23]. Despite different physical scales, the SCF analysis of the observational data show a remarkable similarity between their structures, with all data showing a roughly power-law scaling: $S_2(l) \propto l^{-0.25}$. In contrast, the Offner simulation (S2) shows $S_2(l) \propto l^{-0.6}$ which is markedly different from the observations whereas the Padoan simulation (S1) shows $S_2(l) \propto l^{-0.2}$, similar to the observational data. However, at any given physical scale (e.g., 0.1 pc), the SCF values for the simulations are substantially different from the observations. The SCF analysis shows the utility in comparing observations with the simulated data. The similarity in slopes between the Padoan simulation and the observations suggests a good representation of the correlations present in observational turbulence, but the difference in scales suggests that additional

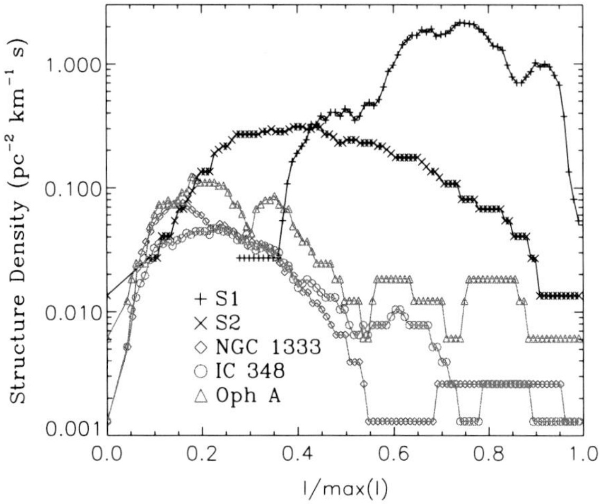

Fig. 35.6 Density of unique structures in the data volume as a function of contour level. The number count at each level is normalized by the volume of the PPV data cube. All observational data show similar behavior, which is different from the two simulations

physical effects may be reducing correlation amplitude. The steep scaling seen for the Offner simulation may imply that more correlation effects through stronger turbulent flows may mimic these observational data better.

I also completed a dendrogram analysis of the five different data cubes. The dendrograms do not provide a statistical comparison per se, but are rather a reduction of the data set, emphasizing the topology of the emission. Subsequent statistical analyses can yield a differential comparison between data sets. Since each point on the dendrogram corresponds to a specific isosurface in the data set, the shape of each of those isosurface can be related to physical quantities such as size, line width and luminosity. Such an analysis of data appears in [29]. In Fig. 35.6, I present a different analysis, emphasizing the topology of the data. For each contour level in the data, the number of unique objects at that level is graphed vs. the contour level, a measured called the *genus*. The number of objects is normalized by the data volume in each simulation, measured in terms of the dimensions of the PPV cube: pc^2 kms^{-1}. Only the behavior of the tree structure containing the brightest emission in each data cube is considered; and smaller, isolated objects are removed. The genus statistic shows a similar behavior for all observational data sets with the density of structures peaking at 20% of the maximum intensity in each data set at a similar density. The Offner simulation (S2) also peaks at a similar brightness, but with a much higher density of structures per data volume. The Padoan simulation (S1) shows a far larger density of structures near the peak brightness, without the contrast in brightness seen in the observations or the other simulation. The inclusion of self-gravity in S2 may explain the increased hierarchical structure seen and hence its similarity to observations.

There are many other possible avenues for using the dendrograms as a statistical tool, including measuring branching ratios between the emission properties of contour levels or number of parent/progeny structures as a function of brightness. In exploring these and other statistics, the largest utility comes from finding a measure such as the SCF or dendrogram-based Genus measure that yield similar results for disparate observed regions. These suggest that the statistics are tracing an intrinsic property of star-forming molecular gas. In such a case, differential measurements made with respect to simulations can offer insights as to which physics must be included for a faithful representation of the observed ISM. By ensuring that the comparisons happen on the common ground of the observational domain, a critical assessment of the faithfulness of simulations is achieved.

References

1. di Francesco, J., Evans, II, N.J., Caselli, P., Myers, P.C., Shirley, Y., Aikawa, Y., Tafalla, M.: An Observational Perspective of Low-Mass Dense Cores I: Internal Physical and Chemical Properties. In: Reipurth, B., Jewitt, D., Keil, K. (eds.) Protostars and Planets V. pp. 17–32 (2007)
2. Dickman, R.L., Kleiner, S.C.: Largescale Structure of the Taurus Molecular Complex - Part Three - Methods for Turbulence. *Astrophys. J.* 295, 479–+ (Aug 1985)
3. Falgarone, E., Lis, D.C., Phillips, T.G., Pouquet, A., Porter, D.H., Woodward, P.R.: Synthesized spectra of turbulent clouds. *Astrophys. J.* 436, 728–740 (Dec 1994)
4. Falgarone, E., Phillips, T.G., Walker, C.K.: The edges of molecular clouds - Fractal boundaries and density structure. *Astrophys. J.* 378, 186–201 (Sep 1991)
5. Ford, H.A., McClure-Griffiths, N.M., Lockman, F.J., Bailin, J., Calabretta, M.R., Kalberla, P.M.W., Murphy, T., Pisano, D.J.: H I Clouds in the Lower Halo. I. The Galactic All-Sky Survey Pilot Region. *Astrophys. J.* 688, 290–305 (Nov 2008)
6. Giovanelli, R., Haynes, M.P., Kent, B.R., Perillat, P., Saintonge, A., Brosch, N., Catinella, B., Hoffman, G.L., Stierwalt, S., Spekkens, K., Lerner, M.S., Masters, K.L., Momjian, E., Rosenberg, J.L., Springob, C.M., Boselli, A., Charmandaris, V., Darling, J.K., Davies, J., Garcia Lambas, D., Gavazzi, G., Giovanardi, C., Hardy, E., Hunt, L.K., Iovino, A., Karachentsev, I.D., Karachentseva, V.E., Koopmann, R.A., Marinoni, C., Minchin, R., Muller, E., Putman, M., Pantoja, C., Salzer, J.J., Scodeggio, M., Skillman, E., Solanes, J.M., Valotto, C., van Driel, W., van Zee, L.: The Arecibo Legacy Fast ALFA Survey. I. Science Goals, Survey Design, and Strategy. *Astron. J.* 130, 2598–2612 (Dec 2005)
7. Goodman, A.A., Rosolowsky, E.W., Borkin, M.A., Foster, J.B., Halle, M., Kauffmann, J., Pineda, J.E.: A role for self-gravity at multiple length scales in the process of star formation. *Nature* 457, 63–66 (Jan 2009)
8. Heyer, M.H., Schloerb, F.P.: Application of Principal Component Analysis to Large-Scale Spectral Line Imaging Studies of the Interstellar Medium. *Astrophys. J.* 475, 173–+ (Jan 1997)
9. Houlahan, P., Scalo, J.: Recognition and characterization of hierarchical interstellar structure. I - Correlation function. *Astrophys. J. Suppl.* 72, 133–152 (Jan 1990)
10. Houlahan, P., Scalo, J.: Recognition and characterization of hierarchical interstellar structure. II - Structure tree statistics. *Astrophys. J.* 393, 172–187 (Jul 1992)
11. Kleiner, S.C., Dickman, R.L.: Large-scale structure of the Taurus molecular complex. II - Analysis of velocity fluctuations and turbulence. III - Methods for turbulence. *Astrophys. J.* 295, 466–484 (Aug 1985)
12. Kolmogorov, A.: The Local Structure of Turbulence in Incompressible Viscous Fluid for Very Large Reynolds' Numbers. Akademiia Nauk SSSR Doklady 30, 301–305 (1941)
13. Kramer, C., Stutzki, J., Rohrig, R., Corneliussen, U.: Clump mass spectra of molecular clouds. *Astron. & Astrophys.* 329, 249–264 (Jan 1998)
14. Lazarian, A., Pogosyan, D.: Velocity Modification of H I Power Spectrum. *Astrophys. J.* 537, 720–748 (Jul 2000)
15. Lazarian, A., Pogosyan, D.: Velocity Modification of the Power Spectrum from an Absorbing Medium. *Astrophys. J.* 616, 943–965 (Dec 2004)
16. Miesch, M.S., Bally, J.: Statistical analysis of turbulence in molecular clouds. *Astrophys. J.* 429, 645–671 (Jul 1994)

17. Nordlund, Å.K., Padoan, P.: The Density PDFs of Supersonic Random Flows. In: J. Franco & A. Carraminana (ed.) Interstellar Turbulence. pp. 218–+ (1999)
18. Offner, S.S.R., Klein, R.I., McKee, C.F.: Driven and Decaying Turbulence Simulations of Low-Mass Star Formation: From Clumps to Cores to Protostars. *Astrophys. J.* 686, 1174–1194 (Oct 2008)
19. Ostriker, E.C., Gammie, C.F., Stone, J.M.: Kinetic and Structural Evolution of Self-gravitating, Magnetized Clouds: 2.5-dimensional Simulations of Decaying Turbulence. *Astrophys. J.* 513, 259–274 (Mar 1999)
20. Padoan, P., Juvela, M., Kritsuk, A., Norman, M.L.: The Power Spectrum of Supersonic Turbulence in Perseus. *Astrophys. J.*l 653, L125–L128 (Dec 2006)
21. Padoan, P., Kim, S., Goodman, A., Staveley-Smith, L.: A New Method to Measure and Map the Gas Scale Height of Disk Galaxies. *Astrophys. J.*l 555, L33–L36 (Jul 2001)
22. Padoan, P., Nordlund, Å.: The Stellar Initial Mass Function from Turbulent Fragmentation. *Astrophys. J.* 576, 870–879 (Sep 2002)
23. Padoan, P., Rosolowsky, E.W., Goodman, A.A.: The Effects of Noise and Sampling on the Spectral Correlation Function. *Astrophys. J.* 547, 862–871 (Feb 2001)
24. Pineda, J.E., Goodman, A.A., Arce, H.G., Caselli, P., Foster, J.B., Myers, P.C., Rosolowsky, E.W.: Direct Observation of a Sharp Transition to Coherence in Dense Cores. *Astrophys. J.*l 712, L116–L121 (Mar 2010)
25. Reeb, G.: Sur les points singuliers d'une forme de pfaff completement intergrable ou d'une fonction numerique. Comptes Rendus Acad. Science Paris 222, 847–849 (1946)
26. Ridge, N.A., Di Francesco, J., Kirk, H., Li, D., Goodman, A.A., Alves, J.F., Arce, H.G., Borkin, M.A., Caselli, P., Foster, J.B., Heyer, M.H., Johnstone, D., Kosslyn, D.A., Lombardi, M., Pineda, J.E., Schnee, S.L., Tafalla, M.: The COMPLETE Survey of Star-Forming Regions: Phase I Data. *Astron. J.* 131, 2921–2933 (Jun 2006)
27. Rosolowsky, E., Leroy, A.: Bias-free Measurement of Giant Molecular Cloud Properties. *Pub. Astro. Soc. Pacific*, 118, 590–610 (Apr 2006)
28. Rosolowsky, E.W., Goodman, A.A., Wilner, D.J., Williams, J.P.: The Spectral Correlation Function: A New Tool for Analyzing Spectral Line Maps. *Astrophys. J.* 524, 887–894 (Oct 1999)
29. Rosolowsky, E.W., Pineda, J.E., Kauffmann, J., Goodman, A.A.: Structural Analysis of Molecular Clouds: Dendrograms. *Astrophys. J.* 679, 1338–1351 (Jun 2008)
30. Scalo, J.M.: Turbulent velocity structure in interstellar clouds. *Astrophys. J.* 277, 556–561 (Feb 1984)
31. Shetty, R., Collins, D.C., Kauffmann, J., Goodman, A.A., Rosolowsky, E.W., Norman, M.L.: The Effect of Projection on Derived Mass-Size and Linewidth-Size Relationships. *Astrophys. J.* 712, 1049–1056 (Apr 2010)
32. Stutzki, J., Bensch, F., Heithausen, A., Ossenkopf, V., Zielinsky, M.: On the fractal structure of molecular clouds. *Astron. & Astrophys.* 336, 697–720 (Aug 1998)
33. Stutzki, J., Güsten, R.: High spatial resolution isotopic CO and CS observations of M17 SW - The clumpy structure of the molecular cloud core. *Astrophys. J.* 356, 513–533 (Jun 1990)
34. Whiting, M.T.: Astronomers! Do You Know Where Your Galaxies are?, Astrophysics and Space Science Proceedings, vol. 2, pp. 343–344. Springer-Verlag (2008)
35. Williams, J.P., de Geus, E.J., Blitz, L.: Determining structure in molecular clouds. *Astrophys. J.* 428, 693–712 (Jun 1994)

Chapter 36
Astronomical Transient Detection Controlling the False Discovery Rate

Nicolle Clements, Sanat K. Sarkar, and Wenge Guo

Abstract Identifying source objects in astronomical observations, in particular with reliable algorithms, is extremely important in large-area surveys. It is of great importance for any source detection algorithm to limit the number of false detections since follow up investigations are timely and costly. In this paper, we consider two new statistical procedures to control the false discovery rate (FDR) for group-dependent data—the two-stage BH method and adaptive two-stage BH method. Motivated by the belief that the spatial dependencies among the hypotheses occur more locally than globally, these procedures test hypotheses in groups that incorporate the local, unknown dependencies. If a group is found significant, further investigation is done to the individual hypotheses within that group. Importantly, these methodologies make no dependence assumption for hypotheses *within* each group. The properties of the two procedures are examined through simulation studies as well as astronomical source detection data.

36.1 Introduction

Detecting, classifying, and monitoring transient sources in the night sky, specifically Type Ia supernovae transients, is an area of astronomical research that receives much attention. Astronomical images represent the intensity of light, or roughly a count

---

N. Clements • S.K. Sarkar
Department of Statistics, Fox School of Business and Management, Temple University, 1810 North 13th Street, Philadelphia, PA 19122
e-mail: tuc37728@temple.edu; sanat@temple.edu

W. Guo
Department of Mathematical Sciences, New Jersey Institute of Technology, University Heights, Newark, NJ 07102
e-mail: wenge.guo@njit.edu

E.D. Feigelson and G.J. Babu (eds.), *Statistical Challenges in Modern Astronomy V*, Lecture Notes in Statistics 209, DOI 10.1007/978-1-4614-3520-4_36,

of the photons at every pixel. However, the number of pixels in each image can be several millions in size, which makes manual source detection impossible.

The term *source pixel* is commonly referred to as a pixel in an image that is above some threshold and thus is part of a true source (transient object). A *source* is a collection of these source pixels that correspond to an astronomical object of interest. The term *background pixel* is an image pixel that does not come from a source. A source, like a supernova transient, is a stellar explosion in the sky that can last for several weeks before fading away. If the host galaxy is reasonably close, then the supernova becomes quite bright. While there is no difficulty in detecting it at peak brightness, the scientific goal is to pick it up as it has just begun to rise and is still very faint. Also, there are many more distant galaxies than bright galaxies, so there are numerous supernovae that will just barely be seen even at peak brightness.

Typically, the data each night are assumed to come from a mixture Gaussian distribution, based on source and background pixels. One issue is that the mean and variance of this Gaussian distribution differs from night to night, due to varying observing conditions, such as cloud coverage and moonlight. The background pixels from the ith night are assumed to be normally distributed with mean μ_i and variance σ_i^2. The source pixels from the ith night and the jth source are normally distributed with mean $\mu_i + \theta_j$, where θ_j can be very small. To detect these sources, we want to test the hypothesis $H_0 : \theta_j = 0$ vs. the alternative $H_1 : \theta_j > 0$. To get around the nightly differences, astronomers standardize the data, also known as computing the signal-to-noise ratio (SNR). One can search for transient sources that exceed some SNR threshold using the standardized data converted to p-values.

It is of great importance for any source detection algorithm to limit the number of false detections. This is because following up new detections is timely and costly. Astronomers want to spend as little of their time as possible viewing what turn out to be vacant regions of sky. Currently, there are several publicly available algorithms for source detection based on sliding cells, Voronoi tessellation, wavelets, and signal-to-noise filtering. Although these algorithms provide some limit to the number of false detections, they cannot provide proof or an upper bound to the number they falsely detect. To give astronomers a source detection procedure that controls a statistically meaningful measure incorporating Type I errors, i.e., false detections, would be a great asset.

36.2 Preliminaries and Background

The False Discovery Rate (FDR) proposed by Benjamini and Hochberg [2], is the expected proportion of Type I errors among all the rejected null hypotheses. It is now a widely accepted notion of error rate to control in large-scale multiple testings arising in modern scientific investigations, including astronomical source detection. Suppose there are N pixels, with P_j, $j = 1,\ldots,N$, being the p-values generated from the observations in those pixels. Then the Benjamini–Hochberg (BH) method controlling the FDR at a level α operates as follows:

The BH Method.

- Order the p-values from the smallest to the largest: $P_{(1)}, \ldots, P_{(N)}$.
- Find $k_{BH} = \max\{j : P_{(j)} \leq j\alpha/N\}$.
- Reject the null hypotheses whose p-values are less than or equal to $P_{(k_{\mathrm{BH}})}$.

The BH method controls the FDR at the desired level α, albeit conservatively, unless there is no real source pixel, only when the p-values are independent or positively dependent (in a certain sense). More specifically, the FDR of the BH method equals $\pi_0\alpha$ when the p-values are independent, and is less than $\pi_0\alpha$ when the p-values are positively dependent [5, 16], where π_0 is the (true) proportion of background pixels. The difference between $\pi_0\alpha$ and the FDR gets larger and larger with increasing (positive) dependence among the p-values.

In absence of knowledge of any specific type of dependence structure among the p-values, the method due to Benjamini and Yekutieli [5], the BY method, is often used. The BY method is an adjusted BH method with α replaced by α/C_N, where $C_N = \sum_{j=1}^{N} j^{-1}$. The BY method is extremely conservative, particularly when N is large, thus is not as powerful as one would hope in detecting true source pixels.

The idea of improving the BH method has been one of the main motivations behind much of the methodological developments taken place in modern multiple testing. This idea has flourished in a number of different directions; for instance, in (a) developing adaptive BH methods incorporating information about π_0 from the data into the BH method or taking an estimation based approach to controlling the FDR [3, 4, 6, 10, 20, 23, 24]; (b) incorporating information about correlations or utilizing the dependence structure into the BH method [7, 14, 25, 26]; and (c) generalizing the notion of FDR to k-FDR by relaxing control over at most $k-1$ false rejections [17–19].

In the context of present astronomical applications, Hopkins et al. [13] suggested a way of improving the BY method by incorporating local dependencies. They argue that astronomical images show some degree of correlation between pixels, but are not *fully* correlated. In other words, the brightness intensity of a given pixel is not influenced by all other $N-1$ pixels, rather it is only *partially* correlated with a smaller number (n) of pixels neighboring it. Any real transient signal should have the spatial shape of the stars covering some adjacent pixels, which is called the telescope 'point spread function' (PSF), and this n is related to the number of pixels representing the PSF. They propose to use the BY method with C_N replaced by $C_n = \sum_{i=1}^{n} i^{-1}$ to account for the local dependencies around the source pixels. This is clearly more powerful than the original BY method, but it can be shown that such adjustment to the BY method may fail to control the FDR when $\pi_0 \approx 1$.

Also in astronomical context, Friedenberg and Genovese [9] considered detecting clusters of pixels, rather than individual pixels, and chose the probability of False Cluster Proportion (FCP) exceeding a certain value as the error rate to control. By relaxing the error rate control to clusters, rather than individuals, there is potential for more powerful procedures due to the reduction in data dimension. However, procedures with cluster-wise control may have some disadvantages compared to individual-wise control, as noted below.

Given the massive influx of data due to large-area surveys, it is crucial to be able to accurately *identify* and *classify* transient sources in real-time data collection. To do so, automated methods must strive to use all the data's available information to first identify and then classify objects (Savage and Oliver 2007). This means using not only clusters of outlying observations as the in the FCP, but also using individual pixels to systematically classify astronomical objects as either point-like (i.e. stars, quasars, supernova, etc.) or extended (i.e. galaxies, nebula, etc.). Currently, many classification methods generate a set of 'features' to determine the type of object discovered. Many of these features are estimated with pixel-wise information, such as source positions, fluxes in a range of apertures, and shapes using radial moments. Another nontrivial problem is deblending or splitting of adjacent sources, typically defined as a number of distinct, adjacent intensity peaks connected above the detection surface brightness threshold (Salzberg et al. 2005; [1]; Henrion et al. 2011). Deblending of nearby objects is nearly impossible with a cluster-wise approach. Because of these classification advantages after identifying new sources, we propose new methodology based on the idea of controlling the rate of false discoveries of individual pixels.

36.3 Proposed Methods

In this paper, we consider using a different idea of incorporating local dependencies and propose an alternative to Hopkins and the BY methods. Our idea is based on the arguments that if the dependencies among the pixels do occur more locally than globally, then by grouping the pixels using an appropriate group size we can make these groups independent of each other. This would be the best scenario where we can apply the BH (more powerful than the BY) method to detect the so called 'potential source groups', which we refer to as the groups containing at least one source pixel. Once a 'potential source group' is identified, we can go back to that group to detect which of the group's individual pixels belong to the source. Based on this general idea of pixel grouping, we propose the following two procedures, by choosing the group size, as in Hopkins et al. [13], related to the PSF of the telescope. In particular, paralleling Hopkins et al.'s choice of n, the number of pixels representing the PSF, we chose our group size S to be this same quantity. Using this argument, the groups containing S 'partially correlated' pixels should behave independently.

Procedure 36.1.

Step 1. Divide the data rectangle into D by D mutually exclusive groups. The group size is $S = D^2$ and the total number of groups is $N/S = G$ (say), with N being the total number of pixels (hypotheses).

Step 2. Find the minimum p-value in each of these G groups. Let $P_{\min}^{(g)}$ be that minimum for the gth group, $g = 1, \ldots, G$. Find $Q_g = SP_{\min}^{(g)}$, for $g = 1, \ldots, G$, which we will call the grouped p-values.

Step 3. Apply the BH method to these grouped p-values to detect the 'potential source groups'. That is, consider the (increasingly) ordered versions of the grouped p-values, $Q_{(1)}, \ldots, Q_{(G)}$, and identify those groups as being potential source groups for which the grouped p-values are less than or equal to $Q_{(k_{\mathrm{BH}}^*)}$, where $k_{\mathrm{BH}}^* = \max\{g : Q_{(g)} \leq g\alpha/G\}$.

Step 4. Identify the jth individual pixel within the gth potential source group as being a source pixel if the corresponding p-value, say P_{gj}, is such that $SP_{gj} \leq k_{\mathrm{BH}}^* \alpha/G$.

Theorem 36.1. *Procedure 36.1 controls the FDR at α if the groups are independent or positively dependent in a certain sense.*

A proof of Theorem 36.1 is provided in the Appendix. Our next procedure is based on the following idea, in addition to that of pixel grouping.

When adapting a multiple testing method to the number of true null hypotheses, say N_0, whether it is for controlling the FDR using the BH method or for controlling the familywise error rate (FWER) using the Bonferroni method (e.g., [8, 11] and Sarkar and Guo 2010), the p-values are modified from P_j to $\tilde{P}_j = \hat{N}_0 P_j$, based on a suitable estimate $\hat{N}_0$ of N_0. One of these estimates is due to Storey et al. [24]:

$$\hat{N}_0 = \frac{W_N(\lambda) + 1}{1 - \lambda}, \tag{36.1}$$

where λ is a tuning parameter and $W_N = \sum_{j=1}^{N} I(P_j > \lambda)$ is the number of p-values exceeding λ and provides an information about the number of true null hypotheses in the data. For instance, in case of the Bonferroni method that rejects H_j if $NP_j \leq \alpha$, its adaptive version would reject the H_j if $\hat{N}_0 P_j \leq \alpha$. This would be potentially more powerful.

Notice that such an adaptive p-value is like a 'shrunken p-value', which gets shrunk towards a smaller value, and thus becomes more significant, if there is evidence of more signals in the data. So, when the p-values are locally dependent and tend to have similar local behaviors in terms of being either significant or non-significant, by doing similar adaptation separately within each group by estimating the number of true group specific signals, one could utilize the dependence within each group and potentially improve Procedure 36.1. With that in mind, we propose our second procedure as follows:

Procedure 36.2.

Step 1. Same as in Procedure 36.1.

Step 2. Find the minimum of the p-values in each of these G groups. Let P_{gj}, $j = 1, \ldots, S$, be the p-values in the gth group, and $P_{\min}^{(g)}$ be the minimum of these

p-values, $g = 1,\ldots,G$. Find $\tilde{Q}_g = \hat{S}_g P_{\min}^{(g)}$, for $g = 1,\ldots,G$, where

$$\hat{S}_g = \min\left\{\frac{\sum_{j=1}^{S} I(P_{gj} > \lambda) + 1}{1-\lambda}, S\right\}, \quad (36.2)$$

which we will call the grouped adaptive p-values.

Step 3. Apply the BH method to these grouped adaptive p-values to detect the 'potential source groups'. That is, consider the (increasingly) ordered versions of the grouped adaptive p-values, $\tilde{Q}_{(1)},\ldots,\tilde{Q}_{(G)}$, and identify those groups as being potential source groups for which the grouped adaptive p-values are less than or equal to $\tilde{Q}_{(\tilde{k}^*_{\text{BH}})}$, where $\tilde{k}^*_{\text{BH}} = \max\{g : \tilde{Q}_{(g)} \leq g\alpha/G\}$.

Step 4. Identify the jth pixel within the gth potential source group as being a source pixel if the corresponding p-value P_{gj} is such that $\hat{S}_g P_{gj} \leq \tilde{k}^*_{\text{BH}}\alpha/G$.

Another adaptive method could also be considered by estimating the number of groups that do not contain any source signal, say G_0, and using the estimate $\hat{G}_0$ in place of G in Procedure 36.1, Steps 3 and 4. However, because of the sparse number of signals in astronomical data, the estimate $\hat{G}_0$ is often just as large or larger than G itself, providing no additional advantage over Procedure 36.1. This type of adaptive group estimation is better suited in data where π_0 is not so close to 1.

36.4 Simulation Study

We ran several simulation studies to examine the FDR control property and the power of our proposed procedures compared to existing methodology. One of the main advantages of the proposed procedures is that there is no dependence assumption of the p-values within each group. Thus, it is only fair to compare our procedures with existing methodology that has such relaxed assumptions (namely, BY and Hopkins).

Since the proposed procedures were developed to control the FDR under arbitrary dependence assumptions within each group, the simulation studies were done under two different dependent scenarios. In the first scenario, each group's p-values are generated from a multivariate normal distribution with common correlation $(0 < \rho < 1)$. In the second scenario, the p-values were also generated from a multivariate normal distribution, but with an autoregressive type of correlation structure within each group, separately for each of the G groups. An autoregressive correlation structure indicates that data collected in a close spatial proximity tend to be more highly correlated than observations taken further apart. For example, let X_{ij} denote an observation in a particular group located in the ith row and jth column. Then, the correlation between two observations in that particular group can be written as $Corr(x_{ij}, x_{i'j'}) = \rho^{\max(|i-i'|,|j-j'|)}$, for any $0 \leq \rho \leq 1$. In other words, the correlation between two observations decreases in value as the absolute spatial distance between (i,i') or (j,j') increases.

Under these two correlation structures, we generated S dependent standard normal random variables independently for each of the G groups. Three of the G groups were chosen randomly for each simulation and one of the values 2, 3 and 4 is added to the variables in each of these three groups. In other words, only three groups were assumed to contain all the signals. Simulation studies with varying number of signal groups (one group to ten groups, instead of three groups) were also computed, but since they yielded similar results, we have decided to restrict the discussion of our simulation studies to three signal groups. The group size S was chosen to be 25, using 5×5 groups ($D = 5$). The number of groups is $G = 900$, totaling $n = 22,500$ individual hypotheses per simulation. Since each simulation contained a fixed three groups of signal each of size 25, the proportion of true null hypotheses $\pi_0 = 1 - \frac{75}{22,500} = 0.996$. Using both correlation structures, we repeated this 1,000 times at each value of ρ.

Four methods were compared—BY, Hopkins, the proposed Two-Stage, and the proposed Adaptive Two-Stage (with $\lambda = 0.5$) procedures. At each simulation, we estimate FDR by the proportion of falsely rejected hypotheses and the power is estimated by the proportion of correctly rejected hypotheses. The simulated FDR and power obtained by averaging these proportions of falsely and correctly rejected hypotheses over all repetitions are shown in Fig. 36.1 for the fixed group correlation and in Fig. 36.2 for the autoregressive case.

When examining the simulated power in the right side of Fig. 36.1, both the Two-Stage and Adaptive Two-Stage Procedures outperform the BY method with the fixed group correlation structure. In other words, these proposed two-stage procedures correctly identify a higher proportion of signals. The Adaptive Two-Stage Procedure has competitive power with Hopkins' procedure and surpasses it when the within group fixed correlation becomes large ($\rho > 0.5$).

The simulated FDR in the left side of Fig. 36.1, reveals a stable Two-Stage Procedure, with the estimated FDR < 0.05 across all fixed group correlations. However, the Adaptive Two-Stage Procedure seems to lose control of the FDR with moderately correlated data within groups ($0.5 < \rho < 0.8$). Although unfortunate, this result is not surprising. Other adaptive methodology also become unstable with large correlation among hypotheses.

Next, we look at the performance of the proposed procedures under the autoregressive within group correlation structure. When examining the simulated power in the right side of Fig. 36.2, both Two-Stage Procedure and Adaptive Two-Stage Procedure outperform the BY method under this group correlation structure. The simulated FDR in the left side of Fig. 36.2, reveals stable Two-Stage Procedure as well as the Adaptive Two-Stage Procedure, with the estimated FDR < 0.05 across all autoregressive group correlation values of ρ.

In conclusion, the simulation study confirms that between the proposed Two-Stage Procedure and the BY method, both of which are theoretically known to control the FDR under arbitrary dependence within the groups, the former is clearly the better choice in terms of controlling the FDR under this dependence situation.

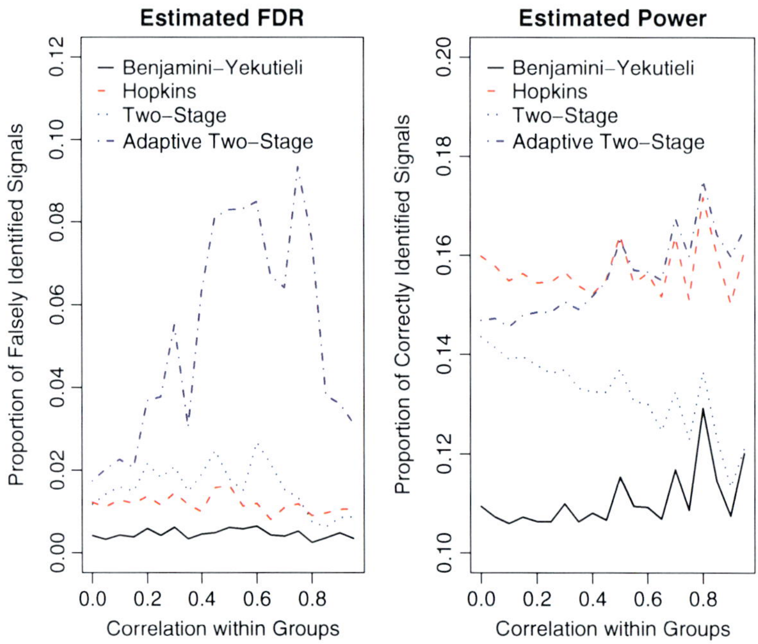

Fig. 36.1 Simulated FDR and power for fixed group correlation structure

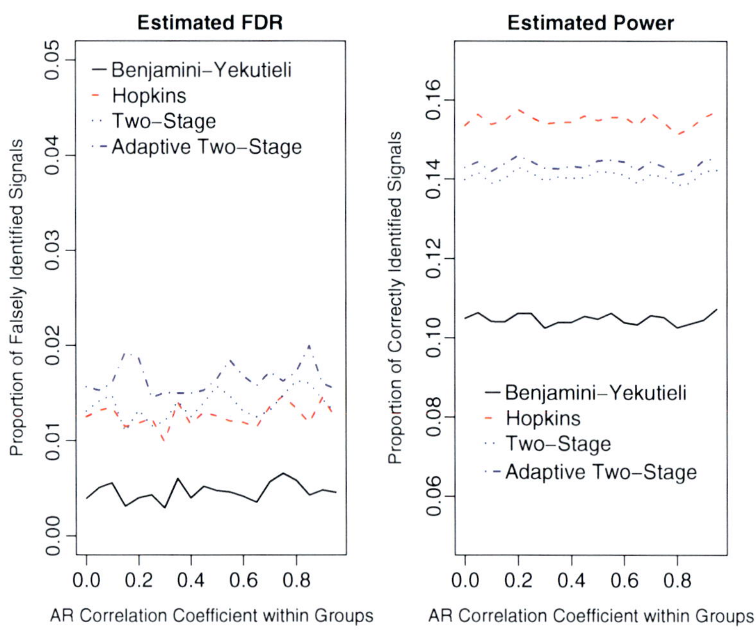

Fig. 36.2 Simulated FDR and power for autoregressive correlation structure

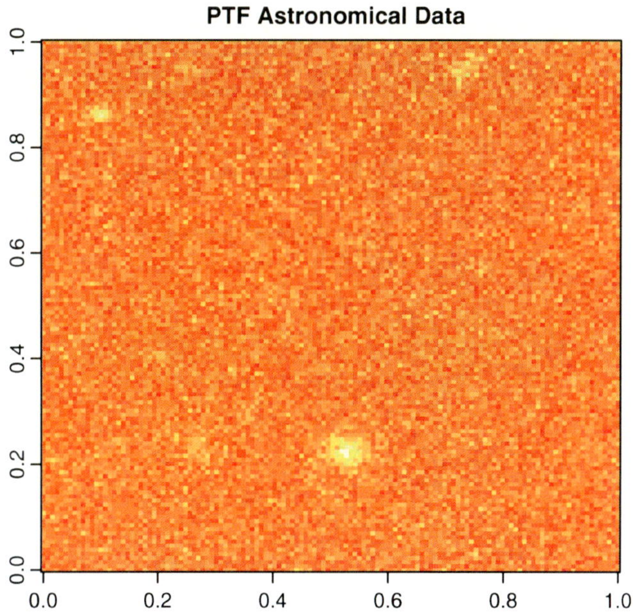

Fig. 36.3 Small portion of a PTF image

Moreover, it is competitive with Hopkins', even though Hopkins' may not control the FDR. The simulation study also seems to indicate that the Adaptive Two-Stage Procedure controls the FDR when the correlation is fixed and small ($0 < \rho < 0.5$), but may become unstable as correlation gets more extreme. Impressively, the Adaptive Two-Stage Procedure under the autoregressive correlation scenario, seems to control the FDR over all positive values of ρ, which is yet to be proved theoretically.

36.5 Application

The astronomical data used to illustrate our procedures comes from Palomar Transient Factory (PTF), one of the mid-size wide-field survey projects currently underway. Each image is $2,048 \times 4,096$ pixels, but a smaller sub-rectangle of noise (130×130) was chosen to apply the methods. The data consist of approximately normally distributed observations with mean $\bar{x} = 721.7$ and variance $s^2 = 476.1$. A heat map of the image can be seen below in Fig. 36.3 and the results in Fig. 36.4. The data were first standardized and converted to p-values. Results of four methods are presented—BY, Hopkins, Two-Stage BH (Procedure 36.1), and Adaptive Two-Stage BH (Procedure 36.2). Again, we have chosen $\lambda = 0.5$ in Procedure 36.2. Applying the BY procedure to the data rejects fourteen pixels and Hopkins' rejects

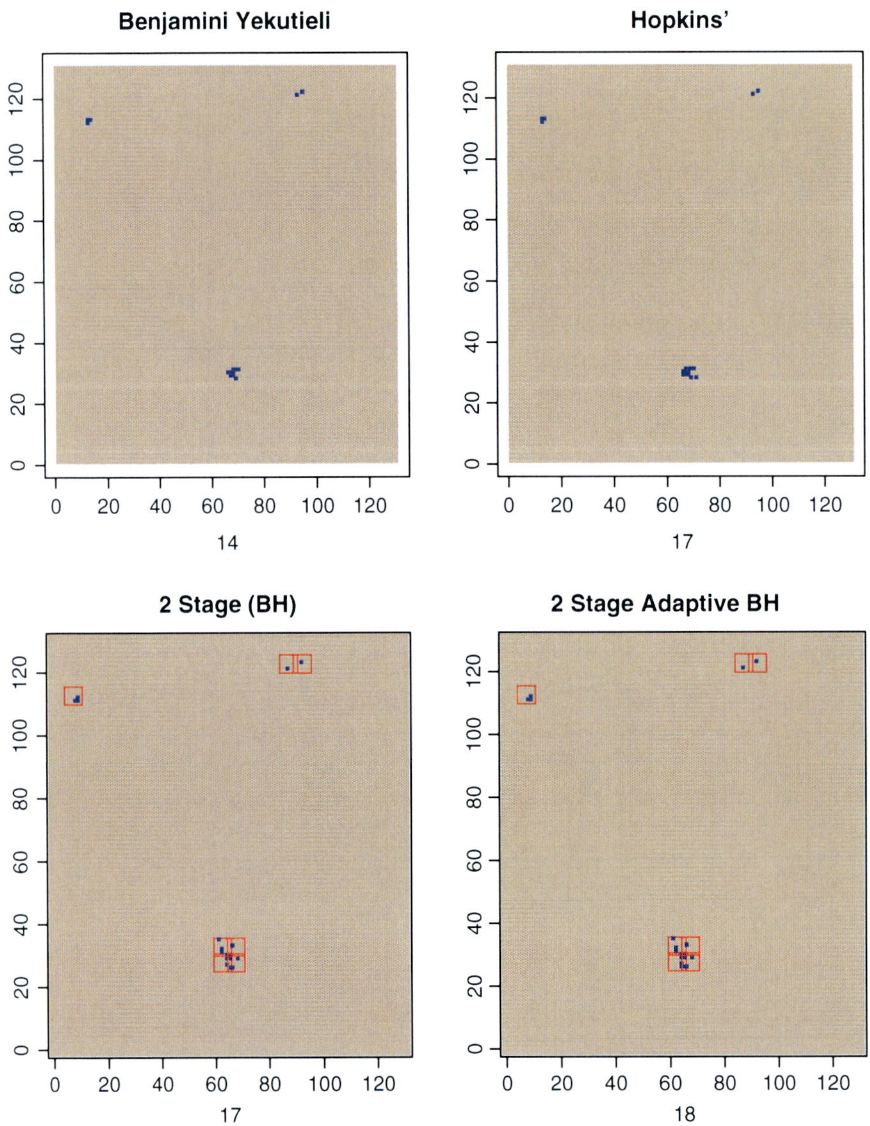

Fig. 36.4 The results of the four methods applied to the PTF astronomical data in Fig. 36.3. The *blue points* represent source pixels and the *red boxes* represent a potential source group. Below each plot is the total number of source pixels found using that method

an additional three pixels. On the other hand, using our Two-Stage BH method, seven potential source groups are found to have seventeen source pixels and the Adaptive Two-Stage BH finds 18 from those seven potential source groups.

36.6 Concluding Remarks

We have proposed, in this research, two new FDR controlling methods to be used in group-dependent data—Two-Stage BH method and Adaptive Two-Stage BH method—and compared them with the existing methods of BY and Hopkins. Both of our proposed methods compare favorably to the BY method in terms of the proportion of detected source pixels. When the group correlation is small ($\rho < 0.5$) or large ($\rho > 0.8$), both of these methods retain control of the FDR; however, when this correlation is moderate ($0.5 < \rho < 0.8$), the adaptive procedure seems to become unstable.

More investigation is needed to estimate the dependence structure of astronomical data to see if the local correlation is small enough to warrant use of adaptive methods. Further simulation studies should be done with larger repetitions, varying π_0, and incorporating other dependence structures.

It would also be interesting to study the astronomical source detection problem differently by adding a third dimension. Since astronomy data is often collected nightly, the assemblage can be thought of as a 'data cube' instead of a 'data matrix', where the first and second dimensions correspond to the spatial location and the third dimension is the date/time of observation. In other words, multiple testing procedures can be adapted to not only search for signals at every ith row and jth column location, but also at every time t. This set up could be explored in both a frequentist and Bayesian contexts.

Acknowledgements The authors would like to thank Eric Feigelson for acclimating us to transient detection methodology and the goals of astronomical research, Peter Nugent for supplying the PTF data, and Peter Freeman for his commentary regarding the False Cluster Proportion methodology. The research of Sarkar and Guo were supported by NSF Grants DMS-1006344 and DMS-1006021 respectively.

Appendix

Proof of Theorem 36.1. We first prove the theorem assuming that the groups are independent. For that we need the following notations:

R: Number of source pixels detected,

V: Number of source pixels falsely detected,

RG: The index of the ordered (in terms of increasing values of grouped p-values) potential source group detected (which is also k^*_{BH}),

$RG^{(-k)}$: The index of the ordered potential source group detected based on the BH method applied to all the groups except the kth one and the critical values $g\alpha/G$, $g = 2, \ldots, G$, and

$J_0(g)$: The set of indices of the p-values in the gth group that correspond to background pixels.

Then,

$$
\begin{aligned}
\text{FDR} &= E\left\{\frac{V}{\max\{R,1\}}\right\} = E\left[E\left\{\frac{V}{\max\{R,1\}}\middle|RG,R\right\}\right] \\
&= \sum_{k=1}^{G}\sum_{j\in J_0(k)}\sum_{g=1}^{G}\sum_{r=1}^{N}\frac{1}{r}Pr\left\{SP_{kj}\le\frac{g}{G}\alpha, RG=g, R=r\right\} \\
&= \sum_{k=1}^{G}\sum_{j\in J_0(k)}\sum_{g=1}^{G}\sum_{r=1}^{N}\frac{1}{r}Pr\left\{P_{kj}\le\frac{g}{N}\alpha, RG^{(-k)}=g-1, R=r\right\} \\
&= \sum_{k=1}^{G}\sum_{j\in J_0(k)}\sum_{g=1}^{G}\sum_{r=1}^{N}\frac{g\alpha}{rN}Pr\left\{RG^{(-k)}=g-1, R=r\middle|P_{kj}\le\frac{g}{N}\alpha\right\} \\
&\le \sum_{k=1}^{G}\sum_{j\in J_0(k)}\sum_{g=1}^{G}\sum_{r=1}^{N}\frac{\alpha}{N}Pr\left\{RG^{(-k)}=g-1, R=r\middle|P_{kj}\le\frac{g}{N}\alpha\right\} \\
&= \sum_{k=1}^{G}\sum_{j\in J_0(k)}\sum_{g=1}^{G}\frac{\alpha}{N}Pr\left\{RG^{(-k)}=g-1\middle|P_{kj}\le\frac{g}{N}\alpha\right\} \\
&= \sum_{k=1}^{G}\sum_{j\in J_0(k)}\sum_{g=1}^{G}\frac{\alpha}{N}Pr\left\{RG^{(-k)}=g-1\right\} \\
&= \sum_{k=1}^{G}\sum_{j\in J_0(k)}\frac{\alpha}{N} = \frac{N_0}{N}\alpha\le\alpha. \qquad (36.3)
\end{aligned}
$$

In (36.3), the fifth equality follows from the assumption that P_{kj} is distributed as $U(0,1)$ when it corresponds to a background pixel, the first inequality follows from the fact that $RG \le R$, and the seventh equality follows from the independence assumption of the groups. This proves the theorem under independence of the groups.

If the groups are not completely independent of each other, we will assume that they are positively dependent in the following sense:

The conditional expectation

$$E\left\{\phi(\mathbf{P}^{(-g)})\middle|P_{gj}=u\right\}, \qquad (36.4)$$

where $\mathbf{P}^{(-g)}$ is the set of p-values corresponding to all pixels except those in the gth group. P_{gj} is the jth p-value corresponding to a background pixel in the gth group, and $\phi(\mathbf{P}^{(-g)})$ is an increasing (coordinatewise) function of all the p-values except those in the gth group, is non-decreasing in $u \in (0,1)$ for each g and j.

From (36.3), we note that

$$
\begin{aligned}
\text{FDR} &\leq \sum_{k=1}^{G} \sum_{j \in J_0(k)} \sum_{g=1}^{G} \frac{\alpha}{N} Pr\left\{RG^{(-k)} = g-1 \middle| P_{kj} \leq \frac{g}{N}\alpha\right\} \\
&= \sum_{k=1}^{G} \sum_{j \in J_0(k)} \sum_{g=1}^{G} \frac{\alpha}{N} \left[Pr\left\{RG^{(-k)} \geq g-1 \middle| P_{kj} \leq \frac{g}{N}\alpha\right\} \right. \\
&\qquad \left. - Pr\left\{RG^{(-k)} \geq g \middle| P_{kj} \leq \frac{g}{N}\alpha\right\} \right] \\
&\leq \sum_{k=1}^{G} \sum_{j \in J_0(k)} \sum_{g=1}^{G} \frac{\alpha}{N} \left[Pr\left\{RG^{(-k)} \geq g-1 \middle| P_{kj} \leq \frac{g-1}{N}\alpha\right\} \right. \\
&\qquad \left. - Pr\left\{RG^{(-k)} \geq g \middle| P_{kj} \leq \frac{g}{N}\alpha\right\} \right] \\
&= \sum_{k=1}^{G} \sum_{j \in J_0(k)} \frac{\alpha}{N} = \frac{N_0 \alpha}{N} \leq \alpha.
\end{aligned}
$$

The second inequality follows from the assumption (36.4) of positive dependence of groups. This completes our proof of Theorem 36.1.

References

1. Becker, A.C (2006). Transient Detection and Classification. *Astronomical Notes* **88**, 789–792.
2. Benjamini, Y & Hochberg, Y. (1995). Controlling the false discovery rate: A practical and powerful approach to multiple testing. *Journal of the Royal Statistical Society, B* **57**, 289–300.
3. Benjamini, Y. & Hochberg, Y. (2000). On the adaptive control of the false discovery rate in multiple testing with independent statistics. *Journal of Educational and Behavioral Statistics.* **25**, 60–83.
4. Benjamini, Y, Krieger, K. & Yekutieli, D. (2006). Adaptive linear step-up procedures that control the false discovery rate. *Biometrika* **93**, 491–507.
5. Benjamini, Y., and Yekutieli D., (2001). The control of the false discovery rate in multiple testing uder dependency. *Annals of Statistics*, **29**, 1165–1188.
6. Blanchard, G. & Roquain, E. (2009). Adaptive FDR control under independence and dependence. *Journal of Machine Learning Research* **10**, 2837–2871.
7. Efron, B. (2007). Correlation and large-scale simultaneous significance testing. *Journal of the American Statistical Association* **102**, 93–103.
8. Finner, H. & Gontscharuk, V. (2009). Controlling the familywise error rate with plug-in estimator for the proportion of true null hypotheses. *Journal of the Royal Statistical Society, B* **71**, 1031–1048.
9. Friedenberg, D. & Genovese, C. (2009) Straight to the Source: Detecting Aggregate Objects in Astronomical Images with Proper Error Control. arXiv: 0910.5449.
10. Gavrilov, Y., Benjamini, Y. & Sarkar, S. K. (2009). An adaptive step-down procedure with proven FDR control. *Annals of Statistics* **37**, 619–629.
11. Guo, W. (2009). A note on adaptive Bonferroni and Holm procedures under dependence. *Biometrika*, **96**, 1012–1018.

12. Henrion, M., Mortlock, D., Hand, D., Gandy, A. (2011). A Bayesian approach to star-galaxy classification. *Monthly Notices of the Royal Astronomical Society*, **412**, 2286–2302.
13. Hopkins, A. M., Miller, C. J., Connolly, A. J., Genovese, C., Nichol, R. C. & Wasserman, L. (2002). A new source detection algorithm using the false discovery rate. *The Astronomical Journal* **123**, 1086–1094.
14. Romano, J. P., Shaikh, A. M. & Wolf, M. (2008). Control of the false discovery rate under dependence using the bootstrap and subsampling. *TEST* **17**, 417–442.
15. Salzberg, S., Chandler, R., Ford, H., Murthy, S., and White, R. (2007). Decision Trees for Automated Identification of Cosmic-Ray Hits in Hubble Space Telescope Images. *The Astronomical Society of the Pacific* **107**, 279–288.
16. Sarkar, S. K. (2002). Some results on false discovery rate in stepwise multiple testing procedures. *Annals of Statistics* **30**, 239–257.
17. Sarkar, S. K. (2007). Stepup procedures controlling generalized FWER and generalized FDR. *Annals of Statistics* **35**, 2405–2420.
18. Sarkar, S. K. & Guo, W. (2009). On a generalized false discovery rate. *Annals of Statistics* **37**, 337–363.
19. Sarkar, S. & Guo, W. (2010). Procedures controlling generalized false discovery rate using bivariate distributions of the null p-values. *Statistica Sinica* **20**, 1227–1238.
20. Sarkar, S. K. (2008). On methods controlling the false discovery rate (with discussions). *Sankhya* **70**, 135–168.
21. Sarkar, S. K., Guo, W. & Finner, H. (2011). On adaptive procedures controlling the familywise error rate. To appear in the Journal of Statistical Planning and Inference.
22. Savage, R. & Oliver, S. (2007). Bayesian Methods of Astronomical Source Extraction. *The Astrophysical Journal* **661**, 1339–1346.
23. Storey, J.D. (2002). A direct approach to false discovery rates. *Journal of the Royal Statistical Society, B* **64**, 479–498.
24. Storey, J. D., Taylor, J. E. & Siegmund, D. (2004). Strong control, conservative point estimation and simultaneous conservative consistency of false discovery rates: a unified approach. *Journal of the Royal Statistical Society, B* **66**, 187–205.
25. Sun, W. & Cai, T. (2009). Large-scale multiple testing under dependence. *Journal of the Royal Statistical Society, B* **71**, 393–424.
26. Yekutieli, D. & Benjamini, Y. (1999) Resampling-based false discovery rate controlling multiple test procedures for correlated test statistics. *Journal of Statistical Planning and Inference* **82**, 171–196.

Chapter 37
Commentary: Astronomical Transient Detection Controlling the False Discovery Rate

Peter E. Freeman

Abstract The two-step, False Discovery Rate-based thresholding procedures presented by Clements and Sarkar in this volume offer a computationally efficient means by which to detect faint sources lurking in collections of megapixel and gigapixel images. We compare Clements and Sarkar's Procedure 36.1 with the False Cluster Proportion-based algorithm of Friedenberg and Genovese (arXiv:0910.5449, 2009). The former employs pixel-wise error control, while the latter employs cluster-wise error control. We find the two techniques yield source lists of similar efficiency (finding ≈50% of the sources detected by a more computationally intensive procedure) and purity (≈100%), if one eliminates single-pixel detections made by the Clements and Sarkar procedure. We propose that the Clements and Sarkar procedure be refined such that only statistically significant clusters are retained in the final source list, mitigating the issue of single-pixel detections and potentially improving the procedure's efficiency.

37.1 Introduction

Rapid advances in telescope technology are allowing us to peer more and more deeply into the Universe and to discover new objects both nearby (e.g., asteroids) and far away (e.g., quasars). *Source detection* is the process of differentiating as-yet-unseen, faint sources from random fluctuations of the astronomical background. It is important for any source detection algorithm to be highly *efficient* (i.e., to detect nearly all sources brighter than some given flux), but it is of even greater importance that it exhibit high *purity* (i.e., to limit the amount of false sources it detects). This is because the time available to follow up detections is limited and astronomers want to

P.E. Freeman (✉)
Department of Statistics, Carnegie Mellon University, Pittsburgh, PA, USA
e-mail: pfreeman@cmu.edu

E.D. Feigelson and G.J. Babu (eds.), *Statistical Challenges in Modern Astronomy V*, Lecture Notes in Statistics 209, DOI 10.1007/978-1-4614-3520-4_37,

spend as little of that time as possible viewing what turn out to be empty patches of sky. One can trivially augment purity by returning to telescopic fields and recording only those sources within them that are repeatedly detected, but important *transient* sources such as Type Ia supernovae may only appear once in a field before fading from view. Thus it is important that the source list generated from any single image be pure.

Commonly used, publicly available algorithms for source detection, like those based on sliding cells,[1] Voronoi tessellation (e.g., [2]), wavelets (e.g., [3]), and signal-to-noise filtering (e.g., [1]) were all created for analyzing megapixel images such as those of the *Chandra* X-ray Observatory. Many of these algorithms follow four basic steps when assessing whether image pixel (i, j) should be associated with a source:

1. Compute background estimate $\hat{B}_{i,j}$;
2. Compute signal estimate $\hat{S}_{i,j}$;
3. Compute p-value estimate, $\hat{p}_{i,j} = \int_{\hat{S}_{i,j}}^{\infty} f(S|\hat{B}_{i,j})dS$, where f denotes the probability distribution function for observing signal strength S given $\hat{B}_{i,j}$; and
4. Compare $\hat{p}_{i,j}$ to a threshold significance α.

If $\hat{p}_{i,j} < \alpha$, we putatively associate pixel (i, j) with an astronomical source.[2] By convention, α is chosen conservatively, with a typical choice being $\alpha = 1/N$, where N is the number of exposed pixels in the image. Using this family-wise error rate (FWER) is akin to performing a Bonferroni correction.

Within the context of these basic steps, the work of Clements and Sarkar in this volume (hereafter C&S) relates directly to step 4: given p-values for each pixel, their procedures based on false detection rate methodology produce threshold values for detection. By applying the Benjamini–Hochberg procedure at two scales-locally, within a box of size similar to that of the telescope's point-spread function (PSF), then globally—they take into account pixel-to-pixel correlations on small scales in a computationally efficient manner while at the same time generating source lists that are more complete than those that would be generated using FWER-based thresholding methods. This is important work, particularly since we will need new, computationally efficient source detection procedures for analyzing gigapixel images. However, we argue in the next section that this work is, in a sense, incomplete: it should be extended beyond the notion of pixel-wise error control to cluster-wise error control so as to help ensure the purity of generated source lists.

[1] See, e.g., http://cxc.harvard.edu/ciao/download/doc/detect_manual/

[2] We note that, depending on the algorithm, a putative source may be rejected before being listed: for instance, in WAVDETECT, putative sources are rejected if they are detected only when the image-smoothing wavelet scale is smaller than that of the telescope's point-spread function (PSF).

37.2 Pixel-Wise Error Control Versus Cluster-Wise Error Control

We compare the performance of C&S's Procedure 36.1 with a method that controls for the proportion of false *clusters* of putative source pixels developed by Friedenberg and Genovese [4] (hereafter F&G). This False Cluster Proportion-based procedure extends work by Perone Pacifico et al. [6] on bounding the rate of false regions within a Gaussian random field by deriving a confidence superset for the location of true nulls (i.e., background regions).

To compare the two procedures, we follow F&G by analyzing a 512×512 pixel subset image of the Chandra Deep Field South (CDFS; see Fig. 37.1). The data in this image are Poisson-distributed counts; by implementing a basic two-pass algorithm (compute the global background using all pixels, detect sources, then recompute the background using only non-source pixels), we estimate the background intensity to be $\hat{\lambda}_b \approx 0.2$ counts/pixel. [5] (hereafter G02), by combining the use of `SExtractor` [1] and `WAVDETECT` [3], ultimately find 27 sources in this image. Of these 27, F&G detect 17, while at the same time detecting no other sources.[3]

To estimate p-values, we compute, for each pixel (i, j),

Fig. 37.1 A 512×512 subimage of the Chandra Deep Field South that we use to compare Procedure 36.1 of Clements and Sarkar with the False Cluster Proportion-based thresholding procedure of Friedenberg and Genovese [4]

[3]F&G run their algorithm assuming a false cluster proportion of 0.1.

Table 37.1 Clements and Sarkar Procedure 36.1: source detection as a function of box size

Box Size	SPD	SPD in G02	MPD	MPD in G02
16	355	–[a]	15 (346)[b]	15
32	27	5	14 (270)	14

SPD single-pixel detections, *MPD* multi-pixel detections

[a] There are too many single pixel sources to determine unambiguous links between them and G02 sources, by eye

[b] Visually identified clumps (total pixels in those clumps)

$$\hat{p}_{i,j} = \sum_{d=d_{i,j}}^{\infty} p(d|\hat{\lambda}_b) = \sum_{d=d_{i,j}}^{\infty} \frac{\hat{\lambda}_b^d e^{-\hat{\lambda}_b}}{d!} .$$

Within the C&S procedure, identification of putative sources depends on the assumed box size. We test box sizes ranging from 4 to 64 pixels. Table 37.1 shows results for box sizes 16 and 32, between which there is a sharp change in the number of detected (i.e., putative source) pixels. In this table, we denote a clump of two or more contiguous or nearly contiguous detected pixels that we identify *by eye* as a multi-pixel detection. All other detected pixels are considered (isolated) single-pixel detections. If we consider the vast majority of single-pixel detections to be false source detections, then we find that for box size 16, the false pixel rate appears much higher than the target of $\alpha = 0.05$, with the C&S procedure coming much closer to performing as advertised at box size 32. The naive conclusion would be to adopt the larger box size. However, we find that for both box sizes, if one eliminates all single-pixel detections and assumes the G02 source list as ground truth, the two C&S source lists have similar efficiency ($\approx$50%) and purity ($\approx$100%) as the source list of F&G. We further note that five of the G02 sources appear to be associated with single pixel detections at box size 32, suggesting that at box size 16 there may be as many or more G02 sources that are associated with clusters of non-contiguous detected pixels whose relative spacing is statistically inconsistent with that expected given the null hypothesis of uniformly distributed false detections. Thus refining the C&S procedure to take into account the relative spacing of detected pixels could potentially lead to an procedure that outperforms the current F&G algorithm.

37.3 Conclusion

The false detection rate-based source thresholding procedures presented by Clements and Sarkar are conceptually simple, computationally efficient, and amenable to parallelization. These features are important to consider as we enter the era of gigapixel image analysis. We test Clements and Sarkar's Procedure 36.1 using Poisson-distributed data from the Chandra Deep Field South. We find that while this procedure limits the number of false pixels, it does not effectively limit the number of false sources. However, we find that by adding a clustering step to the algorithm, i.e., that by requiring that each putative source contain two or more pixels in close proximity, we achieve good cluster-wise error rates. While this is promising, it is obvious that much more work needs to be done before we achieve a robust, computationally efficient detection algorithm for use with gigapixel images that is marked by both high purity and efficiency.

References

1. Bertin, E. and S. Arnouts (1996, June). SExtractor: Software for source extraction. *Astronomy and Astrophysics Supplement Series 117*, 393–404.
2. Ebeling, H. and G. Wiedenmann (1993, January). Detecting structure in two dimensions combining Voronoi tessellation and percolation. *Physical Review E 47*, 704–710.
3. Freeman, P. E., V. Kashyap, R. Rosner, and D. Q. Lamb (2002, January). A Wavelet-Based Algorithm for the Spatial Analysis of Poisson Data. *The Astrophysical Journal Supplement Series 138*, 185–218.
4. Friedenberg, D. A. and C. R. Genovese (2009, October). Straight to the Source: Detecting Aggregate Objects in Astronomical Images with Proper Error Control. *arXiv:0910.5449*.
5. Giacconi, R., A. Zirm, J. Wang, P. Rosati, M. Nonino, P. Tozzi, R. Gilli, V. Mainieri, G. Hasinger, L. Kewley, J. Bergeron, S. Borgani, R. Gilmozzi, N. Grogin, A. Koekemoer, E. Schreier, W. Zheng, and C. Norman (2002, April). Chandra Deep Field South: The 1 Ms Catalog. *The Astrophysical Journal Supplement Series 139*, 369–410.
6. Perone Pacifico, M., C. Genovese, I. Verdinelli, and L. Wasserman (2004). False discovery control for random fields. *Journal of the American Statistical Association 99*(468), 1002–1014.

Chapter 38
Slepian Wavelet Variances for Regularly and Irregularly Sampled Time Series

Debashis Mondal and Donald B. Percival

Abstract We discuss approximate scale-based analysis of variance for Gaussian time series based upon Slepian wavelets. These wavelets arise as eigenfunctions of an energy maximization problem in a pass band of frequencies. Unlike the commonly used Daubechies wavelets, Slepian wavelets have the ability to accommodate both regularly and irregularly sampled data. For regularly sampled Gaussian time series, we derive statistical theory for Slepian-based wavelet variances and show that it is comparable to Daubechies-based variances. For irregularly sampled time series data, we derive a corresponding statistical theory for Slepian-based wavelet variances. We demonstrate its use on X-ray fluctuations from a binary star system and on a light curve from the variable star Z UMa.

38.1 Introduction

Over the past two decades, wavelet variance analysis has become an accepted statistical approach for studying the variability of time series collected at regular time intervals. The wavelet variance (sometime known as the wavelet spectrum) quantifies the variability of a time series with respect to time scales. Wavelet variances at different time scales give rise to a scale-based analysis of variance. In applications arising from astronomy, geophysics, atmospheric sciences, biology, ecology and other areas of science, the wavelet variance has helped practitioners understand quasi-periodic oscillations, small-scale disordered noises, characteristic

D. Mondal (✉)
Department of Statistics, University of Chicago, 5734 S University Ave, Chicago, IL 60637, USA
e-mail: mondal@galton.uchicago.edu

D.B. Percival
Department of Statistics, University of Washington, Box 355640, Seattle, WA 98195, USA
e-mail: dbp@apl.washington.edu

E.D. Feigelson and G.J. Babu (eds.), *Statistical Challenges in Modern Astronomy V*, Lecture Notes in Statistics 209, DOI 10.1007/978-1-4614-3520-4_38,

scales, self-similar behavior, long-range dependence, fractal dimensionality, inhomogeneity and local stationarity that are found in various time series. The reference list for applications of wavelet variance analysis is extensive; see, e.g., [5, 9, 15, 16, 18, 24, 29, 35]. In particular, the work of Scargle et al. [29] is an important early use of the wavelet variance to study time series arising in astronomy. Background material on wavelet-based time series analysis (including the wavelet variance) can be found in [27], and there is a recent review article [26] devoted to the wavelet variance, which includes a basic introduction and discussions on its interpretation and some recent advances.

In astronomy, however, irregularly sampled time series occur more often than not, and their analysis introduces new statistical challenges. Standard wavelet variance analysis is intended to be applied only to regularly sampled time series, and can not easily cope with irregular or unevenly sampled data (a rarely occurring exception would be a time series that could be interpolated onto a regular grid without modification of any of its distributional properties). To date, several approaches have been proposed to adjust this analysis handle unevenly sampled time series, including [7, 8, 36]; however, the statistical properties of these methods have yet to be fully explored. Thus, in this article, we consider irregular sampling schemes, but focus on ones that are based on second-order stationary increment point processes. Indeed, substantial work has been done in other contexts on time series collected under such a sampling scheme; see e.g., [3, 19] and subsequent literature.

Although numerous other possibilities exist, in what follows we exclusively focus on the so-called Slepian wavelets. Slepian wavelets are nonstandard wavelets based on the same ideas leading to Slepian (or discrete prolate spheroidal) sequences [30]. In particular, zeroth-order Slepian sequences are the solutions to an optimization problem in which we seek a regularly spaced sequence whose energy is as concentrated as possible in a band of frequencies centered around zero frequency; i.e., Slepian sequences can be regarded as an optimal approximation to an ideal low-pass filter. Wavelet filters commonly used in the analysis of regularly sampled time series are approximations to ideal band-pass filters. Slepian wavelets are in essence optimal approximations to ideal band-pass filters and have been used previously in a multiwavelet scheme for estimating the wavelet variance [17]. Slepian wavelet filters resemble the familiar least-asymmetric filters due to Daubechies [6], but, unlike the Daubechies filters, the Slepian filters are nonorthogonal and hence decompose the process variance only approximately. Some large sample properties of wavelet variance estimators based on Slepian wavelets are discussed in [21]. The objective of this paper is to extend Slepian wavelets to irregularly sampled data. Bronez [4] introduced the notion of generalized Slepian sequences that can be applied to irregularly sampled time series. Here we focus on an adaptation of Bronez's scheme that yields an estimator of the wavelet variance for irregularly sampled series.

The rest of this article is laid out as follows. We define the Slepian wavelet variance for regularly sampled time series in Sect. 38.2. In Sect. 38.3 we consider estimators for this variance and their corresponding large sample statistical theory, and, in Sect. 38.4 we extend the theory of the Slepian wavelet variance to handle irregularly sampled data. In Sect. 38.5 we provide examples of the use of this methodology on actual time series (a regularly sampled series of X-ray counts and an irregularly sampled series of brightness magnitudes from a variable star). We conclude with some discussion in Sect. 38.5.

38.2 Slepian Wavelets for Regularly Sampled Data

38.2.1 Construction of Slepian Wavelet Filters

For a positive integer j, define the pass-band

$$A_j = \left[-2^{-j}, -2^{-j-1}\right] \cup \left[2^{-j-1}, 2^{-j}\right]. \tag{38.1}$$

Let $\{\psi_m\}_{m=0}^{M-1}$ be the coefficients for a linear filter that approximates a band-pass filter with pass-band A_j, and let ψ be an M-dimensional vector containing these coefficients. Let its Fourier transform be

$$\Psi(f) = \sum_{m=0}^{M-1} \psi_m e^{-i2\pi fm}$$

so that its squared gain function is $|\Psi(f)|^2$. We seek $\{\psi_m\}$ with the following properties: (1) the filter coefficients sum to zero; i.e., $1^T\psi = 0$, where 1^T is a row vector of ones; (2) the sum of the squares of the coefficients satisfies $\psi^T\psi = 2^{-j}$; and (3) the squared gain function is as concentrated as possible within A_j. Note that properties (1) and (2) match those of a jth level Daubechies wavelet filter. The concentration measure for the squared gain function $|\Psi(f)|^2$ within A_j is defined as

$$\lambda(M,j) = \frac{\int_{A_j} |\Psi(f)|^2\,df}{\int_{-1/2}^{1/2} |\Psi(f)|^2\,df} = \frac{\psi^T Q_j \psi}{\psi^T\psi} = 2^j \psi^T Q_j \psi$$

(see, e.g., [17]), where the (s,t)th element of the $M \times M$ matrix Q_j is

$$Q_j(s,t) = \int_{A_j} e^{-i2\pi f(t-s)}\,df = \frac{\sin\left(2^{1-j}\pi(s-t)\right) - \sin\left(2^{-j}\pi(s-t)\right)}{\pi(s-t)} \tag{38.2}$$

(when $s = t$, $Q_j(s,t)$ reduces to 2^{-j}). Maximization of this concentration measure gives rise to the following constrained eigenvalue problem:

$$Q_j\psi = \lambda(M,j)\psi \text{ subject to } 1^T\psi = 0 \text{ and } \psi^T\psi = 2^{-j}. \tag{38.3}$$

Alternatively we can write this eigenvalue problem as

$$C_M Q_j C_M \psi = \lambda(M,j)\psi,$$

where, letting I_M denote the Mth order identity matrix, $C_M = I_M - \frac{1}{M}11^T$ is the centering matrix of order M. To see this, we introduce Lagrangian multipliers a and $\lambda(M,j)$ as in [28, p. 50] and consider the expression

$$\psi^T Q_j \psi - 2a1^T\psi - \lambda(M,j)(\psi^T\psi - 2^{-j})$$

as a function of ψ, a and $\lambda(M,j)$. Equating its partial derivatives to zero yields

$$Q_j\psi - a1 - \lambda(M,j)\psi = 0, \;\; 1^T\psi = 0 \;\text{ and }\; \psi^T\psi = 2^{-j}.$$

Because $C_M 1 = 0$ and $C_M\psi = \psi$, multiplying the first equation by C_M yields

$$C_M Q_j\psi = \lambda(M,j)\psi \;\text{ or, equivalently, }\; C_M Q_j C_M\psi = \lambda(M,j)\psi,$$

which is an eigenvector problem involving a symmetric matrix whose eigenvectors (after proper scaling) satisfy the problem stated in (38.3).

Let $\psi_{j,0}$ be the eigenvector corresponding to the maximum eigenvalue of the above problem. We define $\psi_{j,0}$ to be the Slepian wavelet filter of length M for the level j. We set $M = c2^j$ for level j, where c is a constant independent of j such that $2c$ is an integer. When $c = 1$, the length of the jth level filter matches that of the Haar wavelet filter. We discuss the rationale for setting $c > 1$ in the next section.

38.2.1.1 Continuous Problem

We can elucidate some properties of the Slepian filters by approximating the above eigenvalue problem using a continuous formulation, as follows. Setting $M = 2^j$, we have

$$Q_j(s,t) = \frac{\sin(2\pi(s-t)/M) - \sin(\pi(s-t)/M)}{\pi(s-t)} = \frac{2}{M}\beta(f-f') \tag{38.4}$$

where

$$\beta(u) = \frac{\sin(\pi u) - \sin(\pi u/2)}{\pi u}, \;\; f = \frac{2s}{M} - 1 \;\text{ and }\; f' = \frac{2t}{M} - 1$$

so that $-1 \le (f,f') < 1$ if we assume $0 \le (s,t) \le M-1$. As j tends to infinity, the discrete eigenvalue problem of (38.3) is well approximated by the continuous eigenvalue problem

$$\int_{-1}^{1} \beta(f-f')\psi(f')\,df' = \lambda\psi(f) \tag{38.5}$$

subject to $\int \psi(f)df = 0$ and $\int \psi^2(f)df = 2/M^2$. Note that we can use this eigenfunction $\psi(f)$ to determine the filters for all large j. At level j we would approximate the discrete eigenvector using $\psi(\frac{2s+1}{M} - 1)$, $s = 0, 1, \ldots, M-1$. The continuous formulation suggests that, if matrix computations are too cumbersome for some large j, we can approximate the required eigenvector by applying an interpolation scheme to the eigenvector from a readily computable smaller j. The value of the maximum eigenvalue λ associated with continuous filter $\psi(f)$ represents the fraction of the total energy within the pass-band. We can make this fraction greater by setting $M = c2^j$, where $2c$ is an integer greater than two. The kernel now becomes

$$\beta_c(u) = \frac{\sin(c\pi u) - \sin(c\pi u/2)}{\pi u} \tag{38.6}$$

(the limits of integration in (38.5) remain the same).

Formulating the problem in the continuous domain also allows us to assume that the wavelet filters are either even or odd functions. To see this, we note that, because β_c is an even function,

$$\lambda\,\psi(-f) = \int_{-1}^{1} \beta_c(f + f')\psi(f')df' = \int_{-1}^{1} \beta_c(f - f')\psi(-f')df' \tag{38.7}$$

Thus $\psi(-f)$ is also an eigenfunction with corresponding eigenvalue λ and hence either $\psi(f) = \pm\psi(-f)$ or $(\psi(f) + \psi(-f))/2$ and $(\psi(f) - \psi(-f))/2$ can be taken to be two distinct even and odd eigenfunctions with eigenvalue λ.

38.2.2 Shape and Energy at Different Scales

Plots of Slepian wavelets arising from the kernel β_c for different values of c are shown in Fig. 38.1. For $c = 1$ the Slepian wavelet is S-shaped; for $c = 1.5$ the shape somewhat resembles the Mexican hat wavelet; and for larger values of c, the shape is reminiscent of either the real or imaginary part of the Morlet wavelet popular in geophysics (note that we can flip these wavelets by multiplying them by -1). In Table 38.1 the maximum eigenvalues corresponding to Slepian wavelet filters arising out of matrix Q_j for various values of c are given. Asymptotically, the eigenvectors have a squared gain function of a perfect band-pass filter that is unity inside A_j and zero outside.

Table 38.2 shows the concentration measures for the jth level Haar, D(4), D(6) and LA(8) wavelet filters. The length of these filters is given by $L_j = (2^j - 1)(L-1) + 1$, where $L = 2$, 4, 6 and 8 for, respectively, the Haar, D(4), D(6) and LA(8) filters. By comparison, the length of the Slepian filters is $c2^j$. Setting $c = 2, 3$ and 4 yields Slepian filters that have the same length as, respectively, the D(4), D(6) and LA(8) filters when $j = 1$, but the Slepian filters are shorter when $j \geq 2$ (the $c = 1$ Slepian and Haar filters have the same length for all j). Even though

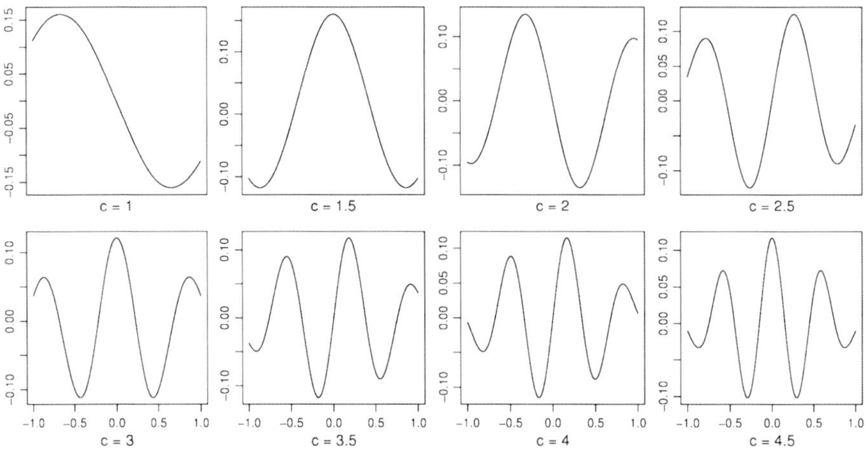

Fig. 38.1 The shape of the first eigenvector for different values of c

Table 38.1 Maximum eigenvalues from (38.3) for different levels j and various c

Level j	$c = 1$	$c = 2$	$c = 3$	$c = 4$
1	0.818	0.989	1.000	1.000
2	0.581	0.787	0.951	0.985
3	0.561	0.781	0.947	0.984
4	0.557	0.779	0.946	0.984
≥ 5	0.556	0.779	0.946	0.984

Table 38.2 Concentration measures for Haar, D(4), D(6) and LA(8) wavelets

Level j	Haar	D(4)	D(6)	LA(8)
1	0.818	0.871	0.895	0.909
2	0.546	0.641	0.696	0.733
3	0.498	0.626	0.691	0.731
4	0.487	0.625	0.691	0.731
≥ 5	0.484	0.625	0.691	0.731

the D(4), D(6) and LA(8) filters are longer than their corresponding Slepian filters for $j \geq 2$, their concentration measures are smaller.

The squared gain functions of the Slepian wavelets at different scales are illustrated in Fig. 38.2. These show how the energy is spread within a band. We also show the spread of energy for an ideal wavelet. Note that for $c = 2$ the energy is more concentrated within the band.

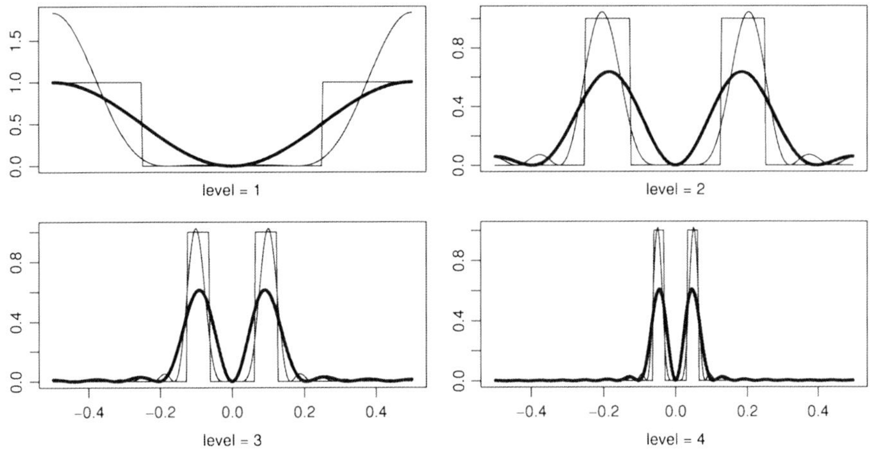

Fig. 38.2 The squared gain functions of the Slepian wavelets at different levels j: *thin curve* is for $c = 2$, *thick curve* is for $c = 1$

38.3 Statistical Theory for Slepian Wavelet Variance

38.3.1 Slepian Wavelet Coefficients and Wavelet Variance

Suppose we have a time series that we regard as a portion $\{X_t\}_{t=0}^{N-1}$ of a stationary process with autocovariance sequence (ACVS) $\{s_{X,\tau}\}_{\tau=-\infty}^{\infty}$ and spectral density function (SDF) denoted by $S(f)$. We denote the Slepian wavelet filter of level j by $\{\psi_{j,u}\}_{u=0}^{M_j-1}$, where $M_j = c2^j$ is the length of the filter. In practice, we restrict ourselves to $c = 1$ or 2.5. We denote the corresponding transfer function by $\Psi_j(f) = \sum_{u=0}^{M_j-1} \psi_{j,u}e^{-i2\pi fu}$ and the associated eigenvalue by λ_j.

Now we define the Slepian wavelet coefficient associated with level j and location or time point t by

$$U_{j,t} = \sum_{u=0}^{M_j-1} \psi_{j,u}X_{t-u}, \quad t = M_j - 1, \ldots, N-1.$$

Because the wavelet filter does not depend on the location t, the stationarity of $\{X_t\}$ and the fact that $\sum_u \psi_{j,u} = 0$ imply that the wavelet coefficients at any fixed level are a portion of a zero mean stationary process with SDF $S_j(f) = |\Psi_j(f)|^2 S(f)$. The Slepian wavelet variance at scale $\tau_j = 2^{j-1}$ is defined as

$$\mu_X^2(\tau_j) = \text{var}(U_{j,t}) = \int_{-\frac{1}{2}}^{\frac{1}{2}} |\Psi_j(f)|^2 S(f)\,df \tag{38.8}$$

38.3.2 Estimation of Slepian Wavelet Variance

Because $\mathrm{e}(U_{j,t}) = 0$, we have $\mathrm{e}(W_{j,t}^2) = \mu_X^2(\tau_j)$, where $E(\cdot)$ denotes the expectation operation. An unbiased estimate of $\mu_X^2(\tau_j)$ is thus given by

$$\hat{\mu}_X^2(\tau_j) = \frac{1}{N_j} \sum_{t=M_j-1}^{N-1} U_{j,t}^2, \tag{38.9}$$

where $N_j = N - M_j + 1$.

We can relate the Slepian wavelet variance to a similar variance based upon the Daubechies wavelets. Under certain reasonable conditions, the Daubechies wavelet variance is approximately equal to

$$\nu_X^2(\tau_j) \approx \int_{A_j} S(f)\,df,$$

where A_j is given by (38.1). Thus the wavelet variance summarizes the information in the SDF using just one value at each scale and also provides the basis for approximating certain SDFs. Moreover, if we assume that SDF is approximately constant over A_j, then the approximation above implies that $S(f) = \nu_X^2(\tau_j)/|A_j| = \nu_X^2(\tau_j)2^j$ in the nominal pass-band A_j. Then we have

$$\mu_X^2(\tau_j) = \int_{-\frac{1}{2}}^{\frac{1}{2}} |\Psi_j(f)|^2 S(f)\,df \approx \int_{A_j} |\Psi_j(f)|^2 \nu_X^2(\tau_j)2^j\,df = \lambda_j \nu_X^2(\tau_j) \approx \nu_X^2(\tau_j), \tag{38.10}$$

where the last approximation holds under the assumption that λ_j is close to unity.

38.3.3 Large Sample Statistical Properties

Theorem 38.1. *Suppose* $\alpha_j = \int_{-\frac{1}{2}}^{\frac{1}{2}} \left(|\Psi_j(f)|^2 S_X(f)\right)^2 df < \infty$. *Then, as* $N_j \to \infty$,

$$N_j^{\frac{1}{2}} \left(\frac{\hat{\mu}_X^2(\tau_j) - \mu_X^2(\tau_j)}{(2\alpha_j)^{\frac{1}{2}}} \right)$$

converges to a Gaussian random variable with zero mean and unit variance.

A proof of Theorem 38.1 is immediate; see e.g., [21]. We note that square integrability of $|\Psi_j(f)|^2 S_X(f)$ holds if the ACVS of $U_{j,t}$ dies down fast enough. In many geophysical application, if $|\Psi_j(f)|^2 S_X(f)$ is not square integrable, it is due to a singularity at $f = 0$, which can be cured by adding additional moment conditions as used in the construction of the higher order Daubechies wavelet filters.

In practice, we need to estimate α_j to make use of the above theorem—see [21, 25, 27] for details.

38.4 Extension to Irregularly Sampled Data

Suppose now that the observed time series consists of values $X(t_0), X(t_1), \ldots X(t_{N-1})$ taken at irregularly spaced time points $t_0, t_1, \ldots t_{N-1}$. Extending the notion of wavelet variance analysis to this scenario is challenging in part because of the lack of an appropriate scale-based wavelet transform for irregularly sampled time series. In particular, while the popular lifting scheme due to Sweldens [32] does define a wavelet transform for such series, its wavelet coefficients cannot be meaningfully associated with specific time scales and hence are not scale-based. By contrast, an attractive property of Slepian wavelets is that they can be generalized to handle irregular sampling in a manner such that maintains each coefficient is associate with a particular scale. Slepian wavelets lead to a statistically tractable wavelet variance analysis applicable to irregularly sampled time series if we assume that the sampling times $\{t_0, t_1, \ldots, t_{N-1}\}$ are a realization of a stationary point process, as we do henceforth. In other words, the sampling intervals $\Delta_1 = t_1 - t_0, \Delta_2 = t_2 - t_1, \ldots, \Delta_{N-1} = t_{N-1} - t_{N-2}$ are a portion of a stationary sequence of positive random variables. Let the marginal density of Δ_k be $p(x)$, $x > 0$, and let μ denote its mean, i.e., the expected sampling interval. The average sampling interval is equal to

$$\bar{\Delta} = \frac{1}{N-1}(\Delta_1 + \Delta_2 + \ldots + \Delta_{N-1}) = \frac{t_{N-1} - t_0}{N-1}, \tag{38.11}$$

and the strong law of large number ensures that, as $N \to \infty$, $\bar{\Delta}$ converges almost surely to μ. Since the average sampling time is $\bar{\Delta}$ rather than being fixed at unity, we must redefine the pass band A_j in (38.1) to be

$$A_j = \left[-2^{-j}/\bar{\Delta}, -2^{-j-1}/\bar{\Delta}\right] \cup \left[2^{-j-1}/\bar{\Delta}, 2^{-j}/\bar{\Delta}\right]. \tag{38.12}$$

Keeping with the tradition of the discrete wavelet transform, we can then consider Slepian wavelet filter constructions at the dyadic scales $\tau_j = 2^{j-1}\bar{\Delta}$ for $j = 1, 2, \ldots$.

38.4.1 *Construction of Adaptive Slepian Wavelet Filters for Irregular Sampling Times*

For each k, we seek a linear filter $\{\psi_{k,m}\}_{m=0}^{M-1}$ that is adapted to time points $t_k, t_{k+1}, \ldots t_{k+M-1}$ and maximizes the energy contained in the pass-band A_j. Thus we consider the Fourier transform

$$\Psi_k(f) = \sum_{m=0}^{M-1} \psi_{k,m} e^{-i2\pi f t_{k+m}}$$

and seek $\{\psi_{k,m}\}$ such that (1) the filter coefficients sum to zero, (2) the sum of the squares of the coefficients are normalized to $2^{-j}/\bar{\Delta}$ and (3) the squared gain function $|\Psi_k(f)|^2$ is as concentrated as possible within A_j. This leads to the following maximization problem:

$$Q_{k,j}\psi_k = \lambda(k,M,j)\psi_k \text{ subject to } 1^T\psi_k = 0, \tag{38.13}$$

where $Q_{k,j}(m,m')$, the (m,m')th element of the $M \times M$ matrix $Q_{k,j}$, is

$$\int_{A_j} e^{-i2\pi f(t_{k+m}-t_{k+m'})}\,df = \frac{\sin\left(\frac{2^{1-j}\pi}{\bar{\Delta}}(t_{k+m}-t_{k+m'})\right) - \sin\left(\frac{2^{-j}\pi}{\bar{\Delta}}(t_{k+m}-t_{k+m'})\right)}{\pi(t_{k+m}-t_{k+m'})}.$$

Set $M = c2^j$. Take $j \to \infty$, and let m, m' be sequences of integers implicitly indexed by j such that $2m/M - 1 \to f$ and $2m'/M - 1 \to f'$ for some $-1 \leq (f, f') \leq 1$. It follows that, for a large class of random sampling schemes that satisfy mild regularity and mixing conditions (renewal process sampling being one example), a functional central limit theorem (see, e.g., [10, 20]) yields

$$\frac{t_{k+m}-t_{k+m'}}{M} \stackrel{\mathrm{d}}{=} \frac{t_m - t_{m'}}{M} = \frac{t_m}{m}\frac{m}{M} - \frac{t_{m'}}{m'}\frac{m'}{M} \to_p \tfrac{1}{2}\mu(f-f')$$

and thus

$$MQ_{k,j}(m,m') \to_p 2\mu\beta_c(f-f')$$

(in the above, '$\stackrel{\mathrm{d}}{=}$' and '$\to_p$' denote equality in distribution and convergence in probability). Consequently, we are led to a continuous eigenvalue problem taking the form

$$\int_{-1}^{1} \mu\beta_c(f-f')\psi_k(f')\,df' = \lambda\psi_k(f), \tag{38.14}$$

which is basically the same eigenvalue problem considered in (38.5) for the regular sampling. As long as M is sufficiently large, we can use continuous Slepian wavelet $\psi(f)$ to obtain the adaptive filters via

$$\psi\left(2\frac{t_{m+k}-t_k}{t_{M+k}-t_k} - 1\right), \quad m = 0, 1, \ldots M-1, \ \ k = 0, 1, \ldots.$$

38.4.2 Adaptive Slepian Wavelet Coefficients and Average Energy

Once we have computed the adaptive wavelet filters $\{\psi_{j,k,m}\}$, we can define Slepian wavelet coefficients indexed by scale τ_j and shift k as follows:

$$U_{j,k} = \sum_{u=0}^{M_j-1} \psi_{j,k,u} X(t_{k+u}), \quad k = 0, 1, \ldots N - M_j.$$

Furthermore, if $X(t_0), \ldots, X(t_{N-1})$ is a realization of a stationary stochastic process, and if the sampling times obey a stationary point process, an immediate consequence is that the adaptive within-scale Slepian wavelet coefficients $U_{j,k}$ form a zero mean stationary time series. Thus we can estimate the overall energy associated with scale τ_j by

$$\hat{v}^2(\tau_j) = \frac{1}{N - M_j + 1} \sum_{k=0}^{N-M_j} U_{j,k}^2 \tag{38.15}$$

We now provide some justification as to why this average energy associated with adaptive Slepian wavelet coefficients gives an approximate estimate of wavelet variance at scale τ_j. Using spectral representation of the stationary time series $X(t)$, we obtain, for any fixed $t_k, \ldots, t_{k+M_j-1}$, conditionally

$$\mathrm{e}(U_{j,k}^2) = \int_{-\infty}^{\infty} \left| \Psi_{j,k}(f) \right|^2 S_X(f)\, df, \quad \Psi_{j,k}(f) = \sum_{m=0}^{M_j-1} \psi_{j,k,m} e^{i2\pi f t_{m+k}}.$$

Now, assuming that the power leakage is negligible for an adaptive Slepian filter, and that the spectral density is approximately constant within the pass band A_j (i.e., $S(f) = v^2(\tau_j) 2^j \bar{\Delta}$), we obtain conditionally

$$\mathrm{e}(U_{j,k}^2) \approx \int_{A_j} \left| \Psi_{j,k}(f) \right|^2 v^2(\tau_j) 2^j \bar{\Delta}\, df = \lambda(k, M_j, j) v^2(\tau_j) \approx v^2(\tau_j).$$

Thus, unconditionally, the average energy $\hat{v}^2(\tau_j)$ provides an approximate estimate of the wavelet variance. As was true in the regularly sampled case, $\hat{v}^2(\tau_j)$ is asymptotically a Gaussian random variable under appropriate regularly conditions. One such set of regularity conditions is given in [1], for which the mean of $\hat{v}^2(\tau_j)$ is asymptotically equal to $v^2(\tau_j) = \mathrm{e}(U_{j,0}^2)$, and its large sample variance is given by $S_{U^2}(0)/(N - M_j + 2)$, where $S_{U^2}(0)$ is the value of the spectral density function of $U_{j,k}^2$ at origin. We can thus apply multitaper method to deduce an estimator [27, 33] of $S_{U^2}(0)$ and construct 95% confidence interval for $v^2(\tau_j)$.

38.5 Applications

As a first example, Fig. 38.3 shows a regularly sampled time series of counts from the X-ray binary system GX 5–1. These $N = 65{,}526$ counts were recorded by the Ginga satellite at successive 1/128 second intervals over 512 s [11, 23].

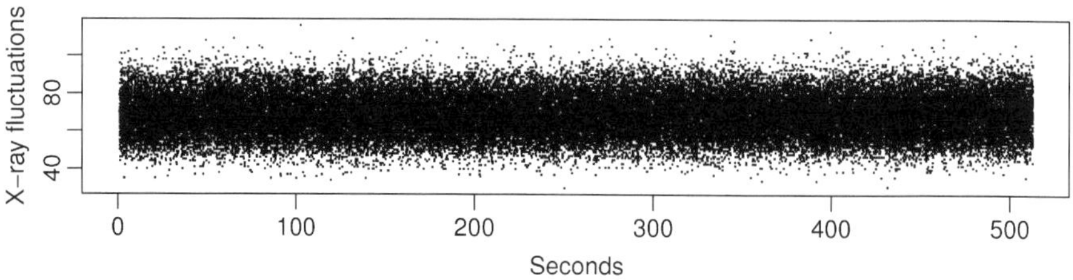

Fig. 38.3 X-ray fluctuations from a binary star system

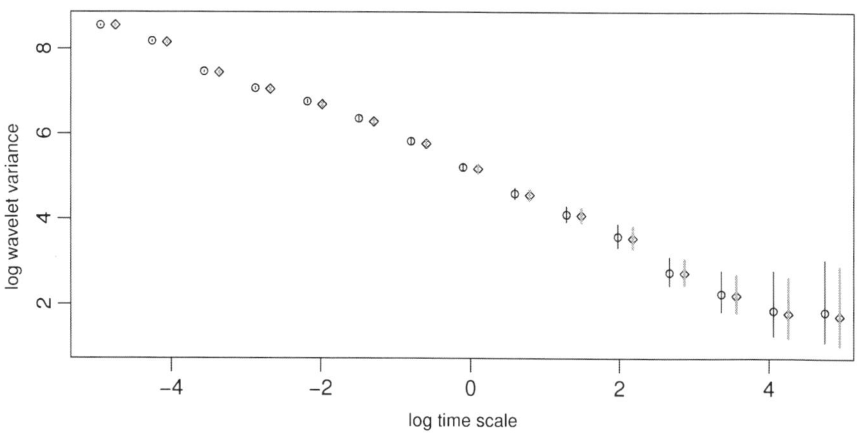

Fig. 38.4 Slepian (*circles* and *black lines*) and Haar (*diamonds* and *gray lines*) wavelet variances of X-ray fluctuations at the log–log scales

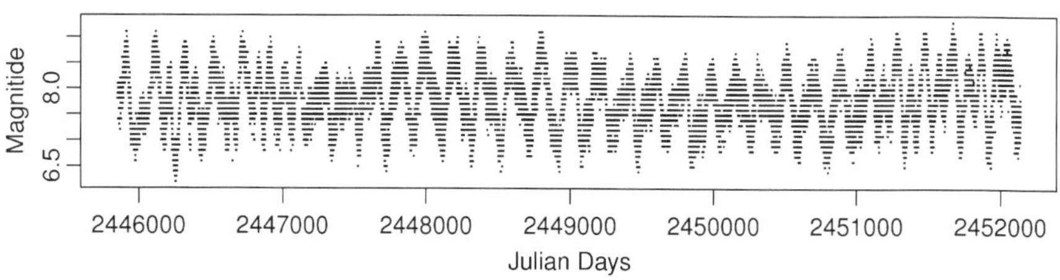

Fig. 38.5 Light curve of Z UMa

A histogram and a quantile-quantile plot of this time series reveal that, despite the count nature of the data, its empirical distribution is well approximated as Gaussian [26]. Figure 38.4 gives the $c = 1$ Slepian and Haar wavelet variances corresponding to $j = 1, \ldots, 15$. The Slepian wavelet variance estimates and their associated uncertainties are in good agreement with the corresponding Haar values.

As a second application, we now consider an irregularly sampled time series. Figure 38.5 displays the light curve data of Z UMa, taken directly from the Web site of American Association of Variable Star Observers. This star is in the constellation

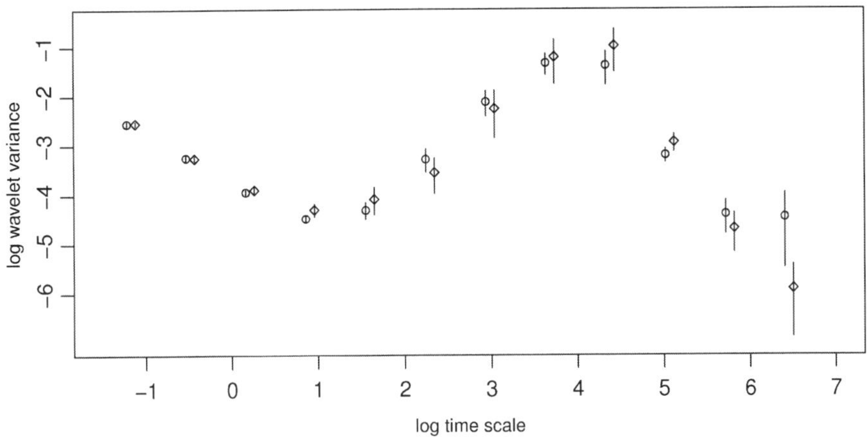

Fig. 38.6 Slepian wavelet variances of Z UMa on a log–log scale. *Circles* (*diamonds*) show the $c = 1$ ($c = 2.5$) estimates, while the *vertical lines* indicate 95% confidence intervals

Ursa Major and is an example of a semi-regular variable star. Its magnitude ranges from 6.2 to 9.4 V, and its pulsating period is about 195.5 days. Observational evidence suggests that Z UMa has more than just one pulsation cycle.

The works of [2, 12, 14, 31] theorize that Z UMa has multiple periods and that the irregularities seen in its light curve are either the result of the superposition of several different pulsation cycles within the star or are driven by the presence of a stellar companion, distorted stellar shapes, rotation, or star spots. The General Catalogue of Variable Stars classifies Z UMa as a semi-regular variable of subtype B. In other words, it has either a poorly defined periodicity or alternating intervals of slow irregular changes.

The time period we consider here ranges over Julian days 2,445,854–2,452,140, which gives us an irregularly sampled light curve data with $N = 20{,}227$ values and an average sampling interval of $\bar{\Delta} = 0.31$ days. Figure 38.6 shows the Slepian wavelet variance estimates for $c = 1$ (circles) and $c = 2.5$ (diamonds). The two estimates agree well at all scales except for the very largest, for which the uncertainty in the estimates is quite large. For $c = 1$ the wavelet variance plot has a peak at scale $\tau_8 = 79.57$ days. A sinusoid with a period of P will show up on a wavelet variance curve as a peak value near scale $P/2$, suggesting a nominal period of 159.14 days, in reasonable agreement with the pulsating period deduced by other methods. By contrast, the $c = 2.5$ plot has its peak at a scale of $\tau_9 = 159.14$ days. The broad-band nature of the fluctuations associated with semi-regular variable stars suggests that it might be more fruitful to view their light curves in terms of a characteristic scale rather than periodicities [13]. Note also that the decrease in the estimated wavelet variance curve at the four smallest scales is approximately linear on the log–log scale, suggesting that a power law might govern the small scale fluctuations of this star.

38.6 Discussion

As an estimator of the wavelet variance, the average energy in (38.15) is inherently biased, largely due the irregular nature of the sampling times. Under certain sampling scheme (e.g., jittered sampling or some other mildly irregular samplings [1, 34]), this bias can be negligible; however, the bias can be substantial for the highly irregular sampling schemes occurring in astrophysical applications. We know from the continuous energy maximization problem that the adaptive Slepian filters at large scales just mimic the continuous wavelet function. Thus the bias problem of bias is not due to the adaptive wavelet filters per se, but is largely due to the nature of the sampling times. In correct this bias, we need to understand the dependence of (38.15) on the distribution of the sampling times. Probabilistic calculations show that the bias depends on the distribution of time intervals $t_l - t_{l'}$ for $l, l' = 0, \ldots M_j - 1$. Correcting for the bias is challenging, but can be attempted at the expense of extensive computations aimed at estimating certain distributions associated with the sampling intervals. Indeed, this is where the work of [22] becomes relevant in that we can essentially generalize their variogram type estimator to obtain asymptotically unbiased quadratic estimate of wavelet variances.

Although multiwavelets scheme have previously been investigated in the context of regularly sampled time series [17], there is scope for development of this scheme, for both the regular and irregular sampling cases. In multitaper spectral estimates [27, 33], the amount of smoothing desired in the resulting estimate is determined by the number of multitapers used to form the estimate. It would be interesting to know how we can optimally determine the value of c and how we can optimally choose the number of Slepian multiwavelets to be included, with the goal of reducing the mean square error as much as possible.

Finally, in our closing remarks, we briefly comment upon the construction of Slepian wavelets in two and higher-dimensions, which can have important applications in variance analysis of random fields. Basically, in higher dimensions, Slepian filters can also be constructed for regular and irregular sampling schemes by maximizing the concentration of energy in an appropriate pass-band of frequencies and solving the related eigenvalue problem. In addition, the use of annular and other elliptical regions as pass-band of frequencies will give rise to genuine higher dimensional wavelet filters, which can have some advantages over the commonly used constructions by tensor products.

Acknowledgements Preparation of this article and the research on irregular data were supported in part by U.S. National Science Foundation Grant No. DMS 0906300. Any opinions, findings and conclusions or recommendations in this article are those of the authors and do not necessarily reflect the views of the National Science Foundation.

References

1. Bardet, J.-M. (2010). A non-parametric estimator of the spectral density of a continuous-time Gaussian process observed at random times. *Scandinavian Journal of Statistics 37*, 458–476.
2. Barnbaum, C., Morris, M., Kahane, C. (1995). Evidence for rapid rotation of the carbon star V Hydrae. *Astrophysical Journal 450*, 862.
3. Beutler, F. (1970). Alias-free randomly timed sampling of stochastic processes. *IEEE Transactions on Information Theory 16*, 147–152.
4. Bronez, T. P. (1998). Spectral estimation of irregularly sampled multidimensional processes by generalized prolate spheroidal sequences. *IEEE Transactions on Acoustics, Speech, and Signal Processing 36*, 1862–1873.
5. Chiann, C., Morettin, P. A. (1998). A wavelet analysis for time series. *Nonparametric Statistics 10*, 1–46.
6. Daubechies, I. (1992). *Ten Lectures on Wavelets*. Philadelphia: SIAM.
7. Foster, G. (1996). Wavelets for period analysis of unevenly sampled time series. *Astronomical Journal 112*, 1709–1729.
8. Frick, P., Grossmann, A., Tchamitchian, P. (1998). Wavelet analysis of signals with gaps. *Journal of Mathematical Physics 39*, 4091–4107.
9. Greenhall, C. A., Howe, D. A., Percival, D. B. (1999). Total variance, an estimator of long-term frequency stability. *IEEE Transactions on Ultrasonics, Ferroelectrics, and Frequency Control 46*, 1183–1191.
10. Herrndorf, N. (1984). A functional central limit theorem for weakly dependent sequences of random variables. *Annals of Probability 12*, 141–153.
11. Hertz, P., Feigelson, E. D. (1997). A sample of astronomical time series. In *Applications of Time Series Analysis in Astronomy and Meteorology*, edited by T. Subba Rao, M. B. Priestley and O. Lessi. London: Chapman & Hall, 340–356.
12. Isles, J. E. (1988). Big Dipper variables. *Sky and Telescope*, 88–100.
13. Keim, M. J., Percival, D. B. (2011). Assessing characteristic scales using wavelets. Submitted.
14. Kiss, L. L., Szatmary, K., Cadmus R. R., Mattei, J. A. (1999). Multiperiodicity in semiregular variables. *Astronomy and Astrophysics 346*, 542–555.
15. Labat, D., Ababou, R., Mangin, A. (2001). Introduction of wavelet analyses to rainfall/runoffs relationship for a karstic basin: the case of licq–atherey karstic system (France). *Ground Water 39*, 605–615.
16. Lark, R. M., Webster, R. (2001). Changes in variance and correlation of soil properties with scale and location: analysis using an adapted maximal overlap discrete wavelet transform. *European Journal of Soil Science 52*, 547–562.
17. Lilly, J. M., Park, J. (1995). Multiwavelet spectral and polarization analyses of seismic records. *Geophysical Journal International, 122*, 1001–1021.
18. Massel, S. R. (2001). Wavelet analysis for processing of ocean surface wave records. *Ocean Engineering 28*, 957–987.
19. Masry, E. (1978). Alias-free sampling: An alternative conceptualization and its applications. *IEEE Transactions on Information Theory 24*, 317–324.
20. Merlevéde, F., Peligrad, M. (2000). The functional central limit theorem under the strong mixing condition. *Annals of Probability 28*, 1336–1356.
21. Mondal, D. (2007). Wavelet variance analysis for time series and random fields. PhD thesis, University of Washington.
22. Mondal, D., Percival, D. B. (2010). Wavelet variance analysis for gappy time series. *Annals of the Institute of Statistical Mathematics 62*, 943–966.
23. Norris, J. P., Hertz, P., Wood, K. S., Vaughan, B. A., Michelson, P. F., Mitsuda, K., T. Dotani, T. (1990). Independence of short time scale fluctuations of quasi-periodic oscillations and low frequency noise in GX 5–1. *Astrophysical Journal 361*, 514–526.

24. Pelgrum, H., Schmugge, T., Rango, A., Ritchie, J., Kustas, B. (2000). Length-scale analysis of surface albedo, temperature, and normalized difference vegetation index in desert grassland. *Water Resources Research 36*, 1757–1766.
25. Percival, D. B. (1995). On estimation of the wavelet variance. *Biometrika 82*, 619–631.
26. Percival, D. B., Mondal, D. (2011). A wavelet variance primer. To appear in *Handbook of statistics Vol 30: Time Series*, edited by T. Subba Rao and C. R. Rao. Chennai: Elsevier.
27. Percival, D. B., Walden, A. T. (2000). *Wavelet Methods for Time Series Analysis*. Cambridge, UK: Cambridge University Press.
28. Rao, C. R. (1973). *Linear Statistical Inference and Its Applications* (Second Edition). New York: John Wiley.
29. Scargle, J. D., Steiman-Cameron, T., Young, K., Donoho, D. L., Crutchfield, J. P., Imamura, J. (1993). The quasi-periodic oscillations and very low frequency noise of Scorpius X1 as transient chaos: a dripping handrail? *Astronomical Journal, 411*, L91–L94.
30. Slepian, D. (1983). Some comments on Fourier analysis, uncertainty and modeling. *SIAM Review, 25*, 379–393.
31. Suchko, M. K. (1980). The periodicities of Z Ursae Majoris. *Journal of the AAVSO 9*, 74–80.
32. Sweldens, W. (1998). The lifting scheme: a construction of second generation wavelets. *SIAM Journal of Mathematical Analysis 29*, 511–546.
33. Thomson, D. J. (1982). Spectrum estimation and harmonic analysis. *Proceedings of the IEEE, 70*, 1055–1096.
34. Thomson, P. J., Robinson, P. M. (1996). Estimation of second-order properties from jittered time series. *Annals of the Institute of Statistical Mathematics 48*, 29–48.
35. Torrence, C., Compo, G. P. (1998). A practical guide to wavelet analysis. *Bulletin of the American Meteorological Society 79*, 61–78.
36. Vio, R., Strohmer, T., Wamsteker, W. (2000). On the reconstruction of irregularly sampled time series. *Publications of the Astronomical Society of the Pacific 112*, 74–90.

Chapter 39
Commentary: Slepian Wavelet Variances for Regularly and Irregularly Samples Time Series

Jeffrey D. Scargle

Abstract This commentary compares the wavelet variance described by Debashis Mondal and Don Percival with the Fourier power spectrum more familar to astronomers. Slepian Wavelets can also be used as tapers for spectral analysis in general, and I briefly describe the corresponding multi-taper estimation of power spectra and time-frequency distributions, demonstrated on the same data analyzed in their paper.

39.1 Characterizing Variability and Its Time-Scale Dependence

While characterizing brightness variations has always been a cornerstone of astronomy, large scale photometric survey programs are now generating a critical need for algorithms to explore massive time series databases. Irregular time sampling, characteristic of many astronomical data streams, may limit the information that classical analysis tools can extract.

In this setting Debashis Mondal and Donald Percival[1] provide an important tool. Their exposition of a method to estimate wavelet variance for irregularly sampled data is a fine example of mathematical expositions of modern time series analysis techniques, in works such as [2,6] to name just two others, that promise rich rewards in a variety of applications. The goal of the next section is to explain the place of wavelet variance in the context of variability analysis in general, largely by comparison with related Fourier methods more familiar to astronomers.

[1]Hereafter deonted **M&P**.

J.D. Scargle (✉)
Planetary Systems Branch, NASA Ames Research Center, Moffett Field, CA, USA
e-mail: Jeffrey.D.Scargle@nasa.gov

E.D. Feigelson and G.J. Babu (eds.), *Statistical Challenges in Modern Astronomy V*, Lecture Notes in Statistics 209, DOI 10.1007/978-1-4614-3520-4_39,

But first let's address the issue of technical conditions needed for the validity of theorems. These mathematical caveats may seem mysterious and difficult to assess in practice, and thus may be barriers to use of the methods by astronomers. **M&P** does not involve many such conditions, but as an example in Sect. 4 we find the assumption "... that the sampling times $t_0, t_1, \ldots, t_{N-1}$ are a realization of a stationary point process" followed by details related to the distribution of sampling intervals. Astronomers do not typically think of sampling as a stochastic process (although of course it is) and would be hard pressed to decide whether any particular mathematical condition is satisfied. The sampling, perhaps determined by weather and telescope time assignment, rather "is what it is." However digging a little deeper what is at play here is the concept of *independence of the samples*—not in the usual sense of statistical independence of the measured quantities but in the sense that the presence of an observation at one time does not affect the probability of an observation at some other time. Unfortunately this condition is rarely satisfied: we have affects ranging from detector dead-time to sampling cadence being changed in mid-stream based on analysis of previous data. So what should we do if we are not sure to what extent such conditions are satisfied by our observations?

Perhaps the most important advice is to begin any analysis by studying the distribution of the time intervals between successive observations. While a simple histogram of these intervals does not address the above independence issue, it almost always provides useful information. Unless something pathological is discovered with this or other exploratory analysis of the sampling it is probably justified to proceed with the analysis without undue concern about whether the technical conditions are fully satisfied. And the sensitivity of an algorithm to statistical conditions and the like can always be studied by analysis of simulated data with known properties.

39.2 Wavelet Variance vs. the Power Spectrum

Wavelets have been enthusiastically taken up by astronomers to implement sophisticated analysis of image and time series data [7]. As described by **M&P** wavelet variance is a basic tool for characterizing variability at different time scales. Wavelet variance as a function of time scale can be called the *wavelet power spectrum*, in justified analogy to the Fourier power spectrum as a function of frequency—a workhorse method for well over a century. The two functions are largely different packaging of essentially the same information. The most fundamental difference lies in the natural independent variable: a logarithmic time-scale for wavelet power *vs.* a frequency scale for Fourier power. Of course the latter can be shown on a logarithmic frequency scale as in Fig. 39.1, which shows the usual noisiness of the Fourier power and its amelioration by smoothing. The unsmoothed spectrum is too noisy to plot; dots and solid lines depict blocks of 16 and 128 points, respectively, averaged logarithmically. In addition the better frequency resolution results in the detection of a quasi-periodic signal that is smoothed over in the wavelet variance

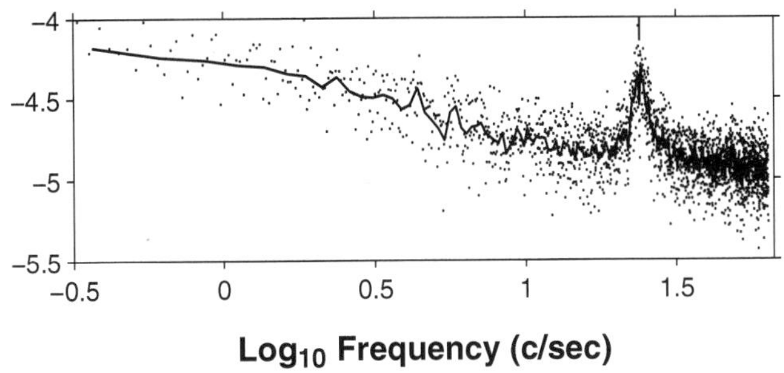

Fig. 39.1 Logarithmic power spectrum of GX 5-1, from the same data as in **M&P**

analysis. The broad peak near 25 c/s (marked at the top of the figure) is the horizontal branch oscillation that ranges from 15 to 50 c/s depending on the physical state of this system (Fig. 2 of [4]).

39.3 Multitaper Time-Scale/Time Frequency Distributions

Another parallel between Fourier and wavelet methods is the connection between time-frequency distributions and the wavelet transform itself. The scalogram, a plot of the magnitude of the wavelet coefficients against the time and scale independent variables, is much like the time-frequency distribution [2]. This very useful quantity presents power as a function of time and frequency, trading off the corresponding resolutions which are of course subject to the uncertainty principle: fine time resolution means coarse frequency resolution and *vice versa.*

To deal with uneven spacing the Lomb-Scargle Periodogram [3, 5, 8, 9, 12], is often used. Here we use an alternative, starting with the correlation algorithm [1] often used in astronomy and well studied in the signal processing literature under the name of *slotted techniques* (e.g. [10, 11]). The basic idea is to construct bins in the lag variable τ and sum the product $x(t_1)x(t_2)$ over all data pairs such that their time difference lies in a given such bin:

$$\rho(\tau_k) = \frac{1}{N_k} \sum_n X(t_n)X(t_m) \tag{39.1}$$

where the sum is over all pairs n, m such that the corresponding time difference $t_n - t_m$ lies within the bin $[\tau_k, \tau_k + \Delta\tau]$, and N_k is the number of such pairs. It is usual to write this formula replacing X_n with $X_n - \mu_X$, where μ_X is the mean value of X, either theoretical or empirical. Here we assume an empirical mean has been subtracted. The average product $x(t_1)x(t_2)$ is taken to describe the degree to which values separated by τ are related (large if positively correlated, large and negative if anti-correlated, and small if uncorrelated).

The role of the factor $\frac{1}{N_k}$ is interesting. In estimating correlation functions for evenly spaced data two variants are used

$$\rho(k) = \frac{1}{N} \sum_n X_n X_{n+k} \tag{39.2}$$

and

$$\rho(k) = \frac{1}{N-k} \sum_n X_n X_{n+k} \tag{39.3}$$

representing a trade-off favoring small variance (with larger bias) or small bias (with larger variance) respectively. Equation 39.1 corresponds to (39.3) since in both cases the denominator is the number of terms contributing to that value of the lag, so the expression is truly an average. If desired the analog of (39.2) could be implemented simply by replacing N_k with a constant.

Even though (39.1) seems a bit abstract it is easily computed in practice. For evenly spaced data with gaps the binning in τ should correspond to the constant sampling interval. The power spectrum can then be computed using the well-known identity that the power spectrum is the Fourier transform of the autocorrelation function (which needs to be evaluated to the maximum lag possible, namely equal to the entire time-span of the observations). Two potential difficulties, the possibility of empty bins or of negative estimated powers, are of no concern here.

With an algorithm in hand to compute the power spectrum (either the procedure just outlined or the Lomb-Scargle periodogram) it is completely straightforward to compute the time-frequency distribution simply by accumulating a matrix of power spectra of the data points in a sequence of windows slid along the observation interval. The most important parameter is the width of the window. A good choice with the present data was found to be about 0.05 times the whole interval.

The final issue has to do with improving the information throughput of the power spectrum procedure by applying a *taper* (sometimes called a *spectral window,* not to be confused with the window discussed in the context of time-frequency distributions). Mulitaper analysis utilizes the solution to the problem of optimizing the ability of the taper to concentrate power into the main lobe of the spectrum, and minimize leakage of power into the side-lobes (cf. **M&P** and [6]). The mathematical solution yields a number of taper shapes that are all approximately optimal. The best one smoothly emphasizes central data at the expense of the data near the ends of the observation interval. The others compensate for this information loss to some extent by reversing this relation. Figure 39.2 shows the time-frequency distribution of the Z UMa data in **M&P**, computed via Edelson and Krolik autocorrelations with a simple Slepian taper. The full spectrum has large power at 1 c/day and an interesting set of aliases, as expected from nightly observations, but in the frequency range shown in the plot we see the comings and goings of the 195.5 day period noted by Isles (reference [12] in **M&P**) and its first harmonic.

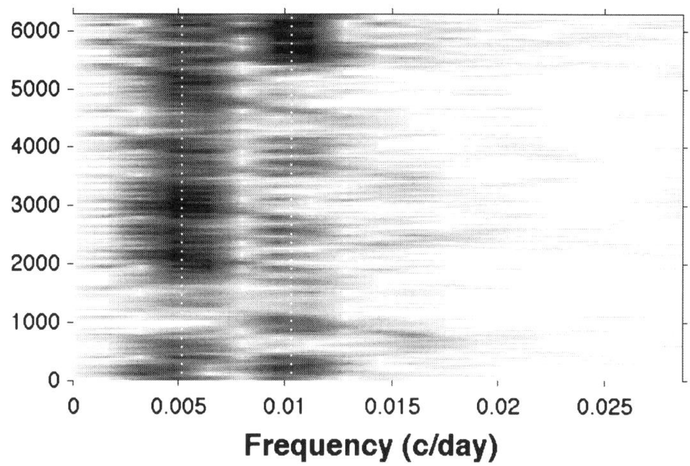

Fig. 39.2 Z UMa time-frequency distribution. *Dotted lines*: 195.5 day period and its harmonic

Wavelet and Fourier power spectra are complementary tools, useful for extracting different aspects of the information contained in time series data. Slepian tapers as optimal solutions to the spectral leakage problem play a similar role in both.

Acknowledgements I am grateful to Debashis Mondal, Don Percival, and Joe Bredekamp and the NASA Applied Information Systems Research Program for encouragement and support.

References

1. Edelson, R. A. and Krolik, J. H. (1988), Astrophysical Journal, 333, 646.
2. Flandrin, P., Time-Frequency/Time-Scale Analysis (Temps-Frquence) (1999), Academic Press: London
3. Gottlieb, Wright and Liller, "Optical studies of UHURU sources. XI. A probable period for Scorpius X-1 = V818 Scorpii," (1975) Astrophysical Journal Letters, Vol. 195, p. L33–L35.
4. Jonker, P. G., van der Klis, M., Homan, J., Mndez, M., Lewin, W. H. G., Wijnands, R., and Zhang, W. (2002), "Low- and high-frequency variability as a function of spectral properties in the bright X-ray binary GX 5-1," Monthly Notices of the Royal Astronomical Society, 333, pp. 665–678.
5. Lomb, N. R. (1976) "Least-squares frequency analysis of unequally spaced data," Astrophysics and Space Science, vol. 39, p. 447
6. Percival, D. B. and Walden A. T. (1993) *Spectral Analysis for Physical Applications: Multitaper and Conventional Univariate Techniques*, Cambridge University Press, Cambridge, UK
7. Percival, D. B. and Walden A. T. (2000) *Wavelet Methods for Time Series Analysis*, Cambridge University Press, Cambridge, UK
8. Scargle, J. D. (1982), "Studies in astronomical time series analysis. II - Statistical aspects of spectral analysis of unevenly spaced data," Astrophysical Journal, **263**, 835–853

9. Scargle, J. D. (1989), "Studies in astronomical time series analysis. III - Fourier transforms, autocorrelation functions, and cross-correlation functions of unevenly spaced data," Astrophysical Journal, **343**, 874–887.
10. Stoica, P. and Sandgren, N. (2006) "Spectral analysis of irregularly-sampled data: Paralleling the regularly-sampled data approaches," Digital Signal Processing, vol. 16, 712.
11. Rehfeld, K., Marwan, N., Heitzig, J. and Kurths, J. (2011) "Comparison of correlation analysis techniques for irregularly sampled time series," Nonlinear Processes in Geophysics, Vol. 18, 389–404
12. Vancek, P. (1971), "Further Development and Properties of the Spectral Analysis by Least-Squares," Astrophysics and Space Science, Vol. 12, p. 10.

Part V
The Future of Astrostatistics

Chapter 40
Astrostatistics in the International Arena

Joseph M. Hilbe

Abstract It was not until the last decades of the twentieth century that computing power and memory allowed the development of statistical software that was rigourous enough to entice astronomers to again become interested in statistics. During the 1990s, personal computers allowed estimation of continually sophisticated iterative statistical routines that could be used to understand astronomical data. During this time small groups of astronomers and statisticians joined together developing both collaborations and conferences to discuss statistical methodology. As computers and accompanying software allowed even more complex models to be developed in the first decade of the twenty-first century, astrostatistics was born as a discipline. Although astrostatistics researchers can now employ memory-intensive Bayesian methods to data that was never possible in earlier years, the vast amount of data being collected by the new data-generating technologies will soon obfuscate current statistical and data mining capabilities. Through the formation of an international astrostatistics network or association and the creation of interdisciplinary astrostatistics degree programs throughout the world, astrostatisticians in the future will collaborate to develop the new mathematics and statistics required to handle the huge amounts of data being gathered. It will also solidify astrostatistics as a profession.

Statistics can be said to have begun with the Babylonians, Egyptians, and in particular with the Greeks, several thousand years ago when scholars, farmers, and businessmen applied such descriptive concepts as the median and range to agricultural and astronomical observations. However, it was not until the first decade of the nineteenth century that simple inferential statistical models were developed.

J.M. Hilbe (✉)
School of Mathematics and Statistical Sciences, Arizona State University,
P.O. Box 871804, Tempe, AZ 85287, USA
e-mail: hilbe@asu.edu

E.D. Feigelson and G.J. Babu (eds.), *Statistical Challenges in Modern Astronomy V*,
Lecture Notes in Statistics 209, DOI 10.1007/978-1-4614-3520-4_40,

The first application of Carl Gauss's method of least squares regression was to the prediction of the apparent position of Ceres as it came into view from its orbit behind the Sun. This occurred in 1801, the first year of the new century. However, Gauss himself did not fully describe the use of least squares regression until 8 years later in a text on Celestial Mechanics.

Although both descriptive statistics and inferential statistics largely began with applications to astronomical data, astronomers from the early-mid 19th to near the end of the last century showed relatively little interest in applying more than basic descriptive statistical methods to astronomical data. There were some exceptions, of course, but astronomers in general turned to non-statistical quantitative methods and later to spectroscopy and differential equations for the understanding of astronomical data throughout the majority of the nineteenth and twentieth centuries. Astronomers have in recent years taken note of the advancements in statistical methodology, and the wide range of capabilities that statisticians now have to understand large and even ill-shaped data situations. These capabilities were not available before. On the other side, the non-statistical methods that were used by most astronomers in the past, which were indeed very effective for what was being analyzed, are becoming no longer satisfactory when attempting to model the large masses of data that are being generated by the new data generating technologies

40.1 Statistical Software for Astronomy

Central to astrostatistical research is the software used for the various analyses being undertaken. The majority of current astrostatisticians now use R, Python, or R and WinBUGS together for their research. R and Python are both freeware software applications to which users may easily add their own functions and scripts. Python, a multi-paradigm programming language, was first released in 1991, with major enhancements in 2000 (ver. 2) and 2008 (ver 3). CPython, the current default implementation of Python, is widely used in the physical sciences. R, on the other hand, was first developed in 1993, went through several years of alpha testing, and was officially released in 2000. Prior to the availability of R and Python, astronomers and astrostatisticians generally used S-Plus or fashioned their own FORTRAN or C functions. S-Plus is similar in structure to R, but is commercial, and has steadily been losing its user-base to software such as Stata and R. R however, because of its low-level programming capabilities, can be used to construct most any statistical model or procedure, making it a powerful analytic tool. The fact that R has become so popular across academic disciplines, thus leading to the creation of a variety of methods and functions which can later be adapted for astronomical research, has made it very attractive as a software application. Astronomers have generally preferred to write their own software rather than rely on commercial statistical packages. R and CPython suit this preference fine.

From the mid-late 1990s to the present, a growing number of astronomers have been turning to Bayesian methodology for the construction of statistical models. However, general Bayesian methods for astronomical research required the development of software appropriate for the types of analyses needed for such work. It was not until WinBUGS was sufficiently programmed during the last decade, though, that it could be used for serious Bayesian astrostatistical modeling. R is also used for Bayesian analysis, and more sophisticated functions are continually being developed, but typically users of R software incorporate WinBUGS into the R environment for difficult modeling tasks. Other Bayesian software exists as well, but is generally not used by astronomers or astrostatisticians. CPython has not had built-in Bayesian modeling capability, and therefore some astronomers have turned to R-WinBUGS as a result. Statisticians have not supported CPython, favoring applications such as R-WinBUGS, Stata, and SAS.

Astrostatistics was first described as such by Babu and Feigelson in the Preface to their seminal work on the subject, *Astrostatistics* [1]. The text was authored just prior to the publication of WinBUGS and to the popularity of R, which was then barely conceived. As of March 2011, however, there are nearly 3,000 R packages residing on CRAN mirror sites located throughout the world. Users may download a core of packages, which consist of the base R program, directly from The R Project for Statistical Computing web site (http://www.r-project.org/). Other packages may be downloaded and installed on one's computer with ease from the Comprehensive R Archive Network (CRAN) folder within the base web site.

Babu and Feigelson's text represents the manner in which astronomers and associated statisticians analyzed data prior to the use of the specialized Bayesian software that made Bayesian analysis feasible for astronomical analysis. Although solid astrostatistical work is still being done using the traditional frequentist approach to statistics, Bayesian methods now predominate in the literature. This trend has particularly grown in the past 5 years.

40.2 Recent Growth of Astrostatistics

Beginning in the mid 1980s, astronomers began to organize small conferences devoted to what we may now call astrostatistics. One of the first was the *Statistical Methods in Astronomy* conference held in Strasbourg in 1983. The *Statistical Challenges in Modern Astronomy* conference has maintained a regular timetable over two decades, held every 5 years since its inception in 1991. Under the direction of Jogesh Babu and Eric Feigelson of the Pennsylvania State University Center for Astrostatistics, the conference has brought together both astronomers and statisticians from around the world for weeklong series of discussions.

During the 1990s several groups were organized consisting of astronomers and statisticians having a common interest in developing new statistical tools for understanding astronomical data. Two of the foremost groups are the California/Boston/Smithsonian Astrostatistics Collaboration (CHASC), headed by David van Dyk of the University of California, Irvine, and the International Computational Astrostatistics (InCA) Group, which is primarily comprised of researchers from

Carnegie Mellon University and the University of Pittsburgh. CHASC, InCA, and the Pennsylvania State University all belong to Large Synoptic Survey Telescope (LSST) Project, which will provide huge amounts of data for analysis. The 8.4 m LSST is currently scheduled to begin surveying activities in 2014.

Several sites in various parts of the world are presently engaged in developing astrostatistics programs and collaborations. An astrostatistics concentration program is being developed by the joint efforts of the departments of Statistics and Astronomy/Astrophysics at Imperial College in London. Conferences on astrostatistics and degree specializations in the discipline are also being developed at the University of Calcutta, and at Pennsylvania State University, the University of Pittsburgh, Carnegie Mellon, Harvard University, University of Florida, University of Birmingham, and other sites.

When astronomers again began to utilize inferential statistical methods into their published research, many of the articles employed inappropriate statistical analyses, or if correct methodology was employed, the analyses generally failed to account for possible violations of the assumptions upon which the research models were based. That is, they did not fully appreciate the statistical theory underlying their analyses. It was certainly not that astronomers lacked mathematical expertise to understand these assumptions; rather it was that many had no special training in statistical estimation. Moreover, many astronomers tended to use only a limited number of statistical procedures. They had not become aware of the vast range of statistical capabilities that had become available to professional statisticians and other researchers (Feigelson and Babu 2004). Of course, there were noted exceptions, but it became readily apparent in the late 20th and during the first decade of the twenty-first centuries that astronomers in general needed to enhance their statistical knowledge. Those astronomers who took up this challenge believed that the best way to address the problem was to conduct conferences and organize collaborative research groups consisting of both astronomers and statisticians. I earlier mentioned some of these groups and conferences.

As of 2009, a relative handful of astronomers and statisticians with an interest in the statistical analysis of astronomical data were associated with collaborative organizations such as CHASC and InCa. Some 100 'astrostatisticians' attended the quint-annual Statistical Challenges conference at Pennsylvania State University. Other conferences have also been ongoing in Europe such as *Astronomical Data Analysis* organized by Jean-Luc Starck and *Cosmostat*. The remaining astrostatisticians have established collaborative associations within their own universities, or within a small group of universities.

40.3 Astrostatistics and the International Statistical Institute

Some excellent work was being done in the area of astrostatistics, but communication between astrostatisticians on a global basis has been rather haphazard. Until recently there has been no overall organization or association for the discipline.

To address this need, in early 2008 I formed an astrostatistics interest group within the fold of the International Statistical Institute (ISI), the world association of statisticians, with headquarters in The Netherlands. As part of my association with NASA's Jet Propulsion Laboratory, I had been on numerous conference calls with the directors of various NASA and JPL projects and missions since 2007, and repeatedly heard that statistical issues were going to be a problem in the analysis of their data. This in turn stimulated me to explore the possibility of forming an association of astrostatisticians that would encourage the global collaboration of statisticians and astronomers with the aim of effecting better statistical research. I was also interested in promoting a professional association for those who considered themselves as astrostatisticians, and not only as a member of the statistical or astronomical communities.

In December 2009, the ISI Executive Committee approved the existence of astrostatistics as a full standing committee of the ISI (http://isi-web.org/com/ast). However, ISI committees consist of no more than 12–15 members. I was receiving numerous inquires about membership from reseachers of both the statistical and astronomical disciplines. As a consequence, the ISI Astrostatistics Network was formed as separate body of researchers, with the ISI astrostatistics committee serving as the Network executive board. Membership has grown to some 130 members from 26 nations and all populated continents.

After 16 months of existence, the Network has established solid relationships with both the ISI and International Astronomical Union, whose leadership has supported the Network and its goals. Network members were awarded an invited papers session and two special topics sessions at this years ISI World Statistics Congress in Dublin. In addition, discussions have been underway with several publishing houses regarding a possible *Journal of Astrostatistics*. The Network will only proceed with such a venture, of course, if it is assured that there will be a rather steady long-term stream of quality submissions made to the journals editorial board. As of this time we are not convinced that this will be the case in the immediate future, but do plan for such a journal in the future.

As a consequence of the initial successes of the Network, in December 2010 Springer Science and Business Media begun a *Springer Series on Astrostatistics*, on which Network members hold the editorial board positions. The series will publish texts and monographs on a wide variety of astrostatistical and astrophysical subjects. A separate Springer astrostatistics e-book series is also being developed to publish the Proceedings of major astrostatistical conferences throughout the world.

It is clear that many in the astrostatistics community believe that the existence of a global association of astrostatisticians is a worthwhile body to support. However, such an association, currently called the *International Astrostatistics Network*, is not aimed to be a governing organization, but rather an association to augment and support the ongoing efforts of established astrostatistics groups and conferences. The Network consists of researchers with a common interest and a resource to help disseminate information regarding astrostatistics related literature, conferences, and research. Most importantly, it can also serve as the professional society for those identifying themselves as astrostatisticians. Astrostatistics as a profession is but in its infancy at this time, but it is hoped that a viable profession will be established within the next 20 years.

40.4 Astrostatistics into the Future

Astrostatistics faces some formidable challenges. The International Virtual Observatory Alliance (IVO, http://www.ivoa.net) is now being constructed which will link archival astronomical databases and catalogues from the many ongoing surveys now being maintained, including LSST. The goal is to make all gathered astronomical data available to astronomers and astrostatisticians for analysis. However, this will involve many petabytes of information. In a relatively short time the amount of data may exceed an Exabyte, or a thousand petabytes. This is a truly huge amount of data. Even when dealing with terabytes, current statistical software is not capable of handling such an amount of information. A regression of a billion observations with ten predictors results in a matrix inversion that far exceeds current and realistically foreseeable capabilities. New methods of statistical analysis will need to be developed to deal with these large datasets, and new statistical methods will need to be created that can evaluate such large amounts of data. There are a host of statistical and data mining problems related to evaluating huge masses of data in the attempt to determine the probability of some proposed outcome or event.

Ultra-large models can be attacked using sequential modeling, saving mean statistics with each iteration, or to partition the data into one-million-observation models, constructing thousands of these models and putting the summary values into a metamodel. The statistics of meta-analysis can help as well. Of course, with more and more data, anomalous observations become dampened out and may be missed in analysis. Other problems exist as well in implementing such methods.

Preferably, statistical analysis should be made on as much data as possible. Researchers are now developing *VOStat* (http://vostat.org), a suite of statistical tools that will hopefully be adequate to evaluate the type of data I have been describing. Statisticians, computation specialists, and astronomers will have to work in concert to deal with these issues.

I believe that astrostatistics will best develop into a mature discipline, capable of handing the looming data and analytic problems, by becoming a profession. This requires developing joint programs in the discipline, sponsored and maintained by the mutual efforts of the departments of statistics and astronomy/astrophysics at leading universities. Graduates will be awarded MS and PhD degrees in astrostatistics, and be trained in statistical analysis, astrophysics, and computer and computational logic. With a new generation of astrostatisticians engaged in handling the problems. I have mentioned here, there is more likelihood that the foremost questions we have of the early universe, of the nature of dark matter and energy, of the likelihood of our existing within a multiverse, as well a host of other queries, can be answered.

References

1. Babu, G.J. and E.D. Feigelson (1996), Astrostatistics, Baton Rouge, FL: Chapman & Hall
2. Feigelson, E.D. and Babu, G.J. (2004) "Statistical challenges in modern astronomy", in *PhyStat2003: Statistical Problems in Particle Physics, Astrophysics, and Cosmology* (L. Lyons, ed.), SLAC, astro-ph/0401404

Chapter 41
The R Statistical Computing Environment

Luke Tierney

Abstract R is a computing environment for data analysis and graphics. R is designed as a high-level programming language that supports complex forms of data analysis as well as the development of new data analysis methodology. In recent years R has become the major framework for providing access to new statistical methodology, and thousands of extension packages are now available. This paper provides a brief introduction to R with examples drawn from astronomical data.

41.1 Introduction

R [9] is a language for data analysis and graphics originally developed by Ross Ihaka and Robert Gentleman at The University of Auckland in New Zealand. R is based on the S language [3, 4] developed by John Chambers and others at Bell Labs. As the primary data analysis framework for the statistics group at Bell Labs the S language placed a strong emphasis on flexibility and the ability to handle and adapt to non-standard problems. R has inherited this design philosophy. R is widely used in the field of statistics and beyond, especially in university environments, and R has become the primary framework for developing and making available new statistical methodology. It is, for example, rare these days for a Ph.D. thesis in statistics or biostatistics not to include and R package implementing the ideas developed in the thesis. Many R extension packages are available through CRAN (R Development Core Team [10]) or similar repositories; the number of packages available on CRAN now exceeds 3,000.

L. Tierney (✉)
Department of Statistics and Actuarial Science, University of Iowa, Iowa City, IA 52240, USA
e-mail: luke-tierney@uiowa.edu

E.D. Feigelson and G.J. Babu (eds.), *Statistical Challenges in Modern Astronomy V*, Lecture Notes in Statistics 209, DOI 10.1007/978-1-4614-3520-4_41,

Table 41.1 R core members and their country of residence

Douglas Bates (USA)	John Chambers (USA)	Peter Dalgaard (Denmark)
Robert Gentleman (USA)	Seth Falcon (USA)	Kurt Hornik (Austria)
Stefano Iacus (Italy)	Ross Ihaka (New Zealand)	Friedrich Leisch (Austria)
Uwe Ligges (Germany)	Thomas Lumley (New Zealand)	Martin Maechler (Switzerland)
Duncan Murdoch (Canada)	Paul Murrell (New Zealand)	Martyn Plummer (France)
Brian Ripley (UK)	Deepayan Sarkar (India)	Duncan Temple Lang (USA)
Luke Tierney (USA)	Simon Urbanek (USA)	

41.1.1 History and Development Model

R is an Open Source project. After a number of yeas of developing R on their own, initially for use in a Macintosh computer lab at the University of Auckland, Ross Ihaka and Robert Gentleman in 1997 established the R core group for developing and maintaining R. This group now consists of 20 researchers from a number of different countries shown in Table 41.1.

An essential component of the success of R is the strong support provided by its community of users. Elements of this support are the many contributed packages made available, contributions to discussion lists and blogs, contributed documentation and tutorials, and *Task Views* provided to help navigate the many available packages that might be useful in different areas of application [16].

41.1.2 Basic Design of R

The S language was originally designed as a framework to support and enable the statistics research group at Bell Labs to handle the rich set of non-standard problems it was confronted with. R retains that basic philosophy of enabling the exploration of new kinds of data, allowing the data to guide the choice of analysis to use, and allowing the analysis tools to be adapted to the data as needed. R is an interactive system, in contrast to batch-oriented systems; this supports and encourages exploratory data analysis. R is also designed as a high level language that can be used to express complex data transformation and analysis steps as well as for implementing new data analysis and display methods. The standard interface to R is a command line interface in which the user enters commands in the R language. This is in contrast to systems designed around a graphical user interface, such as JMP [6], though several extension packages providing graphical user interfaces to R are available [5, 15].

41.1.3 *Extending the R System*

Writing simple R functions is a natural part of working in R. Collections of functions that implement a particular analysis are often best organized into an extension *package*. The R package system provides a framework for developing, documenting, and testing extension code. Packages can include R code as well as foreign code (C, FORTRAN). Many R packages are made available though the CRAN repository [10]; the number of packages available on CRAN recently passed the 3,000 mark.

41.2 Basic Usage and Capabilities

R uses a command line interface, a *read-evaluate-print* loop: The user types an expression, R reads the expression, evaluates it, and prints the result. Some simple examples:

```
> 2 + 3
[1] 5

> exp(-2)
[1] 0.1353353

> log(100, base = 10)
[1] 2
```

A vector containing some uniform random numbers can be created using the `runif` function and assigned to the variable `x` with the *assignment operator* `<-`:

```
> x <- runif(4)
> x
[1] 0.1137034 0.6222994 0.6092747 0.6233794
```

Basic arithmetic operators and functions are *vectorized*: when applied to a vector they are automatically applied one element at a time to produce a vector of results. Scalars are recycled to match the length of the longest vector. Some examples of vectorized operations are

```
> x + 1
[1] 1.113703 1.622299 1.609275 1.623379
> log(x)
[1] -2.1741619 -0.4743339 -0.4954860 -0.4725999
```

41.2.1 *Numerical Summaries*

R provides functions for computing a range of basic numerical summaries such as the mean and standard deviation,

Table 41.2 Functions related to some standard probability distributions

Distribution	Density	CDF	Quantile	Generate
Uniform	dunif	punif	qunif	runif
Normal	dnorm	pnorm	qnorm	rnorm
t	dt	pt	qt	rt
F	df	pf	qf	rf
Gamma	dgamma	pgamma	qgamma	rgamma
Poisson	dpois	ppois	qpois	rpois

```
> mean(x)
[1] 0.4921642
> sd(x)
[1] 0.2523886
```

or the median and inter-quartile range:

```
> median(x)
[1] 0.6157871
> IQR(x)
[1] 0.1371875
```

Functions for sorting and ranking a data vector are also available:

```
> sort(x)
[1] 0.1137034 0.6092747 0.6222994 0.6233794
> rank(x)
[1] 1 3 2 4
```

41.2.2 Probability Distributions

R provides support for computing densities or probability mass functions, cumulative distribution functions, and quantile functions of many standard distributions. Functions for generating random variables are also available. A standard naming convention is used; density functions start with d, cumulative distribution functions with p, quantile functions with q, and random number generators with r. Table 41.2 shows some of the available functions. More distributions are available in extension packages.

41.2.3 Defining Functions

If a method is not available in R it can be implemented by defining a function using the R language. For example, the Pareto distribution is not covered by base R (though it is by several contributed packages). The CDF of the Pareto distribution is

$$F(x) = \begin{cases} 1 - x\left(\frac{x_m}{x}\right)^{\alpha} & \text{for } x \geq x_m \\ 0 & \text{otherwise} \end{cases}$$

for parameters $x_m, \alpha > 0$. R functions are created by the `function` operator and given a name by assigning the result to a variable. A vectorized R function `ppareto` to compute this CDF is defined by

```
ppareto <- function(x, xmin, alpha = 1) {
    stopifnot(all(xmin > 0) \&\&  all(alpha > 0))
    1 - pmin(xmin / x, 1) \^ alpha
}
```

41.2.4 Linear Algebra

Numerical linear algebra computations are at the heart of many statistical methods. R provides functions for computing QR and Cholesky factorizations, singular value decompositions, and PLU decompositions, and for solving triangular systems. Functions for computing eigenvalues and eigenvectors are also available. The implementations are based on the open source LAPACK, LINPACK, and EISPACK libraries. These in turn make use of the basic linear algebra subroutines (BLAS) library. The R source code includes a basic reference implementation of the BLAS, but it is easy to substitute a high performance BLAS such as Atlas [1], OpenBLAS [17] or a vendor BLAS implementation.

The `Matrix` package [2] provides a richer set of facilities for specialized types of matrices, including extensive sparse matrix support.

41.3 Graphics

R contains a rich set of graphical facilities. Several graphics frameworks are available that provide high level functions for creating a variety of statistical graphs. Two frameworks provided in the basic R distribution are *base graphics* and *Lattice graphics*. Two widely used frameworks available as add-on packages are `ggplot2` graphics and the interactive 3D framework `RGL`. A number of other graphics frameworks are available or in development

41.3.1 Base Graphics

Base graphics provides a number of standard graphs, such as dot plots, box plots, histograms, scatter plots, scatter plot matrices, and perspective plots. The base graphics framework is easy to use for simple tasks and supports incremental composition and augmenting of plots.

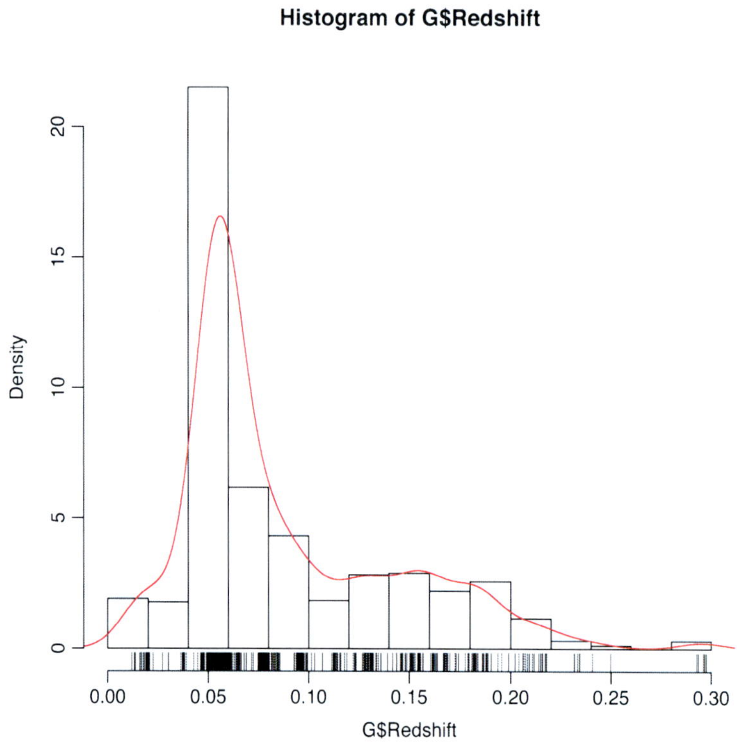

Fig. 41.1 Distribution of redshift values for galaxies in the Abell 85 cluster with redshift values less than 0.3

As an example, adapted from Alastair Sanderson's tutorial pages [11], we can examine data on galaxy cluster Abell 85 from the NASA/IPAC Extragalactic Database [8] and create a plot of the distribution of redshift values for galaxy objects with redshift less than 0.3. First, read in the data and simplify some variable names:

```
A <- read.table("a85\_extended\_NEDsearch.txt",
                sep="|", skip=20, header=TRUE)
colnames(A)[c(2,3,4,5)] <- c("name","ra","dec",
                              "type")
```

Next, create the subset with redshift values less than 0.3 as

```
G <- subset(A,type=="G"&!is.na(Redshift)
        &Redshift<0.3)
```

Finally, create a histogram of the data, superimpose a smooth density estimate, and add a *rug plot* showing the raw data values along the bottom:

```
hist(G\$Redshift, prob = TRUE)
lines(density(G\$Redshift), col = "red")
rug(G\$Redshift)
```

The result is shown in Fig. 41.1

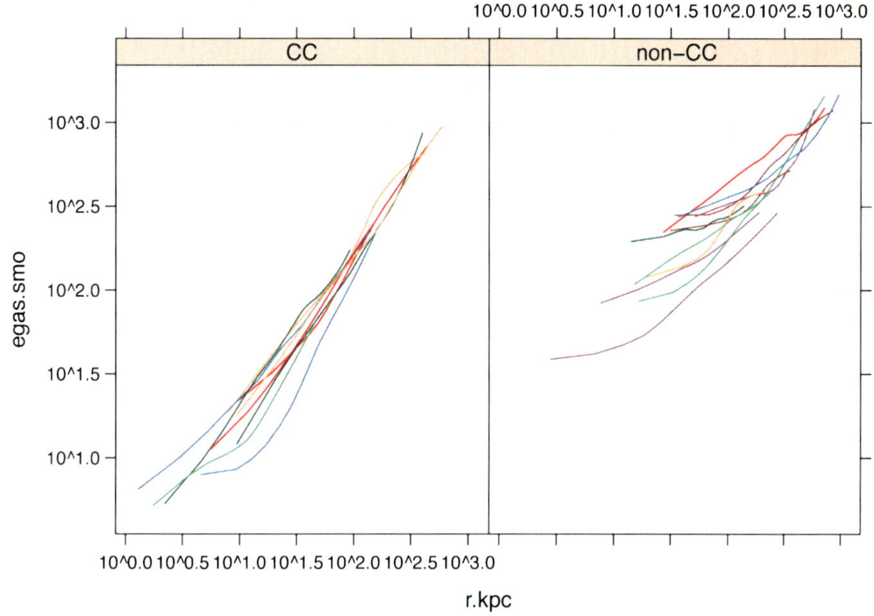

Fig. 41.2 Entropy profiles for cool core and non-cool core galaxy clusters

41.3.2 Lattice Graphics

Lattice graphics [12] are used for creating structured sets of related graphs for understanding multivariate data. Lattice was developed by Deepayan Sarkar based on Cleveland's Trellis system for S. Some advantages of lattice graphics include better, and customizable, default choices for many graphical parameters and layout, a simpler mechanisms for adding annotations, and a richer facility for showing multiple data sets in a single graph or set of graphs. Lattice plots usually display one or two variables given values of additional variables by *grouping*, using separate colors or symbols, or by *conditioning*, using separate plots on identical scales.

As and example, again adapted from Alastair Sanderson's tutorial pages [11], we use a set of smoothed gas entropy measurements at a series of radii of 20 galaxy clusters. The variables in the data set are

`egas.smo`: smoothed gas entropy
`r.kpc`: radius, in kiloparsecs
`cname`: cluster name
`cctype`: cool core or non-cool core

Figure 41.2 shows the entropy profiles for cool core and non-cool core galaxy clusters.

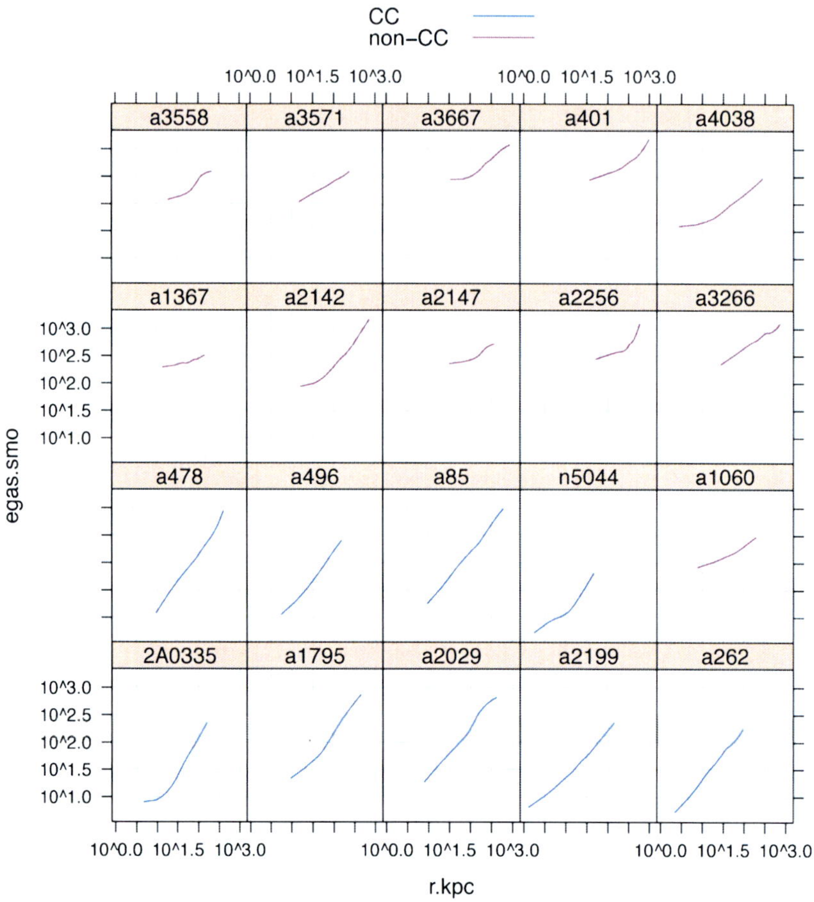

Fig. 41.3 Individual entropy profiles for cool core and non-cool core galaxy clusters

The plot is created by the expression

```
xyplot(egas.smo ~ r.kpc | cctype,  groups=cname,
       scales=list(log=TRUE),
       type=c("g", "l"), aspect = "xy", data=entropy)
```

The *formula* `egas.smo ~ r.kpc | cctype` requests a plot of `egas.smo` against `r.kpc` conditioned on the two values of `cctype`, thus producing two plots with common axes. The `groups` specification establishes the relation between data points within a common galaxy cluster and results in the common color and connected lines used for each galaxy cluster's profile.

A graph showing each profile in its own panel is shown in Fig. 41.3. This plot is created with the expression

```
xyplot(egas.smo ~ r.kpc | reorder(cname, as.numeric(cctype)),
       groups=cctype, type=c("g", "l"), data=entropy,
```

```
        scales=list(log=TRUE),
        auto.key=list(points=FALSE, lines=TRUE))
```

The `reorder` function is used to ensure that the cool core and non-cool core galaxy clusters are shown together.

41.4 Statistical Models

The basic R distribution supports fitting a wide range of statistical models, including linear and non-linear regression models, generalized linear models, mixed models, survival models, time series, and spatial models. Tools for general optimization and maximum likelihood fitting are also provided. Contributed packages add support for many more models and methods.

41.4.1 Common Modeling Function Features

Most modeling functions support a *formula language* for specifying a model. For example, the formula

```
y ~ a + b
```

would be used for specifying a regression of `y` on the variables `a` and `b`. Data is usually taken from a data frame specified as a `data` argument. Functions usually return a model object that can be used to extract coefficients and standard error estimates, compute residuals or fitted values, predict responses at new explanatory variable values, or obtain summary information for the fit. The most basic modeling function is `lm` for fitting linear models.

41.4.2 Linear Model Example

A simple, though not very sensible, model with separate slopes and intercepts for cool core and non-cool core galaxy clusters can be fit to the gas entropy data using the `lm` function. The `summary` function can then be used to print a standard summary of the fit:

```
> summary(lm(log(egas.smo) ~ cctype + cctype * log(r.kpc), data = entropy))
Call:
lm(formula = log(egas.smo) ~ cctype + cctype * log(r.kpc), data = entropy)

Residuals:
     Min       1Q   Median       3Q      Max
-0.87240 -0.15384  0.02201  0.16733  1.05008
```

```
Coefficients:
                           Estimate Std. Error t value Pr(>|t|)
(Intercept)                 0.57021    0.06988    8.16 5.12e-15 ***
cctypenon-CC                2.01888    0.11629   17.36  < 2e-16 ***
log(r.kpc)                  0.97672    0.01700   57.47  < 2e-16 ***
cctypenon-CC:log(r.kpc) -0.32162    0.02520  -12.76  < 2e-16 ***
---
Signif. codes:  0 *** 0.001 ** 0.01 * 0.05 . 0.1   1

Residual standard error: 0.2767 on 375 degrees of freedom
Multiple R-squared: 0.9487, Adjusted R-squared: 0.9482
F-statistic:  2310 on 3 and 375 DF,  p-value: < 2.2e-16
```

41.5 More Language Features

The R language supports a number of fairly standard data types, but also has some unusual language features.

41.5.1 Some R Data Types

R supports a number of different vector data types, including logical vectors, integer vectors, real (double precision) vectors, character vectors, and generic vectors. There are no true scalars, only vectors of length 1. Arrays are vectors with a `dim` attribute specifying the array dimensions.

Generic vectors can have any kind of R data objects as their elements and are building blocks for more general data types. Of these, *data frames* are the most important. They contain the columns of variable values for a data set.

Non-vector data types include functions and environments, which hold bindings between variable names and values.

41.5.2 Some Unusual Language Features

An unusual feature of the R language is that vectors are immutable. Conceptually, the expression

```
x[i] <- y
```

assigns a modified copy of `x` to the variable `x`. The original vector assigned to the variable `x` is not changed. This eliminates errors caused by unintended modifications to data occurring inside of functions. For efficiency reasons R tries to avoid making copies that are not necessary.

Another unusual feature is that all atomic (i.e. non-generic) vector types support missing values. This is very useful for data analysis applications where missing values are all too common.

Arguments to function calls are only evaluated if and when they are needed. This is called *lazy evaluation*. One implication is that even control flow constructs can be viewed as ordinary functions. R functions are also able to obtain the expressions of the arguments with which they were called. This is useful, for example, for creating meaningful default labels for plots. Combined with lazy evaluation, this means functions can implement their own evaluation rules This is called *non-standard evaluation*. This can be very useful, but if not used with care can lead to surprising and confusing results.

41.6 Future Directions

A number of directions for future improvement of R are being explored by the members of the R core group. Efforts I am involved with include work on performance improvement and improved handling of larger data sets.

41.6.1 Future Directions: Performance

One major effort at performance improvement is the development of a byte code compiler for R. The first version of this compiler was released in Spring 2011 and provides significant improvement for scalar-intensive computations. Future work is likely to lead to substantial further performance improvements. Areas to be explored include improved function call performance and possible native code generation.

Given current developments in processor technology, much improvement in performance is likely to come from the use of parallel computing to exploit the availability of multiple processor cores. The ability to use high performance BLAS implementations, including multi-threaded implementations, has already been mentioned in Sect. 41.2.4. Other directions being explored include automatic parallelization of basic arithmetic operations and matrix operations when the sizes of the operands are sufficient to outweigh synchronization overheads. OpenMP is being examined as an implementation framework, and recent work has also made it easier for R package authors to make use of OpenMP in their code.

Other efforts at parallelization include the development of explicit parallelization frameworks such as `snow` [13] and `multicore` [14]. A recent book describes some of the available frameworks [7].

GPU computing is also an area of active research in scientific computing, and several packages that take available of available GPUs are available. Whether it makes sense to integrate GPU computing more directly into the core R engine is not yet clear but is being explored.

Most efforts at parallelization have focused on interpreted code, but it is possible that compilation may help in improving parallel performance as well.

41.6.2 Future Directions: Large Data Sets

Most R functions work on data held in computer memory. A current limitation of R is that integers used to represent vector lengths are stored as 32-bit quantities. As a result, the number of elements in an R array is limited to at most $2^{31} - 1 = 2,147,483,647$. As this corresponds to 16 GB for a numeric array of double precision elements it is not often a limitation yet, but it is becoming more limiting over time. Work over the next year or so will investigate how to remove this limit without forcing a complete rewrite of R and all contributed packages containing C or FORTRAN code.

With storage capacity likely to remain larger than available RAM for the foreseeable future, more methods for handling data sets larger than available memory will continue to be needed. Various approaches can be programmed in R, and several packages are available for operating on very large data stored on disk or other storage media, but more work in this areas is needed.

41.7 Getting Started

R is available from CRAN, the comprehensive R archive network, at http://cran.r-project.com. R is available as source code, as pre-compiled binaries for Windows and Mac OS X, and through the package management systems of many Linux distributions.

Once R is installed the `help` command can be used to accesses the manual included in R. Manuals are also available at http://cran.at.r-project.org. Links to user-contributed documentation, tutorials, and books about R are also available at the CRAN web site. There are several active mailing lists and blogs available. The *Task Views* [16] available at CRAN provide a useful way of learning about relevant contributed packages for different problem domains. As of the time of writing there is no Astronomy task view, but that may change in the near future.

Acknowledgements The author's research was supported in part by NSF grant DMS-09-06398.

References

1. ATLAS 2011, ATLAS Home Page, available at http://math-atlas.sourceforge.net/
2. Bates, D. & Maechler, M. 2011, Matrix: Sparse and Dense Matrix Classes and Methods, R package version 1.0-1, available at http://CRAN.R-project.org/package=Matrix
3. Becker, R. A., Chambers, J. M. & Wilks, A. R. 1988, *The New S Language*, Chaplam & Hall
4. Chambers, J. M. 1998, *Programming with Data: A Guide to the S Language*, Springer
5. Fox, J. 2005, The R Commander: A Basic-Statistics Graphical User Interface to R, *J. Statist. Software*, 14(9), 1–42
6. JMP 2011, JMP Software, available at http://www.jmp.com/

7. McCallum, Q. E. & Weston, S. 2011, *Parallel R*, O'Reilly
8. NED 2011, NASA/IPAC Extragalactic Database, available at http://www.sr.bham.ac.uk/~ajrs/R/r-tutorials.html
9. R Development Core Team 2011a, R: A Language and Environment for Statistical Computing, Vienna Austria
10. R Development Core Team 2011b, CRAN: The Comprehensive Archive Network, available at cran.r-project.org
11. Sanderson, A. 2011, Astronomy-themed R tutorials, available at http://www.sr.bham.ac.uk/~ajrs/R/r-tutorials.html
12. Sarkar, D. 2008, *Lattice: Multivariate Data Visualization with R*, Springer
13. Tierney, L., Rossini, A. J., Li, N. & Sevcikova, H. 2011, snow: Simple Network of Workstations, R package version 0.3-7, available at http://CRAN.R-project.org/package=snow
14. Urbanek, S. 2011, multicore: Parallel processing of R code on machines with multiple cores or CPUs, R package version 0.1-7, available at http://CRAN.R-project.org/package=multicore
15. Verzani, J. 2010, pmg: Poor Man's Gui, R package version 0.9-42, available at http://CRAN.R-project.org/package=pmg
16. Zeileis, A. 2005, CRAN Task Views, *R News*, 5(1), 39–40
17. Zhang, X. & Wang, Q. 2011, OpenBLAS Home Page, available at http://xianyi.github.com/OpenBLAS/

Chapter 42
Panel Discussion: The Future of Astrostatistics

G. Jogesh Babu

Abstract Four experienced astrostatisticians express their views on the promise of astrostatistics in the future: David van Dyk (University College London) discusses massive datasets and complex models, Eric Feigelson (Penn State University) speaks on the past and future of astrostatistics, Thomas Loredo (Cornell University) elucidates fundamentals underlying statistical analysis, and Jeffrey Scargle (NASA-Ames Research Center) presents challenges and opportunities in astrostatistics.

42.1 David A. van Dyk: Understanding Massive Data Sets and Complex Models

Perhaps the most pressing data analytic challenge faced by astronomers is the deluge of massive data sets and data streams. We are in danger of drowning in a torrent of data (Fig. 42.1). The Sloan Digital Sky Survey produced several thousand scientific studies from petabyte photometric and spectroscopic surveys obtained with a modest telescope. This is just the tip of the iceberg. An alphabet soup of other projects are following is Sloan's path: DES, PTF, Pan-STARRS, SN Factory, Kepler, LAMOST, and most spectacularly the planned LSST.

Several presentations during this conference discussed salient aspects of statistical approaches to these mega-data sets.

Domenico Marinucci discussed the extraordinary growth in the size of data sets in cosmology. In some measure this is likely true of all areas of astronomy. (Of course as deeper data becomes available fainter objects become visible and methods for small or low-count data remain important.)

G.J. Babu (✉)
Department of Statistics, Pennsylvania State University, University Park, PA 16802, USA
e-mail: babu@stat.psu.edu

E.D. Feigelson and G.J. Babu (eds.), *Statistical Challenges in Modern Astronomy V*, Lecture Notes in Statistics 209, DOI 10.1007/978-1-4614-3520-4_42,

Fig. 42.1 Today's astronomer may be drowned in a tidal wave of their own massive data sets and complex models

Kirk Borne told us that more data is not just more, it is also qualitatively different data. This may be true, but we must also recognize that the massive growth in the quantity of available data is a major change in and of itself. We must both develop methods to handle large data sets in standard ways, and also realize that entirely new analysis methods are made possible and necessary by the quality and/or quantity of the new data sources.

Alexander Gray presented improvements in the computational efficiency for many standard methods that allow them to be applied to massive data sets.

Ann Lee showed how transforming high-dimensional data to lower dimensional summaries can make them more amenable to standard analysis. Such methods can be very powerful when there are many measurements of the same object.

Joseph Richards discussed an automatic classification procedure for the irregular time series that typically emerge from wide-field surveys. There are dozens of types of variable stars and extragalactic transients that need to be classified without human involvement. Thomas Lee spoke on similar challenges involving the classification of sunspots from imaging data. These procedures can work when reliable training sets for the various classes are available to supervise the classification process. It is, however, difficult to capture the subtlety of classification by humans in any automated procedure.

Statisticians and computer scientists are both developing methodologies for treating large data sets, but from quite different perspectives. Machine learning and artificial intelligence methods are designed to be scalable to large data sets, but often rely on ad hoc foundations and exhibit unpredictable statistical properties. Statistical methods tend to be principled with precise theoretical underpinnings and predictable performance, but are often computationally slow. The challenge is to combine the best of both worlds and develop principled methods that are scaleable to massive data sets.

Fortunately, these challenges are not unique to astronomy, but are being confronted in a broad range of academic fields, including business, industry, economics, biology, and physics. Of course, the models differ in each field, but because many of the computational issues are similar, cross-fertilization can be fruitful. Astronomers can benefit from the experience that statisticians and computer scientists have in treating similar problems in other areas.

The LSST project, in particular, will generate datasets that are not just massive, but also rich and deep. There will be trigonometric parallaxes, proper motions, photometry in six bands, morphology, and time series with 10^2–10^3 irregularly-spaced epochs for billions of objects. Analysis will require both simple but scalable methods and the use of complex models that require sophisticated computing even with small but rich datasets. Such models are sometimes available analytically, but often can only be calculated using sophisticated computer models and/or simulators. An entirely new set of methods are required to incorporate such computer models into principled statistical analyses. This is currently an active area of research in astrostatistics and has come up numerous times during this conference:

Andrew Connolly showed how the observational data from LSST can be understood in detail using sophisticated ray tracing of astronomical populations through the optics of the telescope and accounting for specific properties of the detector.

Vinay Kashyap similarly used principal component analysis to analytically summarize complicated calibration properties using computer models of telescope and detector systems.

David Higdon described how cosmological models, which are extremely time consuming to calculate in full detail, can be emulated as Gaussian processes and fit to statistical summaries (e.g. the galaxy two-point correlation function) of massive datasets.

Chad Schafer introduced ABC, Approximate Bayesian Computation. This method compares data sets simulating under various values of the model parameters with the observed data to deduce what values of the parameter could feasibly have generated the observed data. This is practical when a single run of the model is not extremely expensive computationally.

David van Dyk and, in a contributed paper, Nathan Stein embed computer models into multilevel models in a fully Bayesian setting.

These represent a variety of the strategies for tackling the diverse challenges associated with complex datasets and/or models in astronomy. This rich class of data-analytic problems demands a diverse suite of new statistical methods.

Researchers in all areas of astronomy are turning to computer models to represent complex physical processes that defy analytic formulations. These models pose real challenges even with moderately sized data sets, but as Carlo Graziani noted, "the challenge is acute when complex models are mixed with massive data". Get ready for a computer model near you!

Model fitting is the first of many challenges associated with complex computer models. Model checking, comparison, and improvement pose special challenges when the models under consideration are complex and can only be evaluated numerically. The design of experiments—deciding when and where to make your telescopic observations to optimally constrain your astrophysical model—can also be exceedingly complex when the idiosyncrasies of particular models and particular data sets are considered. The LSST team is now discovering this in their discussions of cadences for repeated observations, even without considering formal statistical optimization. Standard methods for optimal design can guide us when dealing with complex computer models, but are not generally sufficient in and of themselves. These are only a few examples of the challenges posed by complex computer models. Generally speaking however, we cannot expect standard off-the-shelf statistical methods to be sufficient. This is good news for methodologists: there is much work to be done to develop the statistical methods and computational techniques demanded by complex models, especially when they are mixed with massive data.

Let me end by reiterating the urgent need to fully integrate statisticians, computer scientists, and other methodologists into empirical astronomy in general and in the large astronomical survey projects in particular. These data scientists can contribute to the design of data collection, the development of methodology, the actual data analyses, and the linking of empirical results with astrophysical models. These cross-disciplinary colleagues can be integrated within science teams as graduate students and post-docs or brought in as collaborating faculty funded as co-investigators on grants or even hired as faculty members. There is a real thirst among astronomers for statistical help. The current involvement of a handful of statisticians around the world is simply insufficient to meet their needs.

Astrostatistics deserves to grow into an established subdiscipline of astronomy, respected as a full-time research speciality. Statistical subdisiplines are well-established in other fields. Statisticians reside in academic departments of economics, psychology, and biology and publish in prestigious journals of econometrics, psychometrics, and biostatistics. Disciplinary scientists with sophisticated training in statistics have faculty positions in business, engineering, political science, and biology departments. I am pleased to say that my new academic home, Imperial College London, is now growing such an astrostatistics group within an astronomy group in collaboration with its statistics group. I hope this is just the beginning of a worldwide push toward establishing astrostatistics as an academic subdiscipline.

42.2 Eric Feigelson: The Past and Future of Astrostatistics

42.2.1 Astrostatistics Yesterday

Hipparchos arguably started astrostatistics twenty-first centuries ago with his discourse on the length of a year, starting a millennium-long discussion on how to estimate a fixed physical quantity given discrepant measurements [12]. A favorite estimate among the Greeks was the midrange, the mean of the two extrema, which is now considered to be a poor estimate. During Medieval times, some scholars recommended against acquiring repeated measurements; if there is only one measurement, then the value is definitely known! The mean value was advocated by Tycho Brahe and Galileo, but Johannes Kepler inconsistently used arithmetic means, geometric means, and middle values in his work. A consensus among astronomers on the mean emerged only in the eighteenth century, although today many argue that the median is more robust against outlying measurements.

Critical foundations of modern statistics were laid during the nineteenth century by scholars using Newton's laws of motion and gravity to model motions of bodies in the Solar System [11, 28]. This 'celestial mechanics' led some of the greatest mathematicians of the time—Abraham DeMoivre, Adrien-Marie Legendre, Pierre-Simon de Laplace, Siméon-Denis Poisson, Carl Friedrich Gauss—to develop the normal error law, Central Limit Theorem, least-squares regression. Gauss spent most of his career as Professor of Astronomy and Director of the astronomical observatory in Göttingen, Poisson was an astronomer at Paris Bureau des Longitudes, and astronomers later in the century, from John Herschel to Simon Newcomb to Giovanni Schiaparelli, contributed to least squares theory. Thus, for most of the past two millennia, the statisticians were the astronomers and the astronomers were the statisticians.

But the close connection between the fields was sundered during the twentieth century [8]. The statisticians, guided by Frances Galton, Karl Pearson and R. A. Fisher, devoted their talents to serve human affairs: biometrics, demography, economics, political science, and industries ranging from insurance to agriculture. Astronomers found that modern physics derived from terrestrial phenomena—electromagnetism, thermodynamics and fluid mechanics, Einstein's mechanics and gravity, and above all quantum mechanics regulating atomic and nuclear physics—could be powerfully applied to celestial phenomena. The astronomers thus because astrophysicists, seeking to understand the underlying composition and structure of celestial objects and the physical processes underlying their interactions and evolution. Enormously successful models for understanding the physics of stars and cosmology emerged from this enterprise. Thus, by the middle of the twentieth century, astronomical research had largely moved from the statistics of celestial phenomena to astrophysical modeling.

In the 1970s and 1980s there was progress in isolated areas. An appendix of an obscure paper, still under-recognized, by the distinguished British astrophysicist Donald Lynden-Bell [16] made a profound advance in mathematical statistics.

He derived the unique nonparametric maximum likelihood estimator for a randomly truncated univariate dataset, a close analog of the Kaplan–Meier estimator (Kaplan and Meier 1958) for a randomly censored dataset that lies at the foundation of survival analysis. Lynden-Bell's estimator was independently found by Woodroofe [34] who further elucidated its mathematical properties. The Lynden-Bell-Woodroofe estimator should, in my opinion, be used in hundreds of astronomical studies of 'galaxy luminosity functions' and similar distributions derived from flux-limited surveys. Unfortunately, Schmidt's [26] binned V/V_{max} estimator with less desirable properties is still commonly used in its place.

The 1980s also witnessed a great series of papers by Jeffrey Scargle on time series methodology in astronomy, including the derivation of a generalization a Fourier periodogram designed for unevenly spaced data [23]. This Lomb-Scargle periodogram has been extensively used in searches for periodicities, although there is still controversy on reliable estimation of its False Alarm Probabilities. I stumbled across survival analysis and brought some of its methods into common use in astronomy (Feigelson and Nelson 1985). A growing corpus of methodology was also developed to study the statistics of galaxy clustering [17].

42.2.2 Astrostatistics Today

Around 1995–2000, one could see a significant resurgence in interest and activity on advanced statistical methods within the astronomical community. The growth of Bayesian inference during the past decade, well-documented at this conference, is perhaps the most dramatic example. Poisson processes, image processing, time series analysis, likelihood-based modeling, wavelet analysis, neural networks have all witnesses increased applications of sophisticated methodology. We have heard at this meeting about compressive sensing, an amazing approach originating in image restoration and signal processing, which hopefully will become important in the next decade.

However, a serious problem remains with the average astronomical study. While there is a small but growing vanguard of astrostatistical experts in the astronomical community, propelled by a small cadre of statistician collaborators, the great majority of astronomers producing the great majority of research still use a narrow suite of familiar methods. The major astronomical journals publish around 20,000 papers annually, and most use methods such as weighted least squares regression ('minimum χ^2' in the astronomers' lexicon), Kolmogorov-Smirnov nonparametric tests (unaware that the Anderson–Darling test has better performance), linear principal components for multivariate structure (unaware of nonlinear approaches), and the two-point correlation function for point processes (unaware of its relationship to Ripley's K, Baddeley's J and other statistics in spatial statistics). David van Dyk's group showed that even traditional 'likelihood ratio test' is often misused by astronomers [20], and we have pointed out that the Kolmogorov-Smirnov test

probabilities often to not apply to astronomers' problems [2]. Every textbook on unsupervised nonparametric multivariate clustering points out serious deficiencies in 'single linkage agglomerative clustering', yet astronomers continue to use it (the 'friends-of-friends algorithm' in the astronomers' lexicon). A bibliometric accounting of astronomers' use of modern machine learning classification algorithms—such as Support Vector Machines or Random Forests—shows only a handful of studies using these important techniques [7].

The development of the public domain **R** statistical software system with its exponentially growing **CRAN** add-on packages—now numbering over 3,000 packages with over 50,000 statistical functionalities—can dramatically improve the average astronomer's toolkit for statistical analysis ([22], http://www.r-project.org). Extensive Web-based resources and dozens of books are available to inform astronomers about **R**'s capabilities and usage. At Penn State and abroad, we have trained several hundred astronomy graduate students since 2005 in **R** in week-long Summer Schools. Our textbook emerging from these schools, *Modern Statistical Methods for Astronomy with R Applications* [8], may further broaden the **R**-fluent community in astronomy.

42.2.3 Astrostatistics Tomorrow

I suggest that, in addition to learning more methodology, astronomers need to adopt a more sophisticated view for approaching statistical problems arising in data and science analysis. Statistics is not just a collection of mechanical tools that can be quickly applied and that produce scientifically clear results. Rather, the application of statistics to scientific problems requires careful statement of the problem, model formulation, choice of statistical method(s), calculation of statistical functionalities, validation of results, and scientific interpretation.

This is a messier, but much more interesting, process than conducted by most astronomers. The enterprise is further complicated by the vast scope of modern statistics, providing enormous capabilities but make it difficult to find and select methods for a particular problem. Interpreting a statistical result is also not always obvious: in a large sample, a tiny effect of little scientific importance might be statistically significant, while in a small sample, even major scientific effects might be undetectable.

I would like to end with a vision for astrostatistics in 2025:

- The graduate curriculum for astronomers includes a year of statistical methodology tuned to our needs.
- Dozens (out of thousands) of astronomers obtain simultaneous M.S. or Ph.D. degrees in statistics, applied mathematics, and computer science.
- Astronomical papers refer to statistics textbooks, not other astro papers, for methodology.

- Important problems that confront us today are largely solved: multivariate heteroscedastic measurement errors, irregularly spaced time series, faint source detection, etc.
- Astronomers regularly and competently use hundreds of methods in *P*, the public domain statistical software system, successor to *Q* and *R*.
- One to two dozen well-funded research groups in astrostatistics and astroinformatics are active on three continents.
- *Statistical Challenges in Modern Astronomy* conferences are held annually, not every 5 years.

42.3 Thomas Loredo: Statistical Foundations and Statistical Practice

42.3.1 The Frequentist-Bayesian Debate

The future of astrostatistics is linked to the future of statistics as a discipline. The emerging needs of astrostatistics may both motivate and benefit from *fundamental* developments in statistics. This is a two-way street.

Christopher Genovese told us earlier in the conference that, within statistics, the debate between frequentist and Bayesian approaches has largely faded from view. He noted that, although nontrivial philosophical and conceptual differences certainly exist, statisticians recognize that there are situations where each approach has an advantage, and both are used successfully.

My outsider view of contemporary statistics supports the assessment that debate about foundations has faded in recent years. But I do not see this as a positive development, and I disagree with any prescription that fundamentals should not be seriously discussed and researched. Issues at the foundations of statistics are not merely philosophical. Where one comes down on foundational issues has significant implications for statistical practice. I would urge statisticians to think more rather than less about the foundations of their discipline, and to consider doing so in closer partnership with the scientist consumers of their methods. Despite being an outsider to statistics, I take this position emboldened by being in good company from within the discipline, and by the seriousness of the topic. For I see statistics as a kind of theory of the scientific method—at least, that part of the scientific method that may be described with quantitative precision—giving all scientists a vested interest in the field's development.

Prominent statisticians who have contributed enormously to statistical practice continue to embrace the struggle with the foundations and fundamentals of statistical inference. Bradley Efron [5], whose work mostly adopts the frequentist approach, recently lamented the absence of attention to foundations:

> Methodology by itself is an ultimately frustrating exercise. A little statistical philosophy goes a long way but we have had very little in the public forum these days.

In his 2004 American Statistical Association (ASA) Presidential Address [4], he asserted:

> The 250-year debate between Bayesians and frequentists is unusual among philosophical arguments in actually having important practical consequences.... Broadly speaking, Bayesian statistics dominated nineteenth Century statistical practice while the twentieth Century was more frequentist. What's going to happen in the twenty-first Century? ... I strongly suspect that statistics is in for a burst of new theory and methodology, and that this burst will feature a combination of Bayesian and frequentist reasoning.

Efron sees empirical Bayes methods as a promising frequentist/Bayesian hybrid approach ([5]; see the accompanying discussion for critical assessments); I will have more to say about this below.

To cite another example, in his 2005 ASA President's Invited Address, Roderick Little said:

> Pragmatists might argue that good statisticians can get sensible answers under Bayes or frequentist paradigms; indeed maybe two philosophies are better than one, since they provide more tools for the statistician's toolkit.... I am discomforted by this "inferential schizophrenia." Since the Bayesian (B) and frequentist (F) philosophies can differ even on simple problems, at some point decisions seem needed as to which is right. I believe our credibility as statisticians is undermined when we cannot agree on the fundamentals of our subject.

Little, whose work has mostly adopted the Bayesian approach, has recently tried to work out principles for best practices, pulling strengths from each approach. Roughly speaking, his "calibrated Bayes" compromise relies on Bayesian methods for inference under a model, but holds an important role for frequentist methods for model assessment. He feels strongly that Bayesian methods are insufficiently taught to statisticians. But he also criticizes advocates of Bayesian methods for not sufficiently assessing their modeling assumptions.

With such leading lights harping on the need to examine fundamentals, why is there so little of what one might call "foundational self-examination" in statistics? Andrew Gelman [9], in a discussion of the empirical Bayes synthesis of Efron [5], presents three meta-principles of statistics, among them one shedding a bit of light on this question:

> My second meta-principle of statistics is the *methodological attribution problem*, which is that the many useful contributions of a good statistical consultant, or collaborator, will often be attributed to the statistician's methods or philosophy rather than to the artful efforts of the statistician himself or herself. The result is that each of us tends to come away from a collaboration or consulting experience with the warm feeling that our methods really work, and that they represent how scientists really think. In stating this, I am not trying to espouse some sort of empty pluralism.... I think we all have to be careful about attributing too much from our collaborators' and clients' satisfaction with our methods.

The meta-principle speaks to the absence of reflection on foundations: truly talented statisticians adopting different approaches get good work done; the approach they adopt seems not to matter. But Gelman's comment about "empty pluralism" is important. Satisfaction with the current "inferential schizophrenia" in statistics is not justified by past successes. Brilliant analysts can rely on unarticulated intuition, but the rest of us need sound principles, if only they can be uncovered. (We can also benefit from collaboration, but that's another topic!)

Bayarri and Berger [3] provide concrete examples of methodological advances coming from foundational research in their survey, "The Interplay of Bayesian and Frequentist Analysis." They argue that "the debate is far from over and, indeed, should continue, since there are fundamental philosophical and pedagogical issues at stake," with significant implications for practice. They review research that combines frequentist and Bayesian ideas resulting in new directions for statistical practice, including work on frequentist performance of Bayesian procedures, predictive assessment of models, conditional frequentist testing, and so forth. To highlight just one area with practical consequences: Conditional frequentist testing is an alternative to traditional hypothesis testing with p-values (astronomers' "significance levels") that I have found to be appealing to astronomers I work with, because it does what they thought their p-values were doing. It also happens to be closely related to model comparison with Bayes factors, and so serves as a natural bridge between Bayesian and frequentist thinking. Bayarri and Berger also discuss areas where the two approaches seem to fundamentally disagree, such as multiple testing, sequential analysis, and finite population sampling. These topics are important for a variety of astronomical problems, arguing again that work on fundamentals will have practical consequences for astronomers.

The *ISBA Bulletin* from the International Society for Bayesian Analysis, available at http://bayesian.org/, is a good source for occasional informal interchanges on these issues. In a recent issue, ISBA President Michael Jordan [14] polled a number of leading statisticians (including some whose work is largely frequentist) on what they thought were the principal open problems in Bayesian statistics. They noted that Bayesian and frequentist methods often differ considerably on model selection, model misspecification, and model validation. Computation is often seen as difficult; approximate Bayesian computation (ABC) methods, as introduced by Chad Schafer at this conference, may be an important approach. The relationships between frequentist and Bayesian methods need to be elucidated, such as connections between empirical Bayes and the bootstrap and FDR control. Choice of priors continues to be an important issue. Concern was expressed about nonparametric and semiparametric inference where it presently seems safer and easier to use frequentist rather than Bayesian methods; this was discussed by Christopher Genovese earlier in the conference. In all of these areas, clarifying foundations will directly affect practice. And many of them are clearly relevant to current and emerging astrostatistics problems.

42.3.2 Multilevel Models and Multiple Testing

Let me elaborate on one item in Jordan's list as an example of where some struggle at the Bayes/frequentist divide by statisticians and astronomers together might pay dividends: the role of multilevel modeling (empirical or hierarchical Bayes) in multiple testing, where FDR control has become the standard frequentist technique. Statistical research in this area is important for addressing challenges being raised by the astronomy data deluge discussed above by David van Dyk. The deluge coming from synoptic surveys does not just provide astronomers *more* data than we are used to; it also provides a *different kind* of data: collections of modest-sized datasets (such as sparse, irregularly-sampled light curves) for vast numbers of related objects. Astronomers need methods that can accurately and optimally accumulate information, not only within the dataset for a particular object, but also across a population of related objects.

This problem is not unique to astronomy. It is arising in many disciplines, motivating much current statistics research. This research was the main theme of a recent article by Bradley Efron entitled "The future of indirect evidence" in the excellent cross-disciplinary journal *Statistical Science* [5]. Whereas conventional statistical methods accumulate information about an object or process by repeated observations of the same object or process, new data require the ability to pool information across ensembles of related objects or processes—"indirect evidence." Efron advocates empirical Bayes methods as a general framework for using indirect evidence, and False Discovery Rate (FDR) control for the class of problems where the goal is separation of a large ensemble of related observations into discoveries and "nulls." Efron's paper was published with discussion; none of the discussants liked FDR, and neither do I.

For astronomers, a catalog is not just a report of final classifications of candidate sources. Rather, it is a starting point for further analysis and discovery, perhaps the most common goal being estimating population distributions. Catalogs produced using FDR control—say, with the well-known Benjamini–Hochberg (BH) procedure [19]—are ill-suited to this. False discoveries pile up at the low p-values. In typical astronomical settings, the signal-to-noise ratio will be lower for dim sources than for bright ones, so the low p-values will tend to come from dim sources. Applying FDR control methods to this situation will give progressively greater pollution at dimmer fluxes. Simply knowing that you have controlled the FDR at some specified level for the whole catalog does not help you accurately infer the run of $\log N$–$\log S$ (log source counts vs. log flux, i.e., the number-size distribution) or other interesting population-level quantities from the catalog. So BH FDR control addresses a particular question in an almost miraculously beautiful way—nonparametrically, adaptively, and robustly—but it does not provide results that let astronomers answer further, related questions we want to address with the data. Bayesian multilevel modeling can address such questions, via probabilistic "soft" classification rather than thresholding, but the approach requires more care in assessing the impact of modeling assumptions (see, e.g., [27]).

Statisticians themselves are not uniformly enthusiastic about FDR control. Gelman [9] wrote: "To me, the false discovery rate is the latest flavor-of-the-month attempt to make the Bayesian omelette without breaking the Bayesian eggs... it can work fine if the implicit prior is ok... but I really don't like it as an underlying principle." The frequentist literature on multiple testing itself recognizes that FDR control may not address the science questions of interest in a particular study.

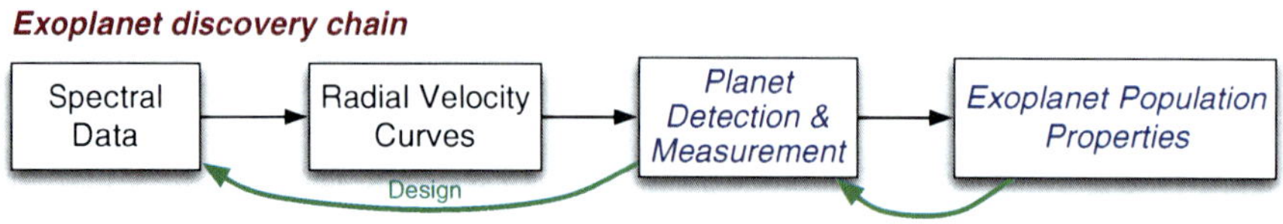

Fig. 42.2 Diagram of a discovery chain whereby exoplanets are found from periodic Doppler shifts in the spectra of the host stars. Progress in detection and characterization of individual planets leads to studies of exoplanet populations, and improved design of the observational experiments and spectral analysis procedures

It includes alternatives to FDR control, such as estimation of confidence bounds on the source fraction advocated by Meinshausen and Rice [18] for some applications. Tighter interaction between astronomers and statisticians is needed to work out how frequentist and Bayesian approaches to multiple testing might interact to produce tools meeting astronomers' needs. For example, can we simultaneously have the robustness offered by BH FDR control and the "soft thresholding" offered by Bayesian multilevel models, enabling a variety of subsequent scientific analyses using the source detection results?

42.3.3 *Statistical Analysis and the Chain of Discovery*

Frequentist methods tend to frame a data analysis task as a monolithic decision, as if addressing that one decision were the sole goal of data-taking. Indeed, this is made explicit in the decision-theoretic formulation of frequentist estimation and testing. But astronomers are seldom seeking to produce a single terminal decision from their data. Instead our observing and cataloging and modeling are all just steps in what one might call unfolding "chains of discovery." An astronomical problem is often first tackled with sequential experimentation and exploration, starting a chain of discovery leading from study of individual objects to study of populations. Figure 42.2 diagrams an example of such a chain for extrasolar planet science using radial velocity data, where planets orbiting other stars are detected from time-dependent Doppler shifts of the spectra of their host stars. Each of the black arrows represents a complicated data analysis problem, converting spectral data into radial velocity curves, modeling these curves to detect planets (as Philip Gregory described at this meeting), and inferring properties of exoplanet populations from the individual planetary measurements. But effective analysis, and even effective data acquisition, requires knowledge from the later steps, so a feedback loop is established. We need a broad statistical approach that facilitates building such chains of discovery.

This notion of a discovery chain is related to the type of problem studied in the branch of statistics known as sequential analysis. A pioneer of this area, Herman

Chernoff, has an intriguing perspective on its relevance to the scientific process more generally:

> I became interested in the notion of experimental design in a much broader context, namely: what's the nature of scientific inference and how do people do science? The thought was not all that unique that it is a sequential procedure.... Although I regard myself as a non-Bayesian, I feel in sequential problems it is rather dangerous to play around with non-Bayesian procedures.... Optimality is, of course, implicit in the Bayesian approach.

An important direction for future fundamental work in statistics would be explicit recognition that most scientific data analysis tasks are just steps in an ongoing sequence of analyses—an unfolding chain of discovery. Efron's "indirect evidence" is a special case of this, where one seeks a framework that can integrate inference about individuals with inference about populations. Given Chernoff's remarks, it is perhaps not surprising that Bayesian ideas are playing an important role in working out how to use indirect evidence, via empirical and hierarchical Bayes methods. I suspect the future of statistics will involve a more thorough integration of Bayesian ideas into statistical practice, if only to enable development of even more elaborate discovery chains. I anticipate that statistical challenges in modern astronomy will be both drivers and beneficiaries of such developments.

42.4 Jeffrey Scargle: Challenges and Opportunities in Astrostatistics

42.4.1 Flawed and Beneficial Statistics

Eric Feigelson gave a somewhat gloomy picture of statistical practice among astronomers in the past, when less-than-perfect methods have been used. In addition to cases where information has been wasted, some catastrophic blunders in various fields have occurred that provide important lessons for astronomers, particularly involving improper treatment of experiment or observation bias. The following are all true stories:

- The president of a major political polling company believes that a selection bias can be eliminated by obtaining larger samples. (Of course this is backward: larger samples only increase the apparent significance of a biased result.)
- A major issue in clinical trials of new drugs or practices, particularly in "meta-analysis" where small clinical studies are combined to get larger samples, is whether *publication bias* is present [21]. This occurs when only studies with high success rates are published, and negative or indeterminate results are kept private. Fifty years ago, a Harvard psychologist presented a statistical formula to evaluate the presence of publication bias, but it has proved to be completely wrong [25], nearly always reporting that no bias can be present. Astronomers must be careful to report all of their findings, not just the positive ones. Some of us remember

when negative stellar parallax measurements were routinely discarded, yielding distances systematically too small.
- A priori analysis of *post facto* 'clusters' of quasar redshifts, or alignments of quasars around bright galaxies, contradicts the cosmological interpretation of redshifts [1] and yields the conclusion that the Earth lies at the center of the Universe [30]. This flawed methodology leads to errors of many orders of magnitude [10, 33, 35].

Astronomical knowledge usually progresses in a more rational fashion. I work at NASA and can give my perspective on the process often followed in space science. Some important scientific questions, or 'mission science goals', drive the design of a new satellite observatory. The instrumentation, data acquisition and processing must be adequate to achieve these goals. The satellite is launched, observations are taken, and science data analysis is pursued to increase our science knowledge relevant to the original science goals. I have seen much improvement in this cycle. When I first joined NASA many years ago, the design of instruments would take little account of the subsequent data processing; data analysis was mostly an afterthought. But today one sees considerable forethought regarding data analysis issues and selection bias in the observations.

42.4.2 A Vision of the Future: Astronomical Time Series Analysis

I would now like to present a vision on how one area of astronomical research—the study of temporal variations of celestial objects—might ideally be pursued. Many major projects are underway to study variability based on multi-epoch wide-field surveys such as the Palomar Transient Factory, the Catalina Real-Time Transient Survey, Pan-STARRS, and the planned Large Synoptic Survey Telescope (LSST). Millions, and soon billions, of time series are emerging from these efforts.

I think there should be a 'Universal Time Series Analysis Machine' into which the data from such surveys can be dumped to give standard analysis products. The machine would permit a variety of input data modes (such as photon events, time-to-spill data, counts in bins, flux measurements) obtained with any observing cadence pattern (evenly spaced, logarithmically spaced, random with periodic gaps, and so forth). The data products would include auto- and cross-correlation functions, Fourier power and phase spectra (with tapering), wavelet scalgrams and scalograms, structure functions, measurement error models, time-scale and time-frequency analysis, and more. These results of automated processing could then be fed into automated machine learning systems, such as the multivariate classifiers described by Joseph Richards at this conference. Advances in machine learning and data mining in astronomy are reviewed in a new volume by Way et al. [32]. While such a Universal TSA Machine may seem ambitious, most of its ingredients are in hand today. I believe that such a Machine can be brought into existence soon.

As part of such a plan, I have been working on an automated and universal algorithm for the construction of histograms. Histograms are used all the time in astronomy for displaying univariate datasets with both small and large samples. But the results are highly dependent on the choice of bin width, number, and phasing. One of the tools that is already available is a histogram scheme based on the Bayesian Blocks algorithm [13, 24], in wide use to construct optimal piecewise-constant time series representations. It turns out that estimating histograms and block representations of event time series are mathematically the same problem. In the former the measurement dimension takes on the role of time as the independent variable in the latter.

In all of the related cases, including time series, spectral analysis, and histograms, the appearance changes considerably as one moves from very coarse to very fine binning. Which one best shows the true distribution? Does coarse binning oversmooth real features (increased bias)? Does the fine binning show real features or just noise (increased variance)? Standard procedure to balance bias and variance can be used (e.g. [31, Chap. 4]) but may miss crucial features of time series with complicated or nonstationary variations. The automatic data-adaptive selection of bin size and location in the Bayesian Blocks algorithm not only resolves these issues but is a countermeasure to the tendency to fiddle with these parameters until the distribution well fits the experimenters predilections.

This algorithm can easily be generalized to higher dimensional problems. For example the multi-scale structure of the 3D positional data in the Sloan Digital Sky Survey redshift survey was revealed through the Bayesian Blocks procedure (Way, Gazis and Scargle). A notable advance is that the "clusters" are not confined to have any particular shape, so that collection of Voronoi cells into blocks reveals the detailed structure of the Cosmic Foam [29].

I will end with an opinion regarding the future that promises to be dominated by data-rich astronomical enterprises. David van Dyk and others spoke of astronomers drowning in the flood of data emerging from large-scale surveys, particularly the planned LSST. I think it is wrong to have a negative attitude towards these challenges. The data flood is not being imposed on us by some external agent, but is emerging from our diverse 'mission science goals' that require very large datasets. They provide wonderful opportunities for new scientific insights. To help improve our attitude, I suggest that this conference, in the future, be renamed *Statistical Opportunities in Modern Astronomy*.

References

1. Arp, H. (1998) *Seeing Red: Redshifts, Cosmology and Academic Science*, C. Roy Keys
2. Babu, G. J. & Feigelson, E. D. (2006) Astrostatistics: Goodness-of-fit and all that!, in *Astronomical Data Analysis Software and Systems XV* (C. Gabriel et al., eds.), ASP Conf. Ser. 351, 127
3. Bayarri, M. J. & Berger, J. O. (2004) The interplay of Bayesian and frequentist analysis, *Statist. Sciences*, 19, 58–80.

4. Efron, B. (2005) Bayesians, frequentists, and scientists, *J. Am. Stat. Assoc.*, 100, 1–5
5. Efron, B. (2010) The future of indirect evidence, *Statistical Science*, 25, 145–157
6. Feigelson, E. D. & Nelson, P. I., (1985) Statistical methods for astronomical data with upper limits I: Univariate distributions, *Astrophys. J.*, 293, 192–206
7. Feigelson, E. D. (2011) Classification in Astronomy: Past and Present, in *Advances in Machine Learning and Data Mining for Astronomy* (K. Ali et al., eds) Chapman & Hall
8. Feigelson, E. D. & Babu G. J. (2012) *Modern Statistical Methods for Astronomy with R Applications*, Cambridge Univ. Press
9. Gelman, A. (2010) Bayesian statistics then and now (discussion of Efron's "The future of indirect evidence"), *Statistical Science*, 25, 162–165
10. Gosset, E., Clowes, R. G., Surdej, J., & Swings, J. P. (1990) A search for quasars in a field around NGC 520, *Mon. Not. Royal Astro. Soc.*, 245, 71
11. Hald, A. (1998) A History of Mathematical Statistics from 1750 to 1930, Springer
12. Hald, A. (2003) A History of Probability and Statistics and Their Applications before 1750, Springer
13. Jackson, B., Scargle, J.D., Barnes, D., Arabhi, S., Alt, A., Gioumousis, P., Gwin, E., Sangtrakulcharoen, P., Tan, L., and Tun Tao Tsai (2005) An algorithm for optimal partitioning of data on an interval, IEEE Signal Processing Letters, 12, 105–108.
14. Jordan, M. (2011) What are the open problems in Bayesian statistics?, *The ISBA Bulletin*, 11(1), 1–4
15. Kaplan, E. L. & Meier, P. (1958) Nonparametric estimation from incomplete observations, *J. Amer. Stat. Assn.*, 53, 457–481
16. Lynden-Bell, D. (1971) A method of allowing for known observational selection in small samples applied to 3CR quasars, *Mon. Not. Royal Astro. Soc.*, 155, 95–118
17. Martínez, V. J. & Saar, E. (2002) *Statistics of the Galaxy Distribution*, Chapman & Hall/CRC
18. Meinshausen, N. & Rice, J. (2006) Estimating the proportion of false null hypotheses among a large number of independently tested hypotheses, *Annals Statistics*, 34, 373–393
19. Miller, C. J., Genovese, C., Nichol, R. C., et al. (2001) Controlling the False-Discovery Rate in astrophysical data analysis, *Astron. J.*, 122, 3492–3505
20. Protassov, R., van Dyk, D. A., Connors, A., Kashyap, V. L., & Siemiginowska, A. (2002) Statistics, handle with care: Detecting multiple model components with the likelihood ratio test, *Astrophys. J.*, 571, 545–559
21. Rothstein, H., Sutton, A. J. & Borenstein, M., eds. (2005) *Publication Bias in Meta-Analysis: Prevention, Assessment, and Adjustments*, Wiley
22. R Development Core Team (2011) R: A Language and Environment for Statistical Computing, R Foundation for Statistical Computing, Vienna Austria
23. Scargle, J. D. (1982) Studies in astronomical time series analysis. II - Statistical aspects of spectral analysis of unevenly spaced data, *Astrophys. J.*, 263, 835–853
24. Scargle, J. D. (1998) Studies in Astronomical Time Series Analysis. V. Bayesian Blocks, a New Method to Analyze Structure in Photon Counting Data, *Astrophys. J.*, 504, 405–418
25. Scargle, J. D. (2000) Publication bias: The "File Drawer" problem in scientific inference, Journal of Scientific Exploration, 14, 91–106.
26. Schmidt, M. (1968) Space Distribution and Luminosity Functions of Quasi-Stellar Radio Sources, *Astrophys. J.*, 151, 393–409
27. Scott, J. G. & Berger, J. O. (2006) An exploration of aspects of Bayesian multiple testing, *J. Statist. Plann. Inference*, 136, 2144–2162.
28. Stigler, S. M. (1986) The History of Statistics: The Measurement of Uncertainty before 1900, Harvard Univ. Press
29. van de Weygaert, R. (2003) The Cosmic Foam: Stochastic Geometry and Spatial Clustering across the Universe, contribution in Proceedings *Statistical Challenges in Modern Astronomy III*, eds. E.D. Feigelson & G.J. Babu, Springer-Verlag, pp. 175–196
30. Varshni, Y. P. (1976), The red shift hypothesis for quasars - Is the earth the center of the Universe, *Astrophysics and Space Science*, 43, 3–8.
31. Wasserman, L. (2006) *All of Nonparametric Statistics*, Springer

32. Way, M. J., Scargle, J. D., Ali, K. & Strvastava, A. N. (2012) *Advanced in Machine Learning and Data Mining for Astronomy*, Chapman & Hall
33. Webster, A. (1982) Quasars, companion galaxies and Poisson statistics, *Mon. Not. Royal Astro. Soc.*, 200, 47P
34. Woodroofe, M. (1985) Estimating a distribution function with truncated data, *Annals Statistics*, 13, 163–177
35. Zuiderwijk, E. J., & de Ruiter, H. R. (1983) On the apparent association of quasars and Arp's companion galaxies, *Mon. Not. Royal Astro. Soc.*, 204, 675

Part VI
Contributed Papers

Chapter 43
Bayesian Estimation of log N — log S

Paul D. Baines, Irina S. Udaltsova, Andreas Zezas, and Vinay L. Kashyap

Abstract The study of source populations is often conducted using the cumulative distribution of the number of sources detected at a given sensitivity. The resulting "$\log(N > S) - \log S$" distribution can be used to compare and evaluate theoretical models for source populations and their evolution. In practice, however, inferring properties of source populations from observational data is complicated by the presence of detector-induced uncertainty and bias. This includes background contamination, uncertainty on both intensity and location of sources, and, most challenging, the issue of non-detections or unobserved sources. Since the probability of a non-detection is a function of the unobserved flux, the missing data mechanism is non-ignorable. We present a computationally efficient Bayesian approach for inferring physical model parameters and the corrected $\log(N > S) \sim \log(S)$ distribution for source populations. Our method extends existing work in allowing for both non-ignorable missing data and an unknown number of unobserved sources. Importantly, our method is also scalable in the number of observed sources, and computationally insensitive to the number of missing sources. By correcting for the non-ignorable missing data mechanism and other detection phenomena, we are able to obtain corrected estimates of the flux and luminosity distribution of source populations.

P.D. Baines (✉) • I.S. Udaltsova
University of California, Davis, CA, USA
e-mail: pdbaines@ucdavis.edu

A. Zezas
Department of Physics, University of Crete, P.O. Box 2208, GR-71003 Heraklion, Greece
e-mail: azezas@physics.uoc.gr

V.L. Kashyap
Harvard-Smithsonian Center for Astrophysics, Cambridge, MA, USA
e-mail: vkashyap@cfa.harvard.edu

E.D. Feigelson and G.J. Babu (eds.), *Statistical Challenges in Modern Astronomy V*, Lecture Notes in Statistics 209, DOI 10.1007/978-1-4614-3520-4_43,

43.1 Overview

The number of sources as a function of flux ("$\log(N) - \log(S)$") is an important tool for describing and investigating the properties of source populations. In practice, observations intended to measure the flux distribution are subject to natural and detector induced uncertainties and biases. The most important consequence of these effects is that a subset of the source population of interest will be unobserved. Since fainter sources are more likely to be unobserved, the missing data mechanism is *non-ignorable* [1]. Failure to account for non-ignorable missing data mechanisms can lead to serious inferential bias. In addition to the missing data, it also necessary to correct for background contamination, and the efficiency of the detector.

To address these challenges we develop a Bayesian method for estimating: (1) the number of sources unobserved due to detector effects, (2) the flux of observed sources, and, (3) the parameters of the $\log(N) - \log(S)$ curve. By modeling the missing data mechanism we correct for detection biases (e.g., Eddington bias) and obtain posterior summaries for the bias-corrected source population.

43.2 Inferring $\log N - \log S$ from Observational Data

The $\log N - \log S$ plot displays $\log_{10}(1 - F(s))$ vs. $\log_{10}(s)$ i.e., the log of the cumulative number of sources detectable at a given sensitivity, as a function of the log-flux. Let the distribution of fluxes in the source population be given by G, with cumulative distribution function F_G. The starting point for probabilistic modeling is that:

$$S_i \overset{iid}{\sim} G \qquad \Rightarrow \qquad \log_{10}(1 - F_G(s)) := H(\log_{10}(s)). \tag{43.1}$$

It can be shown that H in (43.1) is linear if and only if G is a Pareto distribution. Fitting a straight line to the $\log N - \log S$ plot is then seen to be equivalent to fitting a Pareto distribution to a sample of fluxes. Note that more flexible functional forms for the $\log N - \log S$ correspond to more flexible distributions for the true flux population.

In practice the fluxes are not observed directly, so they must be inferred using photon counts contaminated by detector effects and uncertainties. These processes, and any prior information, can naturally be built into our hierarchical model.

To obtain the desired quantities from the posterior distribution, computation is performed using Markov Chain Monte Carlo (MCMC). Figure 43.1 shows posterior samples of the $\log N - \log S$ curve fit using a linear relationship. For this simulation we note that the true $\log N - \log S$ curve is consistent with the posterior inference.

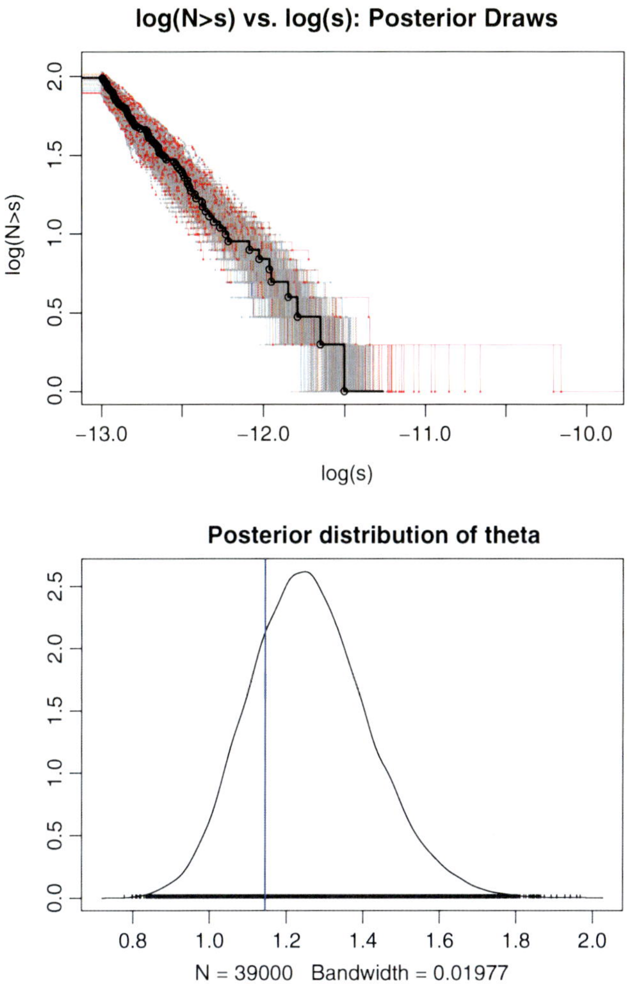

Fig. 43.1 (L) Posterior samples of the $\log N - \log S$ curve for simulated data. The true curve is shown in *bold*. (R) Posterior density of the slope parameter. The true value is shown by the *vertical line*

43.3 Conclusion

By modeling the $\log N - \log S$ relationship within a hierarchical Bayesian framework we achieve flexibility in describing properties of both the source population and the detector induced uncertainties. Our method explicitly corrects for the non-ignorable missing data mechanism that is often ignored by competing methods.

Reference

1. Little, R.J.A., Rubin, D.B.: Statistical Analysis with Missing Data, 2nd Ed. (2002) Wiley-Interscience.

Chapter 44
Techniques for Massive-Data Machine Learning in Astronomy

Nicholas M. Ball

Abstract Important computational algorithms for statistical analysis of massive datasets will require efficient $N\log N$ implementations. A leading group producing these algorithms is the FASTlab group at Georgia Institute of Technology. Substantial speedups over naive algorithms are achieved; for example, from $O(N^3)$ to $O(N)$ for Support Vector Machine classification and from $O(N^n)$ to $O(N^{logn})$ for n-point correlation functions. These methods can be applied to datasets such as the massive image dataset from the Next Generation Virgo Cluster Survey hosted at the Canadian Astronomy Data Centre. Object classification, Virgo Cluster membership, photometric redshifts, catalog cross-matching, and spatial clustering can potentially be achieved with greatly improved efficiency.

44.1 Introduction

Astronomy is increasingly encountering two fundamental truths:

- The field is faced with the task of extracting useful information from extremely large, complex, and high dimensional datasets.
- The techniques of *astroinformatics*[1, 2][1] and *astrostatistics* are the only way to make this tractable, and bring the required level of sophistication to the analysis.

Thus, an approach which provides these tools in a way that scales to these datasets is not just desirable, it is vital. The expertise required spans not just

[1] http://www.ivoa.net/cgi-bin/twiki/bin/view/IVOA/IvoaKDDguide.

N.M. Ball (✉)
National Research Council Herzberg Institute of Astrophysics, 5071 West Saanich Road, Victoria, BC V9E 2E7, Canada
e-mail: nick.ball@nrc-cnrc.gc.ca

E.D. Feigelson and G.J. Babu (eds.), *Statistical Challenges in Modern Astronomy V*, Lecture Notes in Statistics 209, DOI 10.1007/978-1-4614-3520-4_44,

astronomy, but also computer science, statistics, and informatics. We focus here on questions raised by the practical application of various algorithms to real astronomical datasets. That is, what is needed to maximally leverage their potential to improve the science return?

This is not a trivial task. While computing and statistical expertise are required, *so is astronomical expertise*. Precedent has shown that, to-date, the collaborations most productive in producing astronomical science results (e.g., the Sloan Digital Sky Survey), have either involved astronomers expert in computer science and/or statistics, or astronomers involved in close, long-term collaborations with experts in those fields. This does not mean that the astronomers are giving the most important input, but simply that their input is crucial in guiding the effort in the most fruitful directions, and coping with the issues raised by real data. Thus, the tools must be useable and understandable by those whose primary expertise is not computing or statistics, even though they may have quite extensive knowledge of those fields.

'Real' astronomical data are characterized by many issues which differentiate them from ideal data. They may:

- Be large, complex, increasingly high-dimensional, and may be in the time domain
- Contain missing data, such as non-observations or non-detections
- Have heteroscedastic (changing variance), non-Gaussian, or underestimated errors
- Contain outliers, artifacts, false detections, or systematic effects
- Contain correlated inputs
- … and so on

44.2 Relevance of the Algorithms Presented

The algorithms we consider here meet the criteria of being well-known—k-nearest neighbor (kNN), kernel density estimation (KDE), etc.—scalable ($N\log N$ where possible), and useable by astronomers via the software of the FASTlab group at Georgia Institute of Technology[2] directed by Prof. Alex Gray. Some of the well-known algorithms already scale without the work of the group, e.g., mixture of Gaussians, decision tree, linear regression, K-means, and principal components analysis (PCA). However, others, such as all nearest neighbors, KDE, Support Vector Machines, and n-point correlation function (nPCF), do not. What is significant about the results presented here is that they make all of these algorithms scalable. Extensive use is made of the fact that $N\log N$ scaling is achieved when building a *kd*-tree data structure. This and other space-partitioning tree structures are what makes the scaling possible.

[2]FASTlab = Fundamental Algorithmic and Statistical Tools Laboratory, http://www.fast-lab.org.

The relevance of the work of the FASTlab group is two-fold: (a) their results enable scalable versions of the algorithms that do not otherwise scale to be implemented; and (b) they give one the ability to employ more sophisticated variants of the algorithms that do scale. For example, many astronomical phenomena, such as galaxy spectra, are nonlinear, but are often treated by linear analyses such as PCA, or templates. Kernel PCA is a nonlinear extension of PCA, and in the results presented scales as O(N), rather than O(N^3). There are numerous other examples. Both of these points increase the applicability of the algorithms to real astronomical data, i.e., data that contains the issues listed in Sect. 44.1.

44.3 CADC, CANFAR, Petascale Data, and Fast Data Mining Algorithms

The Canadian Advanced Network for Astronomical Research (CANFAR) [3] is a project at the Canadian Astronomy Data Centre (CADC) to provide an infrastructure for data-intensive astronomy projects. It provides those portions of a pipeline that can be usefully supplied in a generic manner, such as access to, processing, storage, and distribution of data, without restricting the analysis that can be performed. The system combines the job scheduling abilities of a batch system with cloud computing resources, and users manage one or more virtual machines, which operate (to them) in the same manner as a desktop machine.

By extension of the arguments for providing a hardware infrastructure and standard software tools within CANFAR, we aim to provide a robust set of generic tools that can be used for data analysis. Given the requirements detailed in Sect. 44.1, that the methods of astroinformatics and astrostatistics are needed for appropriately sophisticated analysis of the data, that such algorithms must scale as $N\log N$ or better to remain tractable in the upcoming petascale regime, and that the aim of the FASTlab group is to implement them such that they may be used on real problems, we are using the software of the group to achieve our aims.

The key point is that, while a given science analysis always specific, *the underlying algorithms are generic*, and it is those that we aim to provide.

44.4 Example: The Next Generation Virgo Cluster Survey

The Next Generation Virgo Cluster Survey (NGVS)[3] is a new 104 square degree survey of the Virgo Cluster, which will provide coverage of this nearby dense environment in the universe to unprecedented depth. The limiting magnitude of the survey is $g_{AB} = 25.7$ (10σ point source), and the 2σ surface brightness limit

[3]https://www.astrosci.ca/NGVS/The_Next_Generation_Virgo_Cluster_Survey.

Table 44.1 NGVS tasks and FASTlab speedups (potential or actual)

Task	Algorithm	Naive speed	FASTlab speed
Object classification	SVM	$O(N^3)$	$O(N)$
Virgo Cluster membership	K-means	$O(N)$	
	PCA	$O(N)$	
	kernel PCA	$O(N^3)$	$O(N)$
Photometric redshifts	NN	$O(N)$	$O(\log N)$
	all NN	$O(N^2)$	$O(N)$
Describing a photo-z PDF	KDE	$O(N^2)$	$O(N)$
Cross-matching multi-wavelength data	nPCF[a]	$O(N^n)$	$O(N^{\log n})$
Clustering of background objects	nPCF	$O(N^n)$	$O(N^{\log n})$

[a]nPCF = n-point correlation function

is $g_{AB} \approx 29$ mag arcsec^{-2}. The data volume of the completed survey will be approximately 50 terabytes. The objects detected span an enormous dynamic range, from the giant elliptical galaxy M87 at $M(B) = -21.6$, to the faintest dwarf ellipticals at $M(B) \approx -6$. Photometry will be available in five broad bands (u^* g' r' i' z'), and the unprecedented depth reveals many complex and previously unseen low surface brightness structures. Some of the survey challenges are given in Table 44.1, together with the relevant machine learning algorithm, and the speedup provided by the results of the FASTlab group.

A typical region of the survey is shown in Fig. 44.1, further exemplifying some of the challenges, and adding others. Many of these, which do not make direct use of the algorithms, but rather of other astronomical software, may be sped up by a linear factor equal to the number of processing cores (currently several hundred) available on the CANFAR system.

Thus, the combination of the fast algorithms provided by the FASTlab group and the CANFAR system enables large datasets to become tractable, while at the same time, for challenges that the algorithms do not directly address, enabling those too to be tackled. Thus, the revolutionary, but nevertheless real and not idealized astronomical data of the NGVS and future surveys, is being tackled in a smart, and scalable way.

44.5 Concluding Questions

These algorithms have excellent potential for improving astronomical analysis. Nevertheless, there are questions one can ask at the interface between astronomical and statistical considerations:

- Will statistical inference (i.e., Bayesian) methods turn out to be more useful for most problems than the prediction-oriented methods presented here?
- Are the approximations introduced in some of the algorithms to enable the speedups (e.g., the kernel methods), unacceptably large?

Fig. 44.1 Typical NGVS survey region, showing several challenges to data mining provided by this survey, including: (1) full-color images, provided by 5-band photometry; (2) *bright stars*, exhibiting halos and bleed trails; (3) large galaxies, showing elliptical light profiles, color gradients, and detailed morphology; (4) complex, irregular, galaxy morphologies—the galaxy on the right is NGC 4438; (5) similar low-surface brightness features, the incidence of which is hugely increased by the survey's unprecedented depth; (6) low surface brightness galaxies, e.g., below NGC 4438; (7) globular clusters and ultra-compact dwarfs—these objects may be unresolved, or partially resolved, and exhibit different light profiles to galaxies, complicating their classification, and the separation of stars (unresolved) and galaxies (resolved); (8) varying sky background, especially near large galaxies, whose light extends to large radii; (9) most objects in the image have no spectroscopy, thus their membership, or not, of the Virgo Cluster, must be deduced by other means

- Will the algorithms be rendered insufficiently useful because of errors on the inputs?
- Are the algorithms limited when the dataset does not fit in memory (either too big, or portions are run in parallel)?
- Will most astronomical data analyses still contain stages that cannot be practically addressed by these algorithms, and that also scale worse than $N\log N$, thus overwhelming even a CANFAR-like parallel computing system?
- Will there be data of high *intrinsic* dimension, that cannot easily be dimension-reduced, thus causing curse-of-dimensionality-type problems that may hamper these algorithms?

- Will novel supercomputing hardware, such as GPGPUs, that enable extremely fast brute-force approaches to problems such as nearest neighbors, prove more practical?
- If the software is licensed, rather than free and open source, will it be practical to deploy it on a distributed computing system for astronomical use?
- Will astronomers require the sophistication of the more advanced algorithms, or will the simple ones that scale remain 'good enough', because the improvements brought by new data still account for most of the new science return?

There are arguments one can make that the answer to all of these is "no". But, as always, if we knew all the answers, it wouldn't be research.

References

1. Ball, N.M., Brunner, R.J.: Data Mining and Machine Learning in Astronomy. Int. J. Mod. Phys. D **19**, 1049–1106 (2010)
2. Borne, K.: Scientific Data Mining in Astronomy. Data Mining and Knowledge Discovery Series, Taylor & Francis: CRC Press, Boca Raton, FL, Ch. 5, pp. 91?114 (2009)
3. Gaudet, S., Hill, N., Armstrong, P., et al. CANFAR: the Canadian Advanced Network for Astronomical Research. In: Radziwill, N.M., Bridger, A. (eds.) Proc. SPIE, Software and Cyberinfrastructure in Astronomy, **7740**, 1L

Chapter 45
A Bayesian Approach to Gravitational Lens Model Selection

Irene Balmès

Abstract Strong gravitational lenses are unique cosmological probes. These produce multiple images of a single source. Whether a single galaxy, a group or a cluster, extracting cosmologically relevant information requires an accurate modeling of the lens mass distribution. A variety of models are available, nevertheless discrimination between them as primarily relied on the quality of fit without accounting for the size of the prior model parameter space. This is a problem of model selection that we address in the Bayesian statistics framework by evaluating Bayes' factors. Using simple test cases, we show that the assumption of more complicate lens models may not be justified given the level of accuracy of the available data.

Images produced by strong gravitational lenses result of different light-paths. If the source behind the lens has a variable luminosity, this will manifest with a time delay between the images. This time delay Δt depends on the gravitational potential of the lens, and the underlying cosmological model. Therefore, we can derive constraints on cosmological parameters (in particular H_0), provided a lens model is assumed. Hence, lens modeling as well as accurate measurements capable of discriminating between models are critical to the study of time delays.

We aim to tackle this problem from the point of view of Bayesian model selection analysis (see e.g. [2]). A large number of lens models have been proposed in a vast literature. Given the fact that observables are limited to the position of the images, their time delay and flux ratio, we restrict our analysis to simple examples characterized by a few parameters. In particular we consider two models for lenses with two images, so called "double" lenses (for a review on lensing, see [1]).

I. Balmès (✉)
Laboratoire Univers et Théories (LUTh), UMR 8102 CNRS, Observatoire de Paris, Université Paris Diderot, 5 place Jules Janssen, 92190 Meudon, France
e-mail: irene.balmes@obspm.fr

E.D. Feigelson and G.J. Babu (eds.), *Statistical Challenges in Modern Astronomy V*, Lecture Notes in Statistics 209, DOI 10.1007/978-1-4614-3520-4_45,

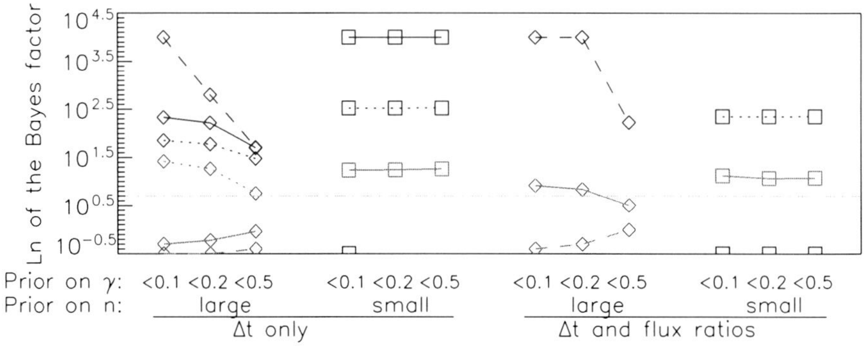

Fig. 45.1 Bayes' factor between model 1 and 2, with different priors. Above the *dot-dashed line*, the evidence in favor of model 1 is strong. Each color represents a different lens

1. Power-law model: assume a density profile $\rho \propto r^{-n}$, with n a free parameter. For $n = 2$, it describes an isothermal lens. In order to assess the dependence on the prior parameter interval we assume two different priors: $0 < n < 3$ (large) and $1 < n < 3$ (small).
2. Power-law model with external shear: assume the previous model with the addition of shear accounting for environmental effect on the lens. This adds two parameters: the strength of the shear γ, and its direction. Expected values for the shear vary up to $\gamma \simeq 0.1$. We assume three different priors on γ: $\gamma < 0.1$, < 0.2 and < 0.5 respectively, testing the shear strength up to unrealistic values.

We performed a likelihood data analysis for a sample of lenses and inferred the Bayes' factor for model 1 and 2 under different priors. Results are summarized in Fig. 45.1. Large Bayes' factors favor the simpler model, model 1. Above a certain threshold (dot-dashed line), the evidence in favor of model 1 is considered strong. In the following, we highlight a few relevant aspects.

- Effect of the prior on n: The lens data set is mainly composed of galaxies, which we expect to be nearly isothermal. Nevertheless, our analysis shows that a large fraction of our sample is accurately described by model 1 if $0 < n < 1$.
- Effect of the prior on γ: In more than half of the cases, allowing higher (unrealistic) shear strength does not change the Bayes' factor. This illustrates the effect of the Occam's razor term in the Bayes' factor: a wider range for a parameter is bound to give a better fit, but this is balanced against a penalty factor.
- Effect of the flux ratios: Time delays depend on the gravitational potential of the lens, whereas flux ratios depend on its second derivative. Furthermore, they are subject to a number of local phenomena that do not affect time delays. As a result, flux ratios require more complex models than time delays. This is consistent with our findings in Fig. 45.1: indeed, adding flux ratios as a constraint leads to having less lenses accurately described by model 2.

Acknowledgements P.-S. Corasaniti provided helpful advice on both this work and this paper. I. Balmès is supported by a scolarship of the "Ministére de l'Éducation Nationale, de la Recherche et de la Technologie" (MENRT).

References

1. C. S. Kochanek. Part 2: Strong gravitational lensing. In G. Meylan, et al., eds., *Saas-Fee Advanced Course 33: Gravitational Lensing: Strong, Weak and Micro*, pages 91–268, 2006.
2. R. Trotta. Bayes in the sky: Bayesian inference and model selection in cosmology. *Contemporary Physics*, 49:71–104, March 2008.

Chapter 46
Identification of Outliers Through Clustering and Semi-supervised Learning for All Sky Surveys

Sharmodeep Bhattacharyya, Joseph W. Richards, John Rice, Dan L. Starr, Nathaniel R. Butler, and Joshua S. Bloom

Abstract Recently there has been a huge surge of data in astronomy, making outlier or novelty detection a crucial step in analyzing these data. Here, we introduce a clustering based semi-supervised approach for outlier detection. The training data, $(X_1, Y_1), \ldots, (X_n, Y_n)$, where $n = 1{,}542$, comes from Hipparcos and Optical Gravitational Lensing Experiment (OGLE) surveys, with, $X_i \in \mathbb{R}^p$ $(p = 64)$ as the features and Y_i is a categorical variable having one of the 25 class labels. The set of 64 periodic and non-periodic features are extracted from the light curves. The test data, $Z_1, \ldots, Z_m$, where $m = 11{,}375$, is the test data, where, $Z_i \in \mathbb{R}^p$. We select these 11,375 low noise variable light sources for our analysis from a set of unlabeled light curves of ∼50,000 variable light sources from All Sky Automated Survey (ASAS). Our goal is to find outlier data points in the unlabeled data set whose labels can not be properly predicted by the information in the labeled data set. We propose a new hierarchical algorithm for outlier detection in this partially labeled setup based on clustering and semi-supervised learning. We apply our method to identify interesting sources in the ASAS data set, with the training data. We present the ASAS light curves of some of these interesting sources, and elaborate on the possible physical mechanisms driving their variability.

46.1 Main Work and Results

We call a collection of data points in the data set *interesting outliers*, if it forms a well-separated cluster and have no known class label with high confidence.

For clustering, we consider two different metrics.

S. Bhattacharyya (✉) • J.W. Richards • J. Rice
Department of Statistics, University of California, Berkeley, CA, USA
e-mail: sharmo@stat.berkeley.edu

J.W. Richards • D.L. Star • N.R. Butler • J.S. Bloom
Department of Astronomy, University of California, Berkeley, CA, USA

E.D. Feigelson and G.J. Babu (eds.), *Statistical Challenges in Modern Astronomy V*, Lecture Notes in Statistics 209, DOI 10.1007/978-1-4614-3520-4_46,

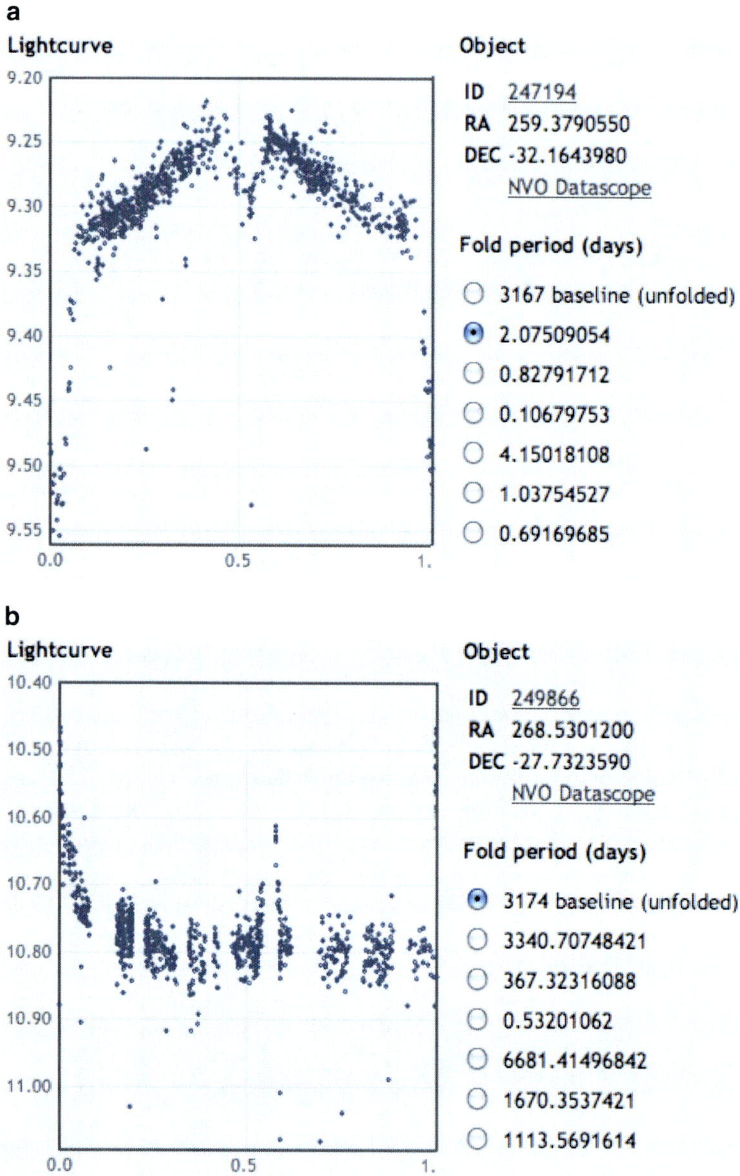

Fig. 46.1 (**a**) Is it an eclipsing Be star? (**b**) Young Emission line star, with no literature

`Weighted Euclidean Metric`: With Random Forest Importance measure of each feature, based on fitting the training data, as the weights.
Random Forest Proximity Metric: See Liaw and Wiener [1] for details about this metric.

The method has parameters (α, L_1, L_2, C, K). We call a cluster $\alpha - outlier$ if it contains less than $(100\alpha)\%$ of the data points from which it has been separated at a given iteration. The parameters L_1, L_2, C and K control how to determine, whether an outlier is *interesting* or not. Here is the method

Step 1 We consider the scaled version of feature space of both labeled and unlabeled data set together, that is, consider, $\mathbf{X} = (X_1, \ldots, X_n, Z_1, \ldots, Z_m)$.
Step 2 Now, cluster the $(n+m)$ data using hierarchical divisive clustering. At any iteration, cluster S divides into S_1 and S_2.
Step 3 At each iteration, if $\alpha.\min(|S_1|, |S_2|) < |S|$ **or** $\min(|S_1|, |S_2|) \leq 10$, we flag the smaller cluster. We stop after a large number of iterations, say K.
Step 4 Now consider each flagged cluster S and the set of labeled data points in S be S_L. If $L_1|S| \leq |S_L|$, then remove flag of S. If $L_2|S| \leq |S_L| < L_1|S|$ $(L_2 < L_1)$, but, more than $C|S_L|$ has same labels, then also, remove the flag of the cluster.
Step 5 Consider all the data points from the unlabeled data set in the flagged clusters as the 'interesting outliers'.

We apply our method to the ASAS data set with $\alpha = 0.01$, $L_1 = 0.5$, $L_2 = 0.25$, $C = 0.75$ and $K = 200$. Below, we present the light curves of two *interesting outliers*, with a note on their peculiarity (Fig. 46.1). This work is described in detail by [2].

References

1. Liaw, A. & Wiener, M. (2002) Classification and Regression by randomForest, *R news*, 2(3), 18–22
2. Richards, J.W., Starr, D.L., Butler, N.R., Bloom, J.S., Brewer, J.M., Crellin-Quick, A., Higgins, J., Kennedy, R. and Rischard, M. (2011) *Astrophys. J.* 733, 10

Chapter 47
Estimation of Moments on the Sphere by Means of Fast Convolution

P. Bielewicz, B.D. Wandelt, and A.J. Banday

Abstract In order to study the statistical properties of large data sets, fast and reliable methods for the estimation of basic statistical quantities, such as moments of the data, are required. We present a method for the estimation of moments on azimuthally symmetric patches defined for data pixelized on the sphere by means of fast convolution. As an example application, we show the results of a search in the WMAP CMB sky maps for regions with anomalous values of the variance, skewness or kurtosis as estimated on a set of concentric rings.

The computation of moments on azimuthally symmetric patches on the sphere can be viewed as the convolution of the data, taken to the appropriate power, with an azimuthally symmetric beam that describes the geometry of the patch. To make this statement more clear, let us consider the example of the computation of variance on the pixelized map ΔT_i for which some regions are excluded by application of the mask M_i. For a region centered on, but not necessarily including, pixel i an estimator of the variance $\mathrm{Var}_{i,r}(\Delta T)$ is given by

$$\mathrm{Var}_{i,r}(\Delta T) = \frac{\sum_j \Delta T_j^2 M_j B_{ij}^r}{\sum_k M_k B_{ik}^r} - \left(\frac{\sum_j \Delta T_j M_j B_{ij}^r}{\sum_k M_k B_{ik}^r}\right)^2 - \frac{\sum_j \sigma_j^2 M_j B_{ij}^r}{\sum_k M_k B_{ik}^r} . \quad (47.1)$$

P. Bielewicz (✉)
University of Toulouse (UPS-OMP), IRAP, Toulouse, France
e-mail: bielewic@cesr.fr

B.D. Wandelt
Institut d'Astrophysique de Paris, 98 bis boulevard Arago, 75014 Paris, France
e-mail: benwandelt@gmail.com

A.J. Banday
Institut de Recherche in Astrophysique et Planétologie, Université de Toulouse, Toulouse, France
e-mail: banday@cesr.fr

E.D. Feigelson and G.J. Babu (eds.), *Statistical Challenges in Modern Astronomy V*, Lecture Notes in Statistics 209, DOI 10.1007/978-1-4614-3520-4_47,

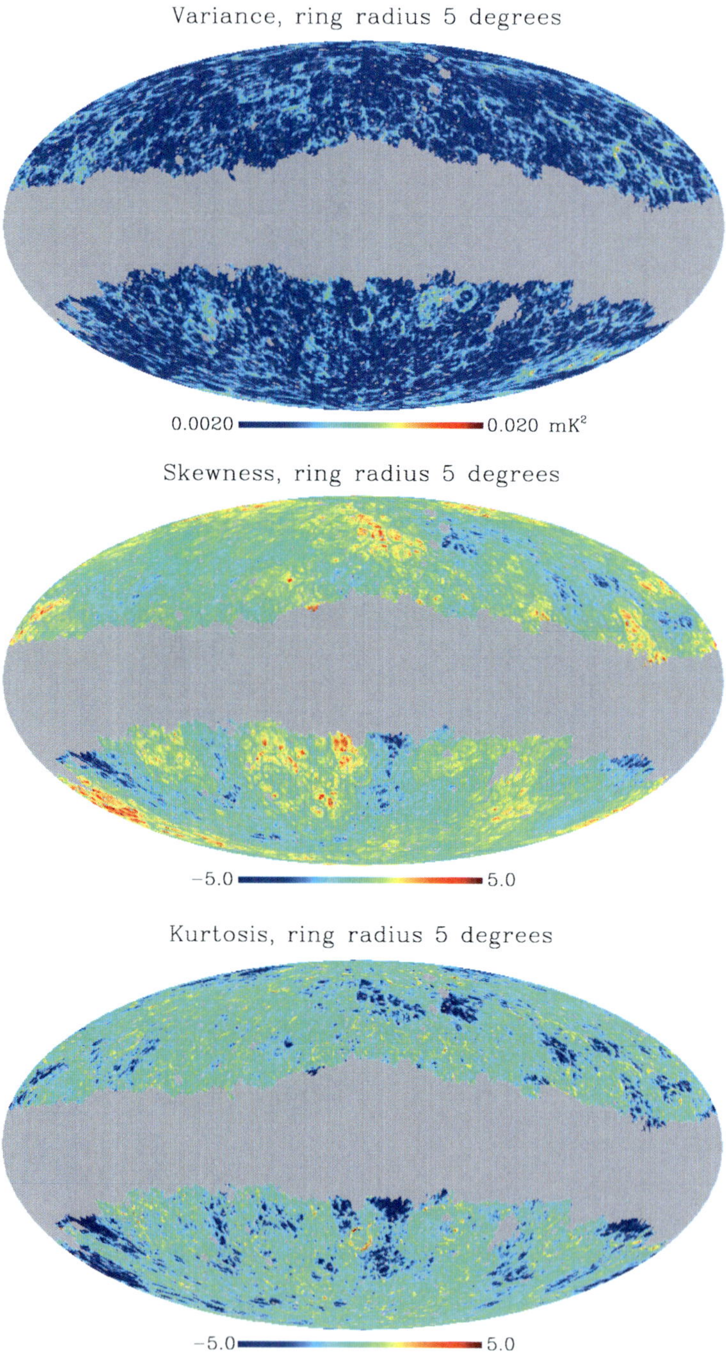

Fig. 47.1 Variance, skewness and kurtosis (from left to right, respectively) estimated on rings with radius of 5° and width of half degree for the foreground corrected V-band WMAP map masked with the KQ75y7 cut

Here, B^r_{ij} denotes the profile of the azimuthally symmetric patch used for estimation of the variance. In the case of our studies of CMB maps presented here, this corresponds to a ring with radius of radius r and width Δr such that

$$B^r_{ij} = \begin{cases} 1 \text{ for } r \leq \arccos(\hat{\mathbf{n}}_i \cdot \hat{\mathbf{n}}_j) < r + \Delta r \\ 0 \text{ otherwise} \end{cases}. \tag{47.2}$$

The last term in (47.1) corresponds to a correction for the bias introduced by the noise variance σ_i^2.

Computation of this estimator is achieved by direct summation over all unmasked pixels j in a given patch centered on pixel i for all possible centers of the patch, and scales as $\mathscr{O}(N_{\text{pix}}^2)$, where N_{pix} is the number of pixels in the map. However, this sum is nothing other than the convolution of the masked map ΔT or ΔT^2 with beam B. Therefore, it can be performed efficiently by decomposition of the data in the basis of spherical harmonics $Y_{\ell m}(\Omega_i)$ functions. Then, for example, the sum of the terms in ΔT_j^2 is given by $\sum_j \Delta T_j^2 M_j B^r_{ij} = \sum_{\ell,m} \tilde{a}_{\ell m} b_\ell Y_{\ell m}(\Omega_i)$, where $\tilde{a}_{\ell m}$ and b_ℓ are the spherical harmonic coefficients of the masked map ΔT^2 and patch, respectively. In the case of azimuthally symmetric patches, we have implicitly utilized the fact that the coefficients b_ℓ can depend only on the multipole order ℓ. Therefore, the complexity of the algorithm for the computation of the variance can be reduced to $\mathscr{O}(N_{\text{pix}} \log N_{\text{pix}})$ operations. This algorithm can also be extended to higher order moments, such as the skewness or kurtosis.

The low complexity of the algorithm makes it well suited for the search for regions with anomalous statistical properties in large data sets such as CMB maps derived from the WMAP or Planck data. We employ this technique to search for regions of the 7-year WMAP data [2] which exhibit anomalous variance, skewness or kurtosis. Examples of the variance, skewness and kurtosis maps are shown in Fig. 47.1. Preliminary results reveal a few interesting regions on the sky, but general consistency with the currently preferred standard ΛCDM model.

Acknowledgements We acknowledge use of CAMB [3] and the HEALPix software [1] analysis packages for deriving these results. We acknowledge the use of data products from the Legacy Archive for Microwave Background Data Analysis (LAMBDA) web site. Support for LAMBDA is provided by the NASA Office of Space Science.

References

1. Górski K. M., Hivon E., Banday A. J., Wandelt B. D., Hansen F. K., Reinecke M., & Bartelmann M., 2005, ApJ, 622, 759
2. Jarosik N., et al., 2011, ApJS, 192, 14
3. Lewis A., Challinor A., Lasenby A., 2000, ApJ, 538, 473

Chapter 48
Variability Detection by Change-Point Analysis

Seo-Won Chang, Yong-Ik Byun, and Jaegyoon Hahm

Abstract We describe a method to detect short-term variability based on the change-point analysis with filtering algorithm using local statistics. The use of cumulative sum scheme and bootstrap rank statistics as a means of detecting a series of change points is discussed. By applying this method to over 30,000 lightcurves from the MMT transit survey data, we found previously unknown evidences about stellar variability (including a total of 606 flare events, 18 eclipsing-like features, and 3 transit-like features). In particular, this approach will be effective in detecting non-periodic events in massive astronomical time series data.

The detection and characterization of variability is often the first step to understand the nature of various cosmic objects. Most variability detection methods require conventional models that are mainly focused on the strictly periodic signals, and are not suitable for the study of arbitrary-shaped, non-periodic, and sporadically occurring variations, especially those of short time scales. Also, in many cases, signal estimation is equated with smoothing of data for de-noising. This sometimes discards vital information in time series data. We introduce a non-parametric method to extract all significant features based on the change-point analysis (CPA) with filtering algorithm using local statistics.

S.-W. Chang (✉) • Y.-k. Byun
Department of Astronomy, Yonsei University, Seoul, Korea
e-mail: seowony@galaxy.yonsei.ac.kr

J. Hahm
Korea Institute of Science and Technology Information Supercomputing Center, Eoeun-dong 52, Yuseong, Daejeon, Korea
e-mail: jaehahm@kisti.re.kr

E.D. Feigelson and G.J. Babu (eds.), *Statistical Challenges in Modern Astronomy V*, Lecture Notes in Statistics 209, DOI 10.1007/978-1-4614-3520-4_48,

48.1 Change-Point Detection Algorithm

Using a combination of cumulative sum scheme and bootstrap rank statistics [3], our method produces a series of estimated change points which correspond to the moments of apparent systematic changes. A given dataset $x_1, x_2, \cdots, x_n$, the estimated change point location $\hat{p}$ is

$$\hat{p} = \underset{p_k \in [1,n]}{\arg\max} \left| S_{p_k} \right|, \tag{48.1}$$

where $S_{p_k} = S_{p_{k+1}} + (x_{p_k} - \bar{x})$, $S_{p_0} = 0$, and the mean $\bar{x}$. The sub-region is successfully segmented into exactly two segments by $\hat{p}\ (\bar{x}_1 = \cdots = \bar{x}_{p_k} \neq \bar{x}_{p_{k+1}} = \bar{x}_n)$. If no change-points could be found at all $(\bar{x}_1 = \bar{x}_2 = \cdots = \bar{x}_n = \bar{x})$, the adjacent sub-region will be considered. Based on N bootstrap samples that are randomly re-ordered original values, we estimate the confidence levels associated with each change point, and then remove some candidates that are not statistically significant anymore. To detect significant features occurring at specific levels in lightcurve, we define a simple criteria similar to Micro-lensing Alert system [1] in the presence of hetero-scedastic measurement errors (w_i):

$$\frac{(x_i - \bar{x}_{p_k} \pm w_i)}{\sigma_{p_k}} \geq N; \quad ConM \geq M, \tag{48.2}$$

where $\sigma_{p_k} = \sqrt{\frac{1}{n-1} \sum_{i=1}^{n} \left(x_i - \bar{x}_{p_k}\right)}$ is the standard deviation of each sub-region and *ConM* is consecutive measurements of x_i that satisfy the selection criteria N. The different combination of N, M values are used for maximize the detection efficiency of significant outlying features.

48.2 Application to MMT Transit Survey Data

By applying this method to over 30,000 lightcurves from the MMT transit survey of M37 open cluster [2], we efficiently identified several hundred instances of abrupt brightness changes **without any smoothing or interpolation of the raw data**.

1. **Flare-like event detection**: The flare detection process is not complete if the quiescent stellar variability remains in the lightcurves. Without any filtering (e.g., moving average) of periodic variability, this method is optimized to detect multiple flare-like features embedded in real astronomical datasets that have unevenly spaced in time and show statistically non-stationary noise behavior.
2. **Eclipsing-like event detection**: Because time series observations often suffer from the incomplete coverage of the entire eclipse periods, eclipsing variables are usually observed repeatedly to form a complete phased lightcurve. This method

is useful to detect the moments of eclipse ingress, center, or egress in cases where the eclipsing pattern is not repeated or the coverage is not sufficient for the detection through conventional period analysis.

Our CPA approach is particularly effective in detecting non-periodic events from data with varying noise as well as short duration events from either non-varying or varying lightcurves.

Acknowledgements This work is supported by Korea Institute of Science and Technology Information under the contract of the commissioned research project, Massive Astronomical Data Applications of Cloud Computation (KISTI-P11020).

References

1. Glicenstein, J. -F. 2001, in *Microlensing 2000: A New Era of Microlensing Astrophysics*, ed. by J. W. Menzies and P. D. Sackett. ASP Conf. Proc., Vol. 239, 28
2. Hartman, J. D., et al. 2008, ApJ, 675, 1233
3. Taylor, W. 2000, in *Change-Point Analyzer 2.0 shareware program*, http://www.variation.com/cpa

Chapter 49
Evolution as a Confounding Parameter in Scaling Relations for Galaxies

Didier Fraix-Burnet

Abstract Early-type galaxies are characterized by many scaling relations. Evolutionary classifications find that some of these correlations are indeed generated by diversification. With a simple mathematical formalism, we show that even the so-called fundamental plane, a relatively tight correlation between three variables, can be easily explained as the artifact of the effect of another parameter influencing all, without any physical hypothesis. In other words, the fundamental plane is probably a confounding correlation, i.e. not physically causal. The complexity of the physics of galaxies and of their evolution suggests that the confounding parameter must be related to the level of diversification reached by the galaxies. Consequently, many scaling relations for galaxies are probably evolutionary correlations, explained by the statistical general evolution of most properties of galaxies.

49.1 The Fundamental Plane as a Confounding Correlation

The fundamental plane for early-type galaxies is a correlation between effective radius, the central velocity dispersion and the surface brightness within the effective radius [1–3]. Let us consider that the effective radius r_e, the central velocity dispersion σ and the luminosity L are all power-law functions of a same generic parameter $\widetilde{X}$:

$$\begin{cases} r_e = A_1 \widetilde{X}^p \\ \sigma = A_2 \widetilde{X}^s \\ L = A_3 \widetilde{X}^t \end{cases} \tag{49.1}$$

D. Fraix-Burnet (✉)
Université Joseph Fourier - Grenoble 1/CNRS, Institut de Planétologie et d'Astrophysique de Grenoble, BP 53, F-38041 Grenoble cedex 9, France
e-mail: fraix@obs.ujf-grenoble.fr

E.D. Feigelson and G.J. Babu (eds.), *Statistical Challenges in Modern Astronomy V*, Lecture Notes in Statistics 209, DOI 10.1007/978-1-4614-3520-4_49,

The surface brightness μ_e can be expressed as $\mu_e = -2.5\log\left(L/4\pi r_e^2\right) + m = (-2.5t + 5p)\log\widetilde{X} + 2.5\log(4\pi) + m$ where m is a constant of normalisation. Any linear correlation of the form

$$\log r_e = a\log\sigma + b\mu_e + c \tag{49.2}$$

translates to

$$\begin{cases} p = sa + (-2.5t + 5p)b \\ \log A_1 = a\log A_2 + b\left(2.5\log\left(4\pi A_1^2/A_3\right) + m\right) + c. \end{cases} \tag{49.3}$$

If a solution can be found for a and b from (49.3), then the equation of the fundamental plane (49.2) is obtained. Conversely, the observations provide a, b and c, so that it is possible to derive p, s and t. There is no need of any further assumption to explain the fundamental plane.

49.2 Evolutionary Correlations

In the course of diversification, many properties of galaxies change, and they tend to statistically change in a more or less monotonous way. It seems difficult to avoid the evolution to act as a confounding factor. It is a well-known problem of comparative methods in phylogeny [4].

We thus propose that the main confounding parameter is $\widetilde{X} = T$ with T an indicator of the level of diversification, being something like an evolutionary clock not necessarily easily related to time or redshift. Indeed, the evolutionary clock, i.e. the factor $\widetilde{X} = T$, can be hidden, not understandable analytically and not directly observable. It is related to an evolutionary classification that gathers objects according to their history. This work is published in [5].

References

1. Djorgovski, S., Davis, M.: Fundamental properties of elliptical galaxies, ApJ, **313**, 59–68 (1987)
2. Dressler, A., Lynden-Bell, D., Burstein, D., Davies, R. L., Faber, S. M., Terlevich, R., Wegner, G.: Spectroscopy and photometry of elliptical galaxies. I - A new distance estimator. ApJ, **313**, 42–58 (1987)
3. Fraix-Burnet, D., Dugué M., Chattopadhyay T., Chattopadhyay A. K., Davoust E.: Structures in the fundamental plane of early-type galaxies, MNRAS, **407**, 2207–2222 (2010) (arXiv:1005.5645)
4. Felsenstein, J.: , Phylogenies and the Comparative Method, The American Naturalist, **125**, 1–15 (1985)
5. Fraix-Burnet, D.: The Fundamental Plane of early-type galaxies as a confounding correlation, MNRAS, **416**, L36-L40, (2011) (arxiv:1106.3154)

Chapter 50
Detecting Galaxy Mergers at High Redshift

P.E. Freeman, R. Izbicki, Ann B. Lee, C. Schafer, D. Slepčev, and J. Newman

Abstract We introduce a new feature of galaxy images, *maxRatio*, and demonstrate its effectiveness at detecting merging galaxies at high redshift.

Mergers play an important role in the development of massive galaxies at redshifts $z \sim 2$, transforming star-forming disks into quenched spheroidal systems. Automated detection of merging systems in this redshift regime is thus critical for testing theories of hierarchical galaxy formation. At low redshifts ($z \sim 0.2$), mergers are efficiently detected by, e.g., extracting estimates of two features from galaxy images: G (the Gini coefficient) and M_{20} [1, 3]. However, Lotz et al. show that the estimators $\hat{G}$ and $\hat{M}_{20}$ become increasingly biased in the low S/N and resolution regimes. We test the efficacy of $\hat{G}$, $\hat{M}_{20}$, and other image features in detecting mergers/interactors (M/I galaxies) at high z by examining 1,653 objects in the H-band GOODS-S ERS2 field [4] whose morphologies were visually identified by members of the CANDELS team [2]. These objects include 236 possible mergers, 70 possible interactors, and 46 tabbed as mergers *and* interactors by different voters.

We introduce a new feature, *maxRatio*, that as shown below is more effective than $\hat{G}$ and $\hat{M}_{20}$ for detecting M/I galaxies at high z. Define a sequence s of levels relative to an object's maximum intensity. For each level i, generate level sets, then compute the ratio R_i of the area of the second-largest set to the largest. *maxRatio*

P.E. Freeman (✉) • R. Izbicki • A.B. Lee • C. Schafer
Department of Statistics, Carnegie Mellon University, Pittsburgh, PA, USA
e-mail: pfreeman@cmu.edu

D. Slepčev
Department of Mathematics, Carnegie Mellon University, Pittsburgh, PA, USA
e-mail: slepcev@math.cmu.edu

J. Newman
Department of Physics and Astronomy, University of Pittsburgh, Pittsburgh, PA, USA
e-mail: janewman@pitt.edu

E.D. Feigelson and G.J. Babu (eds.), *Statistical Challenges in Modern Astronomy V*, Lecture Notes in Statistics 209, DOI 10.1007/978-1-4614-3520-4_50,

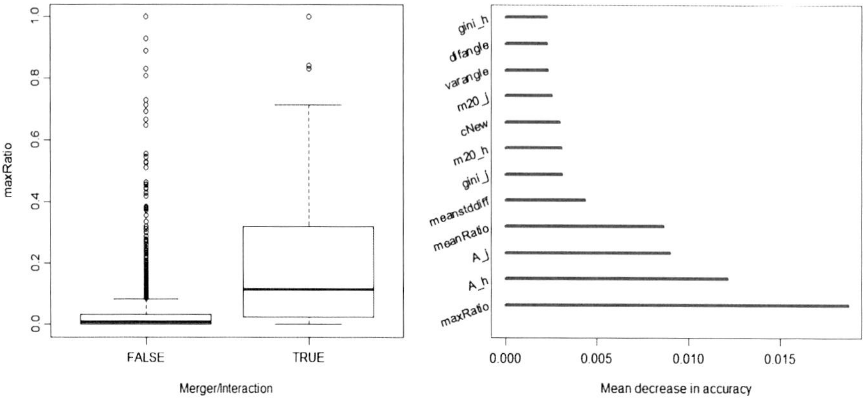

Fig. 50.1 *Left:* boxplots showing the values of *maxRatio* for visually identified non-M/I (*left*) and M/I (*right*). *Right:* relative importance of the 12 most informative features included in our analysis; the most informative is *maxRatio*

is the maximum value of R_i over the sequence s. For non-merging galaxies whose images generally exhibit only one cluster of pixels at each level, *maxRatio* $\rightarrow 0$. For merging systems galaxies with, e.g., two or more distinct nuclei, *maxRatio* $\rightarrow 1$.

We populate a p-dimensional space of features (e.g., $\hat{G}$, *maxRatio*, etc.) with training data, then use various classifiers to predict morphologies (e.g., lasso, random forest, etc.). For instance, lasso selects a sparse set of most important features by zeroing out the coefficients associated with less important features:

$$\min_{\beta}\left(\sum_{i=1}^{n}(Y_i+X_i^T\beta)^2+\lambda||\beta||_1\right), \quad \text{where } ||\beta||_1=\sum_{j=1}^{p}|\beta_j|.$$

The response variable Y_i is the proportion of voters who identify galaxy i with M/I galaxies. In our preliminary work, we identify *maxRatio* is the most important feature of the data for detecting M/I galaxies. See Fig. 50.1.

Acknowledgements This work was supported by NASA AISR grant NNX09AK59G.

References

1. Abraham, R. G., van den Bergh, S. & Nair, P. (2003) A New Approach to Galaxy Morphology. I. Analysis of the Sloan Digital Sky Survey Early Data Release, *Astrophys. J.*, 588, 218–229
2. Grogin, N. A. and others (2011) CANDELS: The Cosmic Assembly Near-infrared Deep Extragalactic Legacy Survey, arXiv:1105.3753
3. Lotz, J. M., Primack, J. & Madau, P. (2004) A New Nonparametric Approach to Galaxy Morphological Classification, *Astron. J.*, 128, 163–182
4. Windhorst, R. A. and others (2011) The HST WFC3 Early Release Science Data: Panchromatic Faint Object Counts for 0.2-2 μm Wavelength, *Astrophys. J. Suppl.*, 193, 27–59

Chapter 51
Multi-component Analysis of a Sample of Bright X-Ray Selected Active Galactic Nuclei

Dirk Grupe

Abstract I report on the statistical analysis of a sample of about 100 Active Galactic Nuclei (AGN) with simultaneous UV and X-ray observations from Swift. I found clear correlations between the X-ray spectral slope α_x, the UV slope α_{UV}, and the optical-to-x-ray spectral slope α_{ox} with the Eddington ratio $L/L_{\rm Edd}$. A major aspect of the statistical analysis will be multi-variant analysis statistical tools such as the Principal Component Analysis (PCA) and cluster analysis. This analysis shows that the main driver of the AGN properties in this sample is the Eddington ratio $L/L_{\rm Edd}$. Although separating Seyfert 1s into Narrow Line Seyfert 1s and Broad Line Seyfert 1s is still a good classification, with the 2,000 km/s cutoff line it is arbitrary. The cluster analysis of this AGN sample suggests that we can separate AGN into those with high and low Eddington ratios and that they form physically distinct groups.

Powered by accretion of surrounding matter onto the central black hole, Active Galactic Nuclei (AGN) are one of the must luminous persistent sources in the Universe. There are several question that we want to answer in AGN research: How do black holes in Active Galactic Nuclei evolve? Are there different phases in the evolution of an AGN? How long does the AGN activity last? How do measurements of the low-redshift Universe relate to high-redshift quasars at the early phases of the Universe?

In order to answer these questions, the key tools are multi-variate statistical methods that can explore the parameter space that is spanned by the AGN emission line and continuum properties. The observed properties of AGN are mainly driven

D. Grupe (✉)
Department of Astronomy and Astrophysics, Pennsylvania State University,
525 Davey Lab, University Park, PA 1682, USA
e-mail: dxg35@psu.edu

E.D. Feigelson and G.J. Babu (eds.), *Statistical Challenges in Modern Astronomy V*, Lecture Notes in Statistics 209, DOI 10.1007/978-1-4614-3520-4_51,

by two parameters: the mass of the central black hole and the accretion rate [1,4,6,8] Both are tied together by the Eddington ratio $L/L_{\rm Edd}$.

The original AGN sample presented here was selected from the ROSAT All-Sky Survey containing 110 AGN [3]. The advantages of this sample is that all sources are bright in X-rays as well as in the Optical/UV. The AGN in this sample are not (strongly) intrinsically absorbed or reddened. Many of these AGN appear to be highly variable at Optical/UV and X-ray energies, which makes studies of the spectral energy distributions of these objects challenging if they are not performed simultaneously. Swift is the most sufficient observatory to provide simultaneous UV and X-ray observations. Half of the AGN observed by Swift are Narrow Line Seyfert 1 Galaxies (NLS1s). NLS1s exhibit extreme properties and the occupy one extreme end in the AGN parameter space. These are AGN with steep X-ray spectra, blue Optical/UV continua, very strong optical FeII emission, and weak emission from the narrow line region.

From the bivariate analysis we found that high $L/L_{\rm Edd}$ AGN have the steepest X-ray spectra, bluest Optical/UV continua and appear to be X-ray weaker than low $L/L_{\rm Edd}$ AGN [5]. Applying a Principal Component Analysis (PCA) to the sample shows that eigenvector 1 which accounts already for 40% of the sample variance, is strongly correlated with $L/L_{\rm Edd}$. The interpretation for eigenvector 2 is that this is the mass of the central black hole. As mentioned earlier, we can also interpret a high L/Ledd as an indicator of the AGN being in an early stage of their development [2, 7]. A cluster analysis using complete linkage shows that the AGN sample can be divided into two groups. We found that these groups are low and high $L/L_{\rm Edd}$ AGN. Consequently this goes together with the usual separation into NLS1s and BLS1s. However, the classical cut-off line at 2,000 km/s turns out to be not always the best way to separate between the two classes. Often we find NLS1s with relatively flat X-ray spectra, etc. and BLS1s that show typical properties of NLS1s but their FWHM(Hβ) is just above the 2,000 km s^{-1} cut-off line. Therefore it may be better to use $L/L_{\rm Edd}$ to characterize AGN [6].

Acknowledgements *Swift* at PSU is supported by NASA contract NAS5-00136. This research was supported by NASA contract NNX07AH67G.

References

1. Boroson, T.A., 2002, ApJ 565, 78
2. Grupe, D., et al., 1999, A&A, 350, 805
3. Grupe, D., et al., 2001, A&A, 367, 470
4. Grupe, D., 2004, AJ, 127, 1799
5. Grupe, D., et al., 2010, ApJS, 187, 64
6. Grupe, D., 2011, PoS(NLS1 004, arXiv:1106.0228
7. Mathur, S., 2000, MNRAS, 314, L17
8. Sulentic, J.W., et al., 2000, ApJ, 536, L5

Chapter 52
Applying the Background-Source Separation Algorithm to Chandra Deep Field South Data

F. Guglielmetti, H. Böhringer, R. Fischer, P. Rosati, and P. Tozzi

Abstract A probabilistic two-component mixture model allows one to separate the diffuse background from the celestial sources within a one-step algorithm without data censoring. The background is modelled with a thin-plate spline combined with the satellite's exposure time. Source probability maps are created in a multi-resolution analysis for revealing faint and extended sources. All detected sources are automatically parametrized to produce a list of source positions, fluxes and morphological parameters. The present analysis is applied to the *Chandra* Deep Field South 2 Ms public released data. Within its 1.884 ks of exposure time and its angular resolution (0.984 arcsec), the *Chandra* Deep Field South data are particularly suited for testing the Background-Source separation algorithm.

An analysis is performed to test the sensitivity and the internal consistency of the Background-Source separation (BSS) algorithm (see [2] and Guglielmetti et al. in this volume) with sources on real fields from pointed observations. The employed field is the *Chandra* Deep Field South (CDF-S) 2 Ms data [3]. The optimal energy

F. Guglielmetti (✉) • H. Böhringer
Max-Planck-Institut für extraterrestrische Physik, Giessenbachstrasse, D-85748, Garching, Germany
e-mail: fabrizia@mpe.mpg.de; hxb@mpe.mpg.de

R. Fischer
Max-Planck-Institute für Plasmaphysik, Boltzmannstrasse 2, D-85748, Garching, Germany
e-mail: Rainer.Fischer@ipp.mpg.de

R. Piero
European Southern Observatory, Karl-Schwarzschild-Strasse 2, D-85748, Garching, Germany
e-mail: prosati@eso.org

P. Tozzi
INAF-OATs, Via Tiepolo 11, I-34143, Trieste, Italy
e-mail: tozzi@oats.inaf.it

E.D. Feigelson and G.J. Babu (eds.), *Statistical Challenges in Modern Astronomy V*, Lecture Notes in Statistics 209, DOI 10.1007/978-1-4614-3520-4_52,

band to detect the emission from both point-like and extended sources is between 0.5 and 2.0 keV. Therefore, this test is concentrated on this energy range. The main advantages of testing real data with respect to simulated ones reside on the fact that real data are characterized by a complex background, a complex point-spread function dependence across the field, source confusion and a wide range of source properties. These characteristics intrinsic to real observations are not easily elaborated with artificial data. Therefore, the CDF-S 2 Ms data are separated in four images of 500 ks exposure time each.

The BSS algorithm is applied on each of the four images. The exponential prior probability density function of the source signal is chosen and 25 pivots equally spaced are used for the background rate estimation. Scales in the range value 0.5–13 arcsec are used in the multiresolution analysis. A threshold value of $P_{\text{source}} \geq 0.9$ is chosen to separate false-positives in source detection from true sources. No contaminations due to steep changes in the exposure time map are seen both in the background map and in the source probability maps. The multiresolution analysis provides for the detection of a wide range of source fluxes and their complex morphologies.

The internal consistency of the BSS algorithm is tested comparing each CDF-S 500 ks data in pairs. Source positions (right ascension and declination), fluxes and extent (i.e. the estimated size of the detected sources) are taken into account. The difference of position estimates are within 1σ (Fig. 52.1a), while the ones of fluxes and extents are within 3σ (Fig. 52.1b). Although 70% of the sources in the CDF-S region are characterized by X-ray variability [4], Poisson fluctuations and contaminations by other sources in the fields can increase the uncertainties estimated for the source flux and extent measurements. The BSS estimates of source parameters are internally consistent.

A sensitivity analysis is performed on the four CDF-S 500 ks data and the results are compared to published ones: CDF-S 1 Ms [1] and 2 Ms [3] data. The information about the sensitivity and the reliability of the survey are described by the sky coverage and the $\log N - \log S$ distribution. The estimated sky coverage and the $\log N - \log S$ distribution depend on the algorithm employed for source detection. The BSS background maps are used to construct the flux limit map of each estimated sky coverage. Hence, vignetting effects and background variations are already accounted in the coverage. The $\log N - \log S$ distributions are computed from the respective sky coverage. It results that the $\log N - \log S$ distributions obtained with the four CDF-S 500 ks data are in agreement with the published ones in Giacconi et al. [1] and Luo et al. [3]: See Fig. 52.1c.

Applying the BSS algorithm to the CDF-S data, we prove that the BSS algorithm provides for a reliable detection of sources and estimation of source and background parameters. An extensive application of the technique is addressed in a forthcoming paper (Guglielmetti et al., in preparation).

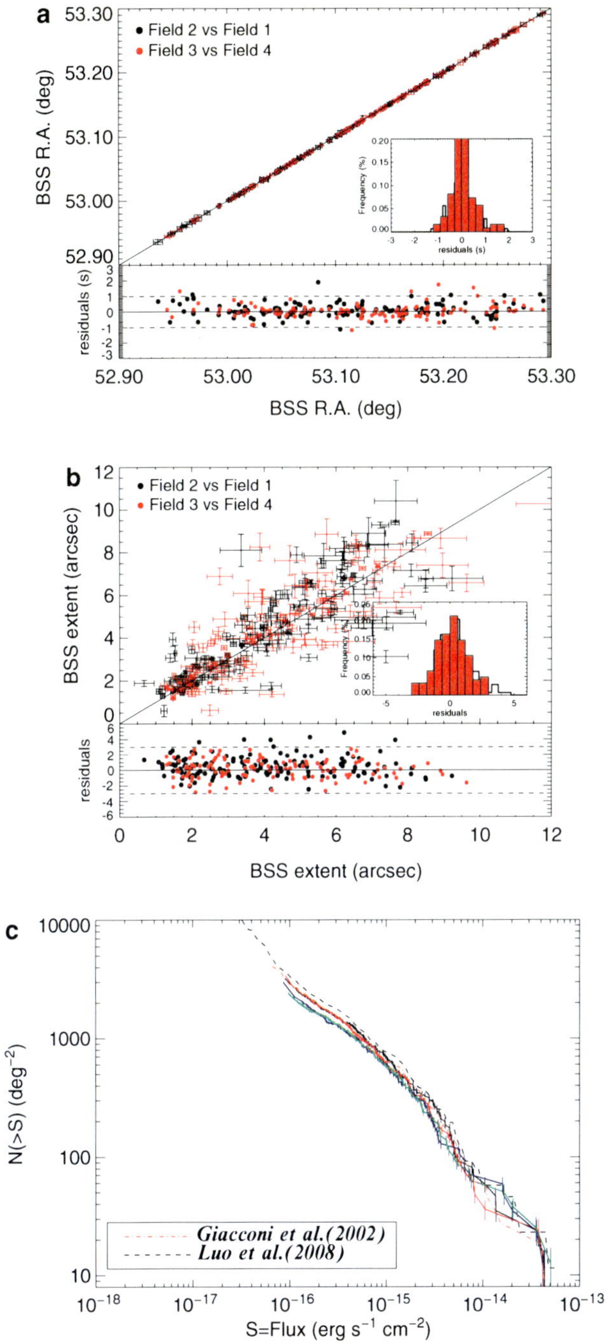

Fig. 52.1 Internal consistency and sensitivity analyses. (**a**): source position (*right ascension*). (**b**): source extent. (**c**): $\log N - \log S$ distribution. In panels **a-b**, field *1-2-3-4* indicate the four analysed CDF-S 500 ks images. In panel **c**, the $\log N - \log S$ distributions obtained from the analysis of the four CDF-S 500 ks images are drawn with a *continuous line* in *red, green, yellow* and *black*

References

1. Giacconi et al.: Chandra Deep Field South: The 1 Ms Catalog. *ApJS* **139**, 369–410 (2002)
2. Guglielmetti, F., Fischer, R., Dose, V.: Background-source separation in astronomical images with Bayesian probability theory - I. The method. MNRAS **396**, 165–190 (2009)
3. Luo, B. et al.: The Chandra Deep Field–South Survey: 2 Ms Source Catalogs. *ApJS* **179**, 19–36 (2008)
4. Paolillo, M. et al.: Prevalence of X-Ray Variability in the Chandra Deep Field-South. *ApJ* **611**, 93–106 (2004)

Chapter 53
Non-Gaussian Physics of the Cosmological Genus Statistic

J. Berian James

Abstract We suggest a technique to calculate the impact of distinct physical processes inducing non-gaussianity on the cosmological density field. The decomposition of the cosmic genus statistic with an orthogonal polynomial sequence allows expression of the scale-dependent evolution of the morphology of large-scale structure, in which effects including galaxy bias, non-linear gravitational evolution and primordial non-gaussianity might be delineated.

53.1 Topology of Cosmic Structures

The study of large-scale cosmological structures with the topological genus statistic promises much from the current generation of galaxy redshift surveys, though links between the physics of structure formation and topological statistics remain unclear. The reason for this seems to be that such measurements are difficult to interpret when the underlying distribution departs from a Gaussian random field [1, 2].

The genus of a (two-dimensional) surface measure its connectedness: the number of holes through the surface less the number of isolated regions; a donut has genus 1, a sphere 0 and two spheres -1. Given a choice of critical density value, the genus quantifies the connection or isolation of structures enclosed by the surface of that density, through the three-dimensional density field. Different choices of density threshold excise filaments, cluster and voids. The curve of genus as a function of density threshold ν has an analytical form when the underlying density field is Gaussian random $g_{\mathrm{GRF}}(\nu) \propto \exp\left(-\frac{\nu^2}{2}\right)\left(1-\nu^2\right)$, with the constant of

---

J.B. James (✉)
Astronomy Department, University of California, Berkeley, CA-94720, USA

Dark Cosmology Centre, Juliane Maries Vej 30, 2100 Copenhagen, Denmark
e-mail: berian@dark-cosmology.dk

E.D. Feigelson and G.J. Babu (eds.), *Statistical Challenges in Modern Astronomy V*, Lecture Notes in Statistics 209, DOI 10.1007/978-1-4614-3520-4_53,

proportionality determined by the power spectrum of the field—the task is to understand the manifestation of non-Gaussian physics in the topology of cosmic structure.

The shape of the curve of genus number as a function of density provides information even in the Gaussian random case: structures at the mean density of the Universe are many times connected, while extreme densities excise many disjoint regions. The balance between these two regimes and the relative abundance of isolated high and low density regions are a scale-dependent probe of many physical processes. Consequently it is desirable to study the surface of genus number as a function of density and scale.

53.2 Hermite Decomposition

To address the manner in which distinct physical processes alter the topology of large-scale structure, we propose that the genus surface (i.e., the genus curve as a function of scale, characterized by λ) be decomposed with an orthogonal basis of Hermite functions ψ_n:

$$g(v;\lambda) = \sum_{n=0}^{\infty} a_n(\lambda)\psi_n(v) \Leftrightarrow a_n(\lambda) = \int_{-\infty}^{\infty} g(v;\lambda)\psi_n(v)dv. \tag{53.1}$$

The evolution of the Hermite modes encode, as a function of scale, the imprint of the physical processes that have modified the field from the Gaussian random form. In this prescription, a Gaussian random field has a trivial decomposition: the base mode of the transform is $n = 2$, i.e., $g_{\mathrm{GRF}}(v) \propto \psi_2(v)$, with other low modes appearing as the field is perturbed from a Gaussian state. Odd-numbered modes will introduce asymmetric features corresponding to an overabundance of either isolated clusters and voids. Quantifying the distortions to the genus curve due to a physical process inducing non-Gaussianity has long been understood to deserve attention. The task is now to map these physical processes onto the Hermite spectrum.

This idea generalises previous theoretical work—two approaches that have led to progress are: (i) using perturbation theory to calculate the impact of small departures from Gaussianity; and (ii) the use of a phenomenological set of derived statistics that quantify differences between a measured genus curve and that for a Gaussian random field. Both of these are naturally expressed using the orthogonal decomposition. Unifying and extending these theoretical directions can provide an over-arching scheme to link the properties of the genus curve to the physics of cosmological structures.

References

1. Y. Choi et al. ApJS 190, 181 (2010)
2. J.B. James, G.F. Lewis, M. Colless, MNRAS 375 128 (2007)

Chapter 54
Modeling Undetectable Flares

Vinay Kashyap, Steve Saar, Jeremy Drake, Kathy Reeves, Jennifer Posson-Brown, and Alanna Connors

Abstract We have developed a fast method for modeling X-ray event list data from stellar coronae as stochastically generated flare distributions. A large portion of the previous algorithm that relied on Monte Carlo simulations has been replaced by analytical computation. This improves upon the speed of previous algorithms by many orders of magnitude. We have verified that the method works by applying it to a star with a previously measured flare distribution.

54.1 The Problem of Small Flares

A fundamental characteristic of solar and stellar flares is that the processes that generate them appear to be scale-free. That is, the distribution of flare energies are power-laws. While there is evidence that the power-law model is invalid at very high and very low energies, this distribution has been verified to hold on the Sun over many orders of magnitude of flare energies and over the range of timescales that are accessible to current high-energy astronomy missions [1]. The number of flares at any given energy range, $(E, E+dE]$, follows a power-law distribution,

$$dN \propto E^{-\alpha}\, dE\,, \tag{54.1}$$

where $\alpha \approx 1.8$ for the Sun.

V. Kashyap (✉) • S. Saar • J. Drake • K. Reeves • J. Posson-Brown
Harvard-Smithsonian Center for Astrophysics, Cambridge MA, USA
e-mail: vkashyap@cfa.harvard.edu; ssaar@cfa.harvard.edu; jdrake@cfa.harvard.edu; kreeves@cfa.harvard.edu; jpbrown@head.cfa.harvard.edu

A. Connors
Eureka Scientific, Oakland CA, USA
e-mail: rpete@head.cfa.harvard.edu

E.D. Feigelson and G.J. Babu (eds.), *Statistical Challenges in Modern Astronomy V*, Lecture Notes in Statistics 209, DOI 10.1007/978-1-4614-3520-4_54,

Similar behavior is suspected on active stars, but some crucial differences exist. The first is that due to sensitivity limitations, we cannot explore the behavior of stellar flares at energies below the so-called milliflare region ($E \sim 10^{29-32}$ erg). Second, the value of α is generally greater than 2 (see, e.g., [2]). The threshold $\alpha = 2$ is critical because beyond that, it is possible to ascribe all of the coronal luminosity to increasingly weaker, but more numerous, flares. It has thus become necessary to systematically study the flare distributions on stars. Unfortunately, current methods to evaluate the flare distribution index α for stars are limited by two factors: they either depend on explicit detections of flares (which limits the analysis to strong flares), or if the flare distribution itself is being modeled, then they are highly computation intensive and are thus slow.

We first developed a method to model the X-ray data directly without resorting to detecting the flares in the first place, by sampling flare energies from an assumed distribution, constructing photon arrival time data stochastically, and comparing the simulated distributions of arrival time differences with that seen in the data [2]. Initial applications of this method were extremely slow because the model distribution of arrival time differences δt had to be empirically generated and the parameter fitting was carried out on a grid.

Here we have speeded up the process considerably by (a) switching to a Markov Chain Monte Carlo fitting method, and (b) computing the model semi-analytically. Because we assume a specific functional form for the flare distribution, their cumulative effect can be easily discerned in the data. Because flare onset is stochastic, it is not feasible to model every feature in a light curve, but rather, they must be modeled only in the aggregate. For a given counting rate R, the probability of finding exactly one event in a duration δt is

$$p(1|R,\delta t) = (R\,\delta t)\,e^{-R\,\delta t} \tag{54.2}$$

and if $R = R(t_i)$ is varying, the overall distribution is the sum of the distributions in the interval $[t_i, t_i + \tau]$, weighted by the expected number of events,

$$f(\delta t) = \sum_i R_i\,\tau \cdot p(1|R_i,\delta t)\,. \tag{54.3}$$

We also achieve a considerable speed increase by discarding the grid-based parameter probability evaluations and using Markov Chain Monte Carlo methods to efficiently explore the parameter space.

We have computed the flare distribution model parameters for the dM3.5 flaring star Ross 154 (see [3]) and have verified that the new method gives the same result as before.

Acknowledgements This work was supported by CXC NASA contract NAS8-39073 and Chandra grant AR0-1101X.

References

1. Aschwanden, M.J., et al., 2000, ApJ, 535, 1047
2. Kashyap, V.L., Drake, J.J., Güdel, M., & Audard, M., 2002, ApJ, 580, 1118
3. Wargelin, B.W., et al., 2008, ApJ, 676, 610

Chapter 55
An F-Statistic Based Multi-detector Veto for Detector Artifacts in Gravitational Wave Data

D. Keitel, R. Prix, M.A. Papa, and M. Siddiqi

Abstract Continuous gravitational waves (CW) are expected from spinning neutron stars with non-axisymmetric deformations. A network of interferometric detectors (LIGO, Virgo and GEO600) is looking for these signals. They are predicted to be very weak and retrievable only by integration over long observation times. One of the standard methods of CW data analysis is the multi-detector $\mathscr{F}$-statistic. In a typical search, the $\mathscr{F}$-statistic is computed over a range in frequency, spin-down and sky position, and the candidates with highest $\mathscr{F}$ values are kept for further analysis. However, this detection statistic is susceptible to a class of noise artifacts, strong monochromatic lines in a single detector. By assuming an extended noise model—standard Gaussian noise plus single-detector lines—we can use a Bayesian odds ratio to derive a generalized detection statistic, the line veto (LV-) statistic. In the absence of lines, it behaves similarly to the $\mathscr{F}$-statistic, but it is more robust against line artifacts. In the past, ad-hoc post-processing vetoes have been implemented in searches to remove these artifacts. Here we provide a systematic framework to develop and benchmark this class of vetoes. We present our results from testing this LV-statistic on simulated data.

In a search for gravitational waves, we are conducting hypothesis tests: at a certain point in parameter space (frequency, spin-down and sky position), is there a signal or not? Assuming Gaussian detector noise only, we have two hypotheses, $\mathscr{H}_\mathrm{G} : \mathbf{x}(t) = \mathbf{n}(t)$ and $\mathscr{H}_\mathrm{S} : \mathbf{x}(t) = \mathbf{n}(t) + \mathbf{h}(t, \mathscr{A})$, where $\mathscr{A}$ are additional signal parameters, like polarization angles. In the Bayesian approach, we compute the odds ratio of the two hypotheses, and we marginalize over the unknown parameters $\mathscr{A}$:

D. Keitel (✉)
Albert-Einstein-Institut, Hannover, Golm, Germany
e-mail: david.keitel@aei.mpg.de

R. Prix • M. A. Papa • M. Siddiqi
Albert-Einstein-Institut, Callinstr. 38, 30167 Hannover, Germany
e-mail: Reinhard.Prix@aei.mpg.de; papa@aei.mpg.de

E.D. Feigelson and G.J. Babu (eds.), *Statistical Challenges in Modern Astronomy V*, Lecture Notes in Statistics 209, DOI 10.1007/978-1-4614-3520-4_55,

$$O_{SG}(\mathbf{x}) \equiv \frac{P(\mathscr{H}_S|\mathbf{x})}{P(\mathscr{H}_G|\mathbf{x})} \propto \int \frac{P(\mathbf{x}|\mathscr{H}_S,\mathscr{A})}{P(\mathbf{x}|\mathscr{H}_G)} P(\mathscr{A}|\mathscr{H}_S)\,\mathrm{d}\mathscr{A} \tag{55.1}$$

The marginalization can be done analytically (for specific priors on $\mathscr{A}$, see [3, 5]). We obtain $O_{SG}(\mathbf{x}) \propto \mathrm{e}^{\mathscr{F}(\mathbf{x})}$, with the standard multi-detector $\mathscr{F}$-statistic [1, 2].

The problem with this approach is that quasi-monochromatic, stationary detector artifacts ("lines") look more like $\mathscr{H}_S$ than $\mathscr{H}_G$ and will result in large values for O_{SG}. So we add an alternative noise hypothesis $\mathscr{H}_L$ that fits lines in single detectors better than the multi-detector coherent $\mathscr{H}_S$, namely $\mathscr{H}_L^X : \mathbf{x}^X(t) = \mathbf{n}^X(t) + \mathbf{h}^X(t,\mathscr{A})$ for a signal in only one detector X, but pure noise $\mathscr{H}_G$ in the others. Again using the $\mathscr{F}$-statistic priors and analytically maximizing over $\mathscr{A}$, we can (e.g. for two detectors $X = 1, 2$) replace the standard $\mathscr{F}$-statistic by a new detection statistic with an extended noise hypothesis:

$$O_{SN}(\mathbf{x}) \equiv \frac{P(\mathscr{H}_S|\mathbf{x})}{P(\mathscr{H}_L|\mathbf{x}) + P(\mathscr{H}_G|\mathbf{x})} \propto \frac{\mathrm{e}^{\mathscr{F}(\mathbf{x})}}{\rho_{\max}^4/70 + l^1\,\mathrm{e}^{\mathscr{F}^1(\mathbf{x}^1)} + l^2\,\mathrm{e}^{\mathscr{F}^2(\mathbf{x}^2)}} \tag{55.2}$$

The new detection statistic downweights candidates which have higher single-detector than multi-detector $\mathscr{F}$-statistics, thereby penalizing lines. The l^X are the prior line probabilities, while the parameter $\rho_{\max}$ from a signal strength prior allows us to tune the detection statistic, determining how much discrepancy between detectors is attributed to Gaussian noise and how soon vetoing sets in. Further work on simulated data is necessary to choose this prior optimally.

In preliminary studies with simulated data, we found the new detection statistic to be much more effective than the standard semi-coherent $\mathscr{F}$-statistic, as seen in the figure below. Especially at low false-alarm rates, which are desirable for GW searches, the new statistic allows for more detections. See [4] for more details.

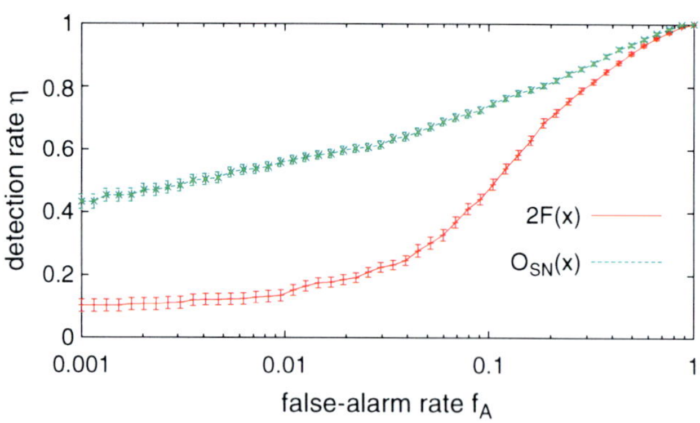

References

1. C. Cutler, B. Schutz, Phys. Rev. D **72**, 063006 (2005)
2. P. Jaranowski, A. Królak, B. Schutz, Phys. Rev. D **58**, 063001 (1998)
3. R. Prix, S. Giampanis, C. Messenger, Phys. Rev. D **84**, 023007 (2011)
4. R. Prix, D. Keitel, M.A. Papa, P. Leaci, M. Siddiqi, in preparation
5. R. Prix, B. Krishnan, Class. Quant. Grav. **26**, 204013 (2009)

Chapter 56
Constrained Probability Distributions of Correlation Functions

D. Keitel and P. Schneider

Abstract Two-point correlation functions are used throughout cosmology as a measure for the statistics of random fields. When used in Bayesian parameter estimation, their likelihood function is usually replaced by a Gaussian approximation. However, this has been shown to be insufficient. For the case of Gaussian random fields, we search for an exact probability distribution of correlation functions, which could improve the accuracy of future data analyses. We use a fully analytic approach, first expanding the random field in its Fourier modes, and then calculating the characteristic function. Finally, we derive the probability distribution function, using integration by residues. The result is strongly non-Gaussian.

For cosmic shear surveys, it was found by Hartlap et al. [2] that a Gaussian likelihood for two-point correlation functions is not a valid approximation. The same is expected for other fields where correlation functions are used for Bayesian or other likelihood-based analyses. In fact, general constraints on correlation functions were analytically derived in [5], which already exclude a Gaussian probability distribution.

Therefore, a better description of the likelihood is necessary. For a Gaussian random field, we can attempt to do this fully analytically. First, we expand the field $g(x)$ in its Fourier components, g_n. Then, the correlation function can be obtained by the estimator

D. Keitel (✉)
Argelander-Institut für Astronomie, Universität Bonn, Auf dem Hügel 71, 53121 Bonn, Germany
Albert-Einstein-Institut, Callinstraße 38, 30167 Hannover, Germany
e-mail: david.keitel@aei.mpg.de

P. Schneider
Argelander-Institut für Astronomie, Auf dem Hügel 71, D-53121 Bonn, Germany
e-mail: peter@astro.uni-bonn.de

E.D. Feigelson and G.J. Babu (eds.), *Statistical Challenges in Modern Astronomy V*, Lecture Notes in Statistics 209, DOI 10.1007/978-1-4614-3520-4_56,

$$\xi(x) = \langle g(y) g^*(x+y) \rangle = 2 \sum_{n=1}^{\infty} |g_n|^2 \cos(k_n x) . \tag{56.1}$$

With each mode g_n having a Gaussian distribution with width σ_n^2 and for different separations x_m, we define the parameters

$$C_{nm} = \sigma_n^2 \cos(k_n x_m) . \tag{56.2}$$

We start our derivation with the characteristic function $\psi(s)$, which for the general multivariate case we obtain, by ensemble averaging, as

$$\psi(s_1, s_2, \ldots, s_k) = \left\langle \exp\left(\mathrm{i} \sum_{n=1}^{k} s_n \xi(x_n) \right) \right\rangle = \prod_{n=1}^{\infty} \left(1 - 2\mathrm{i} \sum_{m=1}^{k} s_m C_{nm} \right)^{-1} . \tag{56.3}$$

Inverse Fourier transformation of the characteristic function yields the distribution function. We solve the integral by the method of residues, since the integrand has a pole for each C_{nm}. For the univariate case, we obtain

$$p(\xi) = \int_{-\infty}^{\infty} \frac{\mathrm{d}s}{2\pi} \mathrm{e}^{-\mathrm{i}s\xi} \prod_{n=1}^{\infty} \frac{1}{1 - 2\mathrm{i}sC_{n1}} = \sum_{n=1}^{\infty} \mathscr{H}_n \mathrm{e}^{-\xi/(2C_{n1})} \frac{1}{2C_{n1}} \prod_{m \neq n} \frac{1}{1 - \frac{C_{m1}}{C_{n1}}} . \tag{56.4}$$

Here, the Heaviside functions in the factor $\mathscr{H}_n = H(\xi)H(C_{n1}) - H(-\xi)H(-C_{n1})$ come from the choice of contours during integration. We find that this distribution is strongly non-Gaussian. For the case of $x = 0$, we have $p(\xi < 0) = 0$, and also for other separations, a Gaussian is a very poor fit both at the peak and in the tails. Also, we find that for reasonable power spectra and field sizes, the mode expansion can be truncated after a few to a few dozen modes. Still, the numerical implementation is quite challenging, since the small numbers involved require either high precision routines or a special reordering of the sum. Also, care needs to be taken if the C_{n1} are not all mutually different.

We obtained the bivariate $p(\xi_1, \xi_2)$ in a similar way, and the full result can be found in [4]. However, for higher multivariates this approach is too cumbersome, and it seems more promising to pursue direct numerical Fourier transformation of the characteristic function, or alternative methods as the one presented by [6]. A similar distribution has also been obtained in the signal processing literature (see e.g. [3]), but is not fully applicable to our case.

In our journal paper [4], we detail the derivation of univariate and bivariate distributions, and we also calculate the moments of these distributions. With them, we can construct an Edgeworth expansion [1] of the distribution, which is only valid up to a few terms, but already presents a large improvement compared to the standard Gaussian approximation. We also found that all our results easily generalize to multi-dimensional Gaussian random fields. Finally, we note that the distribution follows the correlation function constraints derived in [5].

References

1. S. Blinnikov, R. Moessner, Astron. Astrophys. Suppl. Ser. 130, 193–205 (1998)
2. J. Hartlap et al., Astron. & Astrophys. 504(3), 689–703 (2009)
3. D. Hammarwall, M. Bengtsson, B. Ottersten, IEEE Transact. Signal Proc. 56(3), 1188 (2008)
4. D. Keitel, P. Schneider, accepted for Astron. & Astrophys. [arXiv:1105.3672]
5. P. Schneider, J. Hartlap, Astron. & Astrophys. 504(3), 705–717 (2009)
6. P. Wilking, P. Schneider, in *Proc. Stat. Challenges in Modern Astron.*, Springer 2012

Chapter 57
Improving Weak Lensing Reconstructions in 3D Using Sparsity

Adrienne Leonard, François-Xavier Dupé, and Jean-Luc Starck

Abstract Weak gravitational lensing is a powerful tool, which allows us to map the distribution of dark matter in the Universe. With the advent of large, high-resolution and multi-wavelength surveys, it has recently become possible to use photometric redshift information to reconstruct the matter distribution in three dimensions, rather than a two-dimensional projection. This is no easy task, as the inverse problem is ill posed, the data are noise-dominated, and the lensing efficiency kernel is very broad along the line of sight. State-of-the-art linear methods to recover the density distribution typically exhibit a line-of-sight bias in the location of detected peaks, and a broad smearing of the density distribution along the line of sight. We present here a non-linear proximal minimization method incorporating a sparse prior, which allows us to recover the underlying density distribution from lensing measurements with greatly reduced bias and smearing, thus allowing for more accurate mapping of the three-dimensional density distribution.

Weak lensing measures the tiny elliptical distortions induced in images of background galaxies due to the gravitational effects of matter concentrations along the line of sight. Two linear operations map the measured shear γ onto, first, the convergence (dimensionless projected density) κ and then onto the matter overdensity $\delta = \rho/\overline{\rho} - 1$, where $\overline{\rho}$ is the mean matter density in the universe. We can therefore write:

$$\kappa = \mathbf{Q}\delta, \text{ or, equivalently, } \gamma = \mathbf{P}_{\gamma\kappa}\mathbf{Q}\delta. \tag{57.1}$$

A. Leonard (✉) • F.-X. Dupé
Laboratoire AIM, UMR CEA-CNRS-Paris 7, Irfu, SAp/SEDI, Service d'Astrophysique, CEA Saclay, F-91191 GIF-SUR-YVETTE CEDEX, France
e-mail: adrienne.leonard@cea.fr

J.-L. Starck
Service d'Astrophysique, Centre d'Études Atomiques de Saclay, Orme des Merisiers, Gif-sur-Yvette, France

E.D. Feigelson and G.J. Babu (eds.), *Statistical Challenges in Modern Astronomy V*, Lecture Notes in Statistics 209, DOI 10.1007/978-1-4614-3520-4_57,

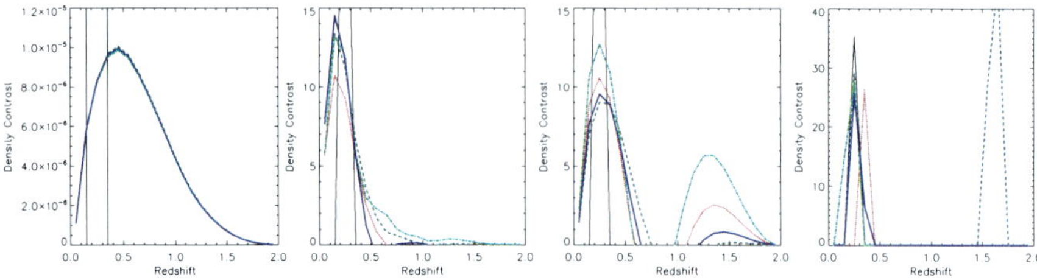

Fig. 57.1 *From left to right*: Reconstructions carried out using the transverse Wiener filter, radial Wiener filter, SVD, and sparse-based methods. The true density is given by the *solid lines*, while the *dashed lines* represent reconstructions along different lines of sight

Our goal in 3D lensing is to invert this equation in the presence of noise to recover a map of the density contrast δ.

We can express the problem as one of the form:

$$\boldsymbol{d} = \mathbf{R}\boldsymbol{s} + \varepsilon, \quad \varepsilon \sim \mathcal{N}(0, \sigma^2)\,, \tag{57.2}$$

with $\boldsymbol{d}$ being the measurements, contaminated by gaussian noise ε, and $\boldsymbol{s}$ being the true underlying density.

Simon et al. [3] and VanderPlas et al. [4] propose to invert this equation using linear methods. The Simon et al. method involves use of a minimum variance filter incorporating Wiener priors (either in the transverse or radial directions), whilst VanderPlas et al. employ an inverse variance filter, which undergoes an SVD decomposition and is then truncated to retain only the largest singular values.

Reconstructions obtained using these methods suffer from the same three fundamental problems: There is a broad smearing of the reconstructed density along the line of sight, there is a bias in the radial location of detected structures and the amplitude of the reconstruction amplitude is damped, sometimes heavily.

We consider the problem in one dimension; i.e. we are concerned with the inversion of the equation $\kappa = \mathbf{Q}\delta + \varepsilon$, and lines of sight are considered independently. We assume that the signal has a ***sparse*** representation in an appropriate dictionary $\boldsymbol{\Phi}$. We therefore wish to solve the following minimisation problem:

$$\min_{\delta \in \mathbb{R}^n} \parallel \boldsymbol{\Phi}^T \delta_1 \parallel \; s.t. \; \tfrac{1}{2} \parallel \kappa - \mathbf{Q}\delta \parallel^2_{\boldsymbol{\Sigma}^{-1}} \leq \varepsilon \,,\; \delta \in \mathscr{C} \tag{57.3}$$

where $\boldsymbol{\Phi}$ is the dictionary, Σ is the covariance matrix of the noise, ε is the *size* of the ℓ_2 ball constraining the data fidelity, and $\mathscr{C}$ is a closed convex set. To do this, we use the primal-dual splitting method proposed by Chambolle and Pock [1] (see [2] for the details).

Figure 57.1 below shows reconstructions of a simulated cluster of galaxies for each of the four methods discussed above. The sparse-based approach clearly outperforms the other three methods, exhibiting no appreciable redshift bias or smearing, and a much smaller damping of the amplitude relative to the true density.

References

1. Chambolle A., Pock T., 2010, *J. Mathematical Imaging & Vision*, 40, 120
2. Leonard A., Dupé F.-X., Starck J.-L., 2012, *Astro. & Astrophys.*, 539, #A85
3. Simon P., Taylor A. N., Hartlap J., 2009, *Mon. Not. Royal Astro. Soc.*, 399, 48
4. VanderPlas J. T., Connolly A. J., Jain B., Jarvis M., 2011, *Astrophys. J.*, 727, 118

Chapter 58
Bayesian Predictions from the Semi-analytic Models of Galaxy Formation

Yu Lu, H.J. Mo, Martin D. Weinberg, and Neal Katz

Abstract The semi-analytic models of galaxy formation (SAMs) have long been criticized because they can not correctly include the model uncertainties into model predictions to make the models testable. We demonstrate that using posterior samples drawn by advanced MCMC algorithms under data constraints we can rigorously establish confidence bounds for model predictions and perform model checks using posterior predictive distributions. We conduct a model inference from the K-band luminosity function of local galaxies and make predictions for galaxies at higher redshifts. The posterior predictive checks show that while the model can reasonably well fit the local galaxy luminosity function, its predictions for the stellar mass function of high redshift galaxies are inconsistent with existing data.

58.1 Introduction

In the conventional implementation of the semi-analytic models of galaxy formation (SAMs), one first "calibrates" the model with a set of observational constraints to find an optimal parameter set, and then uses this parameter set to make predictions for other observations. Such a prediction can not be used to test the model with observational data because of the fact that the model parameters are largely degenerate and the inferential uncertainties are completely ignored [1]. To derive statistically meaningful predictions for further observational tests, one needs to know the joint probability distribution of the model parameters given observational data.

Y. Lu (✉)
Kavli Institute for Particle Astrophysics and Cosmology, Stanford, CA 94309, USA
e-mail: luyu@stanford.edu

H.J. Mo • M.D. Weinberg • N. Katz
Department of Astronomy, University of Massachusetts, Amherst, MA 01003-9305, USA
e-mail: hjmo@astro.umass.edu; weinberg@astro.umass.edu; nsk@astro.umass.edu

E.D. Feigelson and G.J. Babu (eds.), *Statistical Challenges in Modern Astronomy V*, Lecture Notes in Statistics 209, DOI 10.1007/978-1-4614-3520-4_58,

We have developed a Bayesian approach based SAM that allows us to obtain the posterior distribution of the model parameters for given observational data, to rigorously test models, and to make robust predictions taking into account model uncertainties [1]. We use the K-band luminosity function of 2MASS galaxies [2] as the data constraint, and adopt the Tempered Differential Evolution MCMC algorithm [3] included in the UMass Bayesian Inference Engine (BIE)[1] to sample the posterior. We then make Bayesian model predictions and perform numerical model checks using the posterior distribution.

58.2 Bayesian Model Prediction and Posterior Predictive Check

A Bayesian model prediction for an observable $\mathbf{y}'$ is made by marginalizing over the posterior probability distribution of parameter set θ given data $\mathbf{y}$, e.g. $p(\mathbf{y}'|\mathbf{y}) = \int p(\mathbf{y}'|\theta)p(\theta|\mathbf{y})\mathrm{d}\theta$. A usual way to test a model is to check the model predictions with existing data. One can graphically exam the consistency between the predictive distribution of $\mathbf{y}'$ and the data. If the predictive distribution is inconsistent with the data, one should worry about the model assumptions. More importantly, we can quantitatively check a model with numerical posterior predictive checks (PPC) [4]. We first define a χ^2-like test quantity, $\mathscr{T}(\mathbf{y}'_l) = \sum_{i=1}^{N}\left(y'_{l,i} - \overline{y}'_i\right)^2/\sigma_i'^2$, where $y'_{l,j}$ is the prediction of the lth posterior sample for the ith bin of the data, $\overline{y}'_i$ and $\sigma_i'^2$ are the mean and the variance of the predictions for the ith bin, and the summation is over all the bins. Using the reference distribution of the test quantity constructed by the posterior samples, we compute the tail-area probability for the observational data $\mathbf{y}$, $p_B = \frac{1}{L}\sum_{l=1}^{L} I_{\mathscr{T}(\mathbf{y}'_l)\geq\mathscr{T}(\mathbf{y})}$, where I is the indication function. The PPC on the reproduced K-band luminosity function yields $p_B = 0.662$, suggesting the model fits the data fairly well. We then make predictions for the stellar mass function of galaxies at $z \sim 1$, and the PPC with existing data [5] yields $p_B = 0.007$, suggesting the model predictions are inconsistent with the data. The model overpredicts the number of low-mass galaxies and underpredicts the number of high-mass galaxies at higher redshifts, indicating the current model family does not properly describe the redshift evolution of star formation.

References

1. Lu Y., Mo H. J., Weinberg M. D., Katz N. S., 2011, MNRAS, in press
2. Bell E. F., McIntosh D. H., Katz N., Weinberg M. D., 2003, ApJS, 149, 289

[1] http://www.astro.umass.edu/BIE

3. Ter Braak C. J. F., 2006, Stat. Comput., 16, 239
4. Gelman A., Carlin J. B., Stern H. S., Rubin D. B., 2004, Bayesian Data Analysis. 2nd ed. Boca Raton, FL: Chapman and Hall/CRC. xxv, 668 p.
5. Pérez-González P. G., Rieke G. H., Villar V., Barro G., Blaylock M., Egami E., Gallego J., Gil de Paz A., Pascual S., Zamorano J., Donley J. L., 2008, ApJ, 675, 234

Chapter 59
Statistical Issues in Galaxy Cluster Cosmology

Adam Mantz, Steven W. Allen, and David Rapetti

Abstract The number and growth of massive galaxy clusters is a sensitive probe of cosmological structure formation and dark energy. Surveys at various wavelengths can detect clusters to high redshift, but the fact that cluster mass is not directly observable complicates matters, requiring us to simultaneously constrain scaling relations of observable signals with mass. The problem can be cast in the form of a regression, in which the data set is truncated, the (cosmology-dependent) underlying population must be modeled, and strong, complex correlations between measurements often exist.

Simulations of cosmological structure formation provide a robust prediction for the statistical distribution of galaxy clusters in the Universe as a function of mass and redshift. However, they cannot reliably predict the observables used to detect clusters in sky surveys, such as X-ray luminosity, the example we will use throughout this work. Consequently, observers must constrain observable–mass scaling relations using additional data, and use a joint model for the scaling relations and the underlying matter distribution to predict, e.g., the number of clusters as a function of redshift and X-ray luminosity [1]. Here we discuss the features of this

---

A. Mantz (✉)
NASA's Goddard Space Flight Center, Greenbelt, MD 20771, USA
e-mail: amantz@slac.stanford.edu

S.W. Allen
Kavli Institute for Particle Astrophysics and Cosmology, Stanford University,
452 Lomita Mall, Stanford, CA 94305, USA
e-mail: swa@stanford.edu

D. Rapetti
Dark Cosmology Centre, Niels Bohr Institute, University of Copenhagen, Juliane Maries
Vej 30, 2100 Copenhagen, Denmark
e-mail: drapetti@dark-cosmology.dk

E.D. Feigelson and G.J. Babu (eds.), *Statistical Challenges in Modern Astronomy V*, Lecture Notes in Statistics 209, DOI 10.1007/978-1-4614-3520-4_59,

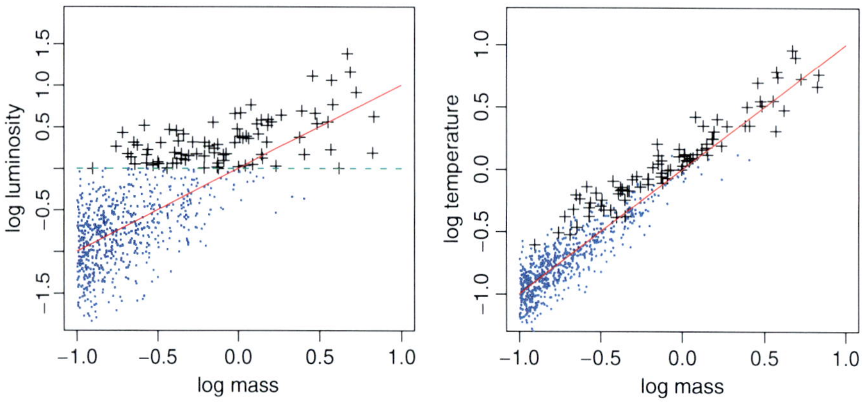

Fig. 59.1 Simulation of a cluster population obeying power-law scaling relations (*solid lines*) with intrinsic scatter. *Black* crosses indicate detected clusters in the simple case where this requires luminosity above a fixed threshold (*dashed line*), with blue dots representing undetected clusters. Temperatures are simulated assuming a strong intrinsic correlation with luminosity at fixed mass

analysis [3,4], described in terms of a Bayesian regression, which touches on several issues presented elsewhere in these proceedings.

Current cluster surveys provide a shallow sample of the full population, so detected clusters are naturally biased high in the detection observable. This is illustrated in Fig. 59.1a, which makes clear that the selection function must be accounted for to recover the true scaling relation. The detectable cluster sample is also sensitive to the underlying distribution in mass, which determines the number of low mass, highly biased detections. The mass distribution consequently must appear in the data likelihood, requiring the scaling relations to be constrained simultaneously with cosmological parameters.

The influence of selection effects can remain even when the observable of immediate interest is not directly involved in cluster detection, as illustrated in Fig. 59.1b. The magnitude of the effect in this case depends on the intrinsic covariance of scatter in cluster luminosity and temperature from the mean values at fixed mass. If this joint scatter has a strong correlation, luminosity selection can effectively imply selection on temperature. As this intrinsic correlation is not known a priori, the best approach is to simultaneously solve for cosmology and the two scaling relations, along with their joint scatter.

In addition to covariance in the intrinsic scatter, measurement errors in different cluster observables are generically correlated, and failing to account for this correlation in the data model can bias the results [2]. Fortunately, the framework of Bayesian analysis provides a straightforward way to addressing this issue (see also Kelly, this volume).

References

1. S. W. Allen, A. E. Evrard, A. B. Mantz. ARA&A, in press, arXiv:1103.4829.
2. A. Mantz S. W. Allen. Submitted, arxiv:1106.4052.
3. A. Mantz, S. W. Allen, D. Rapetti, H. Ebeling. MNRAS, 406:1759–1772.
4. A. Mantz, S. W. Allen, H. Ebeling, D. Rapetti, A. Drlica-Wagner. MNRAS, 406:1773–1795.

Chapter 60
Statistical Analyses to Understand the Relationship Between the Properties of Exoplanets and Their Host Stars

Elizabeth Martínez-Gómez

Abstract The increasing number of exoplanet detections shows that planetary systems around other stars are common in the Universe but also that they may possess a wide range of physical and orbital properties. Physical and statistical models are needed to explain the relation between exoplanets and the characteristics of their host stars. Here we analyze this issue through the application of multivariate statistical techniques. The results show that both the temperature and the magnetic activity of the host star seem to determine the properties of an exoplanet within any system, in particular, of the orbital period. The star's metallicity also appears to be influential in the multiplanet systems.

60.1 Some Remarks About Exoplanets

Until a few years ago the detection techniques (e.g. astrometry, direct imaging, gravitational microlensing, photometry, pulsar timing, radial velocity, and transit method) only offered the possibility to detect mainly Super-Earths above five Earth masses around other stars. Now, the improved methodologies show that planets seem to exist in many possible sizes just as the planets and moons of our own solar system do (e. g. [1]). The increasing number of candidate exoplanets brings some confidence to observed features in statistical distributions of the planet and host star properties. These features can be "fossil traces" of the processes of formation or evolution of these systems and help to constrain the planet–formation models.

In this work we analyze the possible relationship between the observed properties of the exoplanets and their host stars by multivariate statistics.

E. Martínez-Gómez (✉)
Center for Astrostatistics, The Pennsylvania State University, 326 Thomas Building, University Park, PA 16802, USA
e-mail: affabeca@gmail.com

E.D. Feigelson and G.J. Babu (eds.), *Statistical Challenges in Modern Astronomy V*, Lecture Notes in Statistics 209, DOI 10.1007/978-1-4614-3520-4_60,

60.2 Multivariate Statistical Techniques: Application to the Exoplanet Data Explorer

Most of the multivariate techniques rely on the multivariate normality assumption. We transform the variables T, K, T_{eff}, $[Fe/H]$ and log *rhk* of the *Exoplanet Data Explorer* [3] as follows: $Y' = log[Y + (c + |min\,Y|)]$ where c is a constant. Let A a set of p variables X and B a set of q variables Y. We are interested in linear relationships between A and B as well as interrelations among the X's variables in A.

1. *Principal Component Analysis (PCA).* For our data, we get $Z_1 = 0.663\,T - 0.748\,K$, and $Z_2 = -0.748\,T - 0.663\,K$ that account for 52% and 48% of the total variance, respectively. Using the first PC in a multiple regression, we find that T_{eff} and log *rhk* are statistically significant. The PCs can also be used for clustering, in our case, we did not identify any additional groups.
2. *Canonical Correlation Analysis (CCA).* We find two canonical pairs of the type: $U_1 = 0.022T - 0.071K, V_1 = -0.051\,T_{eff} - 0.023\,[Fe/H] - 0.047\,rhk$ with $\hat{\rho}_1^* = 0.302$, and $U_2 = 0.072\,T + 0.024\,K, V_2 = 0.056\,T_{eff} - 0.012\,[Fe/H] - 0.064\,rhk$ with $\hat{\rho}_2^* = 0.191$. The semiamplitude K and the temperature T_{eff} explain most of the variance of the first pair.
3. *Multivariate Analysis of Variance (MANOVA).* We have two groups (simple and multiple planetary systems) which differ from each other significantly, $T^2 = 45.48$ (p–value$\sim$0), that is, the exoplanets in each group could have evolved in different ways.

60.3 Is There a Possible Exoplanet–Host Star Relationship?

Earlier works show the possibility of a relationship (e. g. [2]). We have found that: (1) both the magnetic activity (log *rhk*) and the temperature (T_{eff}) of the host star seem to determine the orbital period of an exoplanet within any system, and (2) the estimated correlation coefficient for multiplanet systems indicates that the metallicity contributes more to the variability of the exoplanet properties.

Acknowledgements The financial support of the Schlumberger Foundation and Consejo Nacional de Ciencia y Tecnología (CONACyT) are gratefully acknowledged.

References

1. Irwin, P. G. J., in *Exoplanets*, ed. by J. W. Mason. (Springer, Heidelberg, 2008), p.1
2. Knutson, H. A., Howard, A. W. & Isaacson, H. 2010, *Astrophys. J.*, **720**, 2, 1569
3. Wright, J. T., Fakhouri, O., Marcy, G. W. et. al. 2011, *Pub. Astro. Soc. Pacific*, **123**, 902, 412

Chapter 61
Identifying High-z Gamma-Ray Burst Candidates Using Random Forest Classification

Adam N. Morgan, James Long, Tamara Broderick, Joseph W. Richards, and Joshua S. Bloom

Abstract The growing number of observed Gamma-ray Bursts (GRBs) necessitates a more efficient use of follow-up resources in order to maximize the expected scientific returns. Studying the most distant (highest redshift) events, for instance, remain a primary goal for many in the field. Toward this goal of optimal resource allocation, we have created the Random Forests Automated Triage Estimator for GRB redshifts (`RATE GRB-`z) to identify high-redshift ($z > 4$) candidates using rapidly available metrics from the *Swift* satellite. Using a training set of 136 GRBs, 17 of which are high-z, our cross-validated performance metrics suggest that following up on just 20% of the GRBs will yield roughly 55% of all high-redshift events.

Data: We collated data on all *Swift* GRBs up to and including GRB 100621A directly from GCN notices and automated pipelines [1, 2] that process and refine the data into more useful metrics. Short bursts ($T_{90} < 2$ s) and bursts without rapid notices from the BAT, XRT, and UVOT were removed from the sample for uniformity. This left 348 events: 136 with known redshift, and 17 with $z > 4$.

Goal: Our primary goal is a decision for each new GRB: Should we devote further telescope observing time to this burst or not?

Methods: The `RATE GRB-`z method uses Random-Forest (RF; [3]) classification, we rank the GRBs in the training set by their out-of-bag probabilities of being in the high-z class. Next, we obtain a probability of high-z for new events by

A.N. Morgan (✉) • J.S. Bloom
Department of Astronomy, University of California, Berkeley, CA 94720-3411, USA
e-mail: amorgan@astro.berkeley.edu; jbloom@astro.berkeley.edu

J. Long • T. Broderick • J.W. Richards
Department of Statistics, University of California, Berkeley, CA 94720-3411, USA
e-mail: jlong@stat.berkeley.edu; tab@stat.berkeley.edu; jwrichar@astro.berkeley.edu

E.D. Feigelson and G.J. Babu (eds.), *Statistical Challenges in Modern Astronomy V*, Lecture Notes in Statistics 209, DOI 10.1007/978-1-4614-3520-4_61,

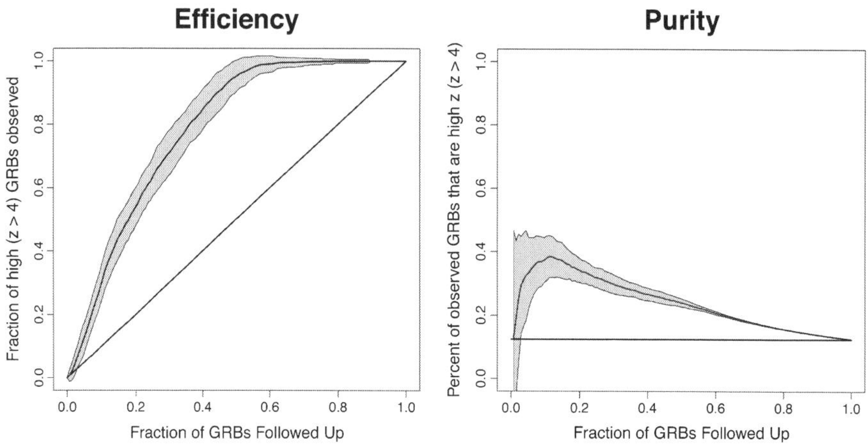

Fig. 61.1 Cross-validated efficiency ($N_{\text{high observed}}/N_{\text{total high}}$; *left panel*) and purity ($N_{\text{high observed}}/N_{\text{total observed}}$; *right panel*) versus fraction of followed-up GRBs. The curve uncertainties are 1 standard deviation from the mean value averaged over 100 RF seeds

inserting them into the RF classifier. If the percentage of training-set bursts with a higher probability of being high-z than the new event is lower than the percentage of GRBs one has telescopic resources to observe, follow-up on the new event is recommended.

Results: Using a total of 12 features in the training set, we used tenfold cross validation [4] to obtain the performance metrics shown in Fig. 61.1. By following up on the top 20% of new GRBs, one can capture ~55% of all high-z (> 4) events. Further, we expect roughly 35% of these followed-up GRBs will be high-z.

References

1. Butler, N. & Kocevski, D.: X-Ray Hardness Evolution in GRB Afterglows and Flares: Late-Time GRB Activity without N_H Variations. The Astrophysical Journal. **663**, 407–419 (2007)
2. Butler, N. et al. A Complete Catalog of Swift Gamma-Ray Burst Spectra and Durations: Demise of a Physical Origin for Pre-Swift High-Energy Correlations. The Astrophysical Journal. **671**, 656–677 (2007)
3. Breiman, L.: Random Forests. Machine Learning. **45** 5–32 (2001)
4. Kohavi, R.: A study of cross-validation and bootstrap for accuracy estimation and model selection. International Joint Conference on Artificial Intelligence, **14** 1137–1145 (1995)

Chapter 62
Fitting Distributions of Points Using τ^2

Tim Naylor

Abstract Fitting datasets which consist of points distributed over a plane where the typical separation between the points is large compared with their uncertainty in position is problematical. Typically this is solved by binning the data, but then the sparsity of the data means that the pixels (if they are to contain several data points) must be larger than the typical uncertainty, thus throwing away precision. Here I present a different solution, developed originally for colour-magnitude diagrams (CMDs) where the model is binned, but the data are not.

62.1 The Problem

The circles in Fig. 62.1 are a typical colour-magnitude dataset for a cluster, with the greyscale showing a typical model. The model distribution is truly two-dimensional, with two curves (single stars and equal mass binaries) separated by a region filled with unequal-mass binaries. Fitting these models to the data is still largely done by eye, but we have developed a technique [1,2], which we [3–6] and others [7–9] have applied to finding ages and distances for clusters.

62.2 The Solution

(1) Simulate a million stars with the parameters of interest (e.g. distance and age), and bin them in colour-magnitude space to create the grey scale model in Fig. 62.1. (2) Assuming the uncertainties for the data points are small, evaluate the model at

T. Naylor (✉)
School of Physics, University of Exeter, Stocker Road, EX4 4QL, Exeter, UK
e-mail: timn@astro.ex.ac.uk

E.D. Feigelson and G.J. Babu (eds.), *Statistical Challenges in Modern Astronomy V*, Lecture Notes in Statistics 209, DOI 10.1007/978-1-4614-3520-4_62,

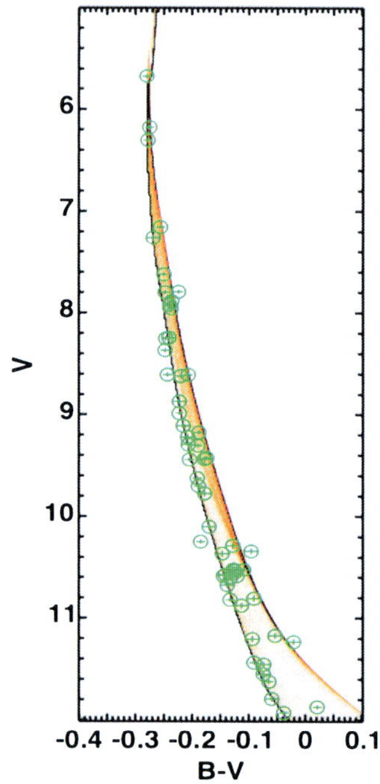

Fig. 62.1 The colour-magnitude data for dereddened members of NGC6530 (*circles*) overlaid on the best fitting model (*greyscale*). The fitting procedure can be thought of qualitatively in the following way, using distance an example fitting parameter. Changing the distance is equivalent to moving the grey scale model up and down over the data points. So for each placement of the model collect the values of the model at the position of each data point. The value of the product of all these values is clearly maximised when the model is at the correct vertical position, i.e. when the distance modulus is correct. The maximum value of the product is a measure of the goodness-of-fit, and the steepness of the maximum a measure of the uncertainty in the distance

the position of each data point i, to obtain P_i for each data point. If the uncertainties are large, convolve the image with the uncertainty for the data point before taking the value. (3) Multiplying these values together gives a goodness-of-fit parameter, though we actually use the log-likelihood $\ln \prod P_i = \tau^2$, since this is related to χ^2. (4) Change the parameters until you find the best (lowest) value of τ^2. There are then techniques for finding probability that this is a good fit, $\Pr(\tau^2)$. If the model is a good fit, one can then determine uncertainties in the parameters in a similar way to χ^2.

References

1. T. Naylor, R.D. Jeffries (2006) MNRAS **373** 1251
2. T. Naylor (2009) MNRAS **399** 432
3. S.P. Littlefair, T. Naylor, N.J. Mayne, E.S. Saunders, R.D. Jeffries (2010), MNRAS **403** 545

4. R.D. Jeffries, T. Naylor, F.M. Walter, M.P. Pozzo, C.R. Devey (2009) MNRAS **393** 538
5. N.J. Mayne, T.Naylor (2008) MNRAS **386** 261
6. R.D. Jeffries, J.M. Oliveira, T. Naylor, N.J. Mayne, S.P. Littlefair (2007) MNRAS **376** 580
7. H. Joshi, B. Kumar, K.P. Singh, R. Sagar, S. Sharma, J.C. Pandey (2008) MNRAS **391** 1279
8. P.A. Cargile, D.J. James (2010) AJ **140** 677
9. N. Da Rio, D.A. Gouliermis, M. Gennaro (2010) ApJ **723** 166

Chapter 63
Theoretical Power Spectrum Estimation from Cosmic Microwave Background Data

Paniez Paykari, Jean-Luc Starck, and M. Jalal Fadili

Abstract The cosmic microwave background (CMB) power spectrum is a powerful cosmological probe as it entails almost all the statistical information of the CMB perturbations. Having access to only one sky, the CMB spectrum measured by our experiments is only a realization of the true underlying angular power spectrum. In this paper we use the sparsity of the CMB spectrum to develop a technique that estimates the true underlying CMB power spectrum from data alone. The developed IDL code, **TOUSI**, for Theoretical pOwer spectrUm using Sparse estImation, will be released with the next version of ISAP.

63.1 Introduction

Measurements of the CMB anisotropies are powerful cosmological probes. In the currently favored cosmological model, with the nearly Gaussian-distributed curvature perturbations, almost all the statistical information are contained in the CMB angular power spectrum. The observed quantity on the sky is generally the CMB temperature anisotropy $\Theta(\mathbf{p})$ in direction $\mathbf{p}$, which is described as $T(\mathbf{p}) = T_{CMB}[1+\Theta(\mathbf{p})]$. This field is expanded on the spherical harmonic functions as

$$\Theta(\mathbf{p}) = \sum_{\ell=0}^{+\infty} \sum_{m=-\ell}^{\ell} a[\ell,m] Y_{\ell m}(\mathbf{p}), \tag{63.1}$$

P. Paykari (✉) • J.-L. Starck
Laboratoire AIM, UMR CEA-CNRS-Paris 7, Irfu, SAp/SEDI, Service d'Astrophysique, CEA Saclay, F-91191 GIF-SUR-YVETTE CEDEX, France
e-mail: paniez.paykari@cea.fr; jeanluc.starck@cea.fr

M.J. Fadili
Groupe de Recherche en Informatique, Image, Automatique et Instrumentation de Caen, École Nationale Superieure d'Ingéntiers, Caen, France
e-mail: Jalal.Fadili@greyc.ensicaen.fr

E.D. Feigelson and G.J. Babu (eds.), *Statistical Challenges in Modern Astronomy V*, Lecture Notes in Statistics 209, DOI 10.1007/978-1-4614-3520-4_63,

$$a[\ell,m] = \int \Theta(\mathbf{p}) Y^{*}_{\ell m}(\mathbf{p}) d\mathbf{p}, \tag{63.2}$$

where ℓ is the multipole moment which is related to the angular size on the sky as $\ell \sim 180^\circ/\theta$ and m is the phase ranging from $-\ell$ to ℓ. For a Gaussian random field, the mean and covariance are sufficient statistics, meaning that they carry all the statistical information of the field. In the case of CMB the mean vanishes and the variance is

$$< |a[\ell,m]|^2 >= C[\ell] > 0. \tag{63.3}$$

The angular power spectrum depends on the cosmological parameters through an angular transfer function $T_\ell(k)$ as

$$C[\ell] = 4\pi \int \frac{dk}{k} T_\ell^2(k) P(k), \tag{63.4}$$

where k defines the scale and $P(k)$ is the primordial power spectrum.

In this paper the sparsity[1] of the CMB power spectrum is used as a key ingredient in order to estimate the theoretical power spectrum without having to know the cosmological parameters; this estimate will not belong to a set of possible theoretical power spectra (i.e. all $C[\ell]$ that can be obtained by CAMB by varying the cosmological parameters). Instead, such an estimation should be useful for other applications, such as:

- Monte Carlo: we may want to make Monte Carlo simulations in some applications without assuming the cosmological parameters.
- Wiener filtering: Wiener filtering is often used to filter the CMB map and it requires the theoretical power spectrum as an input. We may not want to assume any cosmology at this stage of the processing.
- Some estimators (weak lensing, ISW, etc.) require the theoretical power spectrum to be known. Using a data-based estimation of the theoretical $C[\ell]$ could be an interesting alternative, or at least a good first guess in an iterative scheme where the theoretical $C[\ell]$ is required to determine the cosmological parameters.

63.1.1 Which Dictionary for the Theoretical CMB Power Spectrum?

We investigate the sparsity of the CMB power spectrum in two different dictionaries: the Wavelet Transform (WT) and the Discrete Cosine Transform (DCT).

We simulate 100 maps from the theoretical power spectrum and estimate their power spectra. We decompose each realization in the DCT and WT dictionaries

[1] A comprehensive account on sparsity and its applications can be found in the monograph [1].

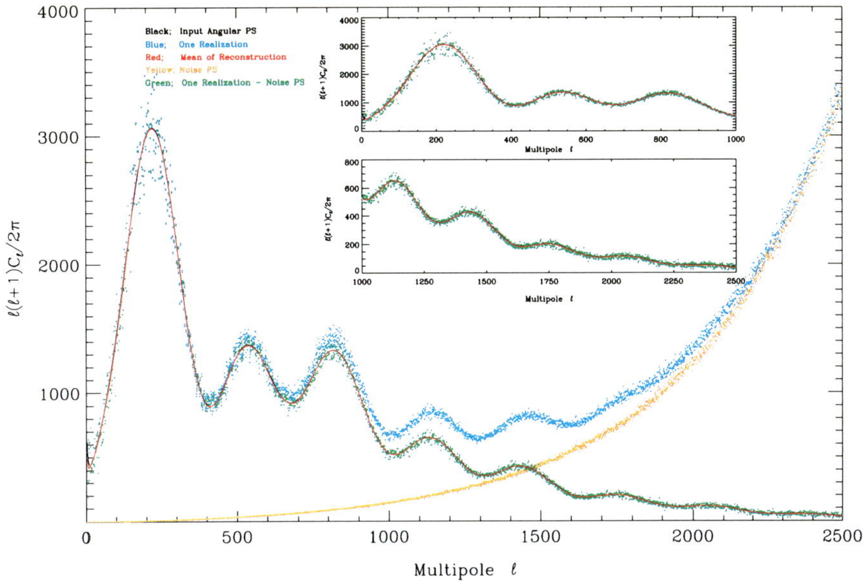

Fig. 63.1 Power spectrum estimation in the presence of instrumental noise. The *blue dots* show the empirical power spectrum of one realization having instrumental noise. *Yellow dots* show the estimated power spectrum of one of the simulated noise maps. *Green dots* show the the spectrum with the noise power spectrum removed. The *black* and *red solid lines* are the input and reconstructed power spectra respectively. The inner plots show a zoomed-in version

and reconstruct them keeping only the significant coefficients. By comparing the reconstructed power spectra to the input one we conclude:

- The CMB power spectrum is very sparse in both the DCT and WT dictionaries, although their sparsifying capabilities are different;
- DCT recovers global features of spectrum (i.e. the peaks and troughs) while WT recovers localized features.

These complementary capabilities of the DCT and WT transforms will be combined to propose a versatile way for adaptively estimating the theoretical power spectrum from a single realization of it.

63.1.2 TOUSI Algorithm on Simulated Noisy CMB Data

Here we present the performance of the TOUSI algorithm in the presence of instrumental noise. The noise maps were simulated using a theoretical (PLANCK level) noise power spectrum. They were added to the CMB maps simulated previously and the power spectra of the combined maps were estimated.

Figure 63.1 shows the reconstruction. The blue dots show the empirical power spectrum of one realization having instrumental noise. Yellow dots show the estimated power spectrum of one of the simulated noise maps. Green dots show the the spectrum with the noise power spectrum removed. The black and red solid lines are the input and reconstructed power spectra respectively. The theoretical power spectrum can be reconstructed up to the point where the structure of the power spectrum has not been destroyed by the instrumental noise. In our case, having PLANCK level noise, this goes to ℓ up to 2,500. It can be seen that TOUSI can do a great job in reconstructing the input power spectrum even in the presence of instrumental noise.

63.2 Conclusion

Measurements of the CMB anisotropies are powerful cosmological probes. In the currently favored cosmological model, with the nearly Gaussian-distributed curvature perturbations, almost all the statistical information are contained in the CMB angular power spectrum. In this paper we have investigated the sparsity of the CMB power spectrum in two dictionaries; DCT and WT. The two dictionaries have different characteristics and can accommodate reconstructing different features of the spectra. The sparsity of the CMB spectrum in these two domains has helped us develop an algorithm, TOUSI, that estimates the true underlying power spectrum from a given realized spectrum. This algorithm uses the sparsity of the CMB power spectrum in both WT and DCT domains and takes the best from both worlds to get a highly accurate estimate from a single realization of the CMB power spectrum. This could be a replacement for CAMB in cases where knowing the cosmological parameters is not necessary. The developed IDL code will be released with the next version of ISAP (Interactive Sparse astronomical data Analysis Packages) via the web site: http://jstarck.free.fr/isap.html

Acknowledgements The authors would like to thank Marian Douspis, Olivier Doré and Amir Hajian for useful discussions. This work is supported by the European Research Council grant SparseAstro (ERC-228261)

Reference

1. J.-L. Starck and F. Murtagh and M.J. Fadili, *Sparse Image and Signal Processing*, Cambridge University Press, 2010.

Chapter 64
Guilt by Association: Finding Cosmic Ray Sources Using Hierarchical Bayesian Clustering

Kunlaya Soiaporn, David Chernoff, Thomas Loredo, David Ruppert, and Ira Wasserman

Abstract The Earth is continuously showered by charged cosmic ray particles, naturally produced atomic nuclei moving with velocity close to the speed of light. Among these are ultra high energy cosmic ray particles with energy exceeding 5×10^{19} eV, which is ten million times more energetic than the most energetic particles produced at the Large Hadron Collider. Astrophysical questions include: what phenomenon accelerates particles to such high energies, and what sort of nuclei are energized? Also, the magnetic deflection of the trajectories of the cosmic rays makes them potential probes of galactic and intergalactic magnetic fields. We develop a Bayesian hierarchical model that can be used to compare different association models between the cosmic rays and source population, using Bayes factors. A measurement model with directional uncertainties and accounting for non-uniform sky exposure is incorporated into the model. The methodology allows us to learn about astrophysical parameters, such as those governing the source luminosity function and the cosmic magnetic field.

K. Soiaporn (✉) • D. Chernoff
Cornell University, Ithaca, NY, USA
e-mail: ks354@cornelll.edu; chernoff@astro.cornell.edu

T. Loredo
Department of Astronomy, Cornell University, Ithaca, NY, USA
e-mail: loredo@astro.cornell.edu

D. Ruppert
School of Operations Research and Information Engineering
Cornell University, Ithaca, NY, USA
e-mail: dr24@cornell.edu

I. Wasserman
Department of Astronomy, Cornell University, 626 Space Sciences Building, Ithaca, NY, USA
e-mail: ira@astro.cornell.edu

E.D. Feigelson and G.J. Babu (eds.), *Statistical Challenges in Modern Astronomy V*, Lecture Notes in Statistics 209, DOI 10.1007/978-1-4614-3520-4_64,

64.1 Introduction

Since the Pierre Auger Observatory (PAO) initiated observations in 2004 it has detected 14 ultra high energy cosmic rays (UHECRs) with energy $\geq$55 Eev in period 1–January 1, 2004 - May 26, 2006, 13 UHECRs in period 2– May 27, 2006 - August 31, 2007, and 42 UHECRs in period 3– September 1, 2007 - December 31, 2009. The energy threshold of 55 Eev was chosen by using period 1 data [1]. These CR particles interact with the cosmic microwave background, and according to GZK limit, CRs with energy $\gtrapprox$60 Eev should come from sources within 200 Mpc [1]. We consider the 17 active galactic nuclei (AGNs) in the volume-complete (to 15 Mpc) catalog of [3] as candidate sources. We use a Bayesian hierarchical model to compare three models, M_0: only isotropic background source, M_1: isotropic background +17 AGNs, M_2: isotropic background +2 AGNs (Centaurus A and NGC 4945–the two closest AGNs) for the UHECRs from the three periods.

64.2 Models and Algorithms

We describe the CR arrival as a Poisson process with rate set by source fluxes and exposure factors, the measurement error as a Fisher distribution with the angular uncertainty of 0.9° and the magnetic deflection as a Fisher distribution with concentration parameter κ. Our hierarchical model has parameters F_0 (flux from isotropic background), F_A (total flux from the AGNs), λ (source label of each UHECR), and κ. We assume AGNs have fixed CR luminosity implying an AGN at distance d generates CR flux $\propto 1/d^2$. We analyze a physically plausible range of deflections $\kappa \in [1, 1000]$. F_0 and F_A have an exponential prior with scale $s \approx 0.063\,\text{km}^{-2}\,\text{year}^{-1}$, based on previous data from CR observatories AGASA and HiRes. Gibbs sampling is performed on the parameters F_A, F_0 and λ to obtain the posterior distributions. We use Chib's method in [2] to estimate the marginal likelihood under each model.

64.3 Results

The Bayes factors as a function of κ are shown in Fig. 64.1. Adopting the log-flat prior for κ, we obtain the overall Bayes factor against the null of 26.10, 5.41 and 0.15 for M_1 and 12.37, 8.27 and 0.11 for M_2, for periods 1, 2 and 3, respectively. The strength of the evidence for AGN association differs markedly from period to period. For M_1 and M_2 we find $\lessapprox$10% of PAO CRs may come from AGN and a significant fraction must originate elsewhere.

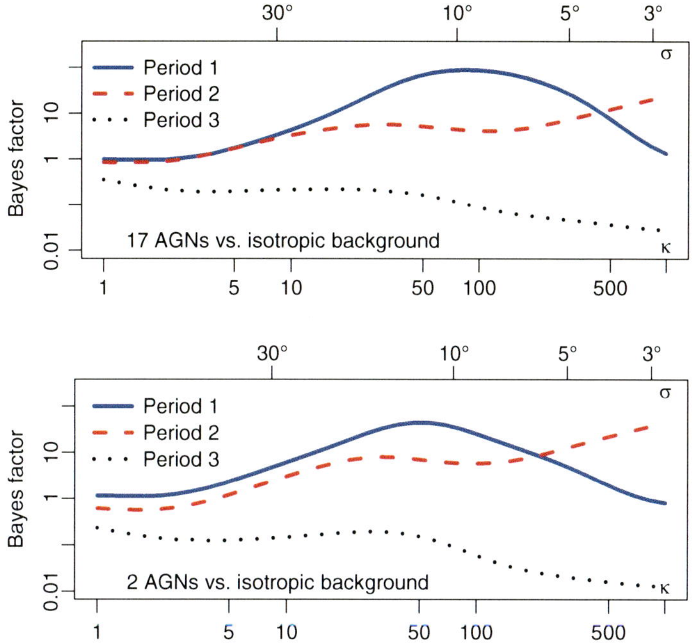

Fig. 64.1 Bayes factors comparing the association model with 17 AGNs (*left*) or 2 AGNs (*right*) to the null isotropic background model. σ is the standard deviation in 2-d Gaussian approximation for the Fisher distribution

References

1. The Pierre Auger Collaboration, Abreu, P., et al. (2010). Update on the Correlation of the Highest Energy Cosmic Rays with Nearby Extragalactic Matter. Astroparticle Physics, 34(5):314-326.
2. Chib, S. (1995). Marginal Likelihood from the Gibbs Output. Journal of the American Statistical Association, 90(432):1313-1321.
3. Goulding, A.D., Alexander, D.M., Lehmer, B., Mullaney, J.R.(2010). Towards a Complete Census of Active Galactic Nuclei in Nearby Galaxies: the Incidence of Growing Black Holes. Monthly Notices of the Royal Astronomical Society, 406(1):597-61.

Chapter 65
Statistical Differences Between Swift Gamma-Ray Burst Classes Based on γ- and X-ray Observations

Dorottya Szécsi, Lajos G. Balázs, Zsolt Bagoly, István Horváth, Attila Mészáros, and Péter Veres

Abstract There are number of evidences that the gamma-ray bursts (GRBs) have a third group beside the short and long ones: the intermediate group. Although at this time, no reasonable physical explanation is known for them. We use discriminant analysis to confirm the former classification and give some further physical properties for the intermediate GRBs.

D. Szécsi (✉)
Eötvös University, H-1117 Budapest, Pázmány P. s. 1/A, Hungary
e-mail: szdpadt@inf.elte.hu

L.G. Balázs
Konkoly Observatory, H-1505 Budapest, POB 67, Hungary
e-mail: balazs@konkoly.hu

Z. Bagoly
Eötvös University, Budapest, Hungary
e-mail: zsolt@yela.elte.hu

I. Horváth
Bolyai Military University, H-1581 Budapest, POB 15, Hungary
e-mail: horvath.istvan@zmne.hu

A. Mészáros
Charles University, V Holešovičkách 2, CZ 180 00 Prague 8, Czech Republic
e-mail: meszaros@cesnet.cz

P. Veres
Pennsylvania State University, 525 Davey Laboratory, University Park, PA 16802, USA
e-mail: puv2@astro.psu.edu

E.D. Feigelson and G.J. Babu (eds.), *Statistical Challenges in Modern Astronomy V*, Lecture Notes in Statistics 209, DOI 10.1007/978-1-4614-3520-4_65,

65.1 Discriminant Analysis: Separation Between Long And Intermediate Groups

We analyse the γ- and X-ray properties observed by the Swift satellite. The variables used in this work are the following: *Fluence* (Fl), *1-sec Peak Photon Flux* (P), *Photon Index* (Pind), *Early X-Flux* (Xfl), *Initial Temporal Decay Index* (Xdec), *Spectral Index* (Xsp) and *HI Column Density* (XNH).

This analysis can confirm the classification of the GRBs based on the hardness-duration joint distribution [1], and also can confirm the separation between the long and the intermediate groups. We can get the discriminant function and the statistical parameters dominating the discriminant function. In our data set, we have 61 intermediate and 123 long GRBs based on the grouping of the hardness-duration joint distribution. We used SPSS[1] in our computations.

In Table 65.1, we compared the means of the variables between the groups using F-statistics. Bold faces mark the variables where the differences in the group means are significant.

In our case, we have two classes (long and intermediate bursts) and one discriminant function. The correlation between the variables and the discriminant function is shown in the last column of Table 65.1. The correlation coefficients marked with bold faces are significant at a very high level. Therefore, the discriminant function is mostly dominated by these variables.

The significance of the differences measured by the discriminant functions is shown in Table 65.2. As the value in the column Significance is .000, we can state that the two groups differ significantly based on the γ- and X-ray observations.

Table 65.1 Variables discriminating long and intermediate bursts

Variable	Wilks' Lambda	F	df1	df2	Sig.	Corr.
Pind	**0.838**	**35.055**	**1**	**182**	**0.000**	**−0.409**
Xdec	0.997	0.476	1	182	0.491	0.048
Xsp	**0.943**	**10.908**	**1**	**182**	**0.001**	**−0.228**
log Fl	**0.634**	**104.841**	**1**	**182**	**0.000**	**0.707**
log P	0.982	3.258	1	182	0.073	0.125
log Xfl	**0.833**	**36.385**	**1**	**182**	**0.000**	**0.417**
log XNH	**0.960**	**7.660**	**1**	**182**	**0.006**	**0.191**

Table 65.2 Discrimination between long and intermediate bursts

Test of function	Wilks' lambda	Chi-square	df	Sig.
1	0.465	136.721	7	**0.000**

[1] SPSS is a registered trademark (http://www.spss.com).

65.2 Conclusion

We confirmed the separation between the long and intermediate groups and gave the variables dominating the discriminant function (Fluence, Early X-Flux, Photon Index, Spectral Index and HI Column Density). It is important constructing or developing a model for the intermediate GRBs.

Acknowledgements This work was supported by OTKA grant K077795, by OTKA/NKTH A08-77719 and A08-77815 grants (Z.B.), by the GAČR grant No. P209/10/0734 (A.M.) and by the Research Program MSM0021620860 of the Ministry of Education of the Czech Republic (A.M.).

Reference

1. Horváth, I. et al.: Detailed Classification of Swift's Gamma-ray Bursts. The Astrophysical Journal **713**, 552 (2010)

Chapter 66
A Quasi-Gaussian Approximation for the Probability Distribution of Correlation Functions

Philipp Wilking and Peter Schneider

Abstract The likelihood function of correlation functions needs to be known whenever they are used for inference about cosmological parameters. It is usually approximated as a multivariate Gaussian, which is not necessarily a good approximation, as can be seen from the existence of constraints on correlation functions (see (Schneider and Hartlap, A&A 504:705–717, 2009))—thus, a better approximation for the likelihood of correlation functions is required. For a 1-D Gaussian field, the univariate and bivariate likelihood has been derived analytically in (Keitel and Schneider Constrained probability distributions of correlation functions. Accepted for A&A [arXiv:1105.3672] 2011) and can deviate very strongly from Gaussians. Based on the constraints and the exact univariate likelihood, we constructed a quasi-Gaussian ansatz for the multi-variate correlation likelihood which (1) strictly obeys the constraints, (2) yields an approximate Gaussian in cases where the Gaussian approximation for the likelihood holds, and, if this is not the case, (3) provides a much better approximation than the Gaussian, as demonstrated with simulations; finally, (4) it provides a significantly better description than the straightforward copula approach.

As shown in [4], correlation functions $\xi(x)$ of a random field cannot take arbitrary values, but are subject to constraints, originating from the non-negativity of the power spectrum $P(k)$. The constraints can be written in terms of the correlation coefficients $r_n \equiv \xi(nx)/\xi(0)$ as $r_{n\mathrm{u}} \leq r_n \leq r_{n\mathrm{u}}$, where the upper and lower bounds

P. Wilking (✉)
Argelander-Institut für Astronomie, Universität Bonn, Auf dem Hügel 71, 53121 Bonn, Germany
e-mail: pwilking@astro.uni-bonn.de

P. Schneider
Argelander-Institut für Astronomie, Auf dem Hügel 71, D-53121 Bonn, Germany
e-mail: peter@astro.uni-bonn.de

E.D. Feigelson and G.J. Babu (eds.), *Statistical Challenges in Modern Astronomy V*, Lecture Notes in Statistics 209, DOI 10.1007/978-1-4614-3520-4_66,

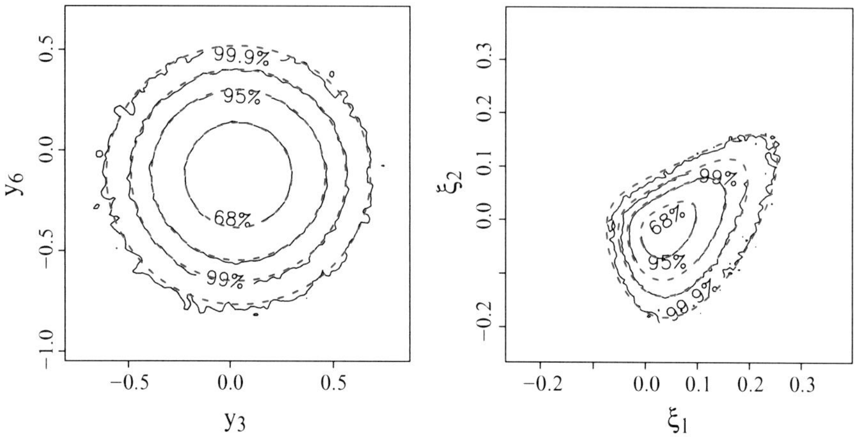

Fig. 66.1 The Gaussian and quasi-Gaussian approximations for $p(y)$ and $p(\xi)$, see text for details

are functions of the r_i with $i < n$. The existence of these constraints shows that the probability distribution of correlation functions cannot be Gaussian.

For a one-dimensional, finite, real Gaussian random field, the exact uni- and bivariate probability distribution functions of its correlation functions have been derived analytically in [1], see also Keitel and Schneider's contribution in these Proceedings. In order to obtain the higher-variate distributions, we can still efficiently construct a new, "quasi-Gaussian" likelihood that also obeys the constraints. For that purpose, we transform the correlation function to a new unbounded quantity

$$y_n = \operatorname{atanh} \frac{2r_n - r_{n\mathrm{u}} - r_{n\mathrm{l}}}{r_{n\mathrm{u}} - r_{n\mathrm{l}}}.$$

Using a Gaussian likelihood for y is by far better justified than for ξ, as illustrated by the left-hand panel of Fig. 66.1 (the solid contours come from simulations, and the dashed ones show the approximation). Transforming the Gaussian back to ξ-space gives a good approximation for the likelihood of ξ (right-hand panel). Details and more quantitative results can be found in our upcoming paper [5], in which we also test how the new likelihood performs in a Bayesian analysis compared to the Gaussian likelihood by constructing a toy model.

As an alternative way to construct likelihood functions, a copula approach (see e.g., [2, 3]) can be used to couple univariate distributions to get a multivariate PDF. However, we showed that coupling the analytical univariate $p(\xi)$ from [1] with a Gaussian copula yields a multi-variate likelihood that is in bad agreement with simulations. Thus our quasi-Gaussian approach should be favored—of course, the accuracy of a copula likelihood might improve with a more realistic coupling.

References

1. Keitel, D., Schneider, P.: Constrained probability distributions of correlation functions. Accepted for A&A [arXiv:1105.3672] (2011)
2. Sato, M., Ichiki, K., Takeuchi, T.T.: Copula cosmology: Constructing a likelihood function. Phys. Rev. D, **83**, 023501 (2011)
3. Scherrer, R.J., Berlind, A.A., Mao, Q., McBride, C.K.: From Finance to Cosmology: The Copula of Large-Scale Structure. ApJ, **708**, L9 (2010)
4. Schneider, P., Hartlap, J.: Constrained correlation functions. A&A **504**, 705–717 (2009)
5. Wilking, P., Schneider, P.: A quasi-Gaussian approximation for the probability distribution of correlation functions. In prep. (2011)

Chapter 67
New Insights into Galaxy Structure from GALPHAT

Ilsang Yoon, Martin Weinberg, and Neal Katz

Abstract We introduce a novel galaxy morphology analysis tool GALPHAT exploiting Bayesian MCMC to provide the full posterior probability distribution of galaxy morphology parameters. Utilizing the full posterior, one can assess a probabilistic significance over the entire parameter space, make parameter inferences with reliable errors and test different hypotheses with statistical confidence levels. GALPHAT provides new insights into galaxy formation and evolution studies based on galaxy morphology and successfully demonstrates the feasibility of a large scale morphology analysis.

67.1 Motivation

The study of galaxy morphologies provides important information to understand galaxy formation and parametric models are widely used to derive galaxy structural parameters. However, an accurate decomposition of galaxy morphology is stymied by degeneracies in the parameter estimation itself. In most previous galaxy decomposition analyses, the correlations of physical properties and structural parameters of galaxies are usually assessed through *scatter* plots of the *best-fit* parameters. Those correlations are subject to strong contamination by underlying systematic correlations of the model parameters. To put galaxy morphology analysis on a rigorous statistical base, we present a novel image decomposition package GALPHAT (GALaxy PHotometric ATtributes [2]), which uses the Bayesian MCMC software package BIE [1], providing full parameter posteriors for reliable parameter estimation and hypothesis testing using a large ensemble of galaxy samples.

I. Yoon (✉) • M. Weinberg • N. Katz
Department of Astronomy, University of Massachusetts, Amherst, MA, USA
e-mail: iyoon@astro.umass.edu; weinberg@astro.umass.edu; nsk@astro.umass.edu

E.D. Feigelson and G.J. Babu (eds.), *Statistical Challenges in Modern Astronomy V*, Lecture Notes in Statistics 209, DOI 10.1007/978-1-4614-3520-4_67,

67.2 Methods

Given a parametric model of a galaxy's surface brightness (i.e. Sérsic), GALPHAT produces a likelihood function for image data by generating the difference with a model image. The BIE samples the posterior for a given prior distribution using a choice of MCMC algorithms. For computational efficiency, GALPHAT pre-generates a table of two-dimensional cumulative distributions of Sérsic profiles by numerical integration over pixels on a scale free grid using a rigorous error tolerance, for many different Sérsic indices n. When generating a model image, GALPHAT interpolates this table and scales using galaxy radius r and axis ratio b/a. Then the image is rotated by its position angle using sequential shear operations carried out in Fourier space, convolved with a given PSF, and combined with an adjustable sky pedestal. By incorporating this fast and accurate image generation algorithm for the likelihood evaluation, GALPHAT can analyze a large number of galaxies (e.g. $\sim$10,000) within two weeks using a Beowulf Linux cluster.

67.3 Results

We simulate a large ensemble of one-component Sérsic and two-component Sérsic bulge/exponential disk profile galaxies with a realistic distribution of galaxy structural parameters for testing the performance of GALPHAT. A summary of our results is.

- Parameter covariance must be fully taken into account using the full posterior to correctly characterize the parameter errors and to avoid biases owing to the parameter covariance, in later inferences.
- A carefully chosen prior significantly improves the inference of galaxy morphology parameters particularly for a low signal-to-noise ratio galaxy.
- Bayes factor model selection enables the reliable classification of a galaxy with statistical confidence, e.g. one- or two-components.

Acknowledgements This work was supported in part by the NSF IIS Program through award 0611948 and by the NASA AISR Program through award NNG06GF25G.

References

1. Weinberg, M. and Moss, E. (2011) Bayesian Inference Engine. *in preparation*. http://www.astro.umass.edu/BIE
2. Yoon, I., Weinberg, M. and Katz, N. (2011) New insights into galaxy structure from GALPHAT-I. Motivation, methodology and benchmarks for Sérsic models. *MNRAS*, 414, 1625–1655. http://sites.google.com/site/galphat/galphat

Index

E.D. Feigelson and G.J. Babu (eds.), *Statistical Challenges in Modern Astronomy V*, Lecture Notes in Statistics 209, DOI 10.1007/978-1-4614-3520-4,

Printed by Publishers' Graphics LLC
SO20120823